I0486286

The Philosophy & Categories of Mathematics

Edited by Paul F. Kisak

Contents

Chapter 1

The Principles of Mathematics

The Principles of Mathematics is a book written by Bertrand Russell in 1903. In it he presented his famous paradox and argued his thesis that mathematics and logic are identical.[1]

The book presents a view of the foundations of mathematics and has become a classic reference. It reported on developments by Giuseppe Peano, Mario Pieri, Richard Dedekind, Georg Cantor, and others. In 1937 Russell prepared a new introduction saying, "Such interest as the book now possesses is historical, and consists in the fact that it represents a certain stage in the development of its subject." Further editions were printed in 1938, 1951, 1996, and 2009.

1.1   Contents

The Principles of Mathematics consists of 59 chapters divided into seven parts: indefinables in mathematics, number, quantity, order, infinity and continuity, space, matter and motion.

In chapter one, "Definition of Pure Mathematics," Russell asserts that:

> The fact that all Mathematics is Symbolic Logic is one of the greatest discoveries of our age; and when this fact has been established, the remainder of the principles of mathematics consists in the analysis of Symbolic Logic itself.[2]

There is an anticipation of relativity physics in the final part as the last three chapters consider Newton's laws of motion, absolute and relative motion, and Hertz's dynamics. However, Russell rejects what he calls "the relational theory", and says on page 489

> For us, since absolute space and time have been admitted, there is no need to avoid absolute motion, and indeed no possibility of doing so.

In his review, G. H. Hardy says "Mr. Russell is a firm believer in absolute position in space and time, a view as much out of fashion nowadays that Chapter [58: Absolute and Relative Motion] will be read with peculiar interest."[3]

1.2   Early reviews

Reviews were prepared by G. E. Moore and Charles Sanders Peirce, but Moore's was never published[4] and that of Peirce was brief and somewhat dismissive. He indicated that he thought it unoriginal, saying that the book "can hardly be called literature" and "Whoever wishes a convenient introduction to the remarkable researches into the logic of mathematics that have been made during the last sixty years [...] will do well to take up this book."[5]

G. H. Hardy wrote a favorable review[3] expecting the book to appeal more to philosophers than mathematicians. But he says

> [I]n spite of its five hundred pages the book is much too short. Many chapters dealing with important questions are compressed into five or six pages, and in some places, especially in the most avowedly controversial parts, the argument is almost too condensed to follow. And the philosopher who attempts to read the book will be especially puzzled by the constant presupposition of a whole philosophical system utterly unlike any of those usually accepted.

In 1904 another review appeared in Bulletin of the American Mathematical Society (11(2):74–93) written by Edwin Bidwell Wilson. He says "The delicacy of the question is such that even the greatest mathematicians and philosophers of to-day have made what seem to be substantial slips of judgement and have shown on occasions an astounding ignorance of the essence of the problem which they were discussing. ... all too frequently it has been the result of a wholly unpardonable disregard of the work already accomplished by others." Wilson recounts the developments of Peano that Russell reports, and takes the occasion to correct Henri Poincaré who had ascribed them to David Hilbert. In praise of Russell, Wilson says "Surely the present work is a monument to patience, perseverance, and thoroughness."(page 88)

1.3   Second edition

In 1938 the book was re-issued with a new preface by Russell. This preface was interpreted as a retreat from the realism of the first edition and a turn toward nominalist philosophy of symbolic logic. James Feibleman, an admirer of the book, thought Russell's new preface went too far into nominalism so he wrote a rebuttal to this introduction.[6] Feibleman says, "It is the first comprehensive treatise on symbolic logic to be written in English; and it gives to that system of logic a realistic interpretation."

1.4   Later reviews

In 1959 Russell wrote *My Philosophical Development*, in which he recalled the impetus to write the *Principles*:

> It was at the International Congress of Philosophy in Paris in the year 1900 that I became aware of the importance of logical reform for the philosophy of mathematics. ... I was impressed by the fact that, in every discussion, [Peano] showed more precision and more logical rigour than was shown by anybody else. ... It was [Peano's works] that gave the impetus to my own views on the principles of mathematics.[7]

Recalling the book after his later work, he provides this evaluation:

> *The Principles of Mathematics*, which I finished on 23 May 1902, turned out to be a crude and rather immature draft of the subsequent work [*Principia Mathematica*], from which, however, it differed in containing controversy with other philosophies of mathematics.[8]

Such self-deprecation from the author after half a century of philosophical growth is understandable. On the other hand, Jules Vuillemin wrote in 1968:

> *The Principles* inaugurated contemporary philosophy. Other works have won and lost the title. Such is not the case with this one. It is serious, and its wealth perseveres. Furthermore, in relation to it, in a deliberate fashion or not, it locates itself again today in the eyes of all those that believe that contemporary science has modified our representation of the universe and through this representation, our relation to ourselves and to others.[9]

When W. V. Quine penned his autobiography, he wrote:[10]

Peano's symbolic notation took Russell by storm in 1900, but Russell's *Principles* was still in unrelieved prose. I was inspired by its profundity [in 1928] and baffled by its frequent opacity. In part it was rough going because of the cumbersomeness of ordinary language as compared with the suppleness of a notation especially devised for these intricate themes. Rereading it years later, I discovered that it had been rough going also because matters were unclear in Russell's own mind in those pioneer days.

The Principles was an early expression of analytic philosophy and thus has come under close examination.[11] Peter Hylton wrote, "The book has an air of excitement and novelty to it ... The salient characteristic of *Principles* is ... the way in which the technical work is integrated into metaphysical argument."[11]:168

Russell's *Principles* looms large in Ivor Grattan-Guinness' study of the roots of modern logic published in 2000.

In 2006, Philip Ehrlich challenged the validity of Russell's analysis of infinitesimals in the Leibniz tradition.[12] A recent study documents the non-sequiturs in Russell's critique of the infinitesimals of Gottfried Leibniz and Hermann Cohen.[13]

1.5 Notes

[1] Russell, Bertrand (1938) [First published 1903]. *Principles of Mathematics* (2nd ed.). W. W. Norton & Company. ISBN 0-393-00249-7. The fundamental thesis of the following pages, that mathematics and logic are identical, is one which I have never since seen any reason to modify. The quotation is from the first page of Russell's introduction to the second (1938) edition.

[2] Bertrand Russell, *Principles of Mathematics* (1903), p.5

[3] G. H. Hardy (18 September 1903) "The Philosophy of Mathematics", Times Literary Supplement #88

[4] Quin, Arthur (1977). *The Confidence of British Philosophers*. p. 221. ISBN 90-04-05397-2.

[5] See the first paragraph of his review of *What is Meaning?* and *The Principles of Mathematics* (1903), *The Nation*, v. 77, n. 1998, p. 308, Google Books Eprint, reprinted in *Collected Papers of Charles Sanders Peirce* v. 8 (1958), paragraph 171 footnote. The review was publicly anonymous like the other reviews (totaling over 300) that Peirce wrote for *The Nation* on a regular basis. Murray Murphy called the review "so brief and cursory that I am convinced that he never read the book." in Murphy, Murray (1993). *The Development of Peirce's Philosophy*. Hackett Pub. Co. p. 241. ISBN 0-87220-231-3. Others such as Norbert Wiener and Christine Ladd-Franklin shared Peirce's view of Russell's work. See Anellis, Irving (1995), "Peirce Rustled, Russell Pierced", *Modern Logic* 5, 270–328.

[6] James Feibleman (1944) Reply to the Introduction of the Second Edition, pages 157 to 174 of *The Philosophy of Bertrand Russell*, P.A. Schilpp, editor, link from HathiTrust

[7] Russell, *My Philosophical Development*, p. 65.

[8] Russell, *My Philosophical Development*, p. 74.

[9] Jules Vuillemin (1968) *Leçons sur la première philosophie de Russell*, page 333, Paris: Colin

[10] W. V. Quine (1985) *The Time of My Life*, page 59, MIT Press ISBN 0-262-17003-5

[11] Peter Hylton (1990) *Russell, Idealism, and the Emergence of Analytic Philosophy*, chapter 5: Russell's *Principles of Mathematics*, pp 167 to 236. Clarendon Press, ISBN 0-19-824626-9

[12] Ehrlich, Philip (2006), "The rise of non-Archimedean mathematics and the roots of a misconception. I. The emergence of non-Archimedean systems of magnitudes", *Archive for History of Exact Sciences* **60** (1): 1–121, doi:10.1007/s00407-005-0102-4

[13] Katz, Mikhail; Sherry, David (2012), "Leibniz's Infinitesimals: Their Fictionality, Their Modern Implementations, and Their Foes from Berkeley to Russell and Beyond", *Erkenntnis*, arXiv:1205.0174, doi:10.1007/s10670-012-9370-y.

1.6 References

- Louis Couturat (1905) *Les Principes des Mathematiques: avec un appendice sur la philosophie des mathématiques de Kant.* Republished 1965, Georg Olms.

- Ivor Grattan-Guinness (2000) *The Search for Mathematical Roots 1870–1940: Logics, Set Theories, and the Foundations of Mathematics from Cantor through Russell to Gödel.* Princeton Univ. Press. ISBN 0-691-05858-X. See pages 292–302 and 310–326.

1.7 External links

- The Principles of Mathematics Online text

- The Principles of Mathematics Full Text at the Internet Archive

Chapter 2

Philosophy of mathematics

The **philosophy of mathematics** is the branch of philosophy that studies the philosophical assumptions, foundations, and implications of mathematics. The aim of the philosophy of mathematics is to provide an account of the nature and methodology of mathematics and to understand the place of mathematics in people's lives. The logical and structural nature of mathematics itself makes this study both broad and unique among its philosophical counterparts.

The terms *philosophy of mathematics* and *mathematical philosophy* are frequently used as synonyms.[1] The latter, however, may be used to refer to several other areas of study. One refers to a project of formalizing a philosophical subject matter, say, aesthetics, ethics, logic, metaphysics, or theology, in a purportedly more exact and rigorous form, as for example the labors of scholastic theologians, or the systematic aims of Leibniz and Spinoza. Another refers to the working philosophy of an individual practitioner or a like-minded community of practicing mathematicians. Additionally, some understand the term "mathematical philosophy" to be an allusion to the approach to the foundations of mathematics taken by Bertrand Russell in his books *The Principles of Mathematics* and *Introduction to Mathematical Philosophy*.

2.1 Recurrent themes

Recurrent themes include:

- What is the role of Mankind in developing mathematics?

- What are the sources of mathematical subject matter?

- What is the ontological status of mathematical entities?

- What does it mean to refer to a mathematical object?

- What is the character of a mathematical proposition?

- What is the relation between logic and mathematics?

- What is the role of hermeneutics in mathematics?

- What kinds of inquiry play a role in mathematics?

- What are the objectives of mathematical inquiry?

- What gives mathematics its hold on experience?

- What are the human traits behind mathematics?

- What is mathematical beauty?

- What is the source and nature of mathematical truth?

- What is the relationship between the abstract world of mathematics and the material universe?

2.2 History

The origin of mathematics is subject to argument. Whether the birth of mathematics was a random happening or induced by necessity duly contingent upon other subjects, say for example physics, is still a matter of prolific debates.

Many thinkers have contributed their ideas concerning the nature of mathematics. Today, some philosophers of mathematics aim to give accounts of this form of inquiry and its products as they stand, while others emphasize a role for themselves that goes beyond simple interpretation to critical analysis. There are traditions of mathematical philosophy in both Western philosophy and Eastern philosophy. Western philosophies of mathematics go as far back as Plato, who studied the ontological status of mathematical objects, and Aristotle, who studied logic and issues related to infinity (actual versus potential).

Greek philosophy on mathematics was strongly influenced by their study of geometry. For example, at one time, the Greeks held the opinion that 1 (one) was not a number, but rather a unit of arbitrary length. A number was defined as a multitude. Therefore, 3, for example, represented a certain multitude of units, and was thus not "truly" a number. At another point, a similar argument was made that 2 was not a number but a fundamental notion of a pair. These views come from the heavily geometric straight-edge-and-compass viewpoint of the Greeks: just as lines drawn in a geometric problem are measured in proportion to the first arbitrarily drawn line, so too are the numbers on a number line measured in proportion to the arbitrary first "number" or "one".

These earlier Greek ideas of numbers were later upended by the discovery of the irrationality of the square root of two. Hippasus, a disciple of Pythagoras, showed that the diagonal of a unit square was incommensurable with its (unit-length) edge: in other words he proved there was no existing (rational) number that accurately depicts the proportion of the diagonal of the unit square to its edge. This caused a significant re-evaluation of Greek philosophy of mathematics. According to legend, fellow Pythagoreans were so traumatized by this discovery that they murdered Hippasus to stop him from spreading his heretical idea. Simon Stevin was one of the first in Europe to challenge Greek ideas in the 16th century. Beginning with Leibniz, the focus shifted strongly to the relationship between mathematics and logic. This perspective dominated the philosophy of mathematics through the time of Frege and of Russell, but was brought into question by developments in the late 19th and early 20th centuries.

2.2.1 20th century

A perennial issue in the philosophy of mathematics concerns the relationship between logic and mathematics at their joint foundations. While 20th century philosophers continued to ask the questions mentioned at the outset of this article, the philosophy of mathematics in the 20th century was characterized by a predominant interest in formal logic, set theory, and foundational issues.

It is a profound puzzle that on the one hand mathematical truths seem to have a compelling inevitability, but on the other hand the source of their "truthfulness" remains elusive. Investigations into this issue are known as the foundations of mathematics program.

At the start of the 20th century, philosophers of mathematics were already beginning to divide into various schools of thought about all these questions, broadly distinguished by their pictures of mathematical epistemology and ontology. Three schools, formalism, intuitionism, and logicism, emerged at this time, partly in response to the increasingly widespread worry that mathematics as it stood, and analysis in particular, did not live up to the standards of certainty and rigor that had been taken for granted. Each school addressed the issues that came to the fore at that time, either attempting to resolve them or claiming that mathematics is not entitled to its status as our most trusted knowledge.

Surprising and counter-intuitive developments in formal logic and set theory early in the 20th century led to new questions concerning what was traditionally called the *foundations of mathematics*. As the century unfolded, the initial focus of concern expanded to an open exploration of the fundamental axioms of mathematics, the axiomatic approach having been taken for granted since the time of Euclid around 300 BCE as the natural basis for mathematics. Notions of axiom, proposition and proof, as well as the notion of a proposition being true of a mathematical object (see Assignment (mathematical logic)), were formalized, allowing them to be treated mathematically. The Zermelo–Fraenkel axioms for set theory were formulated which provided a conceptual framework in which much mathematical discourse would be interpreted. In mathematics, as in physics, new and unexpected ideas had arisen and significant changes were coming. With

Gödel numbering, propositions could be interpreted as referring to themselves or other propositions, enabling inquiry into the consistency of mathematical theories. This reflective critique in which the theory under review "becomes itself the object of a mathematical study" led Hilbert to call such study *metamathematics* or *proof theory*.[2]

At the middle of the century, a new mathematical theory was created by Samuel Eilenberg and Saunders Mac Lane, known as category theory, and it became a new contender for the natural language of mathematical thinking.[3] As the 20th century progressed, however, philosophical opinions diverged as to just how well-founded were the questions about foundations that were raised at the century's beginning. Hilary Putnam summed up one common view of the situation in the last third of the century by saying:

> When philosophy discovers something wrong with science, sometimes science has to be changed—Russell's paradox comes to mind, as does Berkeley's attack on the actual infinitesimal—but more often it is philosophy that has to be changed. I do not think that the difficulties that philosophy finds with classical mathematics today are genuine difficulties; and I think that the philosophical interpretations of mathematics that we are being offered on every hand are wrong, and that "philosophical interpretation" is just what mathematics doesn't need.[4]:169–170

Philosophy of mathematics today proceeds along several different lines of inquiry, by philosophers of mathematics, logicians, and mathematicians, and there are many schools of thought on the subject. The schools are addressed separately in the next section, and their assumptions explained.

2.3 Major themes

2.3.1 Mathematical realism

Mathematical realism, like realism in general, holds that mathematical entities exist independently of the human mind. Thus humans do not invent mathematics, but rather discover it, and any other intelligent beings in the universe would presumably do the same. In this point of view, there is really one sort of mathematics that can be discovered; triangles, for example, are real entities, not the creations of the human mind.

Many working mathematicians have been mathematical realists; they see themselves as discoverers of naturally occurring objects. Examples include Paul Erdős and Kurt Gödel. Gödel believed in an objective mathematical reality that could be perceived in a manner analogous to sense perception. Certain principles (e.g., for any two objects, there is a collection of objects consisting of precisely those two objects) could be directly seen to be true, but the continuum hypothesis conjecture might prove undecidable just on the basis of such principles. Gödel suggested that quasi-empirical methodology could be used to provide sufficient evidence to be able to reasonably assume such a conjecture.

Within realism, there are distinctions depending on what sort of existence one takes mathematical entities to have, and how we know about them. Major forms of mathematical realism include Platonism and empiricism.

2.3.2 Mathematical anti-realism

Mathematical anti-realism generally holds that mathematical statements have truth-values, but that they do not do so by corresponding to a special realm of immaterial or non-empirical entities. Major forms of mathematical anti-realism include Formalism and Fictionalism.

2.4 Contemporary schools of thought

2.4.1 Platonism

Mathematical Platonism is the form of realism that suggests that mathematical entities are abstract, have no spatiotemporal or causal properties, and are eternal and unchanging. This is often claimed to be the view most people have of numbers.

The term *Platonism* is used because such a view is seen to parallel Plato's Theory of Forms and a "World of Ideas" (Greek: *eidos* (εἶδος)) described in Plato's allegory of the cave: the everyday world can only imperfectly approximate an unchanging, ultimate reality. Both *Plato's cave* and *Platonism* have meaningful, not just superficial connections, because Plato's ideas were preceded and probably influenced by the hugely popular *Pythagoreans* of ancient Greece, who believed that the world was, quite literally, generated by numbers.

A major question considered in mathematical platonism is this: precisely where and how do the mathematical entities exist, and how do we know about them? Is there a world, completely separate from our physical one, that is occupied by the mathematical entities? How can we gain access to this separate world and discover truths about the entities? One answer might be the Ultimate Ensemble, which is a theory that postulates all structures that exist mathematically also exist physically in their own universe.

Plato spoke of mathematics by:

> How do you mean?
>
> I mean, as I was saying, that arithmetic has a very great and elevating effect, compelling the soul to reason about abstract number, and rebelling against the introduction of visible or tangible objects into the argument. You know how steadily the masters of the art repel and ridicule any one who attempts to divide absolute unity when he is calculating, and if you divide, they multiply, taking care that one shall continue one and not become lost in fractions.
>
> That is very true.
>
> Now, suppose a person were to say to them: O my friends, what are these wonderful numbers about which you are reasoning, in which, as you say, there is a unity such as you demand, and each unit is equal, invariable, indivisible, --what would they answer?
> — Plato, Chapter 7. "The Republic" (Jowett translation).

In context, chapter 8, of H.D.P. Lee's translation, reports the education of a philosopher contains five mathematical disciplines:

1. mathematics;

2. arithmetic, written in unit fraction "parts" using theoretical unities and abstract numbers;

3. plane geometry and solid geometry also considered the line to be segmented into rational and irrational unit "parts";

4. astronomy

5. harmonics

Translators of the works of Plato rebelled against practical versions of his culture's practical mathematics. However, Plato himself and Greeks had copied 1,500 older Egyptian fraction abstract unities, one being a hekat unity scaled to (64/64) in the Akhmim Wooden Tablet, thereby not getting lost in fractions.

Gödel's Platonism postulates a special kind of mathematical intuition that lets us perceive mathematical objects directly. (This view bears resemblances to many things Husserl said about mathematics, and supports Kant's idea that mathematics is synthetic *a priori*.) Davis and Hersh have suggested in their book *The Mathematical Experience* that most mathematicians act as though they are Platonists, even though, if pressed to defend the position carefully, they may retreat to formalism (see below).

Some mathematicians hold opinions that amount to more nuanced versions of Platonism.

Full-blooded Platonism is a modern variation of Platonism, which is in reaction to the fact that different sets of mathematical entities can be proven to exist depending on the axioms and inference rules employed (for instance, the law of the excluded middle, and the axiom of choice). It holds that all mathematical entities exist, however they may be provable, even if they cannot all be derived from a single consistent set of axioms.

2.4.2 Empiricism

Empiricism is a form of realism that denies that mathematics can be known *a priori* at all. It says that we discover mathematical facts by empirical research, just like facts in any of the other sciences. It is not one of the classical three positions advocated in the early 20th century, but primarily arose in the middle of the century. However, an important early proponent of a view like this was John Stuart Mill. Mill's view was widely criticized, because, according to critics, it makes statements like "2 + 2 = 4" come out as uncertain, contingent truths, which we can only learn by observing instances of two pairs coming together and forming a quartet.

Contemporary mathematical empiricism, formulated by Quine and Putnam, is primarily supported by the indispensability argument: mathematics is indispensable to all empirical sciences, and if we want to believe in the reality of the phenomena described by the sciences, we ought also believe in the reality of those entities required for this description. That is, since physics needs to talk about electrons to say why light bulbs behave as they do, then electrons must exist. Since physics needs to talk about numbers in offering any of its explanations, then numbers must exist. In keeping with Quine and Putnam's overall philosophies, this is a naturalistic argument. It argues for the existence of mathematical entities as the best explanation for experience, thus stripping mathematics of being distinct from the other sciences.

Putnam strongly rejected the term "Platonist" as implying an over-specific ontology that was not necessary to mathematical practice in any real sense. He advocated a form of "pure realism" that rejected mystical notions of truth and accepted much quasi-empiricism in mathematics. Putnam was involved in coining the term "pure realism" (see below).

The most important criticism of empirical views of mathematics is approximately the same as that raised against Mill. If mathematics is just as empirical as the other sciences, then this suggests that its results are just as fallible as theirs, and just as contingent. In Mill's case the empirical justification comes directly, while in Quine's case it comes indirectly, through the coherence of our scientific theory as a whole, i.e. consilience after E.O. Wilson. Quine suggests that mathematics seems completely certain because the role it plays in our web of belief is incredibly central, and that it would be extremely difficult for us to revise it, though not impossible.

For a philosophy of mathematics that attempts to overcome some of the shortcomings of Quine and Gödel's approaches by taking aspects of each see Penelope Maddy's *Realism in Mathematics*. Another example of a realist theory is the embodied mind theory (below). For a modern revision of mathematical empiricism see New Empiricism (below).

For experimental evidence suggesting that human infants can do elementary arithmetic, see Brian Butterworth.

2.4.3 Mathematical monism

Max Tegmark's mathematical universe hypothesis goes further than full-blooded Platonism in asserting that not only do all mathematical objects exist, but nothing else does. Tegmark's sole postulate is: *All structures that exist mathematically also exist physically*. That is, in the sense that "in those [worlds] complex enough to contain self-aware substructures [they] will subjectively perceive themselves as existing in a physically 'real' world".[5][6]

2.4.4 Logicism

Logicism is the thesis that mathematics is reducible to logic, and hence nothing but a part of logic.[7]:41 Logicists hold that mathematics can be known *a priori*, but suggest that our knowledge of mathematics is just part of our knowledge of logic in general, and is thus analytic, not requiring any special faculty of mathematical intuition. In this view, logic is the proper foundation of mathematics, and all mathematical statements are necessary logical truths.

Rudolf Carnap (1931) presents the logicist thesis in two parts:[7]

1. The *concepts* of mathematics can be derived from logical concepts through explicit definitions.

2. The *theorems* of mathematics can be derived from logical axioms through purely logical deduction.

Gottlob Frege was the founder of logicism. In his seminal *Die Grundgesetze der Arithmetik* (*Basic Laws of Arithmetic*) he built up arithmetic from a system of logic with a general principle of comprehension, which he called "Basic Law V" (for concepts F and G, the extension of F equals the extension of G if and only if for all objects a, Fa if and only if Ga), a principle that he took to be acceptable as part of logic.

Frege's construction was flawed. Russell discovered that Basic Law V is inconsistent (this is Russell's paradox). Frege abandoned his logicist program soon after this, but it was continued by Russell and Whitehead. They attributed the paradox to "vicious circularity" and built up what they called ramified type theory to deal with it. In this system, they were eventually able to build up much of modern mathematics but in an altered, and excessively complex form (for example, there were different natural numbers in each type, and there were infinitely many types). They also had to make several compromises in order to develop so much of mathematics, such as an "axiom of reducibility". Even Russell said that this axiom did not really belong to logic.

Modern logicists (like Bob Hale, Crispin Wright, and perhaps others) have returned to a program closer to Frege's. They have abandoned Basic Law V in favor of abstraction principles such as Hume's principle (the number of objects falling under the concept F equals the number of objects falling under the concept G if and only if the extension of F and the extension of G can be put into one-to-one correspondence). Frege required Basic Law V to be able to give an explicit definition of the numbers, but all the properties of numbers can be derived from Hume's principle. This would not have been enough for Frege because (to paraphrase him) it does not exclude the possibility that the number 3 is in fact Julius Caesar. In addition, many of the weakened principles that they have had to adopt to replace Basic Law V no longer seem so obviously analytic, and thus purely logical.

2.4.5 Formalism

Main article: Formalism (mathematics)

Formalism holds that mathematical statements may be thought of as statements about the consequences of certain string manipulation rules. For example, in the "game" of Euclidean geometry (which is seen as consisting of some strings called "axioms", and some "rules of inference" to generate new strings from given ones), one can prove that the Pythagorean theorem holds (that is, one can generate the string corresponding to the Pythagorean theorem). According to formalism, mathematical truths are not about numbers and sets and triangles and the like—in fact, they are not "about" anything at all.

Another version of formalism is often known as deductivism. In deductivism, the Pythagorean theorem is not an absolute truth, but a relative one: *if* one assigns meaning to the strings in such a way that the rules of the game become true (i.e., true statements are assigned to the axioms and the rules of inference are truth-preserving), *then* one must accept the theorem, or, rather, the interpretation one has given it must be a true statement. The same is held to be true for all other mathematical statements. Thus, formalism need not mean that mathematics is nothing more than a meaningless symbolic game. It is usually hoped that there exists some interpretation in which the rules of the game hold. (Compare this position to structuralism.) But it does allow the working mathematician to continue in his or her work and leave such problems to the philosopher or scientist. Many formalists would say that in practice, the axiom systems to be studied will be suggested by the demands of science or other areas of mathematics.

A major early proponent of formalism was David Hilbert, whose program was intended to be a complete and consistent axiomatization of all of mathematics. Hilbert aimed to show the consistency of mathematical systems from the assumption that the "finitary arithmetic" (a subsystem of the usual arithmetic of the positive integers, chosen to be philosophically uncontroversial) was consistent. Hilbert's goals of creating a system of mathematics that is both complete and consistent were dealt a fatal blow by the second of Gödel's incompleteness theorems, which states that sufficiently expressive consistent axiom systems can never prove their own consistency. Since any such axiom system would contain the finitary arithmetic as a subsystem, Gödel's theorem implied that it would be impossible to prove the system's consistency relative to that (since it would then prove its own consistency, which Gödel had shown was impossible). Thus, in order to show that any axiomatic system of mathematics is in fact consistent, one needs to first assume the consistency of a system of mathematics that is in a sense stronger than the system to be proven consistent.

Hilbert was initially a deductivist, but, as may be clear from above, he considered certain metamathematical methods to yield intrinsically meaningful results and was a realist with respect to the finitary arithmetic. Later, he held the opinion that there was no other meaningful mathematics whatsoever, regardless of interpretation.

David Hilbert

Other formalists, such as Rudolf Carnap, Alfred Tarski, and Haskell Curry, considered mathematics to be the investigation of formal axiom systems. Mathematical logicians study formal systems but are just as often realists as they are formalists.

Formalists are relatively tolerant and inviting to new approaches to logic, non-standard number systems, new set theories etc. The more games we study, the better. However, in all three of these examples, motivation is drawn from existing mathematical or philosophical concerns. The "games" are usually not arbitrary.

The main critique of formalism is that the actual mathematical ideas that occupy mathematicians are far removed from the string manipulation games mentioned above. Formalism is thus silent on the question of which axiom systems ought to be studied, as none is more meaningful than another from a formalistic point of view.

Recently, some formalist mathematicians have proposed that all of our *formal* mathematical knowledge should be systematically encoded in computer-readable formats, so as to facilitate automated proof checking of mathematical proofs and the use of interactive theorem proving in the development of mathematical theories and computer software. Because of their close connection with computer science, this idea is also advocated by mathematical intuitionists and constructivists in the "computability" tradition (see below). See QED project for a general overview.

2.4.6   Conventionalism

The French mathematician Henri Poincaré was among the first to articulate a conventionalist view. Poincaré's use of non-Euclidean geometries in his work on differential equations convinced him that Euclidean geometry should not be regarded as *a priori* truth. He held that axioms in geometry should be chosen for the results they produce, not for their apparent coherence with human intuitions about the physical world.

2.4.7   Psychologism

Psychologism in the philosophy of mathematics is the position that mathematical concepts and/or truths are grounded in, derived from or explained by psychological facts (or laws).

John Stuart Mill seems to have been an advocate of a type of logical psychologism, as were many 19th-century German logicians such as Sigwart and Erdmann as well as a number of psychologists, past and present: for example, Gustave Le Bon. Psychologism was famously criticized by Frege in his *The Foundations of Arithmetic*, and many of his works and essays, including his review of Husserl's *Philosophy of Arithmetic*. Edmund Husserl, in the first volume of his *Logical Investigations*, called "The Prolegomena of Pure Logic", criticized psychologism thoroughly and sought to distance himself from it. The "Prolegomena" is considered a more concise, fair, and thorough refutation of psychologism than the criticisms made by Frege, and also it is considered today by many as being a memorable refutation for its decisive blow to psychologism. Psychologism was also criticized by Charles Sanders Peirce and Maurice Merleau-Ponty.

2.4.8   Intuitionism

Main article: Mathematical intuitionism

In mathematics, intuitionism is a program of methodological reform whose motto is that "there are no non-experienced mathematical truths" (L.E.J. Brouwer). From this springboard, intuitionists seek to reconstruct what they consider to be the corrigible portion of mathematics in accordance with Kantian concepts of being, becoming, intuition, and knowledge. Brouwer, the founder of the movement, held that mathematical objects arise from the *a priori* forms of the volitions that inform the perception of empirical objects.[8]

A major force behind intuitionism was L.E.J. Brouwer, who rejected the usefulness of formalized logic of any sort for mathematics. His student Arend Heyting postulated an intuitionistic logic, different from the classical Aristotelian logic; this logic does not contain the law of the excluded middle and therefore frowns upon proofs by contradiction. The axiom of choice is also rejected in most intuitionistic set theories, though in some versions it is accepted. Important work was later done by Errett Bishop, who managed to prove versions of the most important theorems in real analysis within this framework.

In intuitionism, the term "explicit construction" is not cleanly defined, and that has led to criticisms. Attempts have been made to use the concepts of Turing machine or computable function to fill this gap, leading to the claim that only questions regarding the behavior of finite algorithms are meaningful and should be investigated in mathematics. This has led to the study of the computable numbers, first introduced by Alan Turing. Not surprisingly, then, this approach to mathematics is sometimes associated with theoretical computer science.

Constructivism

Main article: Mathematical constructivism

Like intuitionism, constructivism involves the regulative principle that only mathematical entities which can be explicitly constructed in a certain sense should be admitted to mathematical discourse. In this view, mathematics is an exercise of the human intuition, not a game played with meaningless symbols. Instead, it is about entities that we can create directly through mental activity. In addition, some adherents of these schools reject non-constructive proofs, such as a proof by contradiction.

Finitism

Finitism is an extreme form of constructivism, according to which a mathematical object does not exist unless it can be constructed from natural numbers in a finite number of steps. In her book *Philosophy of Set Theory*, Mary Tiles characterized those who allow countably infinite objects as classical finitists, and those who deny even countably infinite objects as strict finitists.

The most famous proponent of finitism was Leopold Kronecker,[9] who said:

> God created the natural numbers, all else is the work of man.

Ultrafinitism is an even more extreme version of finitism, which rejects not only infinities but finite quantities that cannot feasibly be constructed with available resources.

2.4.9   Structuralism

Main article: Mathematical structuralism

Structuralism is a position holding that mathematical theories describe structures, and that mathematical objects are exhaustively defined by their *places* in such structures, consequently having no intrinsic properties. For instance, it would maintain that all that needs to be known about the number 1 is that it is the first whole number after 0. Likewise all the other whole numbers are defined by their places in a structure, the number line. Other examples of mathematical objects might include lines and planes in geometry, or elements and operations in abstract algebra.

Structuralism is an epistemologically realistic view in that it holds that mathematical statements have an objective truth value. However, its central claim only relates to what *kind* of entity a mathematical object is, not to what kind of *existence* mathematical objects or structures have (not, in other words, to their ontology). The kind of existence mathematical objects have would clearly be dependent on that of the structures in which they are embedded; different sub-varieties of structuralism make different ontological claims in this regard.[10]

The *Ante Rem*, or fully realist, variation of structuralism has a similar ontology to Platonism in that structures are held to have a real but abstract and immaterial existence. As such, it faces the usual problems of explaining the interaction between such abstract structures and flesh-and-blood mathematicians.

In Re, or moderately realistic, structuralism is the equivalent of Aristotelian realism. Structures are held to exist inasmuch as some concrete system exemplifies them. This incurs the usual issues that some perfectly legitimate structures might accidentally happen not to exist, and that a finite physical world might not be "big" enough to accommodate some otherwise legitimate structures.

The *Post Res* or eliminative variant of structuralism is anti-realist about structures in a way that parallels nominalism. According to this view mathematical *systems* exist, and have structural features in common. If something is true of a structure, it will be true of all systems exemplifying the structure. However, it is merely convenient to talk of structures being "held in common" between systems: they in fact have no independent existence.

2.4.10   Embodied mind theories

Embodied mind theories hold that mathematical thought is a natural outgrowth of the human cognitive apparatus which finds itself in our physical universe. For example, the abstract concept of number springs from the experience of counting discrete objects. It is held that mathematics is not universal and does not exist in any real sense, other than in human brains. Humans construct, but do not discover, mathematics.

With this view, the physical universe can thus be seen as the ultimate foundation of mathematics: it guided the evolution of the brain and later determined which questions this brain would find worthy of investigation. However, the human mind has no special claim on reality or approaches to it built out of math. If such constructs as Euler's identity are true then they are true as a map of the human mind and cognition.

Embodied mind theorists thus explain the effectiveness of mathematics—mathematics was constructed by the brain in order to be effective in this universe.

The most accessible, famous, and infamous treatment of this perspective is *Where Mathematics Comes From*, by George Lakoff and Rafael E. Núñez. In addition, mathematician Keith Devlin has investigated similar concepts with his book *The Math Instinct*, as has neuroscientist Stanislas Dehaene with his book *The Number Sense*. For more on the philosophical ideas that inspired this perspective, see cognitive science of mathematics.

New empiricism

A more recent empiricism returns to the principle of the English empiricists of the 18th and 19th centuries, in particular John Stuart Mill, who asserted that all knowledge comes to us from observation through the senses. This applies not only to matters of fact, but also to "relations of ideas", as Hume called them: the structures of logic which interpret, organize and abstract observations.

To this principle it adds a materialist connection: all the processes of logic which interpret, organize and abstract observations, are physical phenomena which take place in real time and physical space: namely, in the brains of human beings. Abstract objects, such as mathematical objects, are ideas, which in turn exist as electrical and chemical states of the billions of neurons in the human brain.

This second concept is reminiscent of the social constructivist approach, which holds that mathematics is produced by humans rather than being "discovered" from abstract, *a priori* truths. However, it differs sharply from the constructivist implication that humans arbitrarily construct mathematical principles that have no inherent truth but which instead are created on a conveniency basis. On the contrary, new empiricism shows how mathematics, although constructed by humans, follows rules and principles that will be agreed on by all who participate in the process, with the result that everyone practicing mathematics comes up with the same answer—except in those areas where there is philosophical disagreement on the meaning of fundamental concepts. This is because the new empiricism perceives this agreement as being a physical phenomenon, one which is observed by other humans in the same way that other physical phenomena, like the motions of inanimate bodies, or the chemical interaction of various elements, are observed.

Combining the materialist principle with Millisian epistemology evades the principal difficulty with classical empiricism—that all knowledge comes from the senses. That difficulty lies in the observation that mathematical truths based on logical deduction appear to be more certainly true than knowledge of the physical world itself. (The physical world in this case is taken to mean the portion of it lying outside the human brain.)

Kant argued that the structures of logic which organize, interpret and abstract observations were built into the human mind and were true and valid *a priori*. Mill, on the contrary, said that we believe them to be true because we have enough individual instances of their truth to generalize: in his words, "From instances we have observed, we feel warranted in concluding that what we found true in those instances holds in all similar ones, past, present and future, however numerous they may be".[11] Although the psychological or epistemological specifics given by Mill through which we build our logical apparatus may not be completely warranted, his explanation still nonetheless manages to demonstrate that there is no way around Kant's *a priori* logic. To recant Mill's original idea in an empiricist twist: "Indeed, the very principles of logical deduction are true because we observe that using them leads to true conclusions", which is itself an *a priori* presupposition.

If all this is true, then where do the world senses come in? The early empiricists all stumbled over this point. Hume asserted that all knowledge comes from the senses, and then gave away the ballgame by excepting abstract propositions, which he called "relations of ideas". These, he said, were absolutely true (although the mathematicians who thought them up, being human, might get them wrong). Mill, on the other hand, tried to deny that abstract ideas exist outside the physical world: all numbers, he said, "must be numbers of something: there are no such things as numbers in the abstract". When we count to eight or add five and three we are really counting spoons or bumblebees. "All things possess quantity", he said, so that propositions concerning numbers are propositions concerning "all things whatever". But then in almost a contradiction of himself he went on to acknowledge that numerical and algebraic expressions are not necessarily attached to real world objects: they "do not excite in our minds ideas of any things in particular". Mill's low reputation as a philosopher of logic, and the low estate of empiricism in the century and a half following him, derives from this failed attempt to link abstract thoughts to the physical world, when it may be more plausibly arguable that abstraction consists precisely of separating the thought from its physical foundations.

The conundrum created by our certainty that abstract deductive propositions, if valid (i.e. if we can "prove" them), are true, exclusive of observation and testing in the physical world, gives rise to a further reflection ... What if thoughts themselves, and the minds that create them, are physical objects, existing only in the physical world?

This would reconcile the contradiction between our belief in the certainty of abstract deductions and the empiricist principle that knowledge comes from observation of individual instances. We know that Euler's equation is true because every time a human mind derives the equation, it gets the same result, unless it has made a mistake, which can be acknowledged and corrected. We observe this phenomenon, and we extrapolate to the general proposition that it is always true.

This applies not only to physical principles, like the law of gravity, but to abstract phenomena that we observe only in human brains: in ours and in those of others.

Aristotelian realism

Main article: Aristotle's theory of universals

Similar to empiricism in emphasizing the relation of mathematics to the real world, Aristotelian realism holds that mathematics studies properties such as symmetry, continuity and order that can be literally realized in the physical world (or in any other world there might be). It contrasts with Platonism in holding that the objects of mathematics, such as numbers, do not exist in an "abstract" world but can be physically realized. For example, the number 4 is realized in the relation between a heap of parrots and the universal "being a parrot" that divides the heap into so many parrots.[12] Aristotelian realism is defended by James Franklin and the Sydney School in the philosophy of mathematics and is close to the view of Penelope Maddy that when an egg carton is opened, a set of three eggs is perceived (that is, a mathematical entity realized in the physical world).[13] A problem for Aristotelian realism is what account to give of higher infinities, which may not be realizable in the physical world.

2.4.11   Fictionalism

Fictionalism in mathematics was brought to fame in 1980 when Hartry Field published *Science Without Numbers*, which rejected and in fact reversed Quine's indispensability argument. Where Quine suggested that mathematics was indispensable for our best scientific theories, and therefore should be accepted as a body of truths talking about independently existing entities, Field suggested that mathematics was dispensable, and therefore should be considered as a body of falsehoods not talking about anything real. He did this by giving a complete axiomatization of Newtonian mechanics with no reference to numbers or functions at all. He started with the "betweenness" of Hilbert's axioms to characterize space without coordinatizing it, and then added extra relations between points to do the work formerly done by vector fields. Hilbert's geometry is mathematical, because it talks about abstract points, but in Field's theory, these points are the concrete points of physical space, so no special mathematical objects at all are needed.

Having shown how to do science without using numbers, Field proceeded to rehabilitate mathematics as a kind of useful fiction. He showed that mathematical physics is a conservative extension of his non-mathematical physics (that is, every physical fact provable in mathematical physics is already provable from Field's system), so that mathematics is a reliable process whose physical applications are all true, even though its own statements are false. Thus, when doing mathematics, we can see ourselves as telling a sort of story, talking as if numbers existed. For Field, a statement like "2 + 2 = 4" is just as fictitious as "Sherlock Holmes lived at 221B Baker Street"—but both are true according to the relevant fictions.

By this account, there are no metaphysical or epistemological problems special to mathematics. The only worries left are the general worries about non-mathematical physics, and about fiction in general. Field's approach has been very influential, but is widely rejected. This is in part because of the requirement of strong fragments of second-order logic to carry out his reduction, and because the statement of conservativity seems to require quantification over abstract models or deductions.

2.4.12   Social constructivism or social realism

Social constructivism or *social realism* theories see mathematics primarily as a social construct, as a product of culture, subject to correction and change. Like the other sciences, mathematics is viewed as an empirical endeavor whose results are constantly evaluated and may be discarded. However, while on an empiricist view the evaluation is some sort of comparison with "reality", social constructivists emphasize that the direction of mathematical research is dictated by the fashions of the social group performing it or by the needs of the society financing it. However, although such external forces may change the direction of some mathematical research, there are strong internal constraints—the mathematical traditions, methods, problems, meanings and values into which mathematicians are enculturated—that work to conserve the historically defined discipline.

This runs counter to the traditional beliefs of working mathematicians, that mathematics is somehow pure or objective. But social constructivists argue that mathematics is in fact grounded by much uncertainty: as mathematical practice evolves, the status of previous mathematics is cast into doubt, and is corrected to the degree it is required or desired by the current mathematical community. This can be seen in the development of analysis from reexamination of the calculus of Leibniz and Newton. They argue further that finished mathematics is often accorded too much status, and folk mathematics not enough, due to an overemphasis on axiomatic proof and peer review as practices. However, this might be seen as merely saying that rigorously proven results are overemphasized, and then "look how chaotic and uncertain the rest of it all is!"

The social nature of mathematics is highlighted in its subcultures. Major discoveries can be made in one branch of mathematics and be relevant to another, yet the relationship goes undiscovered for lack of social contact between mathematicians. Social constructivists argue each speciality forms its own epistemic community and often has great difficulty communicating, or motivating the investigation of unifying conjectures that might relate different areas of mathematics. Social constructivists see the process of "doing mathematics" as actually creating the meaning, while social realists see a deficiency either of human capacity to abstractify, or of human's cognitive bias, or of mathematicians' collective intelligence as preventing the comprehension of a real universe of mathematical objects. Social constructivists sometimes reject the search for foundations of mathematics as bound to fail, as pointless or even meaningless.

Contributions to this school have been made by Imre Lakatos and Thomas Tymoczko, although it is not clear that either would endorse the title. More recently Paul Ernest has explicitly formulated a social constructivist philosophy of mathematics.[14] Some consider the work of Paul Erdős as a whole to have advanced this view (although he personally rejected it) because of his uniquely broad collaborations, which prompted others to see and study "mathematics as a social activity", e.g., via the Erdős number. Reuben Hersh has also promoted the social view of mathematics, calling it a "humanistic" approach,[15] similar to but not quite the same as that associated with Alvin White;[16] one of Hersh's co-authors, Philip J. Davis, has expressed sympathy for the social view as well.

A criticism of this approach is that it is trivial, based on the trivial observation that mathematics is a human activity. To observe that rigorous proof comes only after unrigorous conjecture, experimentation and speculation is true, but it is trivial and no-one would deny this. So it's a bit of a stretch to characterize a philosophy of mathematics in this way, on something trivially true. The calculus of Leibniz and Newton was reexamined by mathematicians such as Weierstrass in order to rigorously prove the theorems thereof. There is nothing special or interesting about this, as it fits in with the more general trend of unrigorous ideas which are later made rigorous. There needs to be a clear distinction between the objects of study of mathematics and the study of the objects of study of mathematics. The former doesn't seem to change a great deal; the latter is forever in flux. The latter is what the social theory is about, and the former is what Platonism *et al.* are about.

However, this criticism is rejected by supporters of the social constructivist perspective because it misses the point that the very objects of mathematics are social constructs. These objects, it asserts, are primarily semiotic objects existing in the sphere of human culture, sustained by social practices (after Wittgenstein) that utilize physically embodied signs and give rise to intrapersonal (mental) constructs. Social constructivists view the reification of the sphere of human culture into a Platonic realm, or some other heaven-like domain of existence beyond the physical world, a long-standing category error.

2.4.13 Beyond the traditional schools

Rather than focus on narrow debates about the true nature of mathematical truth, or even on practices unique to mathematicians such as the proof, a growing movement from the 1960s to the 1990s began to question the idea of seeking foundations or finding any one right answer to why mathematics works. The starting point for this was Eugene Wigner's famous 1960 paper *The Unreasonable Effectiveness of Mathematics in the Natural Sciences*, in which he argued that the happy coincidence of mathematics and physics being so well matched seemed to be unreasonable and hard to explain.

The embodied-mind or cognitive school and the social school were responses to this challenge, but the debates raised were difficult to confine to those.

Quasi-empiricism

One parallel concern that does not actually challenge the schools directly but instead questions their focus is the notion of quasi-empiricism in mathematics. This grew from the increasingly popular assertion in the late 20th century that no one foundation of mathematics could be ever proven to exist. It is also sometimes called "postmodernism in mathematics" although that term is considered overloaded by some and insulting by others. Quasi-empiricism argues that in doing their research, mathematicians test hypotheses as well as prove theorems. A mathematical argument can transmit falsity from the conclusion to the premises just as well as it can transmit truth from the premises to the conclusion. Quasi-empiricism was developed by Imre Lakatos, inspired by the philosophy of science of Karl Popper.

Lakatos' philosophy of mathematics is sometimes regarded as a kind of social constructivism, but this was not his intention.

Such methods have always been part of folk mathematics by which great feats of calculation and measurement are sometimes achieved. Indeed, such methods may be the only notion of proof a culture has.

Hilary Putnam has argued that any theory of mathematical realism would include quasi-empirical methods. He proposed that an alien species doing mathematics might well rely on quasi-empirical methods primarily, being willing often to forgo rigorous and axiomatic proofs, and still be doing mathematics—at perhaps a somewhat greater risk of failure of their calculations. He gave a detailed argument for this in *New Directions*.[17]

Popper's "two senses" theory

Realist and constructivist theories are normally taken to be contraries. However, Karl Popper[18] argued that a number statement such as "2 apples + 2 apples = 4 apples" can be taken in two senses. In one sense it is irrefutable and logically true. In the second sense it is factually true and falsifiable. Another way of putting this is to say that a single number statement can express two propositions: one of which can be explained on constructivist lines; the other on realist lines.[19]

Language

Main article: Philosophy of language

Innovations in the philosophy of language during the 20th century renewed interest in whether mathematics is, as is often said, the *language* of science. Although some mathematicians and philosophers would accept the statement "mathematics is a language", linguists believe that the implications of such a statement must be considered. For example, the tools of linguistics are not generally applied to the symbol systems of mathematics, that is, mathematics is studied in a markedly different way than other languages. If mathematics is a language, it is a different type of language than natural languages. Indeed, because of the need for clarity and specificity, the language of mathematics is far more constrained than natural languages studied by linguists. However, the methods developed by Frege and Tarski for the study of mathematical language have been extended greatly by Tarski's student Richard Montague and other linguists working in formal semantics to show that the distinction between mathematical language and natural language may not be as great as it seems.

2.5 Arguments

2.5.1 Indispensability argument for realism

This argument, associated with Willard Quine and Hilary Putnam, is considered by Stephen Yablo to be one of the most challenging arguments in favor of the acceptance of the existence of abstract mathematical entities, such as numbers and sets.[20] The form of the argument is as follows.

1. One must have ontological commitments to *all* entities that are indispensable to the best scientific theories, and to those entities *only* (commonly referred to as "all and only").

2. Mathematical entities are indispensable to the best scientific theories. Therefore,

3. One must have ontological commitments to mathematical entities.[21]

The justification for the first premise is the most controversial. Both Putnam and Quine invoke naturalism to justify the exclusion of all non-scientific entities, and hence to defend the "only" part of "all and only". The assertion that "all" entities postulated in scientific theories, including numbers, should be accepted as real is justified by confirmation holism. Since theories are not confirmed in a piecemeal fashion, but as a whole, there is no justification for excluding any of the entities referred to in well-confirmed theories. This puts the nominalist who wishes to exclude the existence of sets and non-Euclidean geometry, but to include the existence of quarks and other undetectable entities of physics, for example, in a difficult position.[21]

2.5.2 Epistemic argument against realism

The anti-realist "epistemic argument" against Platonism has been made by Paul Benacerraf and Hartry Field. Platonism posits that mathematical objects are *abstract* entities. By general agreement, abstract entities cannot interact causally with concrete, physical entities. ("the truth-values of our mathematical assertions depend on facts involving Platonic entities that reside in a realm outside of space-time"[22]) Whilst our knowledge of concrete, physical objects is based on our ability to perceive them, and therefore to causally interact with them, there is no parallel account of how mathematicians come to have knowledge of abstract objects.[23][24][25] ("An account of mathematical truth ... must be consistent with the possibility of mathematical knowledge."[26]) Another way of making the point is that if the Platonic world were to disappear, it would make no difference to the ability of mathematicians to generate proofs, etc., which is already fully accountable in terms of physical processes in their brains.

Field developed his views into fictionalism. Benacerraf also developed the philosophy of mathematical structuralism, according to which there are no mathematical objects. Nonetheless, some versions of structuralism are compatible with some versions of realism.

The argument hinges on the idea that a satisfactory naturalistic account of thought processes in terms of brain processes can be given for mathematical reasoning along with everything else. One line of defense is to maintain that this is false, so that mathematical reasoning uses some special intuition that involves contact with the Platonic realm. A modern form of this argument is given by Sir Roger Penrose.[27]

Another line of defense is to maintain that abstract objects are relevant to mathematical reasoning in a way that is non-causal, and not analogous to perception. This argument is developed by Jerrold Katz in his book *Realistic Rationalism*.

A more radical defense is denial of physical reality, i.e. the mathematical universe hypothesis. In that case, a mathematician's knowledge of mathematics is one mathematical object making contact with another.

2.6 Aesthetics

Many practicing mathematicians have been drawn to their subject because of a sense of beauty they perceive in it. One sometimes hears the sentiment that mathematicians would like to leave philosophy to the philosophers and get back to mathematics—where, presumably, the beauty lies.

In his work on the divine proportion, H.E. Huntley relates the feeling of reading and understanding someone else's proof of a theorem of mathematics to that of a viewer of a masterpiece of art—the reader of a proof has a similar sense of exhilaration at understanding as the original author of the proof, much as, he argues, the viewer of a masterpiece has a sense of exhilaration similar to the original painter or sculptor. Indeed, one can study mathematical and scientific writings as literature.

Philip J. Davis and Reuben Hersh have commented that the sense of mathematical beauty is universal amongst practicing mathematicians. By way of example, they provide two proofs of the irrationality of $\sqrt{2}$. The first is the traditional proof by contradiction, ascribed to Euclid; the second is a more direct proof involving the fundamental theorem of arithmetic that, they argue, gets to the heart of the issue. Davis and Hersh argue that mathematicians find the second proof more aesthetically appealing because it gets closer to the nature of the problem.

Paul Erdős was well known for his notion of a hypothetical "Book" containing the most elegant or beautiful mathematical proofs. There is not universal agreement that a result has one "most elegant" proof; Gregory Chaitin has argued against this idea.

Philosophers have sometimes criticized mathematicians' sense of beauty or elegance as being, at best, vaguely stated. By the same token, however, philosophers of mathematics have sought to characterize what makes one proof more desirable than another when both are logically sound.

Another aspect of aesthetics concerning mathematics is mathematicians' views towards the possible uses of mathematics for purposes deemed unethical or inappropriate. The best-known exposition of this view occurs in G.H. Hardy's book *A Mathematician's Apology*, in which Hardy argues that pure mathematics is superior in beauty to applied mathematics precisely because it cannot be used for war and similar ends. Some later mathematicians have characterized Hardy's views as mildly dated, with the applicability of number theory to modern-day cryptography.

2.7 See also

2.7.1 Related works

2.7.2 Historical topics

- History and philosophy of science
- History of mathematics
- History of philosophy

2.8 Notes

[1] Maziars, Edward A. (1969). "Problems in the Philosophy of Mathematics (Book Review)". *Philosophy of Science* **36** (3): 325.. For example, when Edward Maziars proposes in a 1969 book review "*to distinguish philosophical mathematics (which is primarily a specialised task for a mathematician) from mathematical philosophy (which ordinarily may be the philosopher's metier)*", he uses the term *mathematical philosophy* as being synonymous with *philosophy of mathematics*.

[2] Kleene, Stephen (1971). *Introduction to Metamathematics*. Amsterdam, Netherlands: North-Holland Publishing Company. p. 5.

[3] Mac Lane, Saunders (1998), *Categories for the Working Mathematician*, 2nd edition, Springer-Verlag, New York, NY.

[4]   • Putnam, Hilary (1967), "Mathematics Without Foundations", *Journal of Philosophy* 64/1, 5-22. Reprinted, pp. 168–184 in W.D. Hart (ed., 1996).

[5] Tegmark, Max (February 2008). "The Mathematical Universe". *Foundations of Physics* **38** (2): 101–150. arXiv:0704.0646. Bibcode:2008FoPh...38..101T. doi:10.1007/s10701-007-9186-9.

[6] Tegmark (1998), p. 1.

[7] Carnap, Rudolf (1931), "Die logizistische Grundlegung der Mathematik", *Erkenntnis* 2, 91-121. Republished, "The Logicist Foundations of Mathematics", E. Putnam and G.J. Massey (trans.), in Benacerraf and Putnam (1964). Reprinted, pp. 41–52 in Benacerraf and Putnam (1983).

[8] Audi, Robert (1999), *The Cambridge Dictionary of Philosophy*, Cambridge University Press, Cambridge, UK, 1995. 2nd edition. Page 542.

[9] From an 1886 lecture at the 'Berliner Naturforscher-Versammlung', according to H. M. Weber's memorial article, as quoted and translated in Gonzalez Cabillon, Julio (2000-02-03). "FOM: What were Kronecker's f.o.m.?". Retrieved 2008-07-19. Gonzalez gives as the sources for the memorial article, the following: 'Weber, H: "Leopold Kronecker". _Jahresberichte der Deutschen Mathematiker Vereinigung_, vol ii (1893) pp 5-31. Cf page 19. See also _Mathematische Annalen_ vol xliii (1893) pp 1-25'.

[10] Brown, James (2008). *Philosophy of Mathematics*. New York: Routledge. ISBN 978-0-415-96047-2.

[11] A System of Logic Ratiocinative and Inductive, The Collected Works of John Stuart Mill published by the University of Toronto Press in 1973. Book II, Chapter vi, Section 2 (Toronto edition 1975, Vol.7, p. 254)

[12] Franklin, James (2014), "An Aristotelian Realist Philosophy of Mathematics", Palgrave Macmillan, Basingstoke; Franklin, James (2011), "Aristotelianism in the philosophy of mathematics," *Studia Neoaristotelica* 8, 3-15.

[13] Maddy, Penelope (1990), *Realism in Mathematics*, Oxford University Press, Oxford, UK.

[14] Ernest, Paul. "Is Mathematics Discovered or Invented?". University of Exeter. Retrieved 2008-12-26.

[15] Hersh, Reuben (February 10, 1997). *What Kind of a Thing is a Number?*. Interview with John Brockman. Edge Foundation. Retrieved 2008-12-26.

[16] "Humanism and Mathematics Education". *Math Forum*. Humanistic Mathematics Network Journal. Retrieved 2008-12-26.

[17] Tymoczko, Thomas (1998), *New Directions in the Philosophy of Mathematics*. ISBN 978-0691034980.

[18] Popper, Karl Raimund (1946) Aristotelian Society Supplementary Volume XX.

[19] Gregory, Frank Hutson (1996) Arithmetic and Reality: A Development of Popper's Ideas. City University of Hong Kong. Republished in Philosophy of Mathematics Education Journal No. 26 (December 2011)

[20] Yablo, S. (November 8, 1998). "A Paradox of Existence".

[21] Putnam, H. *Mathematics, Matter and Method. Philosophical Papers, vol. 1*. Cambridge: Cambridge University Press, 1975. 2nd. ed., 1985.

[22] Field, Hartry, 1989, Realism, Mathematics, and Modality, Oxford: Blackwell, p. 68

[23] "Since abstract objects are outside the nexus of causes and effects, and thus perceptually inaccessible, they cannot be known through their effects on us" Katz, J. *Realistic Rationalism*, p15

[24] .Philosophy Now: *Mathematical_Knowledge_A_Dilemma Mathematical Knowledge: A dilemma*

[25] Standard Encyclopaedia of Philosophy

[26] Benacceraf. 1973, p409

[27] Review of The Emperor's New Mind

2.9 Further reading

- Aristotle, "Prior Analytics", Hugh Tredennick (trans.), pp. 181–531 in *Aristotle, Volume 1*, Loeb Classical Library, William Heinemann, London, UK, 1938.

- Benacerraf, Paul, and Putnam, Hilary (eds., 1983), *Philosophy of Mathematics, Selected Readings*, 1st edition, Prentice-Hall, Englewood Cliffs, NJ, 1964. 2nd edition, Cambridge University Press, Cambridge, UK, 1983.

- Berkeley, George (1734), *The Analyst; or, a Discourse Addressed to an Infidel Mathematician. Wherein It is examined whether the Object, Principles, and Inferences of the modern Analysis are more distinctly conceived, or more evidently deduced, than Religious Mysteries and Points of Faith*, London & Dublin. Online text, David R. Wilkins (ed.), Eprint.

- Bourbaki, N. (1994), *Elements of the History of Mathematics*, John Meldrum (trans.), Springer-Verlag, Berlin, Germany.

- Chandrasekhar, Subrahmanyan (1987), *Truth and Beauty. Aesthetics and Motivations in Science*, University of Chicago Press, Chicago, IL.

- Colyvan, Mark (2004), "Indispensability Arguments in the Philosophy of Mathematics", *Stanford Encyclopedia of Philosophy*, Edward N. Zalta (ed.), Eprint.

- Davis, Philip J. and Hersh, Reuben (1981), *The Mathematical Experience*, Mariner Books, New York, NY.

- Devlin, Keith (2005), *The Math Instinct: Why You're a Mathematical Genius (Along with Lobsters, Birds, Cats, and Dogs)*, Thunder's Mouth Press, New York, NY.

- Dummett, Michael (1991 a), *Frege, Philosophy of Mathematics*, Harvard University Press, Cambridge, MA.

- Dummett, Michael (1991 b), *Frege and Other Philosophers*, Oxford University Press, Oxford, UK.

- Dummett, Michael (1993), *Origins of Analytical Philosophy*, Harvard University Press, Cambridge, MA.

- Ernest, Paul (1998), *Social Constructivism as a Philosophy of Mathematics*, State University of New York Press, Albany, NY.

- George, Alexandre (ed., 1994), *Mathematics and Mind*, Oxford University Press, Oxford, UK.

- Hadamard, Jacques (1949), *The Psychology of Invention in the Mathematical Field*, 1st edition, Princeton University Press, Princeton, NJ. 2nd edition, 1949. Reprinted, Dover Publications, New York, NY, 1954.

- Hardy, G.H. (1940), *A Mathematician's Apology*, 1st published, 1940. Reprinted, C.P. Snow (foreword), 1967. Reprinted, Cambridge University Press, Cambridge, UK, 1992.

- Hart, W.D. (ed., 1996), *The Philosophy of Mathematics*, Oxford University Press, Oxford, UK.

- Hendricks, Vincent F. and Hannes Leitgeb (eds.). *Philosophy of Mathematics: 5 Questions*, New York: Automatic Press / VIP, 2006.

- Huntley, H.E. (1970), *The Divine Proportion: A Study in Mathematical Beauty*, Dover Publications, New York, NY.

- Irvine, A., ed (2009), *The Philosophy of Mathematics*, in *Handbook of the Philosophy of Science* series, North-Holland Elsevier, Amsterdam.

- Klein, Jacob (1968), *Greek Mathematical Thought and the Origin of Algebra*, Eva Brann (trans.), MIT Press, Cambridge, MA, 1968. Reprinted, Dover Publications, Mineola, NY, 1992.

- Kline, Morris (1959), *Mathematics and the Physical World*, Thomas Y. Crowell Company, New York, NY, 1959. Reprinted, Dover Publications, Mineola, NY, 1981.

- Kline, Morris (1972), *Mathematical Thought from Ancient to Modern Times*, Oxford University Press, New York, NY.

- König, Julius (Gyula) (1905), "Über die Grundlagen der Mengenlehre und das Kontinuumproblem", *Mathematische Annalen* 61, 156-160. Reprinted, "On the Foundations of Set Theory and the Continuum Problem", Stefan Bauer-Mengelberg (trans.), pp. 145–149 in Jean van Heijenoort (ed., 1967).

- Körner, Stephan, *The Philosophy of Mathematics, An Introduction*. Harper Books, 1960.

- Lakoff, George, and Núñez, Rafael E. (2000), *Where Mathematics Comes From: How the Embodied Mind Brings Mathematics into Being*, Basic Books, New York, NY.

- Lakatos, Imre 1976 *Proofs and Refutations:The Logic of Mathematical Discovery* (Eds) J. Worrall & E. Zahar Cambridge University Press

- Lakatos, Imre 1978 *Mathematics, Science and Epistemology: Philosophical Papers* Volume 2 (Eds) J.Worrall & G.Currie Cambridge University Press

- Lakatos, Imre 1968 *Problems in the Philosophy of Mathematics* North Holland

- Leibniz, G.W., *Logical Papers* (1666–1690), G.H.R. Parkinson (ed., trans.), Oxford University Press, London, UK, 1966.

- Maddy, Penelope (1997), *Naturalism in Mathematics*, Oxford University Press, Oxford, UK.

- Maziarz, Edward A., and Greenwood, Thomas (1995), *Greek Mathematical Philosophy*, Barnes and Noble Books.

- Mount, Matthew, *Classical Greek Mathematical Philosophy*, .

- Parsons, Charles (2014). *Philosophy of Mathematics in the Twentieth Century: Selected Essays*. Cambridge, MA: Harvard University Press. ISBN 978-0-674-72806-6.

- Peirce, Benjamin (1870), "Linear Associative Algebra", § 1. See *American Journal of Mathematics* 4 (1881).

- Peirce, C.S., *Collected Papers of Charles Sanders Peirce*, vols. 1-6, Charles Hartshorne and Paul Weiss (eds.), vols. 7-8, Arthur W. Burks (ed.), Harvard University Press, Cambridge, MA, 1931 – 1935, 1958. Cited as CP (volume).(paragraph).

- Peirce, C.S., various pieces on mathematics and logic, many readable online through links at the Charles Sanders Peirce bibliography, especially under Books authored or edited by Peirce, published in his lifetime and the two sections following it.

- Plato, "The Republic, Volume 1", Paul Shorey (trans.), pp. 1–535 in *Plato, Volume 5*, Loeb Classical Library, William Heinemann, London, UK, 1930.

- Plato, "The Republic, Volume 2", Paul Shorey (trans.), pp. 1–521 in *Plato, Volume 6*, Loeb Classical Library, William Heinemann, London, UK, 1935.

- Resnik, Michael D. *Frege and the Philosophy of Mathematics*, Cornell University, 1980.

- Resnik, Michael (1997), *Mathematics as a Science of Patterns*, Clarendon Press, Oxford, UK. ISBN 978-0-19-825014-2

- Robinson, Gilbert de B. (1959), *The Foundations of Geometry*. University of Toronto Press, Toronto, Canada, 1940, 1946, 1952, 4th edition 1959.

- Raymond, Eric S. (1993), "The Utility of Mathematics", Eprint.

- Smullyan, Raymond M. (1993), *Recursion Theory for Metamathematics*, Oxford University Press, Oxford, UK.

- Russell, Bertrand (1919), *Introduction to Mathematical Philosophy*, George Allen and Unwin, London, UK. Reprinted, John G. Slater (intro.), Routledge, London, UK, 1993.

- Shapiro, Stewart (2000), *Thinking About Mathematics: The Philosophy of Mathematics*, Oxford University Press, Oxford, UK

- Strohmeier, John, and Westbrook, Peter (1999), *Divine Harmony, The Life and Teachings of Pythagoras*, Berkeley Hills Books, Berkeley, CA.

- Styazhkin, N.I. (1969), *History of Mathematical Logic from Leibniz to Peano*, MIT Press, Cambridge, MA.

- Tait, William W. (1986), "Truth and Proof: The Platonism of Mathematics", *Synthese* 69 (1986), 341-370. Reprinted, pp. 142–167 in W.D. Hart (ed., 1996).

- Tarski, A. (1983), *Logic, Semantics, Metamathematics: Papers from 1923 to 1938*, J.H. Woodger (trans.), Oxford University Press, Oxford, UK, 1956. 2nd edition, John Corcoran (ed.), Hackett Publishing, Indianapolis, IN, 1983.

- Ulam, S.M. (1990), *Analogies Between Analogies: The Mathematical Reports of S.M. Ulam and His Los Alamos Collaborators*, A.R. Bednarek and Françoise Ulam (eds.), University of California Press, Berkeley, CA.

- van Heijenoort, Jean (ed. 1967), *From Frege To Gödel: A Source Book in Mathematical Logic, 1879-1931*, Harvard University Press, Cambridge, MA.

- Wigner, Eugene (1960), "The Unreasonable Effectiveness of Mathematics in the Natural Sciences", *Communications on Pure and Applied Mathematics* **13**(1): 1-14. Eprint

- Wilder, Raymond L. *Mathematics as a Cultural System*, Pergamon, 1980.

- Witzany, Guenther (2011), *Can mathematics explain the evolution of human language?*, Communicative and Integrative Biology, 4(5): 516-520.

2.10 External links

- Philosophy of mathematics at PhilPapers

- Philosophy of mathematics at the Indiana Philosophy Ontology Project

- Philosophy of Mathematics entry by Leon Horsten in the *Stanford Encyclopedia of Philosophy*

- Philosophy of mathematics entry in the *Internet Encyclopedia of Philosophy*

- The London Philosophy Study Guide offers many suggestions on what to read, depending on the student's familiarity with the subject:

  - Philosophy of Mathematics
  - Mathematical Logic
  - Set Theory & Further Logic

- R.B. Jones' philosophy of mathematics page

- Philosophy of mathematics at DMOZ

- The Philosophy of Real Mathematics Blog

- Kaina Stoicheia by C.S. Peirce.

2.10.1 Journals

- Philosophia Mathematica journal

- The Philosophy of Mathematics Education Journal homepage

Chapter 3

Introduction to Mathematical Philosophy

Introduction to Mathematical Philosophy is a book by Bertrand Russell, published in 1919, written in part to exposit in a less technical way the main ideas of his and Whitehead's *Principia Mathematica* (1910–1913), including the theory of descriptions.

Mathematics and logic, historically speaking, have been entirely distinct studies. Mathematics has been connected with science, logic with Greek. But both have developed in modern times: logic has become more mathematical and mathematics has become more logical. The consequence is that it has now become wholly impossible to draw a line between the two; in fact, the two are one. They differ as boy and man: logic is the youth of mathematics and mathematics is the manhood of logic. This view is resented by logicians who, having spent their time in the study of classical texts, are incapable of following a piece of symbolic reasoning, and by mathematicians who have learnt a technique without troubling to inquire into its meaning or justification. Both types are now fortunately growing rarer. So much of modern mathematical work is obviously on the border-line of logic, so much of modern logic is symbolic and formal, that the very close relationship of logic and mathematics has become obvious to every instructed student. The proof of their identity is, of course, a matter of detail: starting with premises which would be universally admitted to belong to logic, and arriving by deduction at results which as obviously belong to mathematics, we find that there is no point at which a sharp line can be drawn, with logic to the left and mathematics to the right. If there are still those who do not admit the identity of logic and mathematics, we may challenge them to indicate at what point, in the successive definitions and deductions of *Principia Mathematica*, they consider that logic ends and mathematics begins. It will then be obvious that any answer must be quite arbitrary. (Russell 1919, 194–195).

3.1 References

- Russell, Bertrand (1919), *Introduction to Mathematical Philosophy*, George Allen and Unwin, London, UK. Reprinted, John G. Slater (intro.), Routledge, London, UK. 1993.

3.2 See also

- Foundations of mathematics

- Philosophy of mathematics

- Introduction to Mathematical Philosophy (free online version created by Kevin C. Klement)

- Introduction to Mathematical Philosophy at the Internet Archive

Chapter 4

Outline of mathematics

The following outline is provided as an overview of and topical guide to mathematics:

Mathematics – the search for fundamental truths in pattern, quantity, and change. For more on the relationship between mathematics and science, refer to the article on science.

4.1 Nature of mathematics

- Definitions of mathematics – Mathematics has no generally accepted definition. Different schools of thought, particularly in philosophy, have put forth radically different definitions, all of which are controversial.

- Philosophy of mathematics – its aim is to provide an account of the nature and methodology of mathematics and to understand the place of mathematics in people's lives.

4.1.1 Mathematics is

- an academic discipline – branch of knowledge that is taught and researched at the college or university level. Disciplines are defined (in part), and recognized by the academic journals in which research is published, and the learned societies and academic departments or faculties to which their practitioners belong.

- a formal science – branch of knowledge concerned with the properties of formal systems based on definitions and rules of inference. Unlike other sciences, the formal sciences are not concerned with the validity of theories based on observations in the real world.

4.1.2 General reference

Classification systems

- Mathematics in the Dewey Decimal Classification system

- *Mathematics Subject Classification* – alphanumerical classification scheme collaboratively produced by staff of and based on the coverage of the two major mathematical reviewing databases, Mathematical Reviews and Zentralblatt MATH.

Reference databases

- *Mathematical Reviews* – journal and online database published by the American Mathematical Society (AMS) that contains brief synopses (and occasionally evaluations) of many articles in mathematics, statistics and theoretical computer science.

- *Zentralblatt MATH* – service providing reviews and abstracts for articles in pure and applied mathematics, published by Springer Science+Business Media. It is a major international reviewing service which covers the entire field of mathematics. It uses the Mathematics Subject Classification codes for organizing their reviews by topic.

4.2   Subjects

4.2.1   Quantity

Quantity –

- Arithmetic –
- Natural numbers –
- Integers –
- Rational numbers –
- Real numbers –
- Complex numbers –
- Hypercomplex numbers –
- Infinity –

4.2.2   Structure

Structure –

- Abstract algebra –
- Linear algebra –
- Number theory –
- Order theory –
- Function (mathematics) –

4.2.3   Space

Space –

- Geometry –
- Algebraic geometry –
- Trigonometry –
- Differential geometry –
- Topology –
- Fractal geometry –

4.2.4 Change

Change –

- Calculus –
- Vector calculus –
- Differential equations –
- Dynamical systems –
- Chaos theory –
- Analysis –

4.2.5 Foundations and philosophy

Foundations of mathematics –

- Philosophy of mathematics –
- Category theory –
- Set theory –
- Type theory –

4.2.6 Mathematical logic

Mathematical logic –

- Model theory –
- Proof theory –
- Recursion theory –
- Set theory –
- Type theory –

4.2.7 Discrete mathematics

Discrete mathematics –

- Combinatorics
- Theory of computation
- Cryptography
- Graph theory

4.2.8 Applied mathematics

Applied mathematics –

- Mathematical physics –
- Analytical mechanics –
- Mathematical fluid dynamics –
- Numerical analysis –
- Mathematical optimization –
- Probability –
- Statistics –
- Mathematical economics –
- Financial mathematics –
- Game theory –
- Mathematical biology –
- Cryptography –
- Operations research –
- Information theory –
- Control theory –
- Dynamical systems –

4.3 History

Main article: History of mathematics

- Babylonian mathematics
- Egyptian mathematics
- Indian mathematics
- Greek mathematics
- Chinese mathematics
    - Abacus
- History of the Hindu-Arabic numeral system
- Islamic mathematics
- Japanese mathematics
- History of algebra

- History of geometry

- History of mathematical notation

- History of trigonometry

- History of writing numbers

4.4 Psychology

- Mathematics education

- Numeracy

- Numerical Cognition

- Subitizing

- Mathematical anxiety

- Dyscalculia

- Acalculia

- Ageometresia

- Number sense

- Numerosity adaptation effect

- Approximate number system

- Mathematical maturity

4.5 Influential mathematicians

See Lists of mathematicians

4.6 Mathematical notation

Main article: Mathematical notation

- List of mathematical abbreviations

- List of mathematical symbols

- List of mathematical symbols by subject

- Table of mathematical symbols by introduction date

- Notation in probability and statistics

- Table of logic symbols

- Physical constants

- Greek letters used in mathematics, science, and engineering

- Latin letters used in mathematics

- Mathematical alphanumeric symbols

- Mathematical operators and symbols in Unicode

- ISO 31-11 (Mathematical signs and symbols for use in physical sciences and technology)

4.7   See also

- Lists of mathematics topics

- Areas of mathematics

- Glossary of areas of mathematics

4.8   External links

- MAA Reviews – The Basic Library List – Mathematical Association of America

- Naoki's Recommended Books, compiled by Naoki Saito, U. C. Davis

- A List of Recommended Books in Topology, compiled by Allen Hatcher, Cornell U.

Chapter 5

Model theory

This article is about the mathematical discipline. For the informal notion in other parts of mathematics and science, see Mathematical model.

In mathematics, **model theory** is the study of classes of mathematical structures (e.g. groups, fields, graphs, universes of set theory) from the perspective of mathematical logic. The objects of study are models of theories in a formal language. We call a set of sentences in a formal language a **theory**; a **model** of a theory is a structure (e.g. an interpretation) that satisfies the sentences of that theory.

Model theory recognises and is intimately concerned with a duality: It examines semantical elements (meaning and truth) by means of syntactical elements (formulas and proofs) of a corresponding language. To quote the first page of Chang & Keisler (1990):[1]

> universal algebra + logic = **model theory**.

Model theory developed rapidly during the 1990s, and a more modern definition is provided by Wilfrid Hodges (1997):

> **model theory** = algebraic geometry − fields,

although model theorists are also interested in the study of fields. Other nearby areas of mathematics include combinatorics, number theory, arithmetic dynamics, analytic functions, and non-standard analysis.

In a similar way to proof theory, model theory is situated in an area of interdisciplinarity among mathematics, philosophy, and computer science. The most prominent professional organization in the field of model theory is the Association for Symbolic Logic.

5.1   Branches of model theory

This article focuses on finitary first order model theory of infinite structures. Finite model theory, which concentrates on finite structures, diverges significantly from the study of infinite structures in both the problems studied and the techniques used. Model theory in higher-order logics or infinitary logics is hampered by the fact that completeness does not in general hold for these logics. However, a great deal of study has also been done in such languages.

Informally, model theory can be divided into classical model theory, model theory applied to groups and fields, and geometric model theory. A missing subdivision is computable model theory, but this can arguably be viewed as an independent subfield of logic.

Examples of early theorems from classical model theory include Gödel's completeness theorem, the upward and downward Löwenheim–Skolem theorems, Vaught's two-cardinal theorem, Scott's isomorphism theorem, the omitting types theorem, and the Ryll-Nardzewski theorem. Examples of early results from model theory applied to fields are Tarski's elimination of quantifiers for real closed fields, Ax's theorem on pseudo-finite fields, and Robinson's development of non-standard analysis. An important step in the evolution of classical model theory occurred with the birth of stability theory (through Morley's theorem on uncountably categorical theories and Shelah's classification program), which developed a calculus of independence and rank based on syntactical conditions satisfied by theories.

During the last several decades applied model theory has repeatedly merged with the more pure stability theory. The result of this synthesis is called geometric model theory in this article (which is taken to include o-minimality, for example, as well as classical geometric stability theory). An example of a theorem from geometric model theory is Hrushovski's proof of the Mordell–Lang conjecture for function fields. The ambition of geometric model theory is to provide a *geography of mathematics* by embarking on a detailed study of definable sets in various mathematical structures, aided by the substantial tools developed in the study of pure model theory.

5.2 Universal algebra

Main article: Universal algebra

Fundamental concepts in universal algebra are signatures σ and σ-algebras. Since these concepts are formally defined in the article on structures, the present article can content itself with an informal introduction which consists in examples of how these terms are used.

> The standard signature of rings is $\sigma_{ring} = \{\times,+,-,0,1\}$, where $\times$ and $+$ are binary, $-$ is unary, and 0 and 1 are nullary.
>
> The standard signature of semirings is $\sigma_{smr} = \{\times,+,0,1\}$, where the arities are as above.
>
> The standard signature of groups (with multiplicative notation) is $\sigma_{grp} = \{\times,^{-1},1\}$, where $\times$ is binary, $^{-1}$ is unary and 1 is nullary.
>
> The standard signature of monoids is $\sigma_{mnd} = \{\times,1\}$.
>
> A ring is a σ_{ring}-structure which satisfies the identities $u + (v + w) = (u + v) + w, u + v = v + u, u + 0 = u, u + (-u) = 0, u \times (v \times w) = (u \times v) \times w, u \times 1 = u, 1 \times u = u, u \times (v + w) = (u \times v) + (u \times w)$ and $(v + w) \times u = (v \times u) + (w \times u)$.
>
> A group is a σ_{grp}-structure which satisfies the identities $u \times (v \times w) = (u \times v) \times w, u \times 1 = u, 1 \times u = u, u \times u^{-1} = 1$ and $u^{-1} \times u = 1$.
>
> A monoid is a σ_{mnd}-structure which satisfies the identities $u \times (v \times w) = (u \times v) \times w, u \times 1 = u$ and $1 \times u = u$.
>
> A semigroup is a $\{\times\}$-structure which satisfies the identity $u \times (v \times w) = (u \times v) \times w$.
>
> A magma is just a $\{\times\}$-structure.

This is a very efficient way to define most classes of algebraic structures, because there is also the concept of σ-homomorphism, which correctly specializes to the usual notions of homomorphism for groups, semigroups, magmas and rings. For this to work, the signature must be chosen well.

Terms such as the σ_{ring}-term $t(u,v,w)$ given by $(u + (v \times w)) + (-1)$ are used to define identities $t = t'$, but also to construct free algebras. An equational class is a class of structures which, like the examples above and many others, is defined as the class of all σ-structures which satisfy a certain set of identities. Birkhoff's theorem states:

> A class of σ-structures is an equational class if and only if it is not empty and closed under subalgebras, homomorphic images, and direct products.

An important non-trivial tool in universal algebra are ultraproducts $\prod_{i \in I} A_i / U$, where I is an infinite set indexing a system of σ-structures Ai, and U is an ultrafilter on I.

While model theory is generally considered a part of mathematical logic, universal algebra, which grew out of Alfred North Whitehead's (1898) work on abstract algebra, is part of algebra. This is reflected by their respective MSC classifications. Nevertheless model theory can be seen as an extension of universal algebra.

5.3   Finite model theory

Main article: Finite model theory

Finite model theory is the area of model theory which has the closest ties to universal algebra. Like some parts of universal algebra, and in contrast with the other areas of model theory, it is mainly concerned with finite algebras, or more generally, with finite σ-structures for signatures σ which may contain relation symbols as in the following example:

> The standard signature for graphs is $\sigma_{grph}=\{E\}$, where E is a binary relation symbol.
>
> A graph is a σ_{grph}-structure satisfying the sentences $\forall u \forall v (uEv \to vEu)$ and $\forall u \neg (uEu)$.

A σ-homomorphism is a map that commutes with the operations and preserves the relations in σ. This definition gives rise to the usual notion of graph homomorphism, which has the interesting property that a bijective homomorphism need not be invertible. Structures are also a part of universal algebra; after all, some algebraic structures such as ordered groups have a binary relation <. What distinguishes finite model theory from universal algebra is its use of more general logical sentences (as in the example above) in place of identities. (In a model-theoretic context an identity $t=t'$ is written as a sentence $\forall u_1 u_2 \ldots u_n (t = t')$.)

The logics employed in finite model theory are often substantially more expressive than first-order logic, the standard logic for model theory of infinite structures.

5.4   First-order logic

Main article: First-order logic

Whereas universal algebra provides the semantics for a signature, logic provides the syntax. With terms, identities and quasi-identities, even universal algebra has some limited syntactic tools; first-order logic is the result of making quantification explicit and adding negation into the picture.

A first-order **formula** is built out of atomic formulas such as $R(f(x,y),z)$ or $y = x + 1$ by means of the Boolean connectives ¬, ∧, ∨, → and prefixing of quantifiers $\forall v$ or $\exists v$. A sentence is a formula in which each occurrence of a variable is in the scope of a corresponding quantifier. Examples for formulas are φ (or φ(x) to mark the fact that at most x is an unbound variable in φ) and ψ defined as follows:

$$\varphi = \forall u \forall v (\exists w (x \times w = u \times v) \to (\exists w (x \times w = u) \lor \exists w (x \times w = v))) \land x \neq 0 \land x \neq 1.$$

$$\psi = \forall u \forall v ((u \times v = x) \to (u = x) \lor (v = x)) \land x \neq 0 \land x \neq 1.$$

(Note that the equality symbol has a double meaning here.) It is intuitively clear how to translate such formulas into mathematical meaning. In the σ_{smr}-structure $\mathcal{N}$ of the natural numbers, for example, an element n **satisfies** the formula φ if and only if n is a prime number. The formula ψ similarly defines irreducibility. Tarski gave a rigorous definition, sometimes called "Tarski's definition of truth", for the satisfaction relation $\models$, so that one easily proves:

$$\mathcal{N} \models \varphi(n) \iff n$$

$$\mathcal{N} \models \psi(n) \iff n$$

A set T of sentences is called a (first-order) theory. A theory is **satisfiable** if it has a **model** $\mathcal{M} \models T$, i.e. a structure (of the appropriate signature) which satisfies all the sentences in the set T. Consistency of a theory is usually defined in a syntactical way, but in first-order logic by the completeness theorem there is no need to distinguish between satisfiability and consistency. Therefore model theorists often use "consistent" as a synonym for "satisfiable".

A theory is called **categorical** if it determines a structure up to isomorphism, but it turns out that this definition is not useful, due to serious restrictions in the expressivity of first-order logic. The Löwenheim–Skolem theorem implies that for every theory $T^{[2]}$ which has an infinite model and for every infinite cardinal number κ, there is a model $\mathcal{M} \models T$ such that the number of elements of $\mathcal{M}$ is exactly κ. Therefore only finitary structures can be described by a categorical theory.

Lack of expressivity (when compared to higher logics such as second-order logic) has its advantages, though. For model theorists, the Löwenheim–Skolem theorem is an important practical tool rather than the source of Skolem's paradox. In a certain sense made precise by Lindström's theorem, first-order logic is the most expressive logic for which both the Löwenheim–Skolem theorem and the compactness theorem hold.

As a corollary (i.e., its contrapositive), the compactness theorem says that every unsatisfiable first-order theory has a finite unsatisfiable subset. This theorem is of central importance in infinite model theory, where the words "by compactness" are commonplace. One way to prove it is by means of ultraproducts. An alternative proof uses the completeness theorem, which is otherwise reduced to a marginal role in most of modern model theory.

5.5 Axiomatizability, elimination of quantifiers, and model-completeness

The first step, often trivial, for applying the methods of model theory to a class of mathematical objects such as groups, or trees in the sense of graph theory, is to choose a signature σ and represent the objects as σ-structures. The next step is to show that the class is an elementary class, i.e. axiomatizable in first-order logic (i.e. there is a theory T such that a σ-structure is in the class if and only if it satisfies T). E.g. this step fails for the trees, since connectedness cannot be expressed in first-order logic. Axiomatizability ensures that model theory can speak about the right objects. Quantifier elimination can be seen as a condition which ensures that model theory does not say too much about the objects.

A theory T has quantifier elimination if every first-order formula $\varphi(x_1,....,xn)$ over its signature is equivalent modulo T to a first-order formula $\psi(x_1,....,xn)$ without quantifiers, i.e. $\forall x_1 \ldots \forall x_n(\phi(x_1,\ldots,x_n) \leftrightarrow \psi(x_1,\ldots,x_n))$ holds in all models of T. For example the theory of algebraically closed fields in the signature $\sigma_{ring}=(\times,+,-,0,1)$ has quantifier elimination because every formula is equivalent to a Boolean combination of equations between polynomials.

A substructure of a σ-structure is a subset of its domain, closed under all functions in its signature σ, which is regarded as a σ-structure by restricting all functions and relations in σ to the subset. An embedding of a σ-structure $\mathcal{A}$ into another σ-structure $\mathcal{B}$ is a map f: A $\rightarrow$ B between the domains which can be written as an isomorphism of $\mathcal{A}$ with a substructure of $\mathcal{B}$. Every embedding is an injective homomorphism, but the converse holds only if the signature contains no relation symbols.

If a theory does not have quantifier elimination, one can add additional symbols to its signature so that it does. Early model theory spent much effort on proving axiomatizability and quantifier elimination results for specific theories, especially in algebra. But often instead of quantifier elimination a weaker property suffices:

A theory T is called model-complete if every substructure of a model of T which is itself a model of T is an elementary substructure. There is a useful criterion for testing whether a substructure is an elementary substructure, called the Tarski–Vaught test. It follows from this criterion that a theory T is model-complete if and only if every first-order formula $\varphi(x_1,....,xn)$ over its signature is equivalent modulo T to an existential first-order formula, i.e. a formula of the following form:

$$\exists v_1 \ldots \exists v_m \psi(x_1,\ldots,x_n,v_1,\ldots,v_m)$$

where ψ is quantifier free. A theory that is not model-complete may or may not have a **model completion**, which is a related model-complete theory that is not, in general, an extension of the original theory. A more general notion is that of **model companions**.

5.6 Categoricity

As observed in the section on first-order logic, first-order theories cannot be categorical, i.e. they cannot describe a unique model up to isomorphism, unless that model is finite. But two famous model-theoretic theorems deal with the weaker notion of κ-categoricity for a cardinal κ. A theory T is called **κ-categorical** if any two models of T that are of cardinality κ are isomorphic. It turns out that the question of κ-categoricity depends critically on whether κ is bigger than the cardinality of the language (i.e. $\aleph_0 + |\sigma|$, where $|\sigma|$ is the cardinality of the signature). For finite or countable signatures this means that there is a fundamental difference between $\aleph_0$-cardinality and κ-cardinality for uncountable κ.

A few characterizations of $\aleph_0$-categoricity include:

For a complete first-order theory T in a finite or countable signature the following conditions are equivalent:

1. T is $\aleph_0$-categorical.
2. For every natural number n, the Stone space $S_n(T)$ is finite.
3. For every natural number n, the number of formulas $\varphi(x_1, ..., x_n)$ in n free variables, up to equivalence modulo T, is finite.

This result, due independently to Engeler, Ryll-Nardzewski and Svenonius, is sometimes referred to as the Ryll-Nardzewski theorem.

Further, $\aleph_0$-categorical theories and their countable models have strong ties with oligomorphic groups. They are often constructed as Fraïssé limits.

Michael Morley's highly non-trivial result that (for countable languages) there is only *one* notion of uncountable categoricity was the starting point for modern model theory, and in particular classification theory and stability theory:

Morley's categoricity theorem

If a first-order theory T in a finite or countable signature is κ-categorical for some uncountable cardinal κ, then T is κ-categorical for all uncountable cardinals κ.

Uncountably categorical (i.e. κ-categorical for all uncountable cardinals κ) theories are from many points of view the most well-behaved theories. A theory that is both $\aleph_0$-categorical and uncountably categorical is called **totally categorical**.

5.7 Model theory and set theory

Set theory (which is expressed in a countable language), if it is consistent, has a countable model; this is known as Skolem's paradox, since there are sentences in set theory which postulate the existence of uncountable sets and yet these sentences are true in our countable model. Particularly the proof of the independence of the continuum hypothesis requires considering sets in models which appear to be uncountable when viewed from *within* the model, but are countable to someone *outside* the model.

The model-theoretic viewpoint has been useful in set theory; for example in Kurt Gödel's work on the constructible universe, which, along with the method of forcing developed by Paul Cohen can be shown to prove the (again philosophically interesting) independence of the axiom of choice and the continuum hypothesis from the other axioms of set theory.

In the other direction, model theory itself can be formalized within ZFC set theory. The development of the fundamentals of model theory (such as the compactness theorem) rely on the axiom of choice, or more exactly the Boolean prime ideal theorem. Other results in model theory depend on set-theoretic axioms beyond the standard ZFC framework. For example, if the Continuum Hypothesis holds then every countable model has an ultrapower which is saturated (in its own cardinality). Similarly, if the Generalized Continuum Hypothesis holds then every model has a saturated elementary extension. Neither of these results are provable in ZFC alone. Finally, some questions arising from model theory (such as compactness for infinitary logics) have been shown to be equivalent to large cardinal axioms.

5.8   Other basic notions of model theory

5.8.1   Reducts and expansions

Main article: Reduct

A field or a vector space can be regarded as a (commutative) group by simply ignoring some of its structure. The corresponding notion in model theory is that of a **reduct** of a structure to a subset of the original signature. The opposite relation is called an *expansion* - e.g. the (additive) group of the rational numbers, regarded as a structure in the signature {+,0} can be expanded to a field with the signature {×,+,1,0} or to an ordered group with the signature {+,0,<}.

Similarly, if σ' is a signature that extends another signature σ, then a complete σ'-theory can be restricted to σ by intersecting the set of its sentences with the set of σ-formulas. Conversely, a complete σ-theory can be regarded as a σ'-theory, and one can extend it (in more than one way) to a complete σ'-theory. The terms reduct and expansion are sometimes applied to this relation as well.

5.8.2   Interpretability

Main article: Interpretation (model theory)

Given a mathematical structure, there are very often associated structures which can be constructed as a quotient of part of the original structure via an equivalence relation. An important example is a quotient group of a group.

One might say that to understand the full structure one must understand these quotients. When the equivalence relation is definable, we can give the previous sentence a precise meaning. We say that these structures are **interpretable**.

A key fact is that one can translate sentences from the language of the interpreted structures to the language of the original structure. Thus one can show that if a structure M interprets another whose theory is undecidable, then M itself is undecidable.

5.8.3   Using the compactness and completeness theorems

Gödel's completeness theorem (not to be confused with his incompleteness theorems) says that a theory has a model if and only if it is consistent, i.e. no contradiction is proved by the theory. This is the heart of model theory as it lets us answer questions about theories by looking at models and vice versa. One should not confuse the completeness theorem with the notion of a complete theory. A complete theory is a theory that contains every sentence or its negation. Importantly, one can find a complete consistent theory extending any consistent theory. However, as shown by Gödel's incompleteness theorems only in relatively simple cases will it be possible to have a complete consistent theory that is also recursive, i.e. that can be described by a recursively enumerable set of axioms. In particular, the theory of natural numbers has no recursive complete and consistent theory. Non-recursive theories are of little practical use, since it is undecidable if a proposed axiom is indeed an axiom, making proof-checking a supertask.

The compactness theorem states that a set of sentences S is satisfiable if every finite subset of S is satisfiable. In the context of proof theory the analogous statement is trivial, since every proof can have only a finite number of antecedents used in the proof. In the context of model theory, however, this proof is somewhat more difficult. There are two well known proofs, one by Gödel (which goes via proofs) and one by Malcev (which is more direct and allows us to restrict the cardinality of the resulting model).

Model theory is usually concerned with first-order logic, and many important results (such as the completeness and compactness theorems) fail in second-order logic or other alternatives. In first-order logic all infinite cardinals look the same to a language which is countable. This is expressed in the Löwenheim–Skolem theorems, which state that any countable theory with an infinite model 𝔄 has models of all infinite cardinalities (at least that of the language) which agree with 𝔄 on all sentences, i.e. they are 'elementarily equivalent'.

5.8.4 Types

Main article: Type (model theory)

Fix an L -structure M , and a natural number n . The set of definable subsets of M^n over some parameters A is a Boolean algebra. By Stone's representation theorem for Boolean algebras there is a natural dual notion to this. One can consider this to be the topological space consisting of maximal consistent sets of formulae over A . We call this the space of (complete) n -types over A , and write $S_n(A)$.

Now consider an element $m \in M^n$. Then the set of all formulae ϕ with parameters in A in free variables $x_1, \dots, x_n$ so that $M \models \phi(m)$ is consistent and maximal such. It is called the *type* of m over A .

One can show that for any n -type p , there exists some elementary extension N of M and some $a \in N^n$ so that p is the type of a over A .

Many important properties in model theory can be expressed with types. Further many proofs go via constructing models with elements that contain elements with certain types and then using these elements.

Illustrative Example: Suppose M is an algebraically closed field. The theory has quantifier elimination . This allows us to show that a type is determined exactly by the polynomial equations it contains. Thus the space of n -types over a subfield A is bijective with the set of prime ideals of the polynomial ring $A[x_1, \dots, x_n]$. This is the same set as the spectrum of $A[x_1, \dots, x_n]$. Note however that the topology considered on the type space is the constructible topology: a set of types is basic open iff it is of the form $\{p : f(x) = 0 \in p\}$ or of the form $\{p : f(x) \neq 0 \in p\}$. This is finer than the Zariski topology.

5.9 History

Model theory as a subject has existed since approximately the middle of the 20th century. However some earlier research, especially in mathematical logic, is often regarded as being of a model-theoretical nature in retrospect. The first significant result in what is now model theory was a special case of the downward Löwenheim–Skolem theorem, published by Leopold Löwenheim in 1915. The compactness theorem was implicit in work by Thoralf Skolem,[3] but it was first published in 1930, as a lemma in Kurt Gödel's proof of his completeness theorem. The Löwenheim–Skolem theorem and the compactness theorem received their respective general forms in 1936 and 1941 from Anatoly Maltsev.

5.10 See also

5.11 Notes

[1] Chang and Keisler, p. 1.

[2] In a countable signature. The theorem has a straightforward generalization to uncountable signatures.

[3] "All three commentators [i.e. Vaught, van Heijenoort and Dreben] agree that both the completeness and compactness theorems were implicit in Skolem 1923...." [Dawson, J. W. (1993). "The compactness of first-order logic:from gödel to lindström". *History and Philosophy of Logic* **14**: 15. doi:10.1080/01445349308837208.]

5.12 References

5.12.1 Canonical textbooks

- Chang, Chen Chung; Keisler, H. Jerome (1990) [1973]. *Model Theory*. Studies in Logic and the Foundations of Mathematics (3rd ed.). Elsevier. ISBN 978-0-444-88054-3.

- Hodges, Wilfrid (1997). *A shorter model theory*. Cambridge: Cambridge University Press. ISBN 978-0-521-58713-6.

- Marker, David (2002). *Model Theory: An Introduction*. Graduate Texts in Mathematics 217. Springer. ISBN 0-387-98760-6.

5.12.2   Other textbooks

- Bell, John L.; Slomson, Alan B. (2006) [1969]. *Models and Ultraproducts: An Introduction* (reprint of 1974 ed.). Dover Publications. ISBN 0-486-44979-3.

- Ebbinghaus, Heinz-Dieter; Flum, Jörg; Thomas, Wolfgang (1994). *Mathematical Logic*. Springer. ISBN 0-387-94258-0.

- Hinman, Peter G. (2005). *Fundamentals of Mathematical Logic*. A K Peters. ISBN 1-56881-262-0.

- Hodges, Wilfrid (1993). *Model theory*. Cambridge University Press. ISBN 0-521-30442-3.

- Manzano, Maria (1999). *Model theory*. Oxford University Press. ISBN 0-19-853851-0.

- Poizat, Bruno (2000). *A Course in Model Theory*. Springer. ISBN 0-387-98655-3.

- Rautenberg, Wolfgang (2010). *A Concise Introduction to Mathematical Logic* (3rd ed.). New York: Springer Science+Business Media. doi:10.1007/978-1-4419-1221-3. ISBN 978-1-4419-1220-6.

- Rothmaler, Philipp (2000). *Introduction to Model Theory* (new ed.). Taylor & Francis. ISBN 90-5699-313-5.

- Ziegler, Martin; Tent, Katrin (2012). *A Course in Model Theory*. Cambridge University Press. ISBN 9780521763240.

5.12.3   Free online texts

- Chatzidakis, Zoe (2001). *Introduction to Model Theory* (PDF). pp. 26 pages.

- Pillay, Anand (2002). *Lecture Notes – Model Theory* (PDF). pp. 61 pages.

- Hazewinkel, Michiel, ed. (2001), "Model theory", *Encyclopedia of Mathematics*, Springer, ISBN 978-1-55608-010-4

- Hodges, Wilfrid, *Model theory*. The Stanford Encyclopedia Of Philosophy, E. Zalta (ed.).

- Hodges, Wilfrid, *First-order Model theory*. The Stanford Encyclopedia Of Philosophy, E. Zalta (ed.).

- Simmons, Harold (2004), *An introduction to Good old fashioned model theory*. Notes of an introductory course for postgraduates (with exercises).

- J. Barwise and S. Feferman (editors), Model-Theoretic Logics, Perspectives in Mathematical Logic, Volume 8, New York: Springer-Verlag, 1985.

Chapter 6

Domain of discourse

In the formal sciences, the **domain of discourse**, also called the **universe of discourse, universal set**, or simply **universe**, is the set of entities over which certain variables of interest in some formal treatment may range.

6.1 Overview

The domain of discourse is usually identified in the preliminaries, so that there is no need in the further treatment to specify each time the range of the relevant variables.[1] Many logicians distinguish, sometimes only tacitly, between the **domain of a science** and the **universe of discourse of a formalization of the science**.[2] Giuseppe Peano formalized number theory (arithmetic of positive integers) taking its domain to be the positive integers and the universe of discourse to include all numbers, not just integers.

6.2 Examples

For example, in an interpretation of first-order logic, the domain of discourse is the set of individuals that the quantifiers range over. In one interpretation, the domain of discourse could be the set of real numbers; in another interpretation, it could be the set of natural numbers. If no domain of discourse has been identified, a proposition such as $\forall x \, (x^2 \neq 2)$ is ambiguous. If the domain of discourse is the set of real numbers, the proposition is false, with $x = \sqrt{2}$ as counterexample; if the domain is the set of naturals, the proposition is true, since 2 is not the square of any natural number.

6.3 Universe of discourse

The term *universe of discourse* generally refers to the collection of objects being discussed in a specific discourse. In model-theoretical semantics, a universe of discourse is the set of entities that a model is based on. The concept *universe of discourse* is generally attributed to Augustus De Morgan (1846) but the name was used for the first time in history by George Boole (1854) on page 42 of his *Laws of Thought* in a long and incisive passage well worth study. Boole's definition is quoted below. The concept, probably discovered independently by Boole in 1847, played a crucial role in his philosophy of logic especially in his stunning principle of wholistic reference.

A database is a model of some aspect of the reality of an organisation. It is conventional to call this reality the "universe of discourse" or "domain of discourse".

6.4 Boole's 1854 definition

In every discourse, whether of the mind conversing with its own thoughts, or of the individual in his intercourse with others, there is an assumed or expressed limit within which the subjects of its operation are confined. The most unfettered discourse is that in which the words we use are understood in the widest possible application, and for them the limits of discourse are co-extensive with those of the universe itself. But more usually we confine ourselves to a less spacious field. Sometimes, in discoursing of men we imply (without expressing the limitation) that it is of men only under certain circumstances and conditions that we speak, as of civilized men, or of men in the vigour of life, or of men under some other condition or relation. Now, whatever may be the extent of the field within which all the objects of our discourse are found, that field may properly be termed the universe of discourse. Furthermore, this universe of discourse is in the strictest sense the ultimate subject of the discourse.[3]

6.5 See also

- Domain of a function
- Domain theory
- Interpretation (logic)
- Term algebra
- Universe (mathematics)

6.6 References

[1] Corcoran, John. *Universe of discourse*. Cambridge Dictionary of Philosophy. Cambridge University Press, 1995, p. 941.

[2] José Miguel Sagüillo, Domains of sciences, universe of discourse, and omega arguments, History and philosophy of logic, vol. 20 (1999), pp. 267–280.

[3] George Boole. 1854/2003. The Laws of Thought, facsimile of 1854 edition, with an introduction by J. Corcoran. Buffalo: Prometheus Books (2003). Reviewed by James van Evra in Philosophy in Review.24 (2004) 167–169.

Chapter 7

Signature (logic)

In logic, especially mathematical logic, a **signature** lists and describes the non-logical symbols of a formal language. In universal algebra, a signature lists the operations that characterize an algebraic structure. In model theory, signatures are used for both purposes.

Signatures play the same role in mathematics as type signatures in computer programming. They are rarely made explicit in more philosophical treatments of logic.

7.1 Definition

Formally, a (single-sorted) **signature** can be defined as a triple $\sigma = (S_{\text{func}}, S_{\text{rel}}, \text{ar})$, where S_{func} and S_{rel} are disjoint sets not containing any other basic logical symbols, called respectively

- *function symbols* (examples: $+$, $\times$, 0, 1) and

- *relation symbols* or *predicates* (examples: $\leq$, $\in$),

and a function ar: $S_{\text{func}} \cup S_{\text{rel}} \to \mathbb{N}_0$ which assigns a non-negative integer called *arity* to every function or relation symbol. A function or relation symbol is called *n*-ary if its arity is *n*. A nullary (*0*-ary) function symbol is called a *constant symbol*.

A signature with no function symbols is called a **relational signature**, and a signature with no relation symbols is called an **algebraic signature**. A **finite signature** is a signature such that S_{func} and S_{rel} are finite. More generally, the **cardinality** of a signature $\sigma = (S_{\text{func}}, S_{\text{rel}}, \text{ar})$ is defined as $|\sigma| = |S_{\text{func}}| + |S_{\text{rel}}|$.

The **language of a signature** is the set of all well formed sentences built from the symbols in that signature together with the symbols in the logical system.

7.2 Other conventions

In universal algebra the word **type** or **similarity type** is often used as a synonym for "signature". In model theory, a signature σ is often called **vocabulary**, or identified with the (first-order) language L to which it provides the non-logical symbols. However, the cardinality of the language L will always be infinite; if σ is finite then $|L|$ will be $\aleph_0$.

As the formal definition is inconvenient for everyday use, the definition of a specific signature is often abbreviated in an informal way, as in:

"The standard signature for abelian groups is $\sigma = (+,-,0)$, where $-$ is a unary operator."

Sometimes an algebraic signature is regarded as just a list of arities, as in:

41

"The similarity type for abelian groups is $\sigma = (2,1,0)$."

Formally this would define the function symbols of the signature as something like f_0 (nullary), f_1 (unary) and f_2 (binary), but in reality the usual names are used even in connection with this convention.

In mathematical logic, very often symbols are not allowed to be nullary, so that constant symbols must be treated separately rather than as nullary function symbols. They form a set S_{const} disjoint from S_{func}, on which the arity function ar is not defined. However, this only serves to complicate matters, especially in proofs by induction over the structure of a formula, where an additional case must be considered. Any nullary relation symbol, which is also not allowed under such a definition, can be emulated by a unary relation symbol together with a sentence expressing that its value is the same for all elements. This translation fails only for empty structures (which are often excluded by convention). If nullary symbols are allowed, then every formula of propositional logic is also a formula of first-order logic.

7.3   Use of signatures in logic and algebra

In the context of first-order logic, the symbols in a signature are also known as the non-logical symbols, because together with the logical symbols they form the underlying alphabet over which two formal languages are inductively defined: The set of *terms* over the signature and the set of (well-formed) *formulas* over the signature.

In a structure, an *interpretation* ties the function and relation symbols to mathematical objects that justify their names: The interpretation of an n-ary function symbol f in a structure A with *domain* A is a function $f^A : A^n \rightarrow A$, and the interpretation of an n-ary relation symbol is a relation $R^A \subseteq A^n$. Here $A^n = A \times A \times ... \times A$ denotes the n-fold cartesian product of the domain A with itself, and so f is in fact an n-ary function, and R an n-ary relation.

7.4   Many-sorted signatures

For many-sorted logic and for many-sorted structures signatures must encode information about the sorts. The most straightforward way of doing this is via **symbol types** that play the role of generalized arities.[1]

Symbol types

Let S be a set (of sorts) not containing the symbols $\times$ or $\rightarrow$.

The symbol types over S are certain words over the alphabet $S \cup \{\times, \rightarrow\}$: the relational symbol types $s_1 \times ... \times sn$, and the functional symbol types $s_1 \times ... \times sn \rightarrow s'$, for non-negative integers n and $s_1, s_2, ..., sn, s' \in S$. (For $n = 0$, the expression $s_1 \times ... \times sn$ denotes the empty word.)

Signature

A (many-sorted) signature is a triple $(S, P, type)$ consisting of

- a set S of sorts,

- a set P of symbols, and

- a map type which associates to every symbol in P a symbol type over S.

7.5   Notes

[1] Many-Sorted Logic, the first chapter in Lecture notes on Decision Procedures, written by Calogero G. Zarba.

7.6 References

- Burris, Stanley N.; Sankappanavar, H.P. (1981). *A Course in Universal Algebra*. Springer. ISBN 3-540-90578-2. Free online edition.

- Hodges, Wilfrid (1997). *A shorter model theory*. Cambridge University Press. ISBN 0-521-58713-1.

7.7 External links

- Stanford Encyclopedia of Philosophy: "Model theory"—by Wilfred Hodges.

- PlanetMath: Entry "Signature" describes the concept for the case when no sorts are introduced.

- Baillie, Jean, "An Introduction to the Algebraic Specification of Abstract Data Types."

Chapter 8

Interpretation (model theory)

For other uses, see Interpretation (disambiguation).

In model theory, **interpretation** of a structure M in another structure N (typically of a different signature) is a technical notion that approximates the idea of representing M inside N. For example every reduct or definitional expansion of a structure N has an interpretation in N.

Many model-theoretic properties are preserved under interpretability. For example if the theory of N is stable and M is interpretable in N, then the theory of M is also stable.

8.1 Definition

An **interpretation** of M in N **with parameters** (or **without parameters**, respectively) is a pair (n, f) where n is a natural number and f is a surjective map from a subset of N^n onto M such that the f-preimage (more precisely the f^k-preimage) of every set $X \subseteq M^k$ definable in M by a first-order formula without parameters is definable (in N) by a first-order formula with parameters (or without parameters, respectively). Since the value of n for an interpretation (n, f) is often clear from context, the map f itself is also called an interpretation.

To verify that the preimage of every definable (without parameters) set in M is definable in N (with or without parameters), it is sufficient to check the preimages of the following definable sets:

- the domain of M;

- the diagonal of M;

- every relation in the signature of M;

- the graph of every function in the signature of M.

In model theory the term *definable* often refers to definability with parameters; if this convention is used, definability without parameters is expressed by the term *0-definable*. Similarly, an interpretation with parameters may be referred to as simply an interpretation, and an interpretation without parameters as a **0-interpretation**.

8.2 Bi-interpretability

If L, M and N are three structures, L is interpreted in M, and M is interpreted in N, then one can naturally construct a composite interpretation of L in N. If two structures M and N are interpreted in each other, then by combining the interpretations in two possible ways, one obtains an interpretation of each of the two structures in itself. This observation permits one to define an equivalence relation among structures, reminiscent of the homotopy equivalence among topological spaces.

Two structures M and N are **bi-interpretable** if there exists an interpretation of M in N and an interpretation of N in M such that the composite interpretations of M in itself and of N in itself are definable in M and in N, respectively (the composite interpretations being viewed as operations on M and on N).

8.3 Example

The partial map f from $\mathbf{Z} \times \mathbf{Z}$ onto $\mathbf{Q}$ which maps (x, y) to x/y provides an interpretation of the field $\mathbf{Q}$ of rational numbers in the ring $\mathbf{Z}$ of integers (to be precise, the interpretation is $(2, f)$). In fact, this particular interpretation is often used to *define* the rational numbers. To see that it is an interpretation (without parameters), one needs to check the following preimages of definable sets in $\mathbf{Q}$:

- the preimage of $\mathbf{Q}$ is defined by the formula $\varphi(x, y)$ given by $\neg\,(y = 0)$;

- the preimage of the diagonal of $\mathbf{Q}$ is defined by the formula $\varphi(x_1, y_1, x_2, y_2)$ given by $x_1 \times y_2 = x_2 \times y_1$;

- the preimages of 0 and 1 are defined by the formulas $\varphi(x, y)$ given by $x = 0$ and $x = y$;

- the preimage of the graph of addition is defined by the formula $\varphi(x_1, y_1, x_2, y_2, x_3, y_3)$ given by $x_1 \times y_2 \times y_3 + x_2 \times y_1 \times y_3 = x_3 \times y_1 \times y_2$;

- the preimage of the graph of multiplication is defined by the formula $\varphi(x_1, y_1, x_2, y_2, x_3, y_3)$ given by $x_1 \times x_2 \times y_3 = x_3 \times y_1 \times y_2$.

8.4 References

- Ahlbrandt, Gisela; Ziegler, Martin (1986), "Quasi finitely axiomatizable totally categorical theories", *Annals of Pure and Applied Logic* **30**: 63–82, doi:10.1016/0168-0072(86)90037-0

- Hodges, Wilfrid (1997), *A shorter model theory*, Cambridge: Cambridge University Press, ISBN 978-0-521-58713-6 (Section 4.3)

- Poizat, Bruno (2000), *A Course in Model Theory*, Springer, ISBN 0-387-98655-3 (Section 9.4)

Chapter 9

Definitions of mathematics

Definitions of mathematics vary widely and different schools of thought, particularly in philosophy, have suggested radically different and controversial accounts.[1][2]

9.1 Early definitions

Aristotle defined mathematics as:

> The science of quantity.

In Aristotle's classification of the sciences, discrete quantities were studied by arithmetic, continuous quantities by geometry.[3]

Auguste Comte's definition tried to explain the role of mathematics in coordinating phenomena in all other fields:[4]

> The science of indirect measurement.[5] Auguste Comte 1851

The "indirectness" in Comte's definition refers to determining quantities that cannot be measured directly, such as the distance to planets or the size of atoms, by means of their relations to quantities that can be measured directly.[6]

9.2 Greater abstraction and competing philosophical schools

The preceding kinds of definitions, which had prevailed since Aristotle's time,[3] were abandoned in the 19th century as new branches of mathematics were developed, which bore no obvious relation to measurement or the physical world, such as group theory, projective geometry,[5] and non-Euclidean geometry.[7] As mathematicians pursued greater rigor and more-abstract foundations, some proposed definitions purely in terms of logic:

> Mathematics is the science that draws necessary conclusions.[8] Benjamin Peirce 1870

> All Mathematics is Symbolic Logic.[7] Bertrand Russell 1903

Peirce did not think that mathematics is the same as logic, since he thought mathematics makes only hypothetical assertions, not categorical ones.[9] Russell's definition, on the other hand, expresses the logicist philosophy of mathematics[10] without reservation. Competing philosophies of mathematics put forth different definitions.

Opposing the completely deductive character of logicism, intuitionism emphasizes the construction of ideas in the mind. Here is an intuitionist definition:[10]

> Mathematics is mental activity which consists in carrying out, one after the other, those mental constructions which are inductive and effective.

meaning that by combining fundamental ideas, one reaches a definite result.

Formalism denies both physical and mental meaning to mathematics, making the symbols and rules themselves the object of study.[10] A formalist definition:

> Mathematics is the manipulation of the meaningless symbols of a first-order language according to explicit, syntactical rules.

Still other approaches emphasize pattern, order, or structure. For example:

> Mathematics is the classification and study of all possible patterns. Walter Warwick Sawyer, 1955

Yet another approach makes abstraction the defining criterion:

> Mathematics is a broad-ranging field of study in which the properties and interactions of idealized objects are examined. Wolfram MathWorld

9.3 General, nonspecialist perspectives

Most contemporary reference works define mathematics mainly by summarizing its main topics and methods and referencing its history:

> The abstract science which investigates deductively the conclusions implicit in the elementary conceptions of spatial and numerical relations, and which includes as its main divisions geometry, arithmetic, and algebra. Oxford English Dictionary, 1933

> I believe maths is concerned with the development of language for expression, validation, falsification, deduction, calculation. This also involves the development of concepts for expression and description of structure and patterns. Ronald Brown

> The study of the measurement, properties, and relationships of quantities and sets, using numbers and symbols. American Heritage Dictionary, 2000

> The science of structure, order, and relation that has evolved from elemental practices of counting, measuring, and describing the shapes of objects.[11] Encyclopaedia Britannica

9.4 Playful, metaphorical, and poetic definitions

Bertrand Russell wrote this famous tongue-in-cheek definition, describing the way all terms in mathematics are ultimately defined by reference to undefined terms:

> The subject in which we never know what we are talking about, nor whether what we are saying is true.[12] Bertrand Russell 1901

Many other attempts to characterize mathematics have led to humor or poetic prose:

> "Mathematics is about making up rules and seeing what happens."[13] Vi Hart

A mathematician is a blind man in a dark room looking for a black cat which isn't there.[14] Charles Darwin

A mathematician, like a painter or poet, is a maker of patterns. If his patterns are more permanent than theirs, it is because they are made with ideas. G. H. Hardy, 1940

Mathematics is the art of giving the same name to different things.[8] Henri Poincaré

Mathematics is the science of skilful operations with concepts and rules invented just for this purpose. [this purpose being the skilful operation ....][15] Eugene Wigner

Mathematics is not a book confined within a cover and bound between brazen clasps, whose contents it needs only patience to ransack; it is not a mine, whose treasures may take long to reduce into possession, but which fill only a limited number of veins and lodes; it is not a soil, whose fertility can be exhausted by the yield of successive harvests; it is not a continent or an ocean, whose area can be mapped out and its contour defined: it is limitless as that space which it finds too narrow for its aspirations; its possibilities are as infinite as the worlds which are forever crowding in and multiplying upon the astronomer's gaze; it is as incapable of being restricted within assigned boundaries or being reduced to definitions of permanent validity, as the consciousness of life, which seems to slumber in each monad, in every atom of matter, in each leaf and bud cell, and is forever ready to burst forth into new forms of vegetable and animal existence.[16] James Joseph Sylvester

What is mathematics? What is it for? What are mathematicians doing nowadays? Wasn't it all finished long ago? How many new numbers can you invent anyway? Is today's mathematics just a matter of huge calculations, with the mathematician as a kind of zookeeper, making sure the precious computers are fed and watered? If it's not, what is it other than the incomprehensible outpourings of superpowered brainboxes with their heads in the clouds and their feet dangling from the lofty balconies of their ivory towers? Mathematics is all of these, and none. Mostly, it's just different. It's not what you expect it to be, you turn your back for a moment and it's changed. It's certainly not just a fixed body of knowledge, its growth is not confined to inventing new numbers, and its hidden tendrils pervade every aspect of modern life.[16] Ian Stewart

9.5   See also

- Philosophy of mathematics

9.6   References

<div class="reflist columns references-column-width" style="-moz-column-width: [1] [2]; -webkit-column-width: [1] [2]; column-width: [1] [2]; list-style-type: decimal;">

[1] Mura, Robert (Dec 1993). "Images of Mathematics Held by University Teachers of Mathematical Sciences". *Educational Studies in Mathematics* **25** (4): 375–385.

[2] Tobies, Renate and Helmut Neunzert (2012). *Iris Runge: A Life at the Crossroads of Mathematics, Science, and Industry.* Springer. pp. 9. ISBN 3-0348-0229-3. It is first necessary to ask what is meant by *mathematics* in general. Illustrious scholars have debated this matter until they were blue in the face, and yet no consensus has been reached about whether mathematics is a natural science, a branch of the humanities, or an art form.

[3] James Franklin, "Aristotelian Realism" in *Philosophy of Mathematics*, ed. A.D. Irvine, p. 104. Elsevier (2009).

[4] Arline Reilein Standley, *Auguste Comte*, p. 61. Twayne Publishers (1981).

[5] Florian Cajori *et al.*, *A History of Mathematics*, 5th ed., p. 285–6. American Mathematical Society (1991).

[6] Auguste Comte, *The Philosophy of Mathematics*, tr. W.M. Gillespie, pp. 17–25. Harper & Brothers, New York (1851).

[7] Bertrand Russell, *The Principles of Mathematics*, p. 5. University Press, Cambridge (1903)

[8] Foundations and fundamental concepts of mathematics By Howard Eves page 150

[9] Carl Boyer, Uta Merzbach, *A History of Mathematics*, p. 426. John Wiley and Sones (2011).

[10] Snapper, Ernst (September 1979), "The Three Crises in Mathematics: Logicism, Intuitionism, and Formalism", *Mathematics Magazine* **52** (4): 207–16, doi:10.2307/2689412, JSTOR 2689412.

[11] "Mathematics. *Encyclopaedia Britannica* from Encyclopaedia Britannica 2006 Ultimate reference Suite DVD.

[12] Russell, Bertrand (1901), "Recent Work on the Principles of Mathematics", *International Monthly* **4**.

[13] 9.999...reasons that .999...=1, Vi Hart

[14] "Pi in the Sky", John Barrow

[15] What is mathematics?

[16] "From Here to Infinity", Ian Stewart

9.7 Further reading

- Courant, Richard; Robbins, Herbert (1996), *What is Mathematics?* (2nd ed.), Oxford University Press, ISBN 978-0-19-510519-3.

- Gowers, Timothy; Barrow-Green, June; Leader, Imre, eds. (2008), *The Princeton Companion to Mathematics*, Princeton University Press, ISBN 978-0-691-11880-2.

- Hersh, Reuben (1999), *What is Mathematics, Really?*, Oxford University Press, ISBN 978-0-19-513087-4.

- Paulos, John Allen (1991), *Beyond Numeracy*, Viking, ISBN 978-0-670-83654-3.

- Stewart, Ian (1996), *From Here to Infinity*, Oxford University Press, ISBN 0-19-283202-6.

Chapter 10

Arithmetic

For the song by Brooke Fraser, see Arithmetic (song).

 Arithmetic or **arithmetics** (from the Greek ἀριθμός *arithmos*, "number") is the oldest[1] and most elementary branch of mathematics. It consists of the study of numbers, especially the properties of the traditional operations between them—addition, subtraction, multiplication and division. Arithmetic is an elementary part of number theory, and number theory is considered to be one of the top-level divisions of modern mathematics, along with algebra, geometry, and analysis. The terms *arithmetic* and *higher arithmetic* were used until the beginning of the 20th century as synonyms for *number theory* and are sometimes still used to refer to a wider part of number theory.[2]

10.1   History

The prehistory of arithmetic is limited to a small number of artifacts which may indicate the conception of addition and subtraction, the best-known being the Ishango bone from central Africa, dating from somewhere between 20,000 and 18,000 BC, although its interpretation is disputed.[3]

The earliest written records indicate the Egyptians and Babylonians used all the elementary arithmetic operations as early as 2000 BC. These artifacts do not always reveal the specific process used for solving problems, but the characteristics of the particular numeral system strongly influence the complexity of the methods. The hieroglyphic system for Egyptian numerals, like the later Roman numerals, descended from tally marks used for counting. In both cases, this origin resulted in values that used a decimal base but did not include positional notation. Complex calculations with Roman numerals required the assistance of a counting board or the Roman abacus to obtain the results.

Early number systems that included positional notation were not decimal, including the sexagesimal (base 60) system for Babylonian numerals and the vigesimal (base 20) system that defined Maya numerals. Because of this place-value concept, the ability to reuse the same digits for different values contributed to simpler and more efficient methods of calculation.

The continuous historical development of modern arithmetic starts with the Hellenistic civilization of ancient Greece, although it originated much later than the Babylonian and Egyptian examples. Prior to the works of Euclid around 300 BC, Greek studies in mathematics overlapped with philosophical and mystical beliefs. For example, Nicomachus summarized the viewpoint of the earlier Pythagorean approach to numbers, and their relationships to each other, in his *Introduction to Arithmetic*.

Greek numerals were used by Archimedes, Diophantus and others in a positional notation not very different from ours. Because the ancient Greeks lacked a symbol for zero (until the Hellenistic period), they used three separate sets of symbols. One set for the unit's place, one for the ten's place, and one for the hundred's. Then for the thousand's place they would reuse the symbols for the unit's place, and so on. Their addition algorithm was identical to ours, and their multiplication algorithm was only very slightly different. Their long division algorithm was the same, and the square root algorithm that was once taught in school was known to Archimedes, who may have invented it. He preferred it to Hero's method of successive approximation because, once computed, a digit doesn't change, and the square roots of perfect squares, such as 7485696, terminate immediately as 2736. For numbers with a fractional part, such as 546.934, they used negative powers of 60 instead of negative powers of 10 for the fractional part 0.934.[4] The ancient Chinese used a similar positional notation. Because they also lacked a symbol for zero, they had one set of symbols for the unit's place, and a second set for the ten's place. For the hundred's place they then reused the symbols for the unit's place, and so on. Their symbols were based on the ancient counting rods. It is a complicated question to determine exactly when the Chinese started calculating with positional representation, but it was definitely before 400 BC.[5] The Bishop of Syria, Severus Sebokht (650 AD), "Indians possess a method of calculation that no word can praise enough. Their rational system of mathematics, or of their method of calculation. I mean the system using nine symbols."[6]

Leonardo of Pisa (Fibonacci) in 1200 AD wrote in *Liber Abaci* "The method of the Indians (Modus Indoram) surpasses any known method to compute. It's a marvelous method. They do their computations using nine figures and symbol zero".[7]

The gradual development of Hindu–Arabic numerals independently devised the place-value concept and positional notation, which combined the simpler methods for computations with a decimal base and the use of a digit representing 0. This allowed the system to consistently represent both large and small integers. This approach eventually replaced all other systems. In the early 6th century AD, the Indian mathematician Aryabhata incorporated an existing version of this system in his work, and experimented with different notations. In the 7th century, Brahmagupta established the use of 0 as a separate number and determined the results for multiplication, division, addition and subtraction of zero and all other numbers, except for the result of division by 0. His contemporary, the Syriac bishop Severus Sebokht described the excellence of this system as "... valuable methods of calculation which surpass description". The Arabs also learned this new method and called it *hesab*.

Although the Codex Vigilanus described an early form of Arabic numerals (omitting 0) by 976 AD, Fibonacci was primarily responsible for spreading their use throughout Europe after the publication of his book *Liber Abaci* in 1202. He considered the significance of this "new" representation of numbers, which he styled the "Method of the Indians" (Latin *Modus Indorum*), so fundamental that all related mathematical foundations, including the results of Pythagoras and the algorism describing the methods for performing actual calculations, were "almost a mistake" in comparison.

In the Middle Ages, arithmetic was one of the seven liberal arts taught in universities.

The flourishing of algebra in the medieval Islamic world and in Renaissance Europe was an outgrowth of the enormous simplification of computation through decimal notation.

Various types of tools exist to assist in numeric calculations. Examples include slide rules (for multiplication, division, and trigonometry) and nomographs in addition to the electrical calculator.

10.2 Arithmetic operations

See also: Algebraic operation

The basic arithmetic operations are addition, subtraction, multiplication and division, although this subject also includes more advanced operations, such as manipulations of percentages, square roots, exponentiation, and logarithmic functions. Arithmetic is performed according to an order of operations. Any set of objects upon which all four arithmetic operations (except division by 0) can be performed, and where these four operations obey the usual laws, is called a field.[8]

10.2.1 Addition (+)

Main article: Addition

Addition is the basic operation of arithmetic. In its simplest form, addition combines two numbers, the *addends* or *terms*, into a single number, the *sum* of the numbers (Such as 2 + 2 = 4 or 3 + 5 = 8).

Adding more than two numbers can be viewed as repeated addition; this procedure is known as summation and includes ways to add infinitely many numbers in an infinite series; repeated addition of the number 1 is the most basic form of counting.

Addition is commutative and associative so the order the terms are added in does not matter. The identity element of addition (the additive identity) is 0, that is, adding 0 to any number yields that same number. Also, the inverse element of addition (the additive inverse) is the opposite of any number, that is, adding the opposite of any number to the number itself yields the additive identity, 0. For example, the opposite of 7 is −7, so 7 + (−7) = 0.

Addition can be given geometrically as in the following example:

> If we have two sticks of lengths 2 and 5, then if we place the sticks one after the other, the length of the stick thus formed is 2 + 5 = 7.

10.2.2   Subtraction (−)

Main article: Subtraction
See also: Method of complements

Subtraction is the inverse of addition. Subtraction finds the *difference* between two numbers, the *minuend* minus the *subtrahend*. If the minuend is larger than the subtrahend, the difference is positive; if the minuend is smaller than the subtrahend, the difference is negative; if they are equal, the difference is 0.

Subtraction is neither commutative nor associative. For that reason, it is often helpful to look at subtraction as addition of the minuend and the opposite of the subtrahend, that is $a - b = a + (-b)$. When written as a sum, all the properties of addition hold.

There are several methods for calculating results, some of which are particularly advantageous to machine calculation. For example, digital computers employ the method of two's complement. Of great importance is the counting up method by which change is made. Suppose an amount P is given to pay the required amount Q, with P greater than Q. Rather than performing the subtraction $P - Q$ and counting out that amount in change, money is counted out starting at Q and continuing until reaching P. Although the amount counted out must equal the result of the subtraction $P - Q$, the subtraction was never really done and the value of $P - Q$ might still be unknown to the change-maker.

10.2.3   Multiplication (× or · or *)

Main article: Multiplication

Multiplication is the second basic operation of arithmetic. Multiplication also combines two numbers into a single number, the *product*. The two original numbers are called the *multiplier* and the *multiplicand*, sometimes both simply called *factors*.

Multiplication may be viewed as a scaling operation. If the numbers are imagined as lying in a line, multiplication by a number, say x, greater than 1 is the same as stretching everything away from 0 uniformly, in such a way that the number 1 itself is stretched to where x was. Similarly, multiplying by a number less than 1 can be imagined as squeezing towards 0. (Again, in such a way that 1 goes to the multiplicand.)

Multiplication is commutative and associative; further it is distributive over addition and subtraction. The multiplicative identity is 1, that is, multiplying any number by 1 yields that same number. Also, the multiplicative inverse is the reciprocal of any number (except 0; 0 is the only number without a multiplicative inverse), that is, multiplying the reciprocal of any number by the number itself yields the multiplicative identity.

The product of a and b is written as $a \times b$ or $a \cdot b$. When a or b are expressions not written simply with digits, it is also written by simple juxtaposition: ab. In computer programming languages and software packages in which one can only use characters normally found on a keyboard, it is often written with an asterisk: $a * b$.

10.2.4   Division (÷ or /)

Main article: Division (mathematics)

Division is essentially the inverse of multiplication. Division finds the *quotient* of two numbers, the *dividend* divided by the *divisor*. Any dividend divided by 0 is undefined. For distinct positive numbers, if the dividend is larger than the divisor, the quotient is greater than 1, otherwise it is less than 1 (a similar rule applies for negative numbers). The quotient multiplied by the divisor always yields the dividend.

Division is neither commutative nor associative. As it is helpful to look at subtraction as addition, it is helpful to look at division as multiplication of the dividend times the reciprocal of the divisor, that is $a \div b = a \times {}^1\!/b$. When written as a product, it obeys all the properties of multiplication.

10.3 Decimal arithmetic

Decimal representation refers exclusively, in common use, to the written numeral system employing arabic numerals as the digits for a radix 10 ("decimal") positional notation; however, any numeral system based on powers of 10, e.g., Greek, Cyrillic, Roman, or Chinese numerals may conceptually be described as "decimal notation" or "decimal representation".

Modern methods for four fundamental operations (addition, subtraction, multiplication and division) were first devised by Brahmagupta of India. This was known during medieval Europe as "Modus Indoram" or Method of the Indians. Positional notation (also known as "place-value notation") refers to the representation or encoding of numbers using the same symbol for the different orders of magnitude (e.g., the "ones place", "tens place", "hundreds place") and, with a radix point, using those same symbols to represent fractions (e.g., the "tenths place", "hundredths place"). For example, 507.36 denotes 5 hundreds (10^2), plus 0 tens (10^1), plus 7 units (10^0), plus 3 tenths (10^{-1}) plus 6 hundredths (10^{-2}).

The concept of 0 as a number comparable to the other basic digits is essential to this notation, as is the concept of 0's use as a placeholder, and as is the definition of multiplication and addition with 0. The use of 0 as a placeholder and, therefore, the use of a positional notation is first attested to in the Jain text from India entitled the *Lokavibhâga*, dated 458 AD and it was only in the early 13th century that these concepts, transmitted via the scholarship of the Arabic world, were introduced into Europe by Fibonacci[9] using the Hindu–Arabic numeral system.

Algorism comprises all of the rules for performing arithmetic computations using this type of written numeral. For example, addition produces the sum of two arbitrary numbers. The result is calculated by the repeated addition of single digits from each number that occupies the same position, proceeding from right to left. An addition table with ten rows and ten columns displays all possible values for each sum. If an individual sum exceeds the value 9, the result is represented with two digits. The rightmost digit is the value for the current position, and the result for the subsequent addition of the digits to the left increases by the value of the second (leftmost) digit, which is always one. This adjustment is termed a *carry* of the value 1.

The process for multiplying two arbitrary numbers is similar to the process for addition. A multiplication table with ten rows and ten columns lists the results for each pair of digits. If an individual product of a pair of digits exceeds 9, the *carry* adjustment increases the result of any subsequent multiplication from digits to the left by a value equal to the second (leftmost) digit, which is any value from 1 to 8 ($9 \times 9 = 81$). Additional steps define the final result.

Similar techniques exist for subtraction and division.

The creation of a correct process for multiplication relies on the relationship between values of adjacent digits. The value for any single digit in a numeral depends on its position. Also, each position to the left represents a value ten times larger than the position to the right. In mathematical terms, the exponent for the radix (base) of 10 increases by 1 (to the left) or decreases by 1 (to the right). Therefore, the value for any arbitrary digit is multiplied by a value of the form 10^n with integer n. The list of values corresponding to all possible positions for a single digit is written as $\{..., 10^2, 10, 1, 10^{-1}, 10^{-2}, ...\}$.

Repeated multiplication of any value in this list by 10 produces another value in the list. In mathematical terminology, this characteristic is defined as closure, and the previous list is described as **closed under multiplication**. It is the basis for correctly finding the results of multiplication using the previous technique. This outcome is one example of the uses of number theory.

10.4   Compound unit arithmetic

Compound[10] unit arithmetic is the application of arithmetic operations to mixed radix quantities such as feet and inches, gallons and pints, pounds shillings and pence, and so on. Prior to the use of decimal-based systems of money and units of measure, the use of compound unit arithmetic formed a significant part of commerce and industry.

10.4.1   Basic arithmetic operations

The techniques used for compound unit arithmetic were developed over many centuries and are well-documented in many textbooks in many different languages.[11][12][13][14] In addition to the basic arithmetic functions encountered in decimal arithmetic, compound unit arithmetic employs three more functions:

- **Reduction** where a compound quantity is reduced to a single quantity, for example conversion of a distance expressed in yards, feet and inches to one expressed in inches.[15]

- **Expansion**, the inverse function to reduction, is the conversion of a quantity that is expressed as a single unit of measure to a compound unit, such as expanding 24 oz to 1 lb, 8 oz.

- **Normalization** is the conversion of a set of compound units to a standard form – for example rewriting "1 ft 13 in" as "2 ft 1 in".

Knowledge of the relationship between the various units of measure, their multiples and their submultiples forms an essential part of compound unit arithmetic.

10.4.2   Principles of compound unit arithmetic

There are two basic approaches to compound unit arithmetic:

- **Reduction–expansion method** where all the compound unit variables are reduced to single unit variables, the calculation performed and the result expanded back to compound units. This approach is suited for automated calculations. A typical example is the handling of time by Microsoft Excel where all time intervals are processed internally as days and decimal fractions of a day.

- **On-going normalization method** in which each unit is treated separately and the problem is continuously normalized as the solution develops. This approach, which is widely described in classical texts, is best suited for manual calculations. An example of the ongoing normalization method as applied to addition is shown below.

10.4.3   Operations in practice

During the 19th and 20th centuries various aids were developed to aid the manipulation of compound units, particularly in commercial applications. The most common aids were mechanical tills which were adapted in countries such as the United Kingdom to accommodate pounds, shillings, pennies and farthings and "Ready Reckoners" – books aimed at traders that catalogued the results of various routine calculations such as the percentages or multiples of various sums of money. One typical booklet[16] that ran to 150 pages tabulated multiples "from one to ten thousand at the various prices from one farthing to one pound".

The cumbersome nature of compound unit arithmetic has been recognized for many years – in 1586, the Flemish mathematician Simon Stevin published a small pamphlet called *De Thiende* ("the tenth")[17] in which he declared that the universal introduction of decimal coinage, measures, and weights to be merely a question of time while in the modern era, many conversion programs, such as that embedded in the calculator supplied as a standard part of the Microsoft Windows 7 operating system display compound units in a reduced decimal format rather than using an expanded format (i.e. "2.5 ft" is displayed rather than "2 ft 6 in").

10.5 Number theory

Main article: Number theory

Until the 19th century, *number theory* was a synonym of "arithmetic". The addressed problems were directly related to the basic operations and concerned primality, divisibility, and the solution of equations in integers, such as Fermat's last theorem. It appeared that most of these problems, although very elementary to state, are very difficult and may not be solved without very deep mathematics involving concepts and methods from many other branches of mathematics. This led to new branches of number theory such as analytic number theory, algebraic number theory, Diophantine geometry and arithmetic algebraic geometry. Wiles' proof of Fermat's Last Theorem is a typical example of the necessity of sophistical methods, which go far beyond the classical methods of arithmetic, for solving problems that can be stated in elementary arithmetic.

10.6 Arithmetic in education

Primary education in mathematics often places a strong focus on algorithms for the arithmetic of natural numbers, integers, fractions, and decimals (using the decimal place-value system). This study is sometimes known as algorism.

The difficulty and unmotivated appearance of these algorithms has long led educators to question this curriculum, advocating the early teaching of more central and intuitive mathematical ideas. One notable movement in this direction was the New Math of the 1960s and 1970s, which attempted to teach arithmetic in the spirit of axiomatic development from set theory, an echo of the prevailing trend in higher mathematics.[18]

Also, arithmetic was used by Islamic Scholars in order to teach application of the rulings related to Zakat and Irth. This was done in a book entitled *The Best of Arithmetic* by Abd-al-Fattah-al-Dumyati.[19]

The book begins with the foundations of mathematics and proceeds to its application in the later chapters.

10.7 See also

- Lists of mathematics topics

- Mathematics

- Outline of arithmetic

10.7.1 Related topics

- Addition of natural numbers

- Additive inverse

- Arithmetic coding

- Arithmetic mean

- Arithmetic progression

- Arithmetic properties

- Associativity

- Commutativity

- Distributivity

- Elementary arithmetic

- Finite field arithmetic

- Integer

- List of important publications in mathematics

- Mental calculation

- Number line

10.8   Notes

[1] "Mathematics". Science Clarified. Retrieved 23 October 2012.

[2] Davenport, Harold, *The Higher Arithmetic: An Introduction to the Theory of Numbers* (7th ed.), Cambridge University Press, Cambridge, UK, 1999, ISBN 0-521-63446-6

[3] Rudman, Peter Strom (2007). *How Mathematics Happened: The First 50,000 Years*. Prometheus Books. p. 64. ISBN 978-1-59102-477-4.

[4] *The Works of Archimedes*, Chapter IV, *Arithmetic in Archimedes*, edited by T.L. Heath, Dover Publications Inc, New York, 2002.

[5] Joseph Needham, *Science and Civilization in China*, Vol. 3, page 9, Cambridge University Press, 1959.

[6] Reference: Revue de l'Orient Chretien by François Nau pp.327-338. (1929)

[7] Reference: Sigler, L., "Fibonacci's Liber Abaci", Springer, 2003.

[8] Tapson, Frank (1996). *The Oxford Mathematics Study Dictionary*. Oxford University Press. ISBN 0 19 914551 2.

[9] Leonardo Pisano - page 3: "Contributions to number theory". *Encyclopædia Britannica* Online, 2006. Retrieved 18 September 2006.

[10] Walkingame, Francis (1860). "The Tutor's Companion; or, Complete Practical Arithmetic" (PDF). Webb, Millington & Co. pp. 24–39.

[11] Palaiseau, JFG (October 1816). *Métrologie universelle, ancienne et moderne: ou rapport des poids et mesures des empires, royaumes, duchés et principautés des quatre parties du monde [Universal, ancient and modern metrology: or report of weights and measurements of empires, kingdoms, duchies and principalities of all parts of the world]* (in French). Bordeaux. Retrieved October 30, 2011.

[12] Jacob de Gelder (1824). *Allereerste Gronden der Cijferkunst [Introduction to Numeracy]* (in Dutch). 's Gravenhage and Amsterdam: de Gebroeders van Cleef. pp. 163–176. Retrieved March 2, 2011.

[13] Malaisé, Ferdinand (1842). *Theoretisch-Praktischer Unterricht im Rechnen für die niederen Classen der Regimentsschulen der Königl. Bayer. Infantrie und Cavalerie [Theoretical and practical instruction in arithmetic for the lower classes of the Royal Bavarian Infantry and Cavalry School]* (in German). Munich. Retrieved 20 March 2012.

[14] *Encyclopædia Britannica*, Vol I, Edinburgh, 1772, Arithmetick

[15] Walkingame, Francis (1860). "The Tutor's Companion; or, Complete Practical Arithmetic" (PDF). Webb, Millington & Co. pp. 43–50.

[16] Thomson, J (1824). *The Ready Reckoner in miniature containing accurate table from one to the thousand at the various prices from one farthing to one pound*. Montreal. Retrieved 25 March 2012.

[17] O'Connor, John J.; Robertson, Edmund F. (January 2004), "Arithmetic", *MacTutor History of Mathematics archive*, University of St Andrews.

[18] Mathematically Correct: Glossary of Terms

[19] Abd-al-Fattah Bin Abd-al-Rahman al-Banna al-Dumyati (1887). "The Best of Arithmetic". *World Digital Library* (in Arabic). Retrieved 30 June 2013.

10.9   References

- Cunnington, Susan, *The Story of Arithmetic: A Short History of Its Origin and Development*, Swan Sonnenschein, London, 1904

- Dickson, Leonard Eugene, *History of the Theory of Numbers* (3 volumes), reprints: Carnegie Institute of Washington, Washington, 1932; Chelsea, New York, 1952, 1966

- Euler, Leonhard, *Elements of Algebra*, Tarquin Press, 2007

- Fine, Henry Burchard (1858–1928), *The Number System of Algebra Treated Theoretically and Historically*, Leach, Shewell & Sanborn, Boston, 1891

- Karpinski, Louis Charles (1878–1956), *The History of Arithmetic*, Rand McNally, Chicago, 1925; reprint: Russell & Russell, New York, 1965

- Ore, Øystein, *Number Theory and Its History*, McGraw–Hill, New York, 1948

- Weil, André, *Number Theory: An Approach through History*, Birkhauser, Boston, 1984; reviewed: Mathematical Reviews 85c:01004

10.10   External links

- MathWorld article about arithmetic

- The New Student's Reference Work/Arithmetic (historical)

- The Great Calculation According to the Indians, of Maximus Planudes – an early Western work on arithmetic at Convergence

- P. H. Vander Weyde (1879). "Arithmetic". *The American Cyclopædia*.

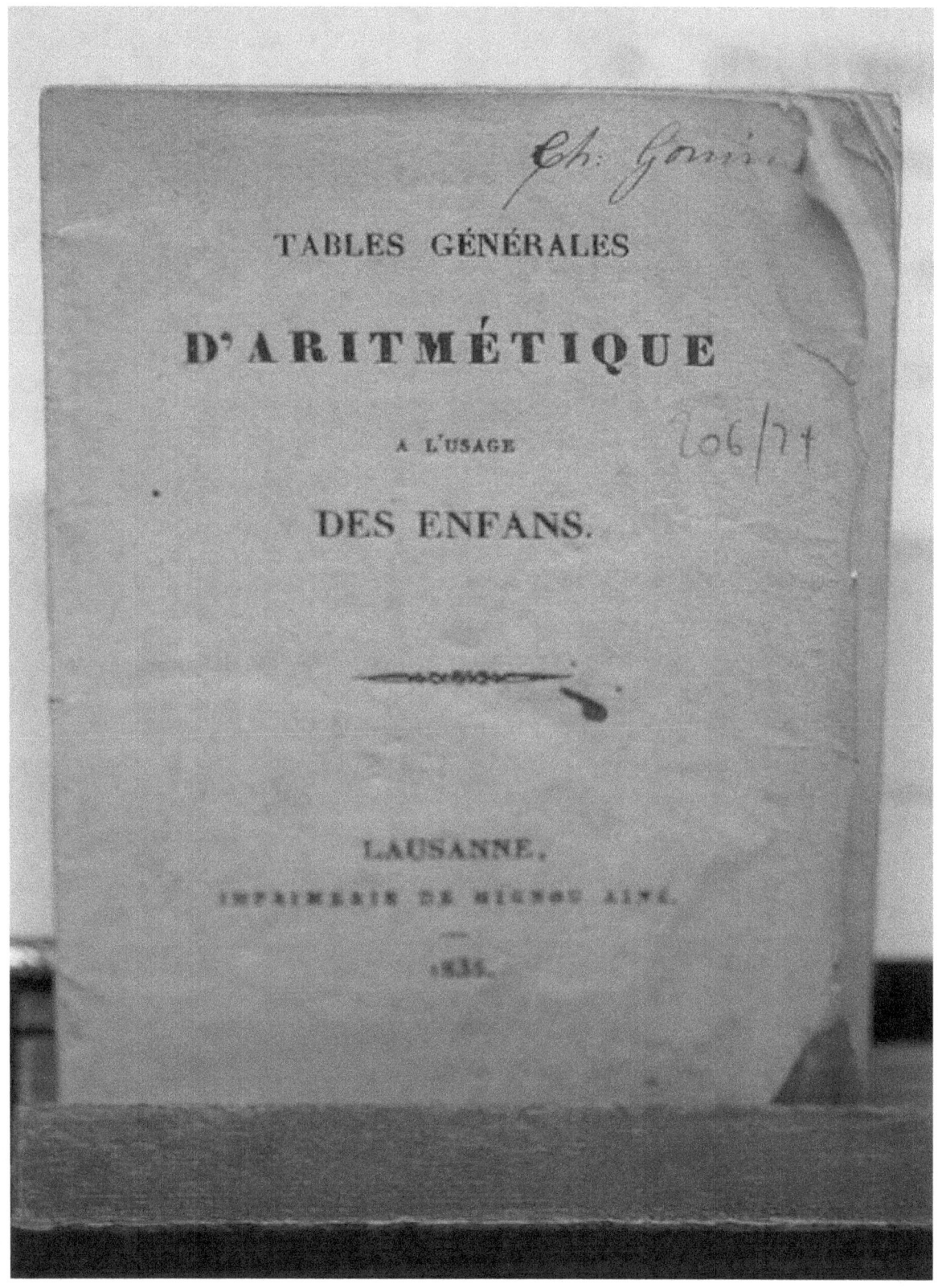

Arithmetic tables for children, Lausanne, 1835

Leibniz's Stepped Reckoner was the first calculator that could perform all four arithmetic operations.

A scale calibrated in imperial units with an associated cost display.

Chapter 11

Natural number

This article is about "positive integers" and "non-negative integers". For all the numbers ..., −2, −1, 0, 1, 2, ..., see Integer. In mathematics, the **natural numbers** (sometimes called the **whole numbers**)[1][2][3][4] are those used for counting (as in "there are *six* coins on the table") and ordering (as in "this is the *third* largest city in the country"). In common language, words used for counting are "cardinal numbers" and words used for ordering are "ordinal numbers".

Another use of natural numbers is for what linguists call nominal numbers, such as the model number of a product, where the "natural number" is used only for naming (as distinct from a serial number where the order properties of the natural numbers distinguish later uses from earlier uses) and generally lacks any meaning of *number* as used in mathematics.

The natural numbers are the basis from which many other number sets may be built by extension: the integers, by including an unresolved negation operation; the rational numbers, by including with the integers an unresolved division operation; the real numbers by including with the rationals the termination of Cauchy sequences; the complex numbers, by including with the real numbers the unresolved square root of minus one; the hyperreal numbers, by including with real numbers the infinitesimal value epsilon; vectors, by including a vector structure with reals; matrices, by having vectors of vectors; the nonstandard integers; and so on.[5][6] Therefore, the natural numbers are canonically embedded (identified) in the other number systems.

Properties of the natural numbers, such as divisibility and the distribution of prime numbers, are studied in number theory. Problems concerning counting and ordering, such as partitioning and enumerations, are studied in combinatorics.

There is no universal agreement about whether to include zero in the set of natural numbers. Some authors begin the natural numbers with 0, corresponding to the **non-negative integers** 0, 1, 2, 3, ..., whereas others start with 1, corresponding to the **positive integers** 1, 2, 3, ...[7][8][9][10] This distinction is of no fundamental concern for the natural numbers as such, since their core construction is the unary operation successor. Including the number 0 just supplies an identity element for the (binary) operation of addition, which makes up together with the multiplication the usual arithmetic in the natural numbers, to be completed within the integers and the rational numbers, only.

In common language, for example in primary school, natural numbers may be called **counting numbers**[11] to distinguish them from the real numbers which are used for measurement.

11.1   History

The most primitive method of representing a natural number is to put down a mark for each object. Later, a set of objects could be tested for equality, excess or shortage, by striking out a mark and removing an object from the set.

The first major advance in abstraction was the use of numerals to represent numbers. This allowed systems to be developed for recording large numbers. The ancient Egyptians developed a powerful system of numerals with distinct hieroglyphs for 1, 10, and all the powers of 10 up to over 1 million. A stone carving from Karnak, dating from around 1500 BC and now at the Louvre in Paris, depicts 276 as 2 hundreds, 7 tens, and 6 ones; and similarly for the number 4,622. The Babylonians had a place-value system based essentially on the numerals for 1 and 10, using base sixty, so that the symbol

Natural numbers can be used for counting (one apple, two apples, three apples, ...)

for sixty was the same as the symbol for one, its value being determined from context.[15]

A much later advance was the development of the idea that 0 can be considered as a number, with its own numeral. The use of a 0 digit in place-value notation (within other numbers) dates back as early as 700 BC by the Babylonians, but they omitted such a digit when it would have been the last symbol in the number.[16] The Olmec and Maya civilizations used 0 as a separate number as early as the 1st century BC, but this usage did not spread beyond Mesoamerica.[17][18] The use of a numeral 0 in modern times originated with the Indian mathematician Brahmagupta in 628. However, 0 had been used as a number in the medieval computus (the calculation of the date of Easter), beginning with Dionysius Exiguus in 525, without being denoted by a numeral (standard Roman numerals do not have a symbol for 0); instead *nulla* (or the genitive form *nullae*) from *nullus*, the Latin word for "none", was employed to denote a 0 value.[19]

The first systematic study of numbers as abstractions is usually credited to the Greek philosophers Pythagoras and Archimedes. Some Greek mathematicians treated the number 1 differently than larger numbers, sometimes even not as a number at all.[20]

Independent studies also occurred at around the same time in India, China, and Mesoamerica.[21]

11.1.1 Modern definitions

In 19th century Europe, there was mathematical and philosophical discussion about the exact nature of the natural numbers. A school of Naturalism stated that the natural numbers were a direct consequence of the human psyche. Henri Poincaré was one of its advocates, as was Leopold Kronecker who summarized "God made the integers, all else is the work of man".

In opposition to the Naturalists, the constructivists saw a need to improve the logical rigor in the foundations of mathematics.[22] In the 1860s, Hermann Grassmann suggested a recursive definition for natural numbers thus stating they were not really natural but a consequence of definitions. Later, two classes of such formal definitions were constructed; later, they were shown to be equivalent in most practical applications.

Set-theoretical definitions of natural numbers were initiated by Frege and he initially defined a natural number as the class of all sets that are in one-to-one correspondence with a particular set, but this definition turned out to lead to paradoxes including Russell's paradox. Therefore, this formalism was modified so that a natural number is defined as a particular set, and any set that can be put into one-to-one correspondence with that set is said to have that number of elements.[23]

The second class of definitions was introduced by Giuseppe Peano and is now called Peano arithmetic. It is based on an axiomatization of the properties of ordinal numbers: each natural number has a successor and every non-zero natural number has a unique predecessor. Peano arithmetic is equiconsistent with several weak systems of set theory. One such system is ZFC with the axiom of infinity replaced by its negation. Theorems that can be proved in ZFC but cannot be proved using the Peano Axioms include Goodstein's theorem.[24]

With all these definitions it is convenient to include 0 (corresponding to the empty set) as a natural number. Including 0 is now the common convention among set theorists[25] and logicians.[26] Other mathematicians also include 0[10] although many have kept the older tradition and take 1 to be the first natural number.[27] Computer scientists often start from zero when enumerating items like loop counters and string- or array- elements.[28][29]

11.2 Notation

Mathematicians use **N** or $\mathbb{N}$ (an N in blackboard bold, displayed as $\mathbb{N}$ in Unicode) to refer to the set of all natural numbers. This set is countably infinite: it is infinite but countable by definition. This is also expressed by saying that the cardinal number of the set is aleph-naught ($\aleph_0$) .[30]

To be unambiguous about whether 0 is included or not, sometimes an index (or superscript) "0" is added in the former case, and a superscript " * " or subscript " 1 " is added in the latter case:

$$\mathbb{N}^0 = \mathbb{N}_0 = \{0, 1, 2, \dots\}$$

$$\mathbb{N}^* = \mathbb{N}^+ = \mathbb{N}_1 = \mathbb{N}_{>0} = \{1, 2, \dots\}.$$

11.3 Properties

11.3.1   Addition

One can recursively define an addition on the natural numbers by setting $a + 0 = a$ and $a + S(b) = S(a + b)$ for all a, b. Here S should be read as "successor". This turns the natural numbers ($\mathbf{N}$, +) into a commutative monoid with identity element 0, the so-called free object with one generator. This monoid satisfies the cancellation property and can be embedded in a group (in the mathematical sense of the word *group*). The smallest group containing the natural numbers is the integers.

If 1 is defined as $S(0)$, then $b + 1 = b + S(0) = S(b + 0) = S(b)$. That is, $b + 1$ is simply the successor of b.

11.3.2   Multiplication

Analogously, given that addition has been defined, a multiplication $\times$ can be defined via $a \times 0 = 0$ and $a \times S(b) = (a \times b) + a$. This turns ($\mathbf{N}^*$, $\times$) into a free commutative monoid with identity element 1; a generator set for this monoid is the set of prime numbers.

11.3.3   Relationship between addition and multiplication

Addition and multiplication are compatible, which is expressed in the distribution law: $a \times (b + c) = (a \times b) + (a \times c)$. These properties of addition and multiplication make the natural numbers an instance of a commutative semiring. Semirings are an algebraic generalization of the natural numbers where multiplication is not necessarily commutative. The lack of additive inverses, which is equivalent to the fact that $\mathbf{N}$ is not closed under subtraction, means that $\mathbf{N}$ is *not* a ring; instead it is a semiring (also known as a *rig*).

If the natural numbers are taken as "excluding 0", and "starting at 1", the definitions of + and $\times$ are as above, except that they begin with $a + 1 = S(a)$ and $a \times 1 = a$.

11.3.4   Order

In this section, juxtaposed variables such as ab indicate the product $a \times b$, and the standard order of operations is assumed.

A total order on the natural numbers is defined by letting $a \leq b$ if and only if there exists another natural number c with $a + c = b$. This order is compatible with the arithmetical operations in the following sense: if a, b and c are natural numbers and $a \leq b$, then $a + c \leq b + c$ and $ac \leq bc$. An important property of the natural numbers is that they are well-ordered: every non-empty set of natural numbers has a least element. The rank among well-ordered sets is expressed by an ordinal number; for the natural numbers this is expressed as ω.

11.3.5   Division

In this section, juxtaposed variables such as ab indicate the product $a \times b$, and the standard order of operations is assumed.

While it is in general not possible to divide one natural number by another and get a natural number as result, the procedure of *division with remainder* is available as a substitute: for any two natural numbers a and b with $b \neq 0$ there are natural numbers q and r such that

$$a = bq + r \text{ and } r < b.$$

The number q is called the *quotient* and r is called the *remainder* of division of a by b. The numbers q and r are uniquely determined by a and b. This Euclidean division is key to several other properties (divisibility), algorithms (such as the Euclidean algorithm), and ideas in number theory.

11.3.6   Algebraic properties satisfied by the natural numbers

The addition (+) and multiplication ($\times$) operations on natural numbers as defined above have several algebraic properties:

- Closure under addition and multiplication: for all natural numbers a and b, both $a + b$ and $a \times b$ are natural numbers.

- Associativity: for all natural numbers a, b, and c, $a + (b + c) = (a + b) + c$ and $a \times (b \times c) = (a \times b) \times c$.

- Commutativity: for all natural numbers a and b, $a + b = b + a$ and $a \times b = b \times a$.

- Existence of identity elements: for every natural number a, $a + 0 = a$ and $a \times 1 = a$.

- Distributivity of multiplication over addition for all natural numbers a, b, and c, $a \times (b + c) = (a \times b) + (a \times c)$.

- No nonzero zero divisors: if a and b are natural numbers such that $a \times b = 0$, then $a = 0$ or $b = 0$.

11.4 Generalizations

Two generalizations of natural numbers arise from the two uses:

- A natural number can be used to express the size of a finite set; more generally a cardinal number is a measure for the size of a set also suitable for infinite sets; this refers to a concept of "size" such that if there is a bijection between two sets they have the same size. The set of natural numbers itself and any other countably infinite set has cardinality aleph-null ($\aleph_0$).

- Linguistic ordinal numbers "first", "second", "third" can be assigned to the elements of a totally ordered finite set, and also to the elements of well-ordered countably infinite sets like the set of natural numbers itself. This can be generalized to ordinal numbers which describe the position of an element in a well-ordered set in general. An ordinal number is also used to describe the "size" of a well-ordered set, in a sense different from cardinality: if there is an order isomorphism between two well-ordered sets they have the same ordinal number. The first ordinal number that is not a natural number is expressed as ω ; this is also the ordinal number of the set of natural numbers itself.

Many well-ordered sets with cardinal number $\aleph_0$ have an ordinal number greater than ω (the latter is the lowest possible). The least ordinal of cardinality $\aleph_0$ (i.e., the initial ordinal) is ω .

For finite well-ordered sets, there is one-to-one correspondence between ordinal and cardinal numbers; therefore they can both be expressed by the same natural number, the number of elements of the set. This number can also be used to describe the position of an element in a larger finite, or an infinite, sequence.

A countable non-standard model of arithmetic satisfying the Peano Arithmetic (i.e., the first-order Peano axioms) was developed by Skolem in 1933. The hypernatural numbers are an uncountable model that can be constructed from the ordinary natural numbers via the ultrapower construction.

Georges Reeb used to claim provocatively that *The naïve integers don't fill up* $\mathbb{N}$. Other generalizations are discussed in the article on numbers.

11.5 Formal definitions

11.5.1 Peano axioms

Main article: Peano axioms

Many properties of the natural numbers can be derived from the Peano axioms.[31][32]

- Axiom One: 0 is a natural number.

- Axiom Two: Every natural number has a successor.

- Axiom Three: 0 is not the successor of any natural number.

- Axiom Four: If the successor of x equals the successor of y, then x equals y.

- Axiom Five (the Axiom of Induction): If a statement is true of 0, and if the truth of that statement for a number implies its truth for the successor of that number, then the statement is true for every natural number.

These are not the original axioms published by Peano, but are named in his honor. Some forms of the Peano axioms have 1 in place of 0. In ordinary arithmetic, the successor of x is x + 1. Replacing Axiom Five by an axiom schema one obtains a (weaker) first-order theory called *Peano Arithmetic*.

11.5.2   Constructions based on set theory

Main article: Set-theoretic definition of natural numbers

In the area of mathematics called set theory, a special case of the von Neumann ordinal construction [33] defines the natural numbers as follows:

Set $0 := \{ \}$, the empty set,

and define $S(a) = a \cup \{a\}$ for every set a. $S(a)$ is the successor of a, and S is called the successor function.

By the axiom of infinity, there exists a set which contains 0 and is closed under the successor function. (Such sets are said to be 'inductive'.) Then the intersection of all inductive sets is defined to be the set of natural numbers. It can be checked that the set of natural numbers satisfies the Peano axioms.

Each natural number is then equal to the set of all natural numbers less than it, so that

- $0 = \{ \}$
- $1 = \{0\} = \{\{ \}\}$
- $2 = \{0, 1\} = \{0, \{0\}\} = \{\{ \}, \{\{ \}\}\}$

:*$3 = \{0, 1, 2\} = \{0, \{0\}, \{0, \{0\}\}\} = \{\{ \}, \{\{ \}\}, \{\{ \}, \{\{ \}\}\}\}$

- $n = \{0, 1, 2, ..., n-2, n-1\} = \{0, 1, 2, ..., n-2\} \cup \{n-1\} = (n-1) \cup \{n-1\} = S(n-1)$

and so on.

With this definition, a natural number n is a particular set with n elements, and $n \leq m$ if and only if n is a subset of m.

Also, with this definition, different possible interpretations of notations like $\mathbf{R}^n$ (n-tuples versus mappings of n into $\mathbf{R}$) coincide.

Even if one does not accept the axiom of infinity and therefore cannot accept that the set of all natural numbers exists, it is still possible to define any one of these sets.

Other constructions

Although the standard construction is useful, it is not the only possible construction. Zermelo's construction goes as follows:

one defines $0 = \{ \}$

and $S(a) = \{a\}$,

producing

- 0 = { }
- 1 = {0} ={{ }}
- 2 = {1} ={{{ }}}, etc.

Each natural number is then equal to the set of the natural number preceding it.

It is also possible to define 0 = {{ }}

and $S(a) = a \cup \{a\}$

producing

- 0 = {{ }}
- 1 = {{ }, 0} = {{ }, {{ }}}
- 2 = {{ }, 0, 1}, etc.

11.6 See also

- Integer

- Set-theoretic definition of natural numbers

- Peano axioms

- Canonical representation of a positive integer

- Countable set

- Number#Classification for other number systems (rational, real, complex etc.)

11.7 Notes

[1] Weisstein, Eric W., "Whole Number", *MathWorld*.

[2] Clapham & Nicholson (2014): "**whole number** An integer, though sometimes it is taken to mean only non-negative integers, or just the positive integers."

[3] James & James (1992) give definitions of "whole number" under several headwords:
INTEGER ... *Syn.* whole number.
NUMBER ... whole number. A nonnegative integer.
WHOLE ... whole number.
(1) One of the integers 0, 1, 2, 3, ...
(2) A positive integer; *i.e.*, a natural number.
(3) An integer, positive, negative, or zero.

[4] The *Common Core State Standards for Mathematics* say: "Whole numbers. The numbers 0, 1, 2, 3, ..." (Glossary, p. 87) (PDF)
Definitions from *The Ontario Curriculum, Grades 1-8: Mathematics*, Ontario Ministry of Education (2005) (PDF)
"**natural numbers.** The counting numbers 1, 2, 3, 4, ..." (Glossary, p. 128)
"**whole number.** Any one of the numbers 0, 1, 2, 3, 4, ..." (Glossary, p. 134)
Musser, Peterson & Burger (2013, p. 57): "As mentioned earlier, the study of the set of whole numbers, $W = \{0, 1, 2, 3, 4, ...\}$, is the foundation of elementary school mathematics."
These pre-algebra books define the *whole numbers*:

- Szczepanski & Kositsky (2008): "Another important collection of numbers is the *whole numbers*, the natural numbers together with zero." (Chapter 1: *The Whole Story*, p. 4). On the inside front cover, the authors say: "We based this book on the state standards for pre-algebra in California, Florida, New York, and Texas, ..."

- Bluman (2010): "When 0 is added to the set of natural numbers, the set is called the whole numbers." (Chapter 1: *Whole Numbers*, p. 1)

  Both books define the *natural numbers* to be: "1, 2, 3, ...".

[5] Mendelson (2008) says: "The whole fantastic hierarchy of number systems is built up by purely set-theoretic means from a few simple assumptions about natural numbers." (Preface, p. x)

[6] Bluman (2010): "Numbers make up the foundation of mathematics." (p. 1)

[7] Weisstein, Eric W., "Natural Number", *MathWorld*.

[8] "natural number", *Merriam-Webster.com* (Merriam-Webster), retrieved 4 October 2014

[9] Carothers (2000) says: "ℕ is the set of natural numbers (positive integers)" (p. 3)

[10] Mac Lane & Birkhoff (1999) include zero in the natural numbers: 'Intuitively, the set ℕ = {0, 1, 2, ... } of all *natural numbers* may be described as follows: ℕ contains an "initial" number 0; ...'. They follow that with their version of the Peano Postulates. (p. 15)

[11] Weisstein, Eric W., "Counting Number", *MathWorld*.

[12] Introduction, Royal Belgian Institute of Natural Sciences, Brussels, Belgium.

[13] Flash presentation, Royal Belgian Institute of Natural Sciences, Brussels, Belgium.

[14] The Ishango Bone, Democratic Republic of the Congo, on permanent display at the Royal Belgian Institute of Natural Sciences, Brussels, Belgium. UNESCO's Portal to the Heritage of Astronomy

[15] Georges Ifrah, *The Universal History of Numbers*, Wiley, 2000. ISBN 0-471-37568-3

[16] "A history of Zero". *MacTutor History of Mathematics*. Retrieved 2013-01-23. ... a tablet found at Kish ... thought to date from around 700 BC, uses three hooks to denote an empty place in the positional notation. Other tablets dated from around the same time use a single hook for an empty place

[17] Mann, Charles C. (2005), *1491: New Revelations Of The Americas Before Columbus*, Knopf, p. 19. ISBN 9781400040063.

[18] Evans, Brian (2014), "Chapter 10. Pre-Columbian Mathematics: The Olmec, Maya, and Inca Civilizations", *The Development of Mathematics Throughout the Centuries: A Brief History in a Cultural Context*, John Wiley & Sons, ISBN 9781118853979.

[19] Michael L. Gorodetsky (2003-08-25). "Cyclus Decemnovennalis Dionysii – Nineteen year cycle of Dionysius". Hbar.phys.msu.ru. Retrieved 2012-02-13.

[20] This convention is used, for example, in Euclid's Elements, see Book VII, definitions 1 and 2.

[21] Morris Kline, *Mathematical Thought From Ancient to Modern Times*, Oxford University Press, 1990 [1972], ISBN 0-19-506135-7

[22] "Much of the mathematical work of the twentieth century has been devoted to examining the logical foundations and structure of the subject." (Eves 1990, p. 606)

[23] Eves 1990, Chapter 15

[24] L. Kirby; J. Paris, *Accessible Independence Results for Peano Arithmetic*, Bulletin of the London Mathematical Society 14 (4): 285. doi:10.1112/blms/14.4.285, 1982.

[25] Bagaria, Joan. "Set Theory". The Stanford Encyclopedia of Philosophy (Winter 2014 Edition).

[26] Goldrei, Derek (1998). "3". *Classic set theory : a guided independent study* (1. ed., 1. print ed.). Boca Raton, Fla. [u.a.]: Chapman & Hall/CRC. p. 33. ISBN 0-412-60610-0.

[27] This is common in texts about Real analysis. See, for example, Carothers (2000, p. 3) or Thomson, Bruckner & Bruckner (2000, p. 2).

[28] Brown, Jim (1978). "In Defense of Index Origin 0". *ACM SIGAPL APL Quote Quad* **9** (2): 7 – 7. doi:10.1145/586050.586053. Retrieved 19 January 2015.

[29] Hui, Roger. "Is Index Origin 0 a Hindrance?". *http://www.jsoftware.com*". *JSoftware / Roger Hui. Retrieved 19 January 2015.*

[30] Weisstein, Eric W., "Cardinal Number", *MathWorld*.

[31] G.E. Mints (originator), "Peano axioms", *Encyclopedia of Mathematics* (Springer, in cooperation with the European Mathematical Society), retrieved 8 October 2014

[32] Hamilton (1988) calls them "Peano's Postulates" and begins with "1. 0 is a natural number." (p. 117f)
Halmos (1960) uses the language of set theory instead of the language of arithmetic for his five axioms. He begins with "(I) $0 \in \omega$ (where, of course, $0 = \varnothing$)" (ω is the set of all natural numbers). (p. 46)
Morash (1991) gives "a two-part axiom" in which the natural numbers begin with 1. (Section 10.1: *An Axiomatization for the System of Positive Integers*)

[33] Von Neumann 1923

11.8 References

- Bluman, Allan (2010), *Pre-Algebra DeMYSTiFieD* (Second ed.), McGraw-Hill Professional

- Carothers, N.L. (2000), *Real analysis*, Cambridge University Press, ISBN 0-521-49756-6

- Clapham, Christopher; Nicholson, James (2014), *The Concise Oxford Dictionary of Mathematics* (Fifth ed.), Oxford University Press

- Dedekind, Richard (1963), *Essays on the Theory of Numbers*, Dover, ISBN 0-486-21010-3

  - Dedekind, Richard (2007), *Essays on the Theory of Numbers*, Kessinger Publishing, LLC, ISBN 0-548-08985-X

- Eves, Howard (1990), *An Introduction to the History of Mathematics* (6th ed.), Thomson, ISBN 978-0-03-029558-4

- Halmos, Paul (1960), *Naive Set Theory*, Springer Science & Business Media

- Hamilton, A. G. (1988), *Logic for Mathematicians* (Revised ed.), Cambridge University Press

- James, Robert C.; James, Glenn (1992), *Mathematics Dictionary* (Fifth ed.), Chapman & Hall

- Landau, Edmund (1966), *Foundations of Analysis* (Third ed.), Chelsea Pub Co, ISBN 0-8218-2693-X

- Mac Lane, Saunders; Birkhoff, Garrett (1999), *Algebra* (3rd ed.), American Mathematical Society

- Mendelson, Elliott (2008) [1973], *Number Systems and the Foundations of Analysis*, Dover Publications

- Morash, Ronald P. (1991), *Bridge to Abstract Mathematics: Mathematical Proof and Structures* (Second ed.), Mcgraw-Hill College

- Musser, Gary L.; Peterson, Blake E.; Burger, William F. (2013), *Mathematics for Elementary Teachers: A Contemporary Approach* (10th ed.), Wiley Global Education, ISBN 978-1118457443

- Szczepanski, Amy F.; Kositsky, Andrew P. (2008), *The Complete Idiot's Guide to Pre-algebra*, Penguin Group

- Thomson, Brian S.; Bruckner, Judith B.; Bruckner, Andrew M. (2008), *Elementary Real Analysis* (Second ed.), ClassicalRealAnalysis.com, ISBN 9781434843678

- Von Neumann, Johann (1923), "Zur Einführung der transfiniten Zahlen", *Acta litterarum ac scientiarum Ragiae Universitatis Hungaricae Francisco-Josephinae, Sectio scientiarum mathematicarum* **1**: 199–208

  - Von Neumann, John (January 2002) [1923], "On the introduction of transfinite numbers", in Jean van Heijenoort, *From Frege to Gödel: A Source Book in Mathematical Logic, 1879-1931* (3rd ed.), Harvard University Press, pp. 346–354, ISBN 0-674-32449-8 - English translation of von Neumann 1923.

11.9   External links

- Hazewinkel, Michiel, ed. (2001), "Natural number", *Encyclopedia of Mathematics*, Springer, ISBN 978-1-55608-010-4

- Axioms and Construction of Natural Numbers

- Essays on the Theory of Numbers by Richard Dedekind at Project Gutenberg

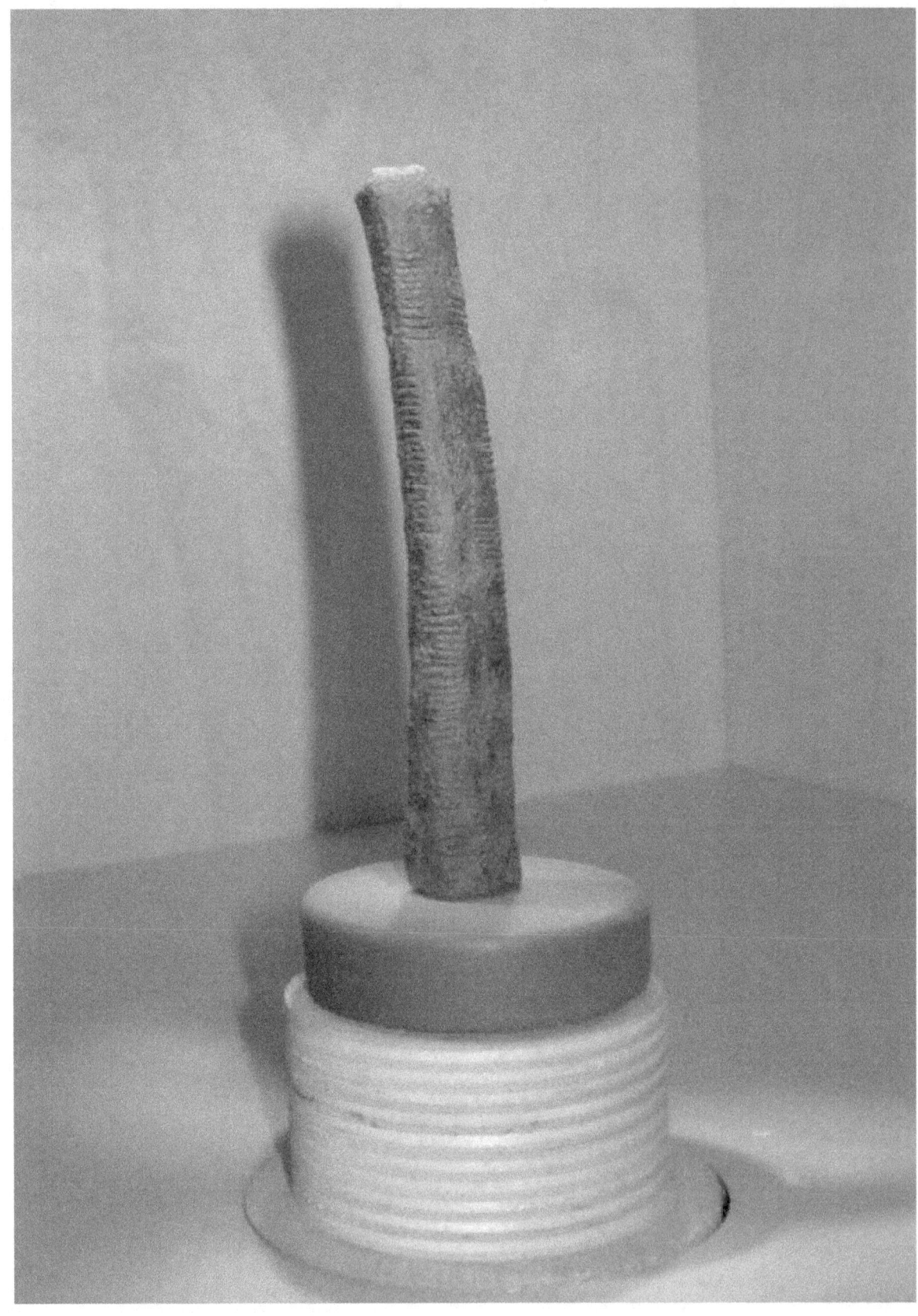

The double-struck capital N symbol, often used to denote the set of all natural numbers (see List of mathematical symbols).

Chapter 12

Integer

This article is about numbers traditionally known as "integers". For computer representations, see integer (computer science). For the concept in algebraic number theory, see integral element.

An **integer** (from the Latin *integer* meaning "whole")[note 1] is a number that can be written without a fractional component. For example, 21, 4, 0, and −2048 are integers, while 9.75, 5½, and √2 are not.

The set of integers consists of zero (0), the natural numbers (1, 2, 3, ...), also called *whole numbers* or *counting numbers*,[1] and their additive inverses (the **negative integers**, i.e. −1, −2, −3, ...). This is often denoted by a boldface Z ("**Z**") or blackboard bold ℤ (Unicode U+2124 ℤ) standing for the German word *Zahlen* (['tsaːlən], "numbers").[2][3] ℤ is a subset of the sets of rational and real numbers and, like the natural numbers, is countably infinite.

The integers form the smallest group and the smallest ring containing the natural numbers. In algebraic number theory, the integers are sometimes called **rational integers** to distinguish them from the more general algebraic integers. In fact, the (rational) integers are the algebraic integers that are also rational numbers.

12.1   Algebraic properties

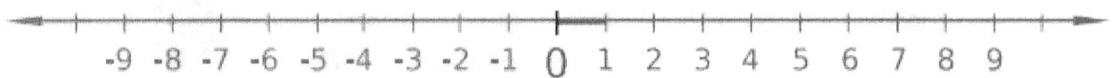

Integers can be thought of as discrete, equally spaced points on an infinitely long number line. In the above, non-negative integers are shown in purple and negative integers in red.

Like the natural numbers, **Z** is closed under the operations of addition and multiplication, that is, the sum and product of any two integers is an integer. However, with the inclusion of the negative natural numbers, and, importantly, 0, **Z** (unlike the natural numbers) is also closed under subtraction. The integers form a unital ring which is the most basic one, in the following sense: for any unital ring, there is a unique ring homomorphism from the integers into this ring. This universal property, namely to be an initial object in the category of rings, characterizes the ring **Z**.

Z is not closed under division, since the quotient of two integers (e.g., 1 divided by 2), need not be an integer. Although the natural numbers are closed under exponentiation, the integers are not (since the result can be a fraction when the exponent is negative).

The following lists some of the basic properties of addition and multiplication for any integers *a*, *b* and *c*.

In the language of abstract algebra, the first five properties listed above for addition say that **Z** under addition is an abelian group. As a group under addition, **Z** is a cyclic group, since every non-zero integer can be written as a finite sum 1 + 1 + ... + 1 or (−1) + (−1) + ... + (−1). In fact, **Z** under addition is the *only* infinite cyclic group, in the sense that any infinite cyclic group is isomorphic to **Z**.

The first four properties listed above for multiplication say that **Z** under multiplication is a commutative monoid. However, not every integer has a multiplicative inverse; e.g. there is no integer x such that $2x = 1$, because the left hand side is even, while the right hand side is odd. This means that **Z** under multiplication is not a group.

All the rules from the above property table, except for the last, taken together say that **Z** together with addition and multiplication is a commutative ring with unity. It is the prototype of all objects of such algebraic structure. Only those equalities of expressions are true in **Z** for all values of variables, which are true in any unital commutative ring. Note that certain non-zero integers map to zero in certain rings.

At last, the property (*) says that the commutative ring **Z** is an integral domain. In fact, **Z** provides the motivation for defining such a structure.

The lack of multiplicative inverses, which is equivalent to the fact that **Z** is not closed under division, means that **Z** is *not* a field. The smallest field with the usual operations containing the integers is the field of rational numbers. The process of constructing the rationals from the integers can be mimicked to form the field of fractions of any integral domain. And back, starting from an algebraic number field (an extension of rational numbers), its ring of integers can be extracted, which includes **Z** as its subring.

Although ordinary division is not defined on **Z**, the division "with remainder" is defined on them. It is called Euclidean division and possesses the following important property: that is, given two integers a and b with $b \neq 0$, there exist unique integers q and r such that $a = q \times b + r$ and $0 \leq r < |b|$, where $|b|$ denotes the absolute value of b. The integer q is called the *quotient* and r is called the *remainder* of the division of a by b. The Euclidean algorithm for computing greatest common divisors works by a sequence of Euclidean divisions.

Again, in the language of abstract algebra, the above says that **Z** is a Euclidean domain. This implies that **Z** is a principal ideal domain and any positive integer can be written as the products of primes in an essentially unique way. This is the fundamental theorem of arithmetic.

12.2   Order-theoretic properties

Z is a totally ordered set without upper or lower bound. The ordering of **Z** is given by:

$$\ldots -3 < -2 < -1 < 0 < 1 < 2 < 3 < \ldots$$

An integer is *positive* if it is greater than zero and *negative* if it is less than zero. Zero is defined as neither negative nor positive.

The ordering of integers is compatible with the algebraic operations in the following way:

1. if $a < b$ and $c < d$, then $a + c < b + d$

2. if $a < b$ and $0 < c$, then $ac < bc$.

It follows that **Z** together with the above ordering is an ordered ring.

The integers are the only nontrivial totally ordered abelian group whose positive elements are well-ordered.[4] This is equivalent to the statement that any Noetherian valuation ring is either a field or a discrete valuation ring.

12.3   Construction

In elementary school teaching, integers are often intuitively defined as the (positive) natural numbers, zero, and the negations of the natural numbers. However, this style of definition leads to many different cases (each arithmetic operation needs to be defined on each combination of types of integer) and makes it tedious to prove that these operations obey

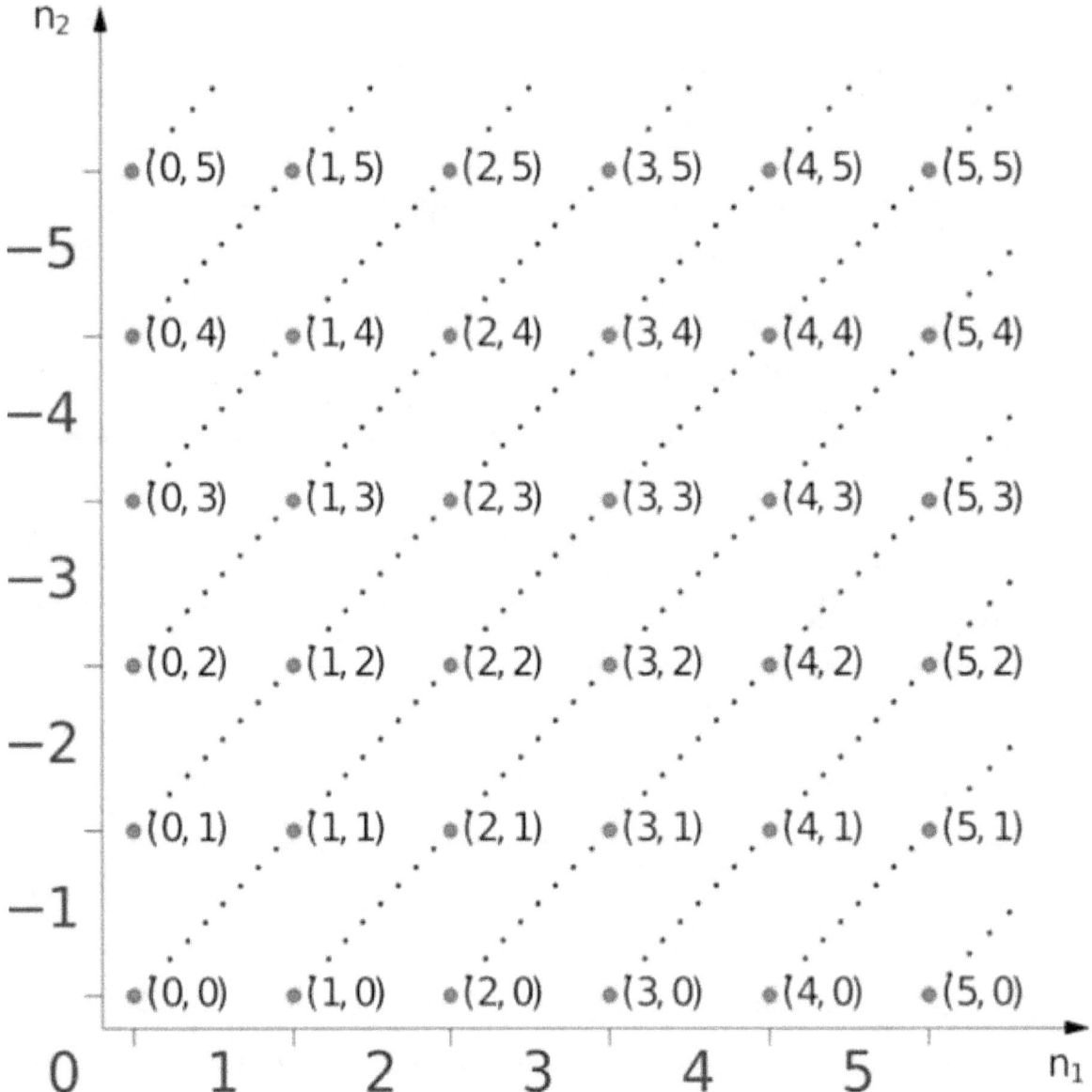

Red points represent ordered pairs of natural numbers. Linked red points are equivalence classes representing the blue integers at the end of the line.

the laws of arithmetic.[5] Therefore, in modern set-theoretic mathematics a more abstract construction,[6] which allows one to define the arithmetical operations without any case distinction, is often used instead.[7] The integers can thus be formally constructed as the equivalence classes of ordered pairs of natural numbers (a,b).[8]

The intuition is that (a,b) stands for the result of subtracting b from a.[8] To confirm our expectation that $1 - 2$ and $4 - 5$ denote the same number, we define an equivalence relation $\sim$ on these pairs with the following rule:

$$(a, b) \sim (c, d)$$

precisely when

$$a + d = b + c.$$

Addition and multiplication of integers can be defined in terms of the equivalent operations on the natural numbers;[8] denoting by $[(a,b)]$ the equivalence class having (a,b) as a member, one has:

$$[(a,b)] + [(c,d)] := [(a+c,b+d)].$$

$$[(a,b)] \cdot [(c,d)] := [(ac+bd,ad+bc)].$$

The negation (or additive inverse) of an integer is obtained by reversing the order of the pair:

$$-[(a,b)] := [(b,a)].$$

Hence subtraction can be defined as the addition of the additive inverse:

$$[(a,b)] - [(c,d)] := [(a+d,b+c)].$$

The standard ordering on the integers is given by:

$$[(a,b)] < [(c,d)] \text{ iff } a+d < b+c.$$

It is easily verified that these definitions are independent of the choice of representatives of the equivalence classes.

Every equivalence class has a unique member that is of the form $(n,0)$ or $(0,n)$ (or both at once). The natural number n is identified with the class $[(n,0)]$ (in other words the natural numbers are embedded into the integers by map sending n to $[(n,0)]$), and the class $[(0,n)]$ is denoted $-n$ (this covers all remaining classes, and gives the class $[(0,0)]$ a second time since $-0 = 0$.

Thus, $[(a,b)]$ is denoted by

$$\begin{cases} a - b, & \text{if } a \geq b \\ -(b-a), & \text{if } a < b. \end{cases}$$

If the natural numbers are identified with the corresponding integers (using the embedding mentioned above), this convention creates no ambiguity.

This notation recovers the familiar representation of the integers as $\{..., -2, -1, 0, 1, 2, ...\}$.

Some examples are:

$$\begin{aligned}
0 &= [(0,0)] &&= [(1,1)] = \cdots &&= [(k,k)] \\
1 &= [(1,0)] &&= [(2,1)] = \cdots &&= [(k+1,k)] \\
-1 &= [(0,1)] &&= [(1,2)] = \cdots &&= [(k,k+1)] \\
2 &= [(2,0)] &&= [(3,1)] = \cdots &&= [(k+2,k)] \\
-2 &= [(0,2)] &&= [(1,3)] = \cdots &&= [(k,k+2)].
\end{aligned}$$

12.4 Computer science

Main article: Integer (computer science)

An integer is often a primitive data type in computer languages. However, integer data types can only represent a subset of all integers, since practical computers are of finite capacity. Also, in the common two's complement representation, the inherent definition of sign distinguishes between "negative" and "non-negative" rather than "negative, positive, and 0". (It is, however, certainly possible for a computer to determine whether an integer value is truly positive.) Fixed length integer approximation data types (or subsets) are denoted *int* or Integer in several programming languages (such as Algol68, C, Java, Delphi, etc.).

Variable-length representations of integers, such as bignums, can store any integer that fits in the computer's memory. Other integer data types are implemented with a fixed size, usually a number of bits which is a power of 2 (4, 8, 16, etc.) or a memorable number of decimal digits (e.g., 9 or 10).

12.5 Cardinality

The cardinality of the set of integers is equal to $\aleph_0$ (aleph-null). This is readily demonstrated by the construction of a bijection, that is, a function that is injective and surjective from **Z** to **N**. If **N** = {0, 1, 2, ...} then consider the function:

$$f(x) = \begin{cases} 2|x|, & \text{if } x \leq 0 \\ 2x - 1, & \text{if } x > 0. \end{cases}$$

{... (−4,8) (−3,6) (−2,4) (−1,2) (0,0) (1,1) (2,3) (3,5) ...}

If **N** = {1, 2, 3, ...} then consider the function:

$$g(x) = \begin{cases} 2|x|, & \text{if } x < 0 \\ 2x + 1, & \text{if } x \geq 0. \end{cases}$$

{... (−4,8) (−3,6) (−2,4) (−1,2) (0,1) (1,3) (2,5) (3,7) ...}

If the domain is restricted to **Z** then each and every member of **Z** has one and only one corresponding member of **N** and by the definition of cardinal equality the two sets have equal cardinality.

12.6 See also

- 0.999...

- Canonical representation of a positive integer

- Hyperinteger

- Integer-valued function

- Integer lattice

- Integer part

- Integer sequence

- Profinite integer

12.7 Notes

[1] *Integer* 's first, literal meaning in Latin is "untouched", from *in* ("not") plus *tangere* ("to touch"). "Entire" derives from the same origin via French (see: Evans, Nick (1995). "A-Quantifiers and Scope". In Bach, Emmon W. *Quantification in Natural Languages*. Dordrecht, The Netherlands; Boston, MA: Kluwer Academic Publishers. p. 262. ISBN 0-7923-3352-7.)

12.8   References

[1] Weisstein, Eric W., "Counting Number", and "Whole Number", *MathWorld*.

[2] Miller, Jeff (2010-08-29). "Earliest Uses of Symbols of Number Theory". Retrieved 2010-09-20.

[3] Peter Jephson Cameron (1998). *Introduction to Algebra*. Oxford University Press. p. 4. ISBN 978-0-19-850195-4.

[4] Warner, Seth (2012), *Modern Algebra*, Dover Books on Mathematics, Courier Corporation, Theorem 20.14, p. 185, ISBN 9780486137094.

[5] Mendelson, Elliott (2008), *Number Systems and the Foundations of Analysis*, Dover Books on Mathematics, Courier Dover Publications, p. 86, ISBN 9780486457925.

[6] Ivorra Castillo: *Álgebra*

[7] Frobisher, Len (1999), *Learning to Teach Number: A Handbook for Students and Teachers in the Primary School*, The Stanley Thornes Teaching Primary Maths Series, Nelson Thornes, p. 126, ISBN 9780748735150.

[8] Campbell, Howard E. (1970). *The structure of arithmetic*. Appleton-Century-Crofts. p. 83. ISBN 0-390-16895-5.

12.9   Sources

• Bell, E.T., *Men of Mathematics*. New York: Simon and Schuster, 1986. (Hardcover; ISBN 0-671-46400-0)/(Paperk; ISBN 0-671-62818-6)

• Herstein, I.N., *Topics in Algebra*, Wiley; 2 edition (June 20, 1975), ISBN 0-471-01090-1.

• Mac Lane, Saunders, and Garrett Birkhoff; *Algebra*, American Mathematical Society; 3rd edition (April 1999). ISBN 0-8218-1646-2.

• Weisstein, Eric W., "Integer", *MathWorld*.

12.10   External links

• Hazewinkel, Michiel, ed. (2001), "Integer", *Encyclopedia of Mathematics*, Springer, ISBN 978-1-55608-010-4

• The Positive Integers — divisor tables and numeral representation tools

• On-Line Encyclopedia of Integer Sequences cf OEIS

This article incorporates material from Integer on PlanetMath, which is licensed under the Creative Commons Attribu-AlikeLicense.

Chapter 13

Rational number

"Rationals" redirects here. For other uses, see Rational (disambiguation).

In mathematics, a **rational number** is any number that can be expressed as the quotient or fraction p/q of two integers, p and q, with the denominator q not equal to zero.[1] Since q may be equal to 1, every integer is a rational number. The set of all rational numbers is usually denoted by a boldface $\mathbf{Q}$ (or blackboard bold $\mathbb{Q}$, Unicode ℚ);[2] it was thus denoted in 1895 by Peano after *quoziente*, Italian for "quotient".

The decimal expansion of a rational number always either terminates after a finite number of digits or begins to repeat the same finite sequence of digits over and over. Moreover, any repeating or terminating decimal represents a rational number. These statements hold true not just for base 10, but also for any other integer base (e.g. binary, hexadecimal).

A real number that is not rational is called irrational. Irrational numbers include $\sqrt{2}$, π, e, and φ. The decimal expansion of an irrational number continues without repeating. Since the set of rational numbers is countable, and the set of real numbers is uncountable, almost all real numbers are irrational.[1]

The rational numbers can be formally defined as the equivalence classes of the quotient set $(\mathbf{Z} \times (\mathbf{Z} \setminus \{0\})) / \sim$, where the cartesian product $\mathbf{Z} \times (\mathbf{Z} \setminus \{0\})$ is the set of all ordered pairs (m,n) where m and n are integers, n is not 0 ($n \neq 0$), and "$\sim$" is the equivalence relation defined by $(m_1,n_1) \sim (m_2,n_2)$ if, and only if, $m_1 n_2 - m_2 n_1 = 0$.

In abstract algebra, the rational numbers together with certain operations of addition and multiplication form the archetypical field of characteristic zero. As such, it is characterized as having no proper subfield or, alternatively, being the field of fractions for the ring of integers. Finite extensions of $\mathbf{Q}$ are called algebraic number fields, and the algebraic closure of $\mathbf{Q}$ is the field of algebraic numbers.[3]

In mathematical analysis, the rational numbers form a dense subset of the real numbers. The real numbers can be constructed from the rational numbers by completion, using Cauchy sequences, Dedekind cuts, or infinite decimals.

Zero divided by any other integer equals zero; therefore, zero is a rational number (but division by zero is undefined).

13.1   Terminology

The term *rational* in reference to the set $\mathbf{Q}$ refers to the fact that a rational number represents a *ratio* of two integers. In mathematics, the adjective *rational* often means that the underlying field considered is the field $\mathbf{Q}$ of rational numbers. Rational polynomial usually, and most correctly, means a polynomial with rational coefficients, also called a "polynomial over the rationals". However, rational function does *not* mean the underlying field is the rational numbers, and a rational algebraic curve is *not* an algebraic curve with rational coefficients.

13.2   Arithmetic

See also: Fraction (mathematics) § Arithmetic with fractions

13.2.1   Embedding of integers

Any integer n can be expressed as the rational number $n/1$.

13.2.2   Equality

$\frac{a}{b} = \frac{c}{d}$ if and only if $ad = bc$.

13.2.3   Ordering

Where both denominators are positive:

$\frac{a}{b} < \frac{c}{d}$ if and only if $ad < bc$.

If either denominator is negative, the fractions must first be converted into equivalent forms with positive denominators, through the equations:

$$\frac{-a}{-b} = \frac{a}{b}$$

and

$$\frac{a}{-b} = \frac{-a}{b}.$$

13.2.4   Addition

Two fractions are added as follows:

$$\frac{a}{b} + \frac{c}{d} = \frac{ad + bc}{bd}.$$

13.2.5   Subtraction

$$\frac{a}{b} - \frac{c}{d} = \frac{ad - bc}{bd}.$$

13.2.6   Multiplication

The rule for multiplication is:

$$\frac{a}{b} \cdot \frac{c}{d} = \frac{ac}{bd}.$$

13.2.7 Division

Where $c \neq 0$:

$$\frac{a}{b} \div \frac{c}{d} = \frac{ad}{bc}.$$

Note that division is equivalent to multiplying by the reciprocal of the divisor fraction:

$$\frac{ad}{bc} = \frac{a}{b} \times \frac{d}{c}.$$

13.2.8 Inverse

Additive and multiplicative inverses exist in the rational numbers:

$$-\left(\frac{a}{b}\right) = \frac{-a}{b} = \frac{a}{-b} \quad \text{and} \quad \left(\frac{a}{b}\right)^{-1} = \frac{b}{a} \text{ if } a \neq 0.$$

13.2.9 Exponentiation to integer power

If n is a non-negative integer, then

$$\left(\frac{a}{b}\right)^{n} = \frac{a^{n}}{b^{n}}$$

and (if $a \neq 0$):

$$\left(\frac{a}{b}\right)^{-n} = \frac{b^{n}}{a^{n}}.$$

13.3 Continued fraction representation

Main article: Continued fraction

A **finite continued fraction** is an expression such as

$$a_0 + \cfrac{1}{a_1 + \cfrac{1}{a_2 + \cfrac{1}{\ddots + \cfrac{1}{a_n}}}},$$

where a_n are integers. Every rational number a/b has two closely related expressions as a finite continued fraction, whose coefficients a_n can be determined by applying the Euclidean algorithm to (a,b).

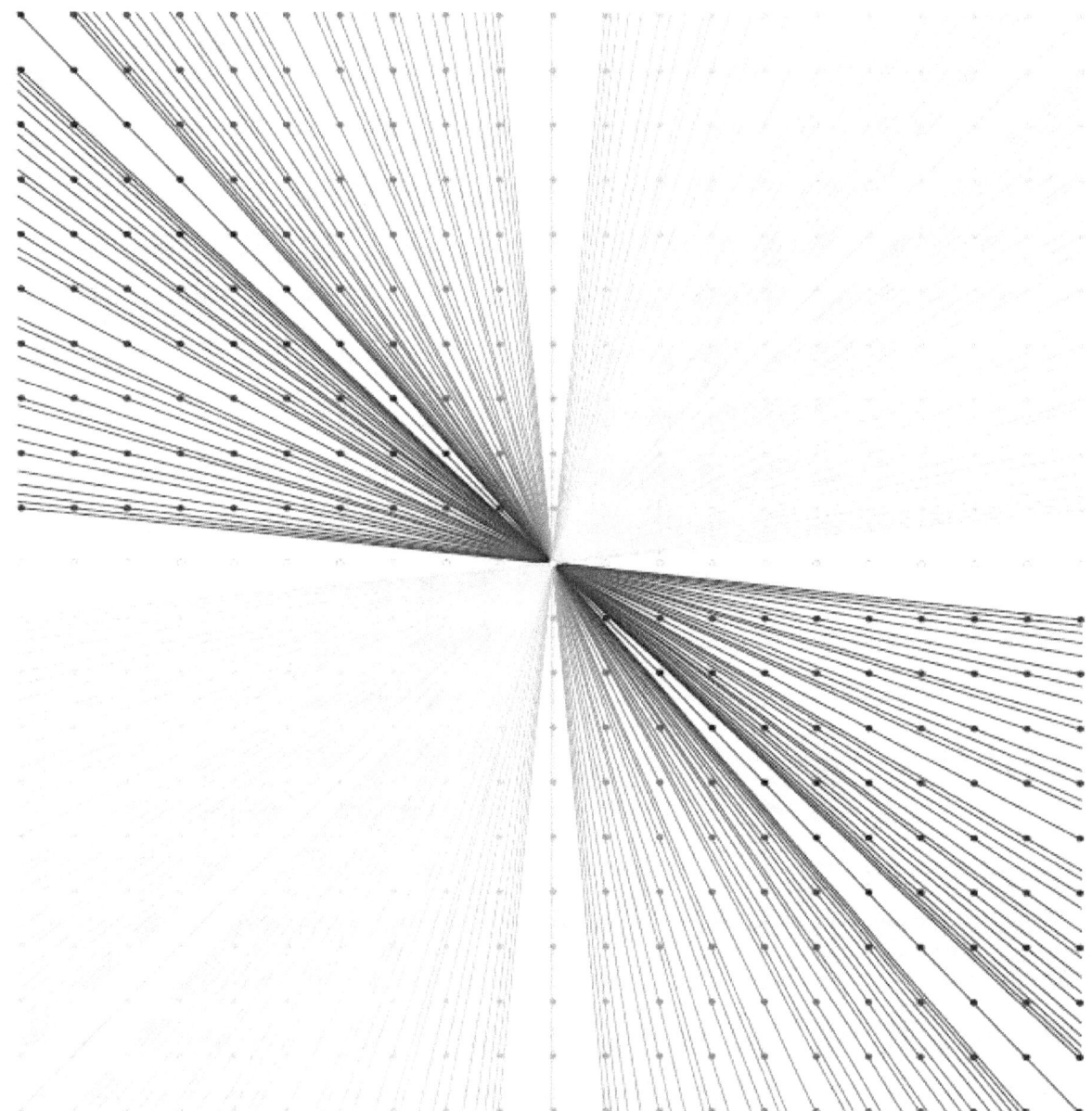

A diagram showing a representation of the equivalent classes of pairs of integers

13.4   Formal construction

Mathematically we may construct the rational numbers as equivalence classes of ordered pairs of integers (m,n), with $n \neq 0$. This space of equivalence classes is the quotient space $(\mathbf{Z} \times (\mathbf{Z} \setminus \{0\})) / \sim$, where $(m_1,n_1) \sim (m_2,n_2)$ if, and only if, $m_1 n_2 - m_2 n_1 = 0$. We can define addition and multiplication of these pairs with the following rules:

$$(m_1, n_1) + (m_2, n_2) \equiv (m_1 n_2 + n_1 m_2, n_1 n_2)$$

$$(m_1, n_1) \times (m_2, n_2) \equiv (m_1 m_2, n_1 n_2)$$

and, if $m_2 \neq 0$, division by

$$\frac{(m_1, n_1)}{(m_2, n_2)} \equiv (m_1 n_2, n_1 m_2).$$

The equivalence relation $(m_1, n_1) \sim (m_2, n_2)$ if, and only if, $m_1 n_2 - m_2 n_1 = 0$ is a congruence relation, i.e. it is compatible with the addition and multiplication defined above, and we may define **Q** to be the quotient set $(\mathbf{Z} \times (\mathbf{Z} \setminus \{0\}))/\sim$, i.e. we identify two pairs (m_1, n_1) and (m_2, n_2) if they are equivalent in the above sense. (This construction can be carried out in any integral domain: see field of fractions.) We denote by $[(m_1, n_1)]$ the equivalence class containing (m_1, n_1). If $(m_1, n_1) \sim (m_2, n_2)$ then, by definition, (m_1, n_1) belongs to $[(m_2, n_2)]$ and (m_2, n_2) belongs to $[(m_1, n_1)]$; in this case we can write $[(m_1, n_1)] = [(m_2, n_2)]$. Given any equivalence class $[(m, n)]$ there are a countably infinite number of representation, since

$$\cdots = [(-2m, -2n)] = [(-m, -n)] = [(m, n)] = [(2m, 2n)] = \cdots .$$

The canonical choice for $[(m, n)]$ is chosen so that n is positive and $\gcd(m, n) = 1$, i.e. m and n share no common factors, i.e. m and n are coprime. For example, we would write $[(1, 2)]$ instead of $[(2, 4)]$ or $[(-12, -24)]$, even though $[(1, 2)] = [(2, 4)] = [(-12, -24)]$.

We can also define a total order on **Q**. Let $\wedge$ be the *and*-symbol and $\vee$ be the *or*-symbol. We say that $[(m_1, n_1)] \leq [(m_2, n_2)]$ if:

$$(n_1 n_2 > 0 \wedge m_1 n_2 \leq n_1 m_2) \vee (n_1 n_2 < 0 \wedge m_1 n_2 \geq n_1 m_2).$$

The integers may be considered to be rational numbers by the embedding that maps m to $[(m, 1)]$.

13.5 Properties

The set **Q**, together with the addition and multiplication operations shown above, forms a field, the field of fractions of the integers **Z**.

The rationals are the smallest field with characteristic zero: every other field of characteristic zero contains a copy of **Q**. The rational numbers are therefore the prime field for characteristic zero.

The algebraic closure of **Q**, i.e. the field of roots of rational polynomials, is the algebraic numbers.

The set of all rational numbers is countable. Since the set of all real numbers is uncountable, we say that almost all real numbers are irrational, in the sense of Lebesgue measure, i.e. the set of rational numbers is a null set.

The rationals are a densely ordered set: between any two rationals, there sits another one, and, therefore, infinitely many other ones. For example, for any two fractions such that

$$\frac{a}{b} < \frac{c}{d}$$

(where b, d are positive), we have

$$\frac{a}{b} < \frac{ad + bc}{2bd} < \frac{c}{d}.$$

Any totally ordered set which is countable, dense (in the above sense), and has no least or greatest element is order isomorphic to the rational numbers.

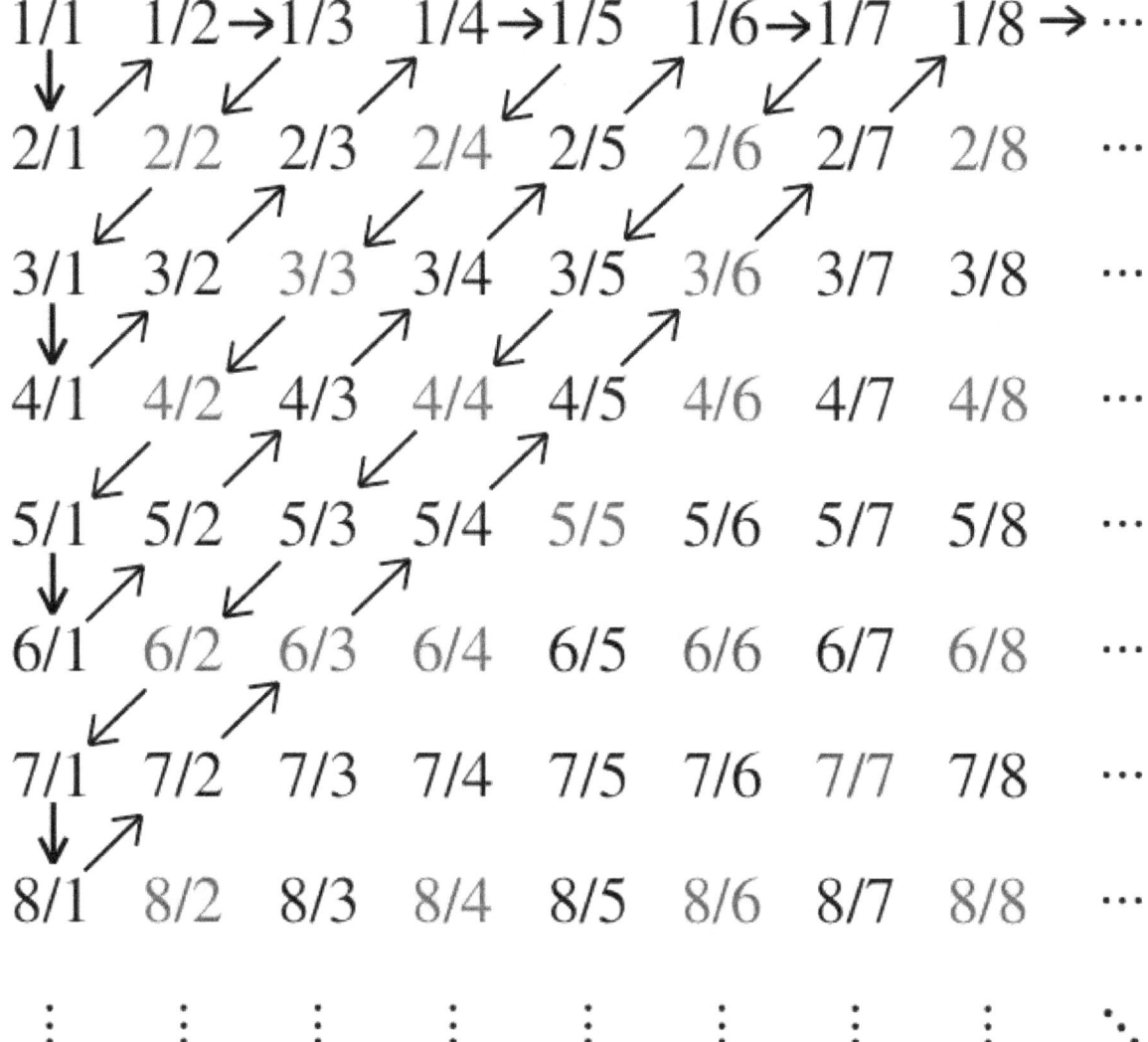

A diagram illustrating the countability of the rationals

13.6   Real numbers and topological properties

The rationals are a dense subset of the real numbers: every real number has rational numbers arbitrarily close to it. A related property is that rational numbers are the only numbers with finite expansions as regular continued fractions.

By virtue of their order, the rationals carry an order topology. The rational numbers, as a subspace of the real numbers, also carry a subspace topology. The rational numbers form a metric space by using the absolute difference metric $d(x,y) = |x - y|$, and this yields a third topology on **Q**. All three topologies coincide and turn the rationals into a topological field. The rational numbers are an important example of a space which is not locally compact. The rationals are characterized topologically as the unique countable metrizable space without isolated points. The space is also totally disconnected. The rational numbers do not form a complete metric space; the real numbers are the completion of **Q** under the metric $d(x,y) = |x - y|$, above.

13.7   *p*-adic numbers

See also: p-adic Number

In addition to the absolute value metric mentioned above, there are other metrics which turn **Q** into a topological field:

Let *p* be a prime number and for any non-zero integer *a*, let |a|p = p^{-n}, where p^n is the highest power of *p* dividing *a*.

In addition set |0|p = 0. For any rational number *a/b*, we set |a/b|p = |a|p / |b|p.

Then dp(*x*,*y*) = |*x* − *y*|p defines a metric on **Q**.

The metric space (**Q**,dp) is not complete, and its completion is the *p*-adic number field **Q**p. Ostrowski's theorem states that any non-trivial absolute value on the rational numbers **Q** is equivalent to either the usual real absolute value or a *p*-adic absolute value.

13.8   See also

- Floating point
- Ford circles
- Niven's theorem
- Rational data type

13.9   References

[1] Rosen, Kenneth (2007). *Discrete Mathematics and its Applications* (6th ed.). New York, NY: McGraw-Hill. pp. 105,158–160. ISBN 978-0-07-288008-3.

[2] Rouse, Margaret. "Mathematical Symbols". Retrieved 1 April 2015.

[3] Gilbert, Jimmie; Linda, Gilbert (2005). *Elements of Modern Algebra* (6th ed.). Belmont, CA: Thomson Brooks/Cole. pp. 243–244. ISBN 0-534-40264-X.

13.10   External links

- Hazewinkel, Michiel, ed. (2001), "Rational number", *Encyclopedia of Mathematics*, Springer, ISBN 978-1-55608-010-4
- "Rational Number" From MathWorld – A Wolfram Web Resource

Chapter 14

Real number

For the real numbers used in descriptive set theory, see Baire space (set theory). For the computing datatype, see Floating-point number.

In mathematics, a **real number** is a value that represents a quantity along a continuous line. The adjective *real* in this context was introduced in the 17th century by Descartes, who distinguished between real and imaginary roots of polynomials.

The real numbers include all the rational numbers, such as the integer −5 and the fraction 4/3, and all the irrational numbers such as $\sqrt{2}$ (1.41421356..., the square root of two, an irrational algebraic number) and all transcendental numbers such as π (3.14159265..., a transcendental number). Real numbers can be thought of as points on an infinitely long line called the number line or real line, where the points corresponding to integers are equally spaced. Any real number can be determined by a possibly infinite decimal representation such as that of 8.632, where each consecutive digit is measured in units one tenth the size of the previous one. The real line can be thought of as a part of the complex plane, and complex numbers include real numbers.

These descriptions of the real numbers are not sufficiently rigorous by the modern standards of pure mathematics. The discovery of a suitably rigorous definition of the real numbers – indeed, the realization that a better definition was needed – was one of the most important developments of 19th century mathematics. The currently standard axiomatic definition is that real numbers form the unique Archimedean complete totally ordered field ($\mathbf{R}$; + ; · ; <), up to an isomorphism,[1] whereas popular constructive definitions of real numbers include declaring them as equivalence classes of Cauchy sequences of rational numbers, Dedekind cuts, or certain infinite "decimal representations", together with precise interpretations for the arithmetic operations and the order relation. These definitions are equivalent in the realm of classical mathematics.

The reals are uncountable; that is: while both the set of all natural numbers and the set of all real numbers are infinite sets, there can be no one-to-one function from the real numbers to the natural numbers: the cardinality of the set of all real numbers (denoted c and called cardinality of the continuum) is strictly greater than the cardinality of the set of all natural numbers (denoted $\aleph_0$). The statement that there is no subset of the reals with cardinality strictly greater than $\aleph_0$ and strictly smaller than c is known as the continuum hypothesis (CH). It is known to be neither provable nor refutable using the axioms of Zermelo–Fraenkel set theory (ZFC), the standard foundation of modern mathematics, in the sense that some models of ZFC satisfy CH, while others violate it.

14.1 History

Simple fractions have been used by the Egyptians around 1000 BC; the Vedic "Sulba Sutras" ("The rules of chords") in, c. 600 BC, include what may be the first "use" of irrational numbers. The concept of irrationality was implicitly accepted by early Indian mathematicians since Manava (c. 750–690 BC), who were aware that the square roots of certain numbers such as 2 and 61 could not be exactly determined.[2] Around 500 BC, the Greek mathematicians led by Pythagoras realized the need for irrational numbers, in particular the irrationality of the square root of 2.

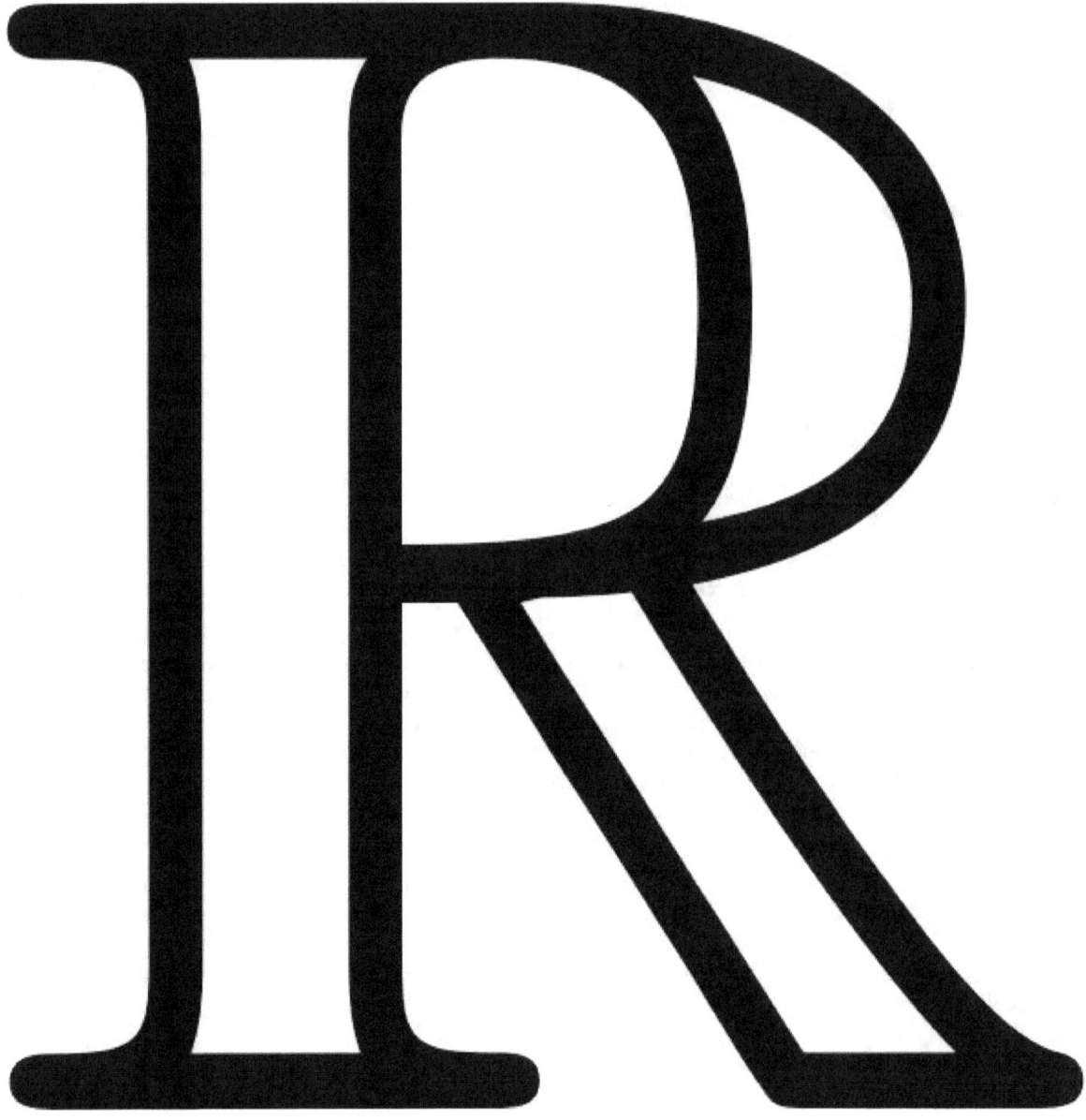

*A symbol of the set of **real numbers** ($\mathbb{R}$)*

The Middle Ages brought the acceptance of zero, negative, integral, and fractional numbers, first by Indian and Chinese mathematicians, and then by Arabic mathematicians, who were also the first to treat irrational numbers as algebraic objects,[3] which was made possible by the development of algebra. Arabic mathematicians merged the concepts of "number" and "magnitude" into a more general idea of real numbers.[4] The Egyptian mathematician Abū Kāmil Shujā ibn Aslam (c. 850–930) was the first to accept irrational numbers as solutions to quadratic equations or as coefficients in an equation, often in the form of square roots, cube roots and fourth roots.[5]

In the 16th century, Simon Stevin created the basis for modern decimal notation, and insisted that there is no difference between rational and irrational numbers in this regard.

In the 17th century, Descartes introduced the term "real" to describe roots of a polynomial, distinguishing them from "imaginary" ones.

In the 18th and 19th centuries there was much work on irrational and transcendental numbers. Johann Heinrich Lambert (1761) gave the first flawed proof that π cannot be rational; Adrien-Marie Legendre (1794) completed the proof,[6]

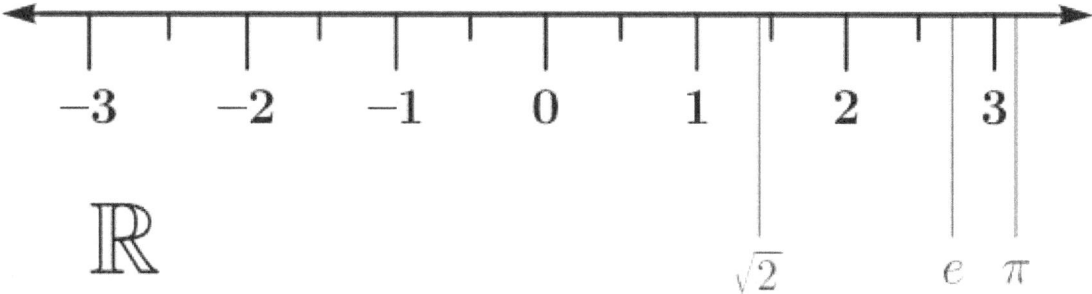

Real numbers can be thought of as points on an infinitely long number line

and showed that π is not the square root of a rational number.[7] Paolo Ruffini (1799) and Niels Henrik Abel (1842) both constructed proofs of the Abel–Ruffini theorem: that the general quintic or higher equations cannot be solved by a general formula involving only arithmetical operations and roots.

Évariste Galois (1832) developed techniques for determining whether a given equation could be solved by radicals, which gave rise to the field of Galois theory. Joseph Liouville (1840) showed that neither e nor e^2 can be a root of an integer quadratic equation, and then established the existence of transcendental numbers; Georg Cantor (1873) extended and greatly simplified this proof.[8] Charles Hermite (1873) first proved that e is transcendental, and Ferdinand von Lindemann (1882), showed that π is transcendental. Lindemann's proof was much simplified by Weierstrass (1885), still further by David Hilbert (1893), and has finally been made elementary by Adolf Hurwitz and Paul Gordan.

The development of calculus in the 18th century used the entire set of real numbers without having defined them cleanly. The first rigorous definition was given by Georg Cantor in 1871. In 1874 he showed that the set of all real numbers is uncountably infinite but the set of all algebraic numbers is countably infinite. Contrary to widely held beliefs, his first method was not his famous diagonal argument, which he published in 1891. See Cantor's first uncountability proof.

14.2   Definition

Main article: Construction of the real numbers

The real number system $(\mathbb{R}; +; \cdot; <)$ can be defined axiomatically up to an isomorphism, which is described hereafter. There are also many ways to construct "the" real number system, for example, starting from natural numbers, then defining rational numbers algebraically, and finally defining real numbers as equivalence classes of their Cauchy sequences or as Dedekind cuts, which are certain subsets of rational numbers. Another possibility is to start from some rigorous axiomatization of Euclidean geometry (Hilbert, Tarski, etc.) and then define the real number system geometrically. From the structuralist point of view all these constructions are on equal footing.

14.2.1   Axiomatic approach

Let $\mathbb{R}$ denote the set of all real numbers. Then:

- The set $\mathbb{R}$ is a field, meaning that addition and multiplication are defined and have the usual properties.

- The field $\mathbb{R}$ is ordered, meaning that there is a total order $\geq$ such that, for all real numbers x, y and z:

    - if $x \geq y$ then $x + z \geq y + z$;
    - if $x \geq 0$ and $y \geq 0$ then $xy \geq 0$.

- The order is Dedekind-complete; that is: every non-empty subset S of $\mathbb{R}$ with an upper bound in $\mathbb{R}$ has a least upper bound (also called supremum) in $\mathbb{R}$.

The last property is what differentiates the reals from the rationals. For example, the set of rationals with square less than 2 has a rational upper bound (e.g., 1.5) but no rational least upper bound, because the square root of 2 is not rational.

The real numbers are uniquely specified by the above properties. More precisely, given any two Dedekind-complete ordered fields $\mathbb{R}_1$ and $\mathbb{R}_2$, there exists a unique field isomorphism from $\mathbb{R}_1$ to $\mathbb{R}_2$, allowing us to think of them as essentially the same mathematical object.

For another axiomatization of $\mathbb{R}$, see Tarski's axiomatization of the reals.

14.2.2 Construction from the rational numbers

The real numbers can be constructed as a completion of the rational numbers in such a way that a sequence defined by a decimal or binary expansion like (3; 3.1; 3.14; 3.141; 3.1415; ...) converges to a unique real number, in this case π. For details and other constructions of real numbers, see construction of the real numbers.

14.3 Properties

14.3.1 Basic properties

A real number may be either rational or irrational; either algebraic or transcendental; and either positive, negative, or zero. Real numbers are used to measure continuous quantities. They may be expressed by decimal representations that have an infinite sequence of digits to the right of the decimal point; these are often represented in the same form as 324.823122147... The ellipsis (three dots) indicates that there would still be more digits to come.

More formally, real numbers have the two basic properties of being an ordered field, and having the least upper bound property. The first says that real numbers comprise a field, with addition and multiplication as well as division by non-zero numbers, which can be totally ordered on a number line in a way compatible with addition and multiplication. The second says that, if a non-empty set of real numbers has an upper bound, then it has a real least upper bound. The second condition distinguishes the real numbers from the rational numbers: for example, the set of rational numbers whose square is less than 2 is a set with an upper bound (e.g. 1.5) but no (rational) least upper bound: hence the rational numbers do not satisfy the least upper bound property.

14.3.2 Completeness

Main article: Completeness of the real numbers

A main reason for using real numbers is that the reals contain all limits. More precisely, every sequence of real numbers having the property that consecutive terms of the sequence become arbitrarily close to each other necessarily has the property that after some term in the sequence the remaining terms are arbitrarily close to some specific real number. In mathematical terminology, this means that the reals are complete (in the sense of metric spaces or uniform spaces, which is a different sense than the Dedekind completeness of the order in the previous section). This is formally defined in the following way:

A sequence (x_n) of real numbers is called a *Cauchy sequence* if for any $\varepsilon > 0$ there exists an integer N (possibly depending on ε) such that the distance $|x_n - x_m|$ is less than ε for all n and m that are both greater than N. In other words, a sequence is a Cauchy sequence if its elements x_n eventually come and remain arbitrarily close to each other.

A sequence (x_n) *converges to the limit* x if for any $\varepsilon > 0$ there exists an integer N (possibly depending on ε) such that the distance $|x_n - x|$ is less than ε provided that n is greater than N. In other words, a sequence has limit x if its elements eventually come and remain arbitrarily close to x.

Notice that every convergent sequence is a Cauchy sequence. The converse is also true:

> Every Cauchy sequence of real numbers is convergent to a real number.

That is: the reals are complete.

Note that the rationals are not complete. For example, the sequence (1; 1.4; 1.41; 1.414; 1.4142; 1.41421...), where each term adds a digit of the decimal expansion of the positive square root of 2, is Cauchy but it does not converge to a rational number. (In the real numbers, in contrast, it converges to the positive square root of 2.)

The existence of limits of Cauchy sequences is what makes calculus work and is of great practical use. The standard numerical test to determine if a sequence has a limit is to test if it is a Cauchy sequence, as the limit is typically not known in advance.

For example, the standard series of the exponential function

$$e^x = \sum_{n=0}^{\infty} \frac{x^n}{n!}$$

converges to a real number because for every x the sums

$$\sum_{n=N}^{M} \frac{x^n}{n!}$$

can be made arbitrarily small by choosing N sufficiently large. This proves that the sequence is Cauchy, so we know that the sequence converges even if the limit is not known in advance.

14.3.3   "The complete ordered field"

The real numbers are often described as "the complete ordered field", a phrase that can be interpreted in several ways.

First, an order can be lattice-complete. It is easy to see that no ordered field can be lattice-complete, because it can have no largest element (given any element z, $z + 1$ is larger), so this is not the sense that is meant.

Additionally, an order can be Dedekind-complete, as defined in the section **Axioms**. The uniqueness result at the end of that section justifies using the word "the" in the phrase "complete ordered field" when this is the sense of "complete" that is meant. This sense of completeness is most closely related to the construction of the reals from Dedekind cuts, since that construction starts from an ordered field (the rationals) and then forms the Dedekind-completion of it in a standard way.

These two notions of completeness ignore the field structure. However, an ordered group (in this case, the additive group of the field) defines a uniform structure, and uniform structures have a notion of completeness (topology); the description in the previous section **Completeness** is a special case. (We refer to the notion of completeness in uniform spaces rather than the related and better known notion for metric spaces, since the definition of metric space relies on already having a characterization of the real numbers.) It is not true that **R** is the *only* uniformly complete ordered field, but it is the only uniformly complete *Archimedean field*, and indeed one often hears the phrase "complete Archimedean field" instead of "complete ordered field". Every uniformly complete Archimedean field must also be Dedekind-complete (and vice versa, of course), justifying using "the" in the phrase "the complete Archimedean field". This sense of completeness is most closely related to the construction of the reals from Cauchy sequences (the construction carried out in full in this article), since it starts with an Archimedean field (the rationals) and forms the uniform completion of it in a standard way.

But the original use of the phrase "complete Archimedean field" was by David Hilbert, who meant still something else by it. He meant that the real numbers form the *largest* Archimedean field in the sense that every other Archimedean field is a subfield of **R**. Thus **R** is "complete" in the sense that nothing further can be added to it without making it no longer an Archimedean field. This sense of completeness is most closely related to the construction of the reals from surreal numbers, since that construction starts with a proper class that contains every ordered field (the surreals) and then selects from it the largest Archimedean subfield.

14.3.4   Advanced properties

See also: Real line

The reals are uncountable; that is: there are strictly more real numbers than natural numbers, even though both sets are infinite. In fact, the cardinality of the reals equals that of the set of subsets (i.e. the power set) of the natural numbers, and Cantor's diagonal argument states that the latter set's cardinality is strictly greater than the cardinality of **N**. Since the set of algebraic numbers is countable, almost all real numbers are transcendental. The non-existence of a subset of the reals with cardinality strictly between that of the integers and the reals is known as the continuum hypothesis. The continuum hypothesis can neither be proved nor be disproved; it is independent from the axioms of set theory.

As a topological space, the real numbers are separable. This is because the set of rationals, which is countable, is dense in the real numbers. The irrational numbers are also dense in the real numbers, however they are uncountable and have the same cardinality as the reals.

The real numbers form a metric space: the distance between x and y is defined as the absolute value $|x - y|$. By virtue of being a totally ordered set, they also carry an order topology; the topology arising from the metric and the one arising from the order are identical, but yield different presentations for the topology – in the order topology as ordered intervals, in the metric topology as epsilon-balls. The Dedekind cuts construction uses the order topology presentation, while the Cauchy sequences construction uses the metric topology presentation. The reals are a contractible (hence connected and simply connected), separable and complete metric space of Hausdorff dimension 1. The real numbers are locally compact but not compact. There are various properties that uniquely specify them; for instance, all unbounded, connected, and separable order topologies are necessarily homeomorphic to the reals.

Every nonnegative real number has a square root in **R**, although no negative number does. This shows that the order on **R** is determined by its algebraic structure. Also, every polynomial of odd degree admits at least one real root: these two properties make **R** the premier example of a real closed field. Proving this is the first half of one proof of the fundamental theorem of algebra.

The reals carry a canonical measure, the Lebesgue measure, which is the Haar measure on their structure as a topological group normalized such that the unit interval [0;1] has measure 1. There exist sets of real numbers that are not Lebesgue measurable, e.g. Vitali sets.

The supremum axiom of the reals refers to subsets of the reals and is therefore a second-order logical statement. It is not possible to characterize the reals with first-order logic alone: the Löwenheim–Skolem theorem implies that there exists a countable dense subset of the real numbers satisfying exactly the same sentences in first-order logic as the real numbers themselves. The set of hyperreal numbers satisfies the same first order sentences as **R**. Ordered fields that satisfy the same first-order sentences as **R** are called nonstandard models of **R**. This is what makes nonstandard analysis work; by proving a first-order statement in some nonstandard model (which may be easier than proving it in **R**), we know that the same statement must also be true of **R**.

The field **R** of real numbers is an extension field of the field **Q** of rational numbers, and **R** can therefore be seen as a vector space over **Q**. Zermelo–Fraenkel set theory with the axiom of choice guarantees the existence of a basis of this vector space: there exists a set B of real numbers such that every real number can be written uniquely as a finite linear combination of elements of this set, using rational coefficients only, and such that no element of B is a rational linear combination of the others. However, this existence theorem is purely theoretical, as such a base has never been explicitly described.

The well-ordering theorem implies that the real numbers can be well-ordered if the axiom of choice is assumed: there exists a total order on **R** with the property that every non-empty subset of **R** has a least element in this ordering. (The standard ordering ≤ of the real numbers is not a well-ordering since e.g. an open interval does not contain a least element in this ordering.) Again, the existence of such a well-ordering is purely theoretical, as it has not been explicitly described. If V=L is assumed in addition to the axioms of ZF, a well ordering of the real numbers can be shown to be explicitly definable by a formula.[9]

14.4   Applications and connections to other areas

14.4.1   Real numbers and logic

The real numbers are most often formalized using the Zermelo–Fraenkel axiomatization of set theory, but some mathematicians study the real numbers with other logical foundations of mathematics. In particular, the real numbers are also studied in reverse mathematics and in constructive mathematics.[10]

The hyperreal numbers as developed by Edwin Hewitt, Abraham Robinson and others extend the set of the real numbers by introducing infinitesimal and infinite numbers, allowing for building infinitesimal calculus in a way closer to the original intuitions of Leibniz, Euler, Cauchy and others.

Edward Nelson's internal set theory enriches the Zermelo–Fraenkel set theory syntactically by introducing a unary predicate "standard". In this approach, infinitesimals are (non-"standard") elements of the set of the real numbers (rather than being elements of an extension thereof, as in Robinson's theory).

The continuum hypothesis posits that the cardinality of the set of the real numbers is $\aleph_1$; i.e. the smallest infinite cardinal number after $\aleph_0$, the cardinality of the integers. Paul Cohen proved in 1963 that it is an axiom independent of the other axioms of set theory; that is: one may choose either the continuum hypothesis or its negation as an axiom of set theory, without contradiction.

14.4.2   In physics

In the physical sciences, most physical constants such as the universal gravitational constant, and physical variables, such as position, mass, speed, and electric charge, are modeled using real numbers. In fact, the fundamental physical theories such as classical mechanics, electromagnetism, quantum mechanics, general relativity and the standard model are described using mathematical structures, typically smooth manifolds or Hilbert spaces, that are based on the real numbers, although actual measurements of physical quantities are of finite accuracy and precision.

In some recent developments of theoretical physics stemming from the holographic principle, the Universe is seen fundamentally as an information store, essentially zeroes and ones, organized in much less geometrical fashion and manifesting itself as space-time and particle fields only on a more superficial level. This approach removes the real number system from its foundational role in physics and even prohibits the existence of infinite precision real numbers in the physical universe by considerations based on the Bekenstein bound.[11]

14.4.3   In computation

With some exceptions, most calculators do not operate on real numbers. Instead, they work with finite-precision approximations called floating-point numbers. In fact, most scientific computation uses floating-point arithmetic. Real numbers satisfy the usual rules of arithmetic, but floating-point numbers do not.

Computers cannot directly store arbitrary real numbers with infinitely many digits.

The precision is limited by the number of bits allocated to store a number, whether as floating-point numbers or arbitrary precision numbers. However, computer algebra systems can operate on irrational quantities exactly by manipulating formulas for them (such as $\sqrt{2}$, $\arcsin\left(\frac{2}{23}\right)$, or $\int_0^1 x^x\ dx$) rather than their rational or decimal approximation.[12] however, it is not in general possible to determine whether two such expressions are equal (the constant problem).

A real number is called *computable* if there exists an algorithm that yields its digits. Because there are only countably many algorithms,[13] but an uncountable number of reals, almost all real numbers fail to be computable. Moreover, the equality of two computable numbers is an undecidable problem. Some constructivists accept the existence of only those reals that are computable. The set of definable numbers is broader, but still only countable.

14.4.4 "Reals" in set theory

In set theory, specifically descriptive set theory, the Baire space is used as a surrogate for the real numbers since the latter have some topological properties (connectedness) that are a technical inconvenience. Elements of Baire space are referred to as "reals".

14.5 Vocabulary and notation

Mathematicians use the symbol **R**, or, alternatively, $\mathbb{R}$, the letter "R" in blackboard bold (encoded in Unicode as U+211D $\mathbb{R}$ double-struck capital r (HTML ℝ)), to represent the set of all real numbers. As this set is naturally endowed with the structure of a field, the expression *field of real numbers* is frequently used when its algebraic properties are under consideration.

The sets of positive real numbers and negative real numbers are often noted $\mathbf{R}^+$ and $\mathbf{R}^-$,[14] respectively; $\mathbf{R}_+$ and $\mathbf{R}_-$ are also used.[15] The non-negative real numbers can be noted $\mathbf{R}_{\geq 0}$ but one often sees this set noted $\mathbf{R}^+ \cup \{0\}$.[14] In French mathematics, the *positive real numbers* and *negative real numbers* commonly include zero, and these sets are noted respectively $\mathbb{R}_+$ and $\mathbb{R}_-$.[15] In this understanding, the respective sets without zero are called strictly positive real numbers and strictly negative real numbers, and are noted $\mathbb{R}_+^*$ and $\mathbb{R}_-^*$.[15]

The notation $\mathbf{R}^n$ refers to the cartesian product of n copies of **R**, which is an n-dimensional vector space over the field of the real numbers; this vector space may be identified to the n-dimensional space of Euclidean geometry as soon as a coordinate system has been chosen in the latter. For example, a value from $\mathbf{R}^3$ consists of three real numbers and specifies the coordinates of a point in 3-dimensional space.

In mathematics, *real* is used as an adjective, meaning that the underlying field is the field of the real numbers (or *the real field*). For example, *real matrix*, *real polynomial* and *real Lie algebra*. The word is also used as a noun, meaning a real number (as in "the set of all reals").

14.6 Generalizations and extensions

The real numbers can be generalized and extended in several different directions:

- The complex numbers contain solutions to all polynomial equations and hence are an algebraically closed field unlike the real numbers. However, the complex numbers are not an ordered field.

- The affinely extended real number system adds two elements $+\infty$ and $-\infty$. It is a compact space. It is no longer a field, not even an additive group, but it still has a total order; moreover, it is a complete lattice.

- The real projective line adds only one value ∞. It is also a compact space. Again, it is no longer a field, not even an additive group. However, it allows division of a non-zero element by zero. It has cyclic order described by a separation relation.

- The long real line pastes together $\aleph_1^* + \aleph_1$ copies of the real line plus a single point (here $\aleph_1^*$ denotes the reversed ordering of $\aleph_1$) to create an ordered set that is "locally" identical to the real numbers, but somehow longer; for instance, there is an order-preserving embedding of $\aleph_1$ in the long real line but not in the real numbers. The long real line is the largest ordered set that is complete and locally Archimedean. As with the previous two examples, this set is no longer a field or additive group.

- Ordered fields extending the reals are the hyperreal numbers and the surreal numbers; both of them contain infinitesimal and infinitely large numbers and are therefore non-Archimedean ordered fields.

- Self-adjoint operators on a Hilbert space (for example, self-adjoint square complex matrices) generalize the reals in many respects: they can be ordered (though not totally ordered), they are complete, all their eigenvalues are real and they form a real associative algebra. Positive-definite operators correspond to the positive reals and normal operators correspond to the complex numbers.

14.7   See also

- Complex number

- Continued fraction

- Hypercomplex number

- Imaginary number

- Limit of a sequence

- Natural number

- Real analysis

14.8   Notes

[1] More precisely, given two complete totally ordered fields, there is a *unique* isomorphism between them. This implies that the identity is the unique field automorphism of the reals that is compatible with the ordering.

[2] T. K. Puttaswamy, "The Accomplishments of Ancient Indian Mathematicians", pp. 410–1. In: Selin, Helaine; D'Ambrosio, Ubiratan, eds. (2000), *Mathematics Across Cultures: The History of Non-western Mathematics*, Springer, ISBN 1-4020-0260-2.

[3] O'Connor, John J.; Robertson, Edmund F., "Arabic mathematics: forgotten brilliance?", *MacTutor History of Mathematics archive*, University of St Andrews.

[4] Matvievskaya, Galina (1987), "The Theory of Quadratic Irrationals in Medieval Oriental Mathematics", *Annals of the New York Academy of Sciences* **500**: 253–277 [254], doi:10.1111/j.1749-6632.1987.tb37206.x

[5] Jacques Sesiano, "Islamic mathematics", p. 148, in Selin, Helaine; D'Ambrosio, Ubiratan (2000), *Mathematics Across Cultures: The History of Non-western Mathematics*, Springer, ISBN 1-4020-0260-2

[6] Beckmann, Petr (1993), *A History of Pi*, Dorset Classic Reprints, Barnes & Noble Publishing, p. 170, ISBN 9780880294188.

[7] Arndt, Jörg; Haenel, Christoph (2001), *Pi Unleashed*, Springer, p. 192, ISBN 9783540665724.

[8] Dunham, William (2015), *The Calculus Gallery: Masterpieces from Newton to Lebesgue*, Princeton University Press, p. 127, ISBN 9781400866793, Cantor found a remarkable shortcut to reach Liouville's conclusion with a fraction of the work

[9] Moschovakis, Yiannis N. Descriptive set theory. Studies in Logic and the Foundations of Mathematics, 100. North-Holland Publishing Co., Amsterdam - New York, 1980. xii+637 pp. ISBN 0-444-85305-7. Chapter V.

[10] Bishop, Errett; Bridges, Douglas (1985), *Constructive analysis*, Grundlehren der Mathematischen Wissenschaften [Fundamental Principles of Mathematical Sciences] **279**, Berlin, New York: Springer-Verlag, ISBN 978-3-540-15066-4, chapter 2.

[11] Scott Aaronson, *NP-complete Problems and Physical Reality*, ACM SIGACT News, vol. 36, no. 1. (March 2005), pp. 30–52.

[12] Cohen, Joel S. (2002), *Computer algebra and symbolic computation: elementary algorithms* **1**, A K Peters, p. 32, ISBN 978-1-56881-158-1

[13] James L. Hein, *Discrete Structures, Logic, and Computability*, 3rd edition (Jones and Bartlett Publishers, Sudbury, Massachusetts, USA), section 14.1.1 (2010).

[14] Schumacher 1996, pp. 114-115

[15] École Normale Supérieure of Paris, "Nombres réels" ("Real numbers"), p. 6

14.9   References

- Georg Cantor, 1874, "Über eine Eigenschaft des Inbegriffes aller reellen algebraischen Zahlen", *Journal für die Reine und Angewandte Mathematik*, volume 77, pages 258–262.

- Solomon Feferman, 1989, *The Number Systems: Foundations of Algebra and Analysis*, AMS Chelsea, ISBN 0-8218-2915-7.

- Robert Katz, 1964, *Axiomatic Analysis*, D. C. Heath and Company.

- Edmund Landau, 2001, ISBN 0-8218-2693-X, *Foundations of Analysis*, American Mathematical Society.

- Howie, John M., *Real Analysis*, Springer, 2005, ISBN 1-85233-314-6.

- Schumacher, Carol (1996), *ChapterZero / Fundamental Notions of Abstract Mathematics*, Addison-Wesley, ISBN 0-201-82653-4.

14.10   External links

- Hazewinkel, Michiel, ed. (2001), "Real number", *Encyclopedia of Mathematics*, Springer, ISBN 978-1-55608-010-4

- The real numbers: Pythagoras to Stevin

- The real numbers: Stevin to Hilbert

- The real numbers: Attempts to understand

- What are the "real numbers," really?

Chapter 15

Complex number

A **complex number** is a number that can be expressed in the form $a + bi$, where a and b are real numbers and i is the imaginary unit, that satisfies the equation $i^2 = -1$.[1] In this expression, a is the *real part* and b is the *imaginary part* of the complex number.

Complex numbers extend the concept of the one-dimensional number line to the two-dimensional complex plane (also called Argand plane) by using the horizontal axis for the real part and the vertical axis for the imaginary part. The complex number $a + bi$ can be identified with the point (a, b) in the complex plane. A complex number whose real part is zero is said to be purely imaginary, whereas a complex number whose imaginary part is zero is a real number. In this way, the complex numbers contain the ordinary real numbers while extending them in order to solve problems that cannot be solved with real numbers alone.

As well as their use within mathematics, complex numbers have practical applications in many fields, including physics, chemistry, biology, economics, electrical engineering, and statistics. The Italian mathematician Gerolamo Cardano is the first known to have introduced complex numbers. He called them "fictitious" during his attempts to find solutions to cubic equations in the 16th century.[2]

15.1 Overview

Complex numbers allow for solutions to certain equations that have no solutions in real numbers. For example, the equation

$$(x + 1)^2 = -9$$

has no real solution, since the square of a real number cannot be negative. Complex numbers provide a solution to this problem. The idea is to extend the real numbers with the imaginary unit i where $i^2 = -1$, so that solutions to equations like the preceding one can be found. In this case the solutions are $-1 + 3i$ and $-1 - 3i$, as can be verified using the fact that $i^2 = -1$:

$$((-1 + 3i) + 1)^2 = (3i)^2 = (3^2)(i^2) = 9(-1) = -9.$$

$$((-1 - 3i) + 1)^2 = (-3i)^2 = (-3)^2(i^2) = 9(-1) = -9.$$

According to the fundamental theorem of algebra, all polynomial equations with real or complex coefficients in a single variable have a solution in complex numbers.

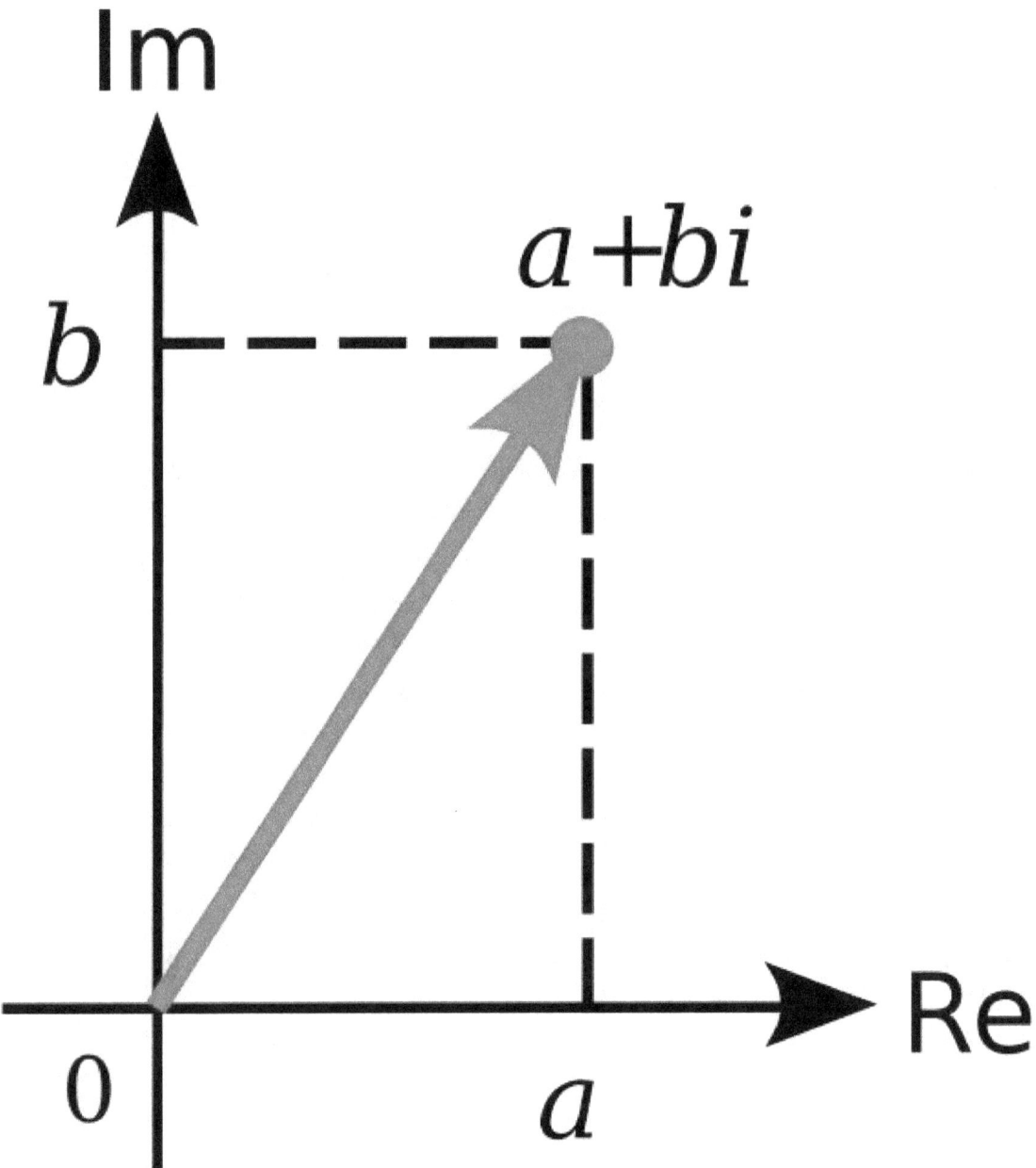

A complex number can be visually represented as a pair of numbers (a, b) forming a vector on a diagram called an Argand diagram, representing the complex plane. "Re" is the real axis, "Im" is the imaginary axis, and i is the imaginary unit which satisfies $i^2 = -1$.

15.1.1   Definition

A complex number is a number of the form $a + bi$, where a and b are real numbers and i is an *imaginary unit*, satisfying $i^2 = -1$. For example, $-3.5 + 2i$ is a complex number.

The real number a is called the *real part* of the complex number $a + bi$; the real number b is called the *imaginary part* of $a + bi$. By this convention the imaginary part does not include the imaginary unit: hence b, not bi, is the imaginary part.[3][4] The real part of a complex number z is denoted by $\operatorname{Re}(z)$ or $\Re(z)$; the imaginary part of a complex number z is denoted by $\operatorname{Im}(z)$ or $\Im(z)$. For example,

$$\mathrm{Re}(-3.5 + 2i) = -3.5$$
$$\mathrm{Im}(-3.5 + 2i) = 2.$$

Hence, in terms of its real and imaginary parts, a complex number z is equal to $\mathrm{Re}(z) + \mathrm{Im}(z) \cdot i$. This expression is sometimes known as the Cartesian form of z.

A real number a can be regarded as a complex number $a + 0i$ whose imaginary part is 0. A purely imaginary number bi is a complex number $0 + bi$ whose real part is zero. It is common to write a for $a + 0i$ and bi for $0 + bi$. Moreover, when the imaginary part is negative, it is common to write $a - bi$ with $b > 0$ instead of $a + (-b)i$, for example $3 - 4i$ instead of $3 + (-4)i$.

The set of all complex numbers is denoted by $\mathbb{C}$, **C** or $\mathbb{C}$.

15.1.2   Notation

Some authors[5] write $a + ib$ instead of $a + bi$. In some disciplines, in particular electromagnetism and electrical engineering, j is used instead of i,[6] since i is frequently used for electric current. In these cases complex numbers are written as $a + bj$ or $a + jb$.

15.1.3   Complex plane

Main article: Complex plane

A complex number can be viewed as a point or position vector in a two-dimensional Cartesian coordinate system called the complex plane or Argand diagram (see Pedoe 1988 and Solomentsev 2001), named after Jean-Robert Argand. The numbers are conventionally plotted using the real part as the horizontal component, and imaginary part as vertical (see Figure 1). These two values used to identify a given complex number are therefore called its *Cartesian*, *rectangular*, or *algebraic form*.

A position vector may also be defined in terms of its magnitude and direction relative to the origin. These are emphasized in a complex number's *polar form*. Using the polar form of the complex number in calculations may lead to a more intuitive interpretation of mathematical results. Notably, the operations of addition and multiplication take on a very natural geometric character when complex numbers are viewed as position vectors: addition corresponds to vector addition while multiplication corresponds to multiplying their magnitudes and adding their arguments (i.e. the angles they make with the x axis). Viewed in this way the multiplication of a complex number by i corresponds to rotating the position vector counterclockwise by a quarter turn (90°) about the origin: $(a+bi)i = ai+bi^2 = -b+ai$.

15.1.4   History in brief

Main section: History

The solution in radicals (without trigonometric functions) of a general cubic equation contains the square roots of negative numbers when all three roots are real numbers, a situation that cannot be rectified by factoring aided by the rational root test if the cubic is irreducible (the so-called casus irreducibilis). This conundrum led Italian mathematician Gerolamo Cardano to conceive of complex numbers in around 1545, though his understanding was rudimentary.

Work on the problem of general polynomials ultimately led to the fundamental theorem of algebra, which shows that with complex numbers, a solution exists to every polynomial equation of degree one or higher. Complex numbers thus form an algebraically closed field, where any polynomial equation has a root.

Many mathematicians contributed to the full development of complex numbers. The rules for addition, subtraction, multiplication, and division of complex numbers were developed by the Italian mathematician Rafael Bombelli.[7] A more abstract formalism for the complex numbers was further developed by the Irish mathematician William Rowan Hamilton, who extended this abstraction to the theory of quaternions.

15.2 Relations

15.2.1 Equality

Two complex numbers are equal if and only if both their real and imaginary parts are equal. In symbols:

$$z_1 = z_2 \leftrightarrow (\mathrm{Re}(z_1) = \mathrm{Re}(z_2) \wedge \mathrm{Im}(z_1) = \mathrm{Im}(z_2)).$$

15.2.2 Ordering

Because complex numbers are naturally thought of as existing on a two-dimensional plane, there is no natural linear ordering on the set of complex numbers.[8]

There is no linear ordering on the complex numbers that is compatible with addition and multiplication. Formally, we say that the complex numbers cannot have the structure of an ordered field. This is because any square in an ordered field is at least 0, but $i^2 = -1$.

15.3 Elementary operations

15.3.1 Conjugation

Main article: Complex conjugate
The *complex conjugate* of the complex number $z = x + yi$ is defined to be $x - yi$. It is denoted $\bar{z}$ or z^*.

Formally, for any complex number z:

$$\bar{z} = \mathrm{Re}(z) - \mathrm{Im}(z) \cdot i.$$

Geometrically, $\bar{z}$ is the "reflection" of z about the real axis. Conjugating twice gives the original complex number: $\bar{\bar{z}} = z$.

The real and imaginary parts of a complex number z can be extracted using the conjugate:

$$\mathrm{Re}\,(z) = \tfrac{1}{2}(z + \bar{z}),$$

$$\mathrm{Im}\,(z) = \tfrac{1}{2i}(z - \bar{z}).$$

Moreover, a complex number is real if and only if it equals its conjugate.

Conjugation distributes over the standard arithmetic operations:

$$\overline{z + w} = \bar{z} + \bar{w},$$

$$\overline{z - w} = \bar{z} - \bar{w},$$

$$\overline{zw} = \bar{z}\bar{w},$$

$$\overline{(z/w)} = \bar{z}/\bar{w}.$$

The reciprocal of a nonzero complex number $z = x + yi$ is given by

$$\frac{1}{z} = \frac{\bar{z}}{z\bar{z}} = \frac{\bar{z}}{x^2 + y^2}.$$

This formula can be used to compute the multiplicative inverse of a complex number if it is given in rectangular coordinates. Inversive geometry, a branch of geometry studying reflections more general than ones about a line, can also be expressed in terms of complex numbers. In the network analysis of electrical circuits, the complex conjugate is used in finding the equivalent impedance when the maximum power transfer theorem is used.

15.3.2   Addition and subtraction

Complex numbers are added by adding the real and imaginary parts of the summands. That is to say:

$$(a + bi) + (c + di) = (a + c) + (b + d)i.$$

Similarly, subtraction is defined by

$$(a + bi) - (c + di) = (a - c) + (b - d)i.$$

Using the visualization of complex numbers in the complex plane, the addition has the following geometric interpretation: the sum of two complex numbers A and B, interpreted as points of the complex plane, is the point X obtained by building a parallelogram three of whose vertices are O, A and B. Equivalently, X is the point such that the triangles with vertices O, A, B, and X, B, A, are congruent.

15.3.3   Multiplication and division

The multiplication of two complex numbers is defined by the following formula:

$$(a + bi)(c + di) = (ac - bd) + (bc + ad)i.$$

In particular, the square of the imaginary unit is -1:

$$i^2 = i \times i = -1.$$

The preceding definition of multiplication of general complex numbers follows naturally from this fundamental property of the imaginary unit. Indeed, if i is treated as a number so that di means d times i, the above multiplication rule is identical to the usual rule for multiplying two sums of two terms.

$(a + bi)(c + di) = ac + bci + adi + bidi$ (distributive law)

$\qquad = ac + bidi + bci + adi$ (commutative law of addition—the order of the summands can be changed)

$\qquad = ac + bdi^2 + (bc + ad)i$ (commutative and distributive laws)

$\qquad = (ac - bd) + (bc + ad)i$ (fundamental property of the imaginary unit).

The division of two complex numbers is defined in terms of complex multiplication, which is described above, and real division. When at least one of c and d is non-zero, we have

$$\frac{a + bi}{c + di} = \left(\frac{ac + bd}{c^2 + d^2}\right) + \left(\frac{bc - ad}{c^2 + d^2}\right)i.$$

Division can be defined in this way because of the following observation:

$$\frac{a + bi}{c + di} = \frac{(a + bi) \cdot (c - di)}{(c + di) \cdot (c - di)} = \left(\frac{ac + bd}{c^2 + d^2}\right) + \left(\frac{bc - ad}{c^2 + d^2}\right)i.$$

As shown earlier, $c - di$ is the complex conjugate of the denominator $c + di$. At least one of the real part c and the imaginary part d of the denominator must be nonzero for division to be defined. This is called "rationalization" of the denominator (although the denominator in the final expression might be an irrational real number).

15.3.4 Square root

See also: Square roots of negative and complex numbers

The square roots of $a + bi$ (with $b \neq 0$) are $\pm(\gamma + \delta i)$, where

$$\gamma = \sqrt{\frac{a + \sqrt{a^2 + b^2}}{2}}$$

and

$$\delta = \mathrm{sgn}(b)\sqrt{\frac{-a + \sqrt{a^2 + b^2}}{2}}.$$

where sgn is the signum function. This can be seen by squaring $\pm(\gamma + \delta i)$ to obtain $a + bi$.[9][10] Here $\sqrt{a^2 + b^2}$ is called the modulus of $a + bi$, and the square root sign indicates the square root with non-negative real part, called the **principal square root**; also $\sqrt{a^2 + b^2} = \sqrt{z\bar{z}}$, where $z = a + bi$.[11]

15.4 Polar form

Main article: Polar coordinate system

15.4.1 Absolute value and argument

An alternative way of defining a point P in the complex plane, other than using the x- and y-coordinates, is to use the distance of the point from O, the point whose coordinates are $(0, 0)$ (the origin), together with the angle subtended between the positive real axis and the line segment OP in a counterclockwise direction. This idea leads to the polar form of complex numbers.

The *absolute value* (or *modulus* or *magnitude*) of a complex number $z = x + yi$ is

$$r = |z| = \sqrt{x^2 + y^2}.$$

If z is a real number (i.e., $y = 0$), then $r = |x|$. In general, by Pythagoras' theorem, r is the distance of the point P representing the complex number z to the origin. The square of the absolute value is

$$|z|^2 = z\bar{z} = x^2 + y^2.$$

where $\bar{z}$ is the complex conjugate of z .

The *argument* of z (in many applications referred to as the "phase") is the angle of the radius OP with the positive real axis, and is written as $\arg(z)$. As with the modulus, the argument can be found from the rectangular form $x + yi$:[12]

$$\varphi = \arg(z) = \begin{cases} \arctan(\frac{y}{x}) & \text{if } x > 0 \\ \arctan(\frac{y}{x}) + \pi & \text{if } x < 0 \text{ and } y \geq 0 \\ \arctan(\frac{y}{x}) - \pi & \text{if } x < 0 \text{ and } y < 0 \\ \frac{\pi}{2} & \text{if } x = 0 \text{ and } y > 0 \\ -\frac{\pi}{2} & \text{if } x = 0 \text{ and } y < 0 \\ \text{indeterminate} & \text{if } x = 0 \text{ and } y = 0. \end{cases}$$

The value of φ is expressed in radians in this article. It can increase by any integer multiple of 2π and still give the same angle. Hence, the arg function is sometimes considered as multivalued. Normally, as given above, the principal value in the interval $(-\pi, \pi]$ is chosen. Values in the range $[0, 2\pi)$ are obtained by adding 2π if the value is negative. The polar angle for the complex number 0 is indeterminate, but arbitrary choice of the angle 0 is common.

The value of φ equals the result of atan2: $\varphi = \text{atan2}(\text{imaginary}, \text{real})$.

Together, r and φ give another way of representing complex numbers, the *polar form*, as the combination of modulus and argument fully specify the position of a point on the plane. Recovering the original rectangular co-ordinates from the polar form is done by the formula called *trigonometric form*

$$z = r(\cos\varphi + i\sin\varphi).$$

Using Euler's formula this can be written as

$$z = re^{i\varphi}.$$

Using the cis function, this is sometimes abbreviated to

$$z = r\,\text{cis}\,\varphi.$$

In angle notation, often used in electronics to represent a phasor with amplitude r and phase φ, it is written as[13]

$$z = r\angle\varphi.$$

15.4.2   Multiplication and division in polar form

Formulas for multiplication, division and exponentiation are simpler in polar form than the corresponding formulas in Cartesian coordinates. Given two complex numbers $z_1 = r_1(\cos\varphi_1 + i\sin\varphi_1)$ and $z_2 = r_2(\cos\varphi_2 + i\sin\varphi_2)$, because of the well-known trigonometric identities

$$\cos(a)\cos(b) - \sin(a)\sin(b) = \cos(a+b)$$

$$\cos(a)\sin(b) + \sin(a)\cos(b) = \sin(a+b)$$

we may derive

$$z_1 z_2 = r_1 r_2(\cos(\varphi_1 + \varphi_2) + i\sin(\varphi_1 + \varphi_2)).$$

In other words, the absolute values are multiplied and the arguments are added to yield the polar form of the product. For example, multiplying by i corresponds to a quarter-turn counter-clockwise, which gives back $i^2 = -1$. The picture at the right illustrates the multiplication of

$$(2+i)(3+i) = 5 + 5i.$$

Since the real and imaginary part of $5 + 5i$ are equal, the argument of that number is 45 degrees, or $\pi/4$ (in radian). On the other hand, it is also the sum of the angles at the origin of the red and blue triangles are arctan(1/3) and arctan(1/2), respectively. Thus, the formula

$$\frac{\pi}{4} = \arctan\frac{1}{2} + \arctan\frac{1}{3}$$

holds. As the arctan function can be approximated highly efficiently, formulas like this—known as Machin-like formulas—are used for high-precision approximations of π.

Similarly, division is given by

$$\frac{z_1}{z_2} = \frac{r_1}{r_2}\left(\cos(\varphi_1 - \varphi_2) + i\sin(\varphi_1 - \varphi_2)\right).$$

15.5 Exponentiation

15.5.1 Euler's formula

Euler's formula states that, for any real number x,

$$e^{ix} = \cos x + i\sin x$$

where e is the base of the natural logarithm. This can be proved through induction by observing that

$$i^0 = 1, \quad i^1 = i, \quad i^2 = -1, \quad i^3 = -i,$$
$$i^4 = 1, \quad i^5 = i, \quad i^6 = -1, \quad i^7 = -i,$$

and so on, and by considering the Taylor series expansions of e^x, $cos(x)$ and $sin(x)$:

$$e^{ix} = 1 + ix + \frac{(ix)^2}{2!} + \frac{(ix)^3}{3!} + \frac{(ix)^4}{4!} + \frac{(ix)^5}{5!} + \frac{(ix)^6}{6!} + \frac{(ix)^7}{7!} + \frac{(ix)^8}{8!} + \cdots$$

$$= 1 + ix - \frac{x^2}{2!} - \frac{ix^3}{3!} + \frac{x^4}{4!} + \frac{ix^5}{5!} - \frac{x^6}{6!} - \frac{ix^7}{7!} + \frac{x^8}{8!} + \cdots$$

$$= \left(1 - \frac{x^2}{2!} + \frac{x^4}{4!} - \frac{x^6}{6!} + \frac{x^8}{8!} - \cdots \right) + i \left(x - \frac{x^3}{3!} + \frac{x^5}{5!} - \frac{x^7}{7!} + \cdots \right)$$

$$= \cos x + i \sin x \; .$$

The rearrangement of terms is justified because each series is absolutely convergent.

15.5.2   Natural logarithm

Euler's formula allows us to observe that, for any complex number

$$z = r(\cos \varphi + i \sin \varphi).$$

where r is a non-negative real number, one possible value for z's natural logarithm is

$$\ln(z) = \ln(r) + \varphi i$$

Because cos and sin are periodic functions, the natural logarithm may be considered a multi-valued function, with:

$$\ln(z) = \{\ln(r) + (\varphi + 2\pi k)i \mid k \in \mathbb{Z}\}$$

15.5.3   Integer and fractional exponents

We may use the identity

$$\ln(a^b) = b \ln(a)$$

to define complex exponentiation, which is likewise multi-valued:

$$\ln(z^n) = \ln((r(\cos \varphi + i \sin \varphi))^n)$$

$$= n \ln(r(\cos \varphi + i \sin \varphi))$$

$$= \{n(\ln(r) + (\varphi + k2\pi)i) \mid k \in \mathbb{Z}\}$$

$$= \{n \ln(r) + n\varphi i + nk2\pi i \mid k \in \mathbb{Z}\}.$$

When n is an integer, this simplifies to de Moivre's formula:

$$z^n = (r(\cos \varphi + i \sin \varphi))^n = r^n (\cos n\varphi + i \sin n\varphi).$$

The nth roots of z are given by

$$\sqrt[n]{z} = \sqrt[n]{r}\left(\cos\left(\frac{\varphi + 2k\pi}{n}\right) + i\sin\left(\frac{\varphi + 2k\pi}{n}\right)\right)$$

for any integer k satisfying $0 \le k \le n - 1$. Here $\sqrt[n]{r}$ is the usual (positive) nth root of the positive real number r. While the nth root of a positive real number r is chosen to be the *positive* real number c satisfying $c^n = x$ there is no natural way of distinguishing one particular complex nth root of a complex number. Therefore, the nth root of z is considered as a multivalued function (in z), as opposed to a usual function f, for which $f(z)$ is a uniquely defined number. Formulas such as

$$\sqrt[n]{z^n} = z$$

(which holds for positive real numbers), do in general not hold for complex numbers.

15.6 Properties

15.6.1 Field structure

The set **C** of complex numbers is a field. Briefly, this means that the following facts hold: first, any two complex numbers can be added and multiplied to yield another complex number. Second, for any complex number z, its additive inverse $-z$ is also a complex number; and third, every nonzero complex number has a reciprocal complex number. Moreover, these operations satisfy a number of laws, for example the law of commutativity of addition and multiplication for any two complex numbers z_1 and z_2:

$$z_1 + z_2 = z_2 + z_1.$$

$$z_1 z_2 = z_2 z_1.$$

These two laws and the other requirements on a field can be proven by the formulas given above, using the fact that the real numbers themselves form a field.

Unlike the reals, **C** is not an ordered field, that is to say, it is not possible to define a relation $z_1 < z_2$ that is compatible with the addition and multiplication. In fact, in any ordered field, the square of any element is necessarily positive, so $i^2 = -1$ precludes the existence of an ordering on **C**.

When the underlying field for a mathematical topic or construct is the field of complex numbers, the topic's name is usually modified to reflect that fact. For example: complex analysis, complex matrix, complex polynomial, and complex Lie algebra.

15.6.2 Solutions of polynomial equations

Given any complex numbers (called coefficients) $a_0, \dots, an$, the equation

$$a_n z^n + \cdots + a_1 z + a_0 = 0$$

has at least one complex solution z, provided that at least one of the higher coefficients $a_1, \dots, an$ is nonzero. This is the statement of the *fundamental theorem of algebra*. Because of this fact, **C** is called an algebraically closed field. This property does not hold for the field of rational numbers **Q** (the polynomial $x^2 - 2$ does not have a rational root, since $\sqrt{2}$ is not a rational number) nor the real numbers **R** (the polynomial $x^2 + a$ does not have a real root for $a > 0$, since the square of x is positive for any real number x).

There are various proofs of this theorem, either by analytic methods such as Liouville's theorem, or topological ones such as the winding number, or a proof combining Galois theory and the fact that any real polynomial of *odd* degree has at least one real root.

Because of this fact, theorems that hold *for any algebraically closed field*, apply to **C**. For example, any non-empty complex square matrix has at least one (complex) eigenvalue.

15.6.3   Algebraic characterization

The field **C** has the following three properties: first, it has characteristic 0. This means that $1 + 1 + \cdots + 1 \neq 0$ for any number of summands (all of which equal one). Second, its transcendence degree over **Q**, the prime field of **C**, is the cardinality of the continuum. Third, it is algebraically closed (see above). It can be shown that any field having these properties is isomorphic (as a field) to **C**. For example, the algebraic closure of $\mathbf{Q}p$ also satisfies these three properties, so these two fields are isomorphic. Also, **C** is isomorphic to the field of complex Puiseux series. However, specifying an isomorphism requires the axiom of choice. Another consequence of this algebraic characterization is that **C** contains many proper subfields that are isomorphic to **C**.

15.6.4   Characterization as a topological field

The preceding characterization of **C** describes only the algebraic aspects of **C**. That is to say, the properties of nearness and continuity, which matter in areas such as analysis and topology, are not dealt with. The following description of **C** as a topological field (that is, a field that is equipped with a topology, which allows the notion of convergence) does take into account the topological properties. **C** contains a subset P (namely the set of positive real numbers) of nonzero elements satisfying the following three conditions:

- P is closed under addition, multiplication and taking inverses.

- If x and y are distinct elements of P, then either $x - y$ or $y - x$ is in P.

- If S is any nonempty subset of P, then $S + P = x + P$ for some x in **C**.

Moreover, **C** has a nontrivial involutive automorphism $x \mapsto x^*$ (namely the complex conjugation), such that $x\,x^*$ is in P for any nonzero x in **C**.

Any field F with these properties can be endowed with a topology by taking the sets $B(x, p) = \{\ y \mid p - (y - x)(y - x)^* \in P\ \}$ as a base, where x ranges over the field and p ranges over P. With this topology F is isomorphic as a *topological* field to **C**.

The only connected locally compact topological fields are **R** and **C**. This gives another characterization of **C** as a topological field, since **C** can be distinguished from **R** because the nonzero complex numbers are connected, while the nonzero real numbers are not.

15.7   Formal construction

15.7.1   Formal development

Above, complex numbers have been defined by introducing i, the imaginary unit, as a symbol. More rigorously, the set **C** of complex numbers can be defined as the set $\mathbf{R}^2$ of ordered pairs (a, b) of real numbers. In this notation, the above formulas for addition and multiplication read

$$(a,b) + (c,d) = (a+c, b+d)$$
$$(a,b) \cdot (c,d) = (ac - bd, bc + ad).$$

It is then just a matter of notation to express (a, b) as $a + bi$.

Though this low-level construction does accurately describe the structure of the complex numbers, the following equivalent definition reveals the algebraic nature of $\mathbf{C}$ more immediately. This characterization relies on the notion of fields and polynomials. A field is a set endowed with addition, subtraction, multiplication and division operations that behave as is familiar from, say, rational numbers. For example, the distributive law

$$(x + y)z = xz + yz$$

must hold for any three elements x, y and z of a field. The set $\mathbf{R}$ of real numbers does form a field. A polynomial $p(X)$ with real coefficients is an expression of the form

$$a_n X^n + \cdots + a_1 X + a_0$$

where the a_0, ..., a_n are real numbers. The usual addition and multiplication of polynomials endows the set $\mathbf{R}[X]$ of all such polynomials with a ring structure. This ring is called polynomial ring.

The quotient ring $\mathbf{R}[X]/(X^2 + 1)$ can be shown to be a field. This extension field contains two square roots of -1, namely (the cosets of) X and $-X$, respectively. (The cosets of) 1 and X form a basis of $\mathbf{R}[X]/(X^2 + 1)$ as a real vector space, which means that each element of the extension field can be uniquely written as a linear combination in these two elements. Equivalently, elements of the extension field can be written as ordered pairs (a, b) of real numbers. Moreover, the above formulas for addition etc. correspond to the ones yielded by this abstract algebraic approach—the two definitions of the field $\mathbf{C}$ are said to be isomorphic (as fields). Together with the above-mentioned fact that $\mathbf{C}$ is algebraically closed, this also shows that $\mathbf{C}$ is an algebraic closure of $\mathbf{R}$.

15.7.2 Matrix representation of complex numbers

Complex numbers $a + bi$ can also be represented by 2×2 matrices that have the following form:

$$\begin{pmatrix} a & -b \\ b & a \end{pmatrix}.$$

Here the entries a and b are real numbers. The sum and product of two such matrices is again of this form, and the sum and product of complex numbers corresponds to the sum and product of such matrices. The geometric description of the multiplication of complex numbers can also be expressed in terms of rotation matrices by using this correspondence between complex numbers and such matrices. Moreover, the square of the absolute value of a complex number expressed as a matrix is equal to the determinant of that matrix:

$$|z|^2 = \begin{vmatrix} a & -b \\ b & a \end{vmatrix} = (a^2) - ((-b)(b)) = a^2 + b^2.$$

The conjugate $\bar{z}$ corresponds to the transpose of the matrix.

Though this representation of complex numbers with matrices is the most common, many other representations arise from matrices *other than* $\begin{pmatrix} 0 & -1 \\ 1 & 0 \end{pmatrix}$ that square to the negative of the identity matrix. See the article on 2×2 real matrices for other representations of complex numbers.

15.8 Complex analysis

Main article: Complex analysis

The study of functions of a complex variable is known as complex analysis and has enormous practical use in applied mathematics as well as in other branches of mathematics. Often, the most natural proofs for statements in real analysis or even number theory employ techniques from complex analysis (see prime number theorem for an example). Unlike real functions, which are commonly represented as two-dimensional graphs, complex functions have four-dimensional graphs and may usefully be illustrated by color-coding a three-dimensional graph to suggest four dimensions, or by animating the complex function's dynamic transformation of the complex plane.

15.8.1 Complex exponential and related functions

The notions of convergent series and continuous functions in (real) analysis have natural analogs in complex analysis. A sequence of complex numbers is said to converge if and only if its real and imaginary parts do. This is equivalent to the (ε, δ)-definition of limits, where the absolute value of real numbers is replaced by the one of complex numbers. From a more abstract point of view, $\mathbf{C}$, endowed with the metric

$$d(z_1, z_2) = |z_1 - z_2|$$

is a complete metric space, which notably includes the triangle inequality

$$|z_1 + z_2| \le |z_1| + |z_2|$$

for any two complex numbers z_1 and z_2.

Like in real analysis, this notion of convergence is used to construct a number of elementary functions: the *exponential function* $\exp(z)$, also written e^z, is defined as the infinite series

$$\exp(z) := 1 + z + \frac{z^2}{2 \cdot 1} + \frac{z^3}{3 \cdot 2 \cdot 1} + \cdots = \sum_{n=0}^{\infty} \frac{z^n}{n!}.$$

and the series defining the real trigonometric functions sine and cosine, as well as hyperbolic functions such as sinh also carry over to complex arguments without change. *Euler's identity* states:

$$\exp(i\varphi) = \cos(\varphi) + i\sin(\varphi)$$

for any real number φ, in particular

$$\exp(i\pi) = -1$$

Unlike in the situation of real numbers, there is an infinitude of complex solutions z of the equation

$$\exp(z) = w$$

for any complex number $w \ne 0$. It can be shown that any such solution z—called complex logarithm of a—satisfies

$\log(x + iy) = \ln |w| + i \arg(w)$.

where arg is the argument defined above, and ln the (real) natural logarithm. As arg is a multivalued function, unique only up to a multiple of 2π, log is also multivalued. The principal value of log is often taken by restricting the imaginary part to the interval $(-\pi,\pi]$.

Complex exponentiation z^ω is defined as

$$z^\omega = \exp(\omega \log z).$$

Consequently, they are in general multi-valued. For $\omega = 1/n$, for some natural number n, this recovers the non-uniqueness of nth roots mentioned above.

Complex numbers, unlike real numbers, do not in general satisfy the unmodified power and logarithm identities, particularly when naïvely treated as single-valued functions; see failure of power and logarithm identities. For example, they do not satisfy

$$a^{bc} = (a^b)^c.$$

Both sides of the equation are multivalued by the definition of complex exponentiation given here, and the values on the left are a subset of those on the right.

15.8.2 Holomorphic functions

A function $f : \mathbf{C} \to \mathbf{C}$ is called holomorphic if it satisfies the Cauchy–Riemann equations. For example, any **R**-linear map $\mathbf{C} \to \mathbf{C}$ can be written in the form

$$f(z) = az + b\bar{z}$$

with complex coefficients a and b. This map is holomorphic if and only if $b = 0$. The second summand $b\bar{z}$ is real-differentiable, but does not satisfy the Cauchy–Riemann equations.

Complex analysis shows some features not apparent in real analysis. For example, any two holomorphic functions f and g that agree on an arbitrarily small open subset of **C** necessarily agree everywhere. Meromorphic functions, functions that can locally be written as $f(z)/(z - z_0)^n$ with a holomorphic function f, still share some of the features of holomorphic functions. Other functions have essential singularities, such as $\sin(1/z)$ at $z = 0$.

15.9 Applications

Complex numbers have essential concrete applications in a variety of scientific and related areas such as signal processing, control theory, electromagnetism, fluid dynamics, quantum mechanics, cartography, and vibration analysis. Some applications of complex numbers are:

15.9.1 Control theory

In control theory, systems are often transformed from the time domain to the frequency domain using the Laplace transform. The system's poles and zeros are then analyzed in the *complex plane*. The root locus, Nyquist plot, and Nichols plot techniques all make use of the complex plane.

In the root locus method, it is especially important whether the poles and zeros are in the left or right half planes, i.e. have real part greater than or less than zero. If a linear, time-invariant (LTI) system has poles that are

- in the right half plane, it will be unstable,

- all in the left half plane, it will be stable,

- on the imaginary axis, it will have marginal stability.

If a system has zeros in the right half plane, it is a nonminimum phase system.

15.9.2   Improper integrals

In applied fields, complex numbers are often used to compute certain real-valued improper integrals, by means of complex-valued functions. Several methods exist to do this; see methods of contour integration.

15.9.3   Fluid dynamics

In fluid dynamics, complex functions are used to describe potential flow in two dimensions.

15.9.4   Dynamic equations

In differential equations, it is common to first find all complex roots r of the characteristic equation of a linear differential equation or equation system and then attempt to solve the system in terms of base functions of the form $f(t) = e^{rt}$. Likewise, in difference equations, the complex roots r of the characteristic equation of the difference equation system are used, to attempt to solve the system in terms of base functions of the form $f(t) = r^t$.

15.9.5   Electromagnetism and electrical engineering

Main article: Alternating current

In electrical engineering, the Fourier transform is used to analyze varying voltages and currents. The treatment of resistors, capacitors, and inductors can then be unified by introducing imaginary, frequency-dependent resistances for the latter two and combining all three in a single complex number called the impedance. This approach is called phasor calculus.

In electrical engineering, the imaginary unit is denoted by j, to avoid confusion with I, which is generally in use to denote electric current, or, more particularly, i, which is generally in use to denote instantaneous electric current.

Since the voltage in an AC circuit is oscillating, it can be represented as

$$V(t) = V_0 e^{j\omega t} = V_0 \left(\cos \omega t + j \sin \omega t \right).$$

To obtain the measurable quantity, the real part is taken:

$$v(t) = \text{Re}(V) = \text{Re}\left[V_0 e^{j\omega t} \right] = V_0 \cos \omega t.$$

The complex-valued signal $V(t)$ is called the analytic representation of the real-valued, measurable signal $v(t)$. [114]

15.9.6   Signal analysis

Complex numbers are used in signal analysis and other fields for a convenient description for periodically varying signals. For given real functions representing actual physical quantities, often in terms of sines and cosines, corresponding complex functions are considered of which the real parts are the original quantities. For a sine wave of a given frequency, the absolute value $|z|$ of the corresponding z is the amplitude and the argument $\arg(z)$ is the phase.

If Fourier analysis is employed to write a given real-valued signal as a sum of periodic functions, these periodic functions are often written as complex valued functions of the form

$$x(t) = Re\{X(t)\}$$

and

$$X(t) = Ae^{i\omega t} = ae^{i\phi}e^{i\omega t} = ae^{i(\omega t + \phi)}$$

where ω represents the angular frequency and the complex number A encodes the phase and amplitude as explained above.

This use is also extended into digital signal processing and digital image processing, which utilize digital versions of Fourier analysis (and wavelet analysis) to transmit, compress, restore, and otherwise process digital audio signals, still images, and video signals.

Another example, relevant to the two side bands of amplitude modulation of AM radio, is:

$$
\begin{aligned}
\cos((\omega + \alpha)t) + \cos((\omega - \alpha)t) &= \mathrm{Re}\left(e^{i(\omega+\alpha)t} + e^{i(\omega-\alpha)t}\right) \\
&= \mathrm{Re}\left((e^{i\alpha t} + e^{-i\alpha t}) \cdot e^{i\omega t}\right) \\
&= \mathrm{Re}\left(2\cos(\alpha t) \cdot e^{i\omega t}\right) \\
&= 2\cos(\alpha t) \cdot \mathrm{Re}\left(e^{i\omega t}\right) \\
&= 2\cos(\alpha t) \cdot \cos(\omega t) \ .
\end{aligned}
$$

15.9.7 Quantum mechanics

The complex number field is intrinsic to the mathematical formulations of quantum mechanics, where complex Hilbert spaces provide the context for one such formulation that is convenient and perhaps most standard. The original foundation formulas of quantum mechanics—the Schrödinger equation and Heisenberg's matrix mechanics—make use of complex numbers.

15.9.8 Relativity

In special and general relativity, some formulas for the metric on spacetime become simpler if one takes the time component of the spacetime continuum to be imaginary. (This approach is no longer standard in classical relativity, but is used in an essential way in quantum field theory.) Complex numbers are essential to spinors, which are a generalization of the tensors used in relativity.

15.9.9 Geometry

Fractals

Certain fractals are plotted in the complex plane, e.g. the Mandelbrot set and Julia sets.

Triangles

Every triangle has a unique Steiner inellipse—an ellipse inside the triangle and tangent to the midpoints of the three sides of the triangle. The foci of a triangle's Steiner inellipse can be found as follows, according to Marden's theorem:[15][16] Denote the triangle's vertices in the complex plane as $a = xA + yAi$, $b = xB + yBi$, and $c = xC + yCi$. Write the cubic equation $(x-a)(x-b)(x-c)=0$, take its derivative, and equate the (quadratic) derivative to zero. Marden's Theorem says that the solutions of this equation are the complex numbers denoting the locations of the two foci of the Steiner inellipse.

15.9.10   Algebraic number theory

As mentioned above, any nonconstant polynomial equation (in complex coefficients) has a solution in **C**. A fortiori, the same is true if the equation has rational coefficients. The roots of such equations are called algebraic numbers – they are a principal object of study in algebraic number theory. Compared to **Q**, the algebraic closure of **Q**, which also contains all algebraic numbers, **C** has the advantage of being easily understandable in geometric terms. In this way, algebraic methods can be used to study geometric questions and vice versa. With algebraic methods, more specifically applying the machinery of field theory to the number field containing roots of unity, it can be shown that it is not possible to construct a regular nonagon using only compass and straightedge – a purely geometric problem.

Another example are Gaussian integers, that is, numbers of the form $x + iy$, where x and y are integers, which can be used to classify sums of squares.

15.9.11   Analytic number theory

Main article: Analytic number theory

Analytic number theory studies numbers, often integers or rationals, by taking advantage of the fact that they can be regarded as complex numbers, in which analytic methods can be used. This is done by encoding number-theoretic information in complex-valued functions. For example, the Riemann zeta function $\zeta(s)$ is related to the distribution of prime numbers.

15.10   History

The earliest fleeting reference to square roots of negative numbers can perhaps be said to occur in the work of the Greek mathematician Hero of Alexandria in the 1st century AD, where in his *Stereometrica* he considers, apparently in error, the volume of an impossible frustum of a pyramid to arrive at the term $\sqrt{81-144}=3i\sqrt{7}$ in his calculations, although negative quantities were not conceived of in Hellenistic mathematics and Heron merely replaced it by its positive ($\sqrt{144-81}=3\sqrt{7}$).[17]

The impetus to study complex numbers proper first arose in the 16th century when algebraic solutions for the roots of cubic and quartic polynomials were discovered by Italian mathematicians (see Niccolò Fontana Tartaglia, Gerolamo Cardano). It was soon realized that these formulas, even if one was only interested in real solutions, sometimes required the manipulation of square roots of negative numbers. As an example, Tartaglia's formula for a cubic equation of the form $x^3=px+q$ [18] gives the solution to the equation $x^3 = x$ as

$$\frac{1}{\sqrt{3}}\left((\sqrt{-1})^{1/3} + \frac{1}{(\sqrt{-1})^{1/3}}\right).$$

At first glance this looks like nonsense. However formal calculations with complex numbers show that the equation $z^3 = i$ has solutions $-i$, $\frac{\sqrt{3}}{2}+\frac{1}{2}i$ and $-\frac{\sqrt{3}}{2}+\frac{1}{2}i$. Substituting these in turn for $\sqrt{-1}^{1/3}$ in Tartaglia's cubic formula and simplifying, one gets 0, 1 and −1 as the solutions of $x^3 - x = 0$. Of course this particular equation can be solved at sight but it does illustrate that when general formulas are used to solve cubic equations with real roots then, as later mathematicians showed rigorously, the use of complex numbers is unavoidable. Rafael Bombelli was the first to explicitly address these seemingly paradoxical solutions of cubic equations and developed the rules for complex arithmetic trying to resolve these issues.

The term "imaginary" for these quantities was coined by René Descartes in 1637, although he was at pains to stress their imaginary nature[19]

> [...] sometimes only imaginary, that is one can imagine as many as I said in each equation, but sometimes there exists no quantity that matches that which we imagine.
>
> *([...] quelquefois seulement imaginaires c'est-à-dire que l'on peut toujours en imaginer autant que j'ai dit en chaque équation, mais qu'il n'y a quelquefois aucune quantité qui corresponde à celle qu'on imagine.)*

A further source of confusion was that the equation $\sqrt{-1}^2 = \sqrt{-1}\sqrt{-1} = -1$ seemed to be capriciously inconsistent with the algebraic identity $\sqrt{a}\sqrt{b} = \sqrt{ab}$, which is valid for non-negative real numbers a and b, and which was also used in complex number calculations with one of a, b positive and the other negative. The incorrect use of this identity (and the related identity $\frac{1}{\sqrt{a}} = \sqrt{\frac{1}{a}}$) in the case when both a and b are negative even bedeviled Euler. This difficulty eventually led to the convention of using the special symbol i in place of $\sqrt{-1}$ to guard against this mistake. Even so, Euler considered it natural to introduce students to complex numbers much earlier than we do today. In his elementary algebra text book, Elements of Algebra, he introduces these numbers almost at once and then uses them in a natural way throughout.

In the 18th century complex numbers gained wider use, as it was noticed that formal manipulation of complex expressions could be used to simplify calculations involving trigonometric functions. For instance, in 1730 Abraham de Moivre noted that the complicated identities relating trigonometric functions of an integer multiple of an angle to powers of trigonometric functions of that angle could be simply re-expressed by the following well-known formula which bears his name, de Moivre's formula:

$$(\cos\theta + i\sin\theta)^n = \cos n\theta + i\sin n\theta.$$

In 1748 Leonhard Euler went further and obtained Euler's formula of complex analysis:

$$\cos\theta + i\sin\theta = e^{i\theta}$$

by formally manipulating complex power series and observed that this formula could be used to reduce any trigonometric identity to much simpler exponential identities.

The idea of a complex number as a point in the complex plane (above) was first described by Caspar Wessel in 1799, although it had been anticipated as early as 1685 in Wallis's *De Algebra tractatus*.

Wessel's memoir appeared in the Proceedings of the Copenhagen Academy but went largely unnoticed. In 1806 Jean-Robert Argand independently issued a pamphlet on complex numbers and provided a rigorous proof of the fundamental theorem of algebra. Gauss had earlier published an essentially topological proof of the theorem in 1797 but expressed his doubts at the time about "the true metaphysics of the square root of −1". It was not until 1831 that he overcame these doubts and published his treatise on complex numbers as points in the plane, largely establishing modern notation and terminology. The English mathematician G. H. Hardy remarked that Gauss was the first mathematician to use complex numbers in 'a really confident and scientific way' although mathematicians such as Niels Henrik Abel and Carl Gustav Jacob Jacobi were necessarily using them routinely before Gauss published his 1831 treatise.[20] Augustin Louis Cauchy and Bernhard Riemann together brought the fundamental ideas of complex analysis to a high state of completion, commencing around 1825 in Cauchy's case.

The common terms used in the theory are chiefly due to the founders. Argand called $\cos\phi + i\sin\phi$ the *direction factor*, and $r = \sqrt{a^2 + b^2}$ the *modulus*; Cauchy (1828) called $\cos\phi + i\sin\phi$ the *reduced form* (l'expression réduite) and apparently introduced the term *argument*; Gauss used i for $\sqrt{-1}$, introduced the term *complex number* for $a + bi$, and called $a^2 + b^2$ the *norm*. The expression *direction coefficient*, often used for $\cos\phi + i\sin\phi$, is due to Hankel (1867), and *absolute value*, for *modulus*, is due to Weierstrass.

Later classical writers on the general theory include Richard Dedekind, Otto Hölder, Felix Klein, Henri Poincaré, Hermann Schwarz, Karl Weierstrass and many others.

15.11   Generalizations and related notions

The process of extending the field **R** of reals to **C** is known as Cayley–Dickson construction. It can be carried further to higher dimensions, yielding the quaternions **H** and octonions **O** which (as a real vector space) are of dimension 4 and 8, respectively.

However, just as applying the construction to reals loses the property of ordering, more properties familiar from real and complex numbers vanish with increasing dimension. The quaternions are only a skew field, i.e. for some x, y: $x \cdot y \neq y \cdot x$ for two quaternions, the multiplication of octonions fails (in addition to not being commutative) to be associative: for some x, y, z: $(x \cdot y) \cdot z \neq x \cdot (y \cdot z)$.

Reals, complex numbers, quaternions and octonions are all normed division algebras over **R**. However, by Hurwitz's theorem they are the only ones. The next step in the Cayley–Dickson construction, the sedenions, in fact fails to have this structure.

The Cayley–Dickson construction is closely related to the regular representation of **C**, thought of as an **R**-algebra (an **R**-vector space with a multiplication), with respect to the basis $(1, i)$. This means the following: the **R**-linear map

$$\mathbb{C} \to \mathbb{C}, z \mapsto wz$$

for some fixed complex number w can be represented by a 2×2 matrix (once a basis has been chosen). With respect to the basis $(1, i)$, this matrix is

$$\begin{pmatrix} \mathrm{Re}(w) & -\mathrm{Im}(w) \\ \mathrm{Im}(w) & \mathrm{Re}(w) \end{pmatrix}$$

i.e., the one mentioned in the section on matrix representation of complex numbers above. While this is a linear representation of **C** in the 2×2 real matrices, it is not the only one. Any matrix

$$J = \begin{pmatrix} p & q \\ r & -p \end{pmatrix}, \quad p^2 + qr + 1 = 0$$

has the property that its square is the negative of the identity matrix: $J^2 = -I$. Then

$$\{z = aI + bJ : a, b \in R\}$$

is also isomorphic to the field **C**, and gives an alternative complex structure on $\mathbf{R}^2$. This is generalized by the notion of a linear complex structure.

Hypercomplex numbers also generalize **R**, **C**, **H**, and **O**. For example, this notion contains the split-complex numbers, which are elements of the ring $\mathbf{R}[x]/(x^2 - 1)$ (as opposed to $\mathbf{R}[x]/(x^2 + 1)$). In this ring, the equation $a^2 = 1$ has four solutions.

The field **R** is the completion of **Q**, the field of rational numbers, with respect to the usual absolute value metric. Other choices of metrics on **Q** lead to the fields $\mathbf{Q}p$ of p-adic numbers (for any prime number p), which are thereby analogous to **R**. There are no other nontrivial ways of completing **Q** than **R** and $\mathbf{Q}p$, by Ostrowski's theorem. The algebraic closure $\overline{\mathbf{Q}_p}$ of $\mathbf{Q}p$ still carry a norm, but (unlike **C**) are not complete with respect to it. The completion $\mathbf{C}_p$ of $\overline{\mathbf{Q}_p}$ turns out to be algebraically closed. This field is called p-adic complex numbers by analogy.

The fields **R** and **Q**p and their finite field extensions, including **C**, are local fields.

15.12   See also

- Algebraic surface

- Circular motion using complex numbers

- Complex base systems

- Complex geometry

- Complex square root

- Domain coloring

- Eisenstein integer

- Euler's identity

- Gaussian integer

- Mandelbrot set

- Quaternion

- Riemann sphere (extended complex plane)

- Root of unity

- Unit complex number

15.13   Notes

[1] Charles P. McKeague (2011), *Elementary Algebra*, Brooks/Cole, p. 524, ISBN 978-0-8400-6421-9

[2] Burton (1995, p. 294)

[3] Complex Variables (2nd Edition), M.R. Spiegel, S. Lipschutz, J.J. Schiller, D. Spellman, Schaum's Outline Series, Mc Graw Hill (USA), ISBN 978-0-07-161569-3

[4] Aufmann, Richard N.; Barker, Vernon C.; Nation, Richard D. (2007), "Chapter P", *College Algebra and Trigonometry* (6 ed.), Cengage Learning, p. 66, ISBN 0-618-82515-0

[5] For example Ahlfors (1979).

[6] Brown, James Ward; Churchill, Ruel V. (1996), *Complex variables and applications* (6th ed.), New York: McGraw-Hill, p. 2, ISBN 0-07-912147-0, In electrical engineering, the letter j is used instead of i.

[7] Katz (2004, §9.1.4)

[8] http://mathworld.wolfram.com/ComplexNumber.html

[9] Abramowitz, Milton; Stegun, Irene A. (1964), *Handbook of mathematical functions with formulas, graphs, and mathematical tables*, Courier Dover Publications, p. 17, ISBN 0-486-61272-4, Section 3.7.26, p. 17

[10] Cooke, Roger (2008), *Classical algebra: its nature, origins, and uses*, John Wiley and Sons, p. 59, ISBN 0-470-25952-3, Extract: page 59

[11] Ahlfors (1979, p. 3)

[12] Kasana, H.S. (2005), "Chapter 1", *Complex Variables: Theory And Applications* (2nd ed.), PHI Learning Pvt. Ltd, p. 14, ISBN 81-203-2641-5

[13] Nilsson, James William; Riedel, Susan A. (2008), "Chapter 9", *Electric circuits* (8th ed.), Prentice Hall, p. 338, ISBN 0-13-198925-1

[14] Electromagnetism (2nd edition), I.S. Grant, W.R. Phillips, Manchester Physics Series, 2008 ISBN 0-471-92712-0

[15] Kalman, Dan (2008a), "An Elementary Proof of Marden's Theorem", *The American Mathematical Monthly* **115**: 330–38, ISSN 0002-9890

[16] Kalman, Dan (2008b), "The Most Marvelous Theorem in Mathematics", *Journal of Online Mathematics and its Applications*

[17] Nahin, Paul J. (2007), *An Imaginary Tale: The Story of* $\sqrt{-1}$, Princeton University Press, ISBN 978-0-691-12798-9, retrieved 20 April 2011

[18] In modern notation, Tartaglia's solution is based on expanding the cube of the sum of two cube roots: $(\sqrt[3]{u}+\sqrt[3]{v})^3 = 3\sqrt[3]{uv}(\sqrt[3]{u}+\sqrt[3]{v}) + u+v$ With $x = \sqrt[3]{u}+\sqrt[3]{v}$, $p = 3\sqrt[3]{uv}$, $q = u+v$, u and v can be expressed in terms of p and q as $u = q/2 + \sqrt{(q/2)^2 - (p/3)^3}$ and $v = q/2 - \sqrt{(q/2)^2 - (p/3)^3}$, respectively. Therefore, $x = \sqrt[3]{q/2 + \sqrt{(q/2)^2 - (p/3)^3}} + \sqrt[3]{q/2 - \sqrt{(q/2)^2 - (p/3)^3}}$. When $(q/2)^2 - (p/3)^3$ is negative (casus irreducibilis), the second cube root should be regarded as the complex conjugate of the first one.

[19] Descartes, René (1954) [1637], *La Géométrie | The Geometry of René Descartes with a facsimile of the first edition*, Dover Publications, ISBN 0-486-60068-8, retrieved 20 April 2011

[20] Hardy, G. H.; Wright, E. M. (2000) [1938], *An Introduction to the Theory of Numbers*, OUP Oxford, p. 189 (fourth edition), ISBN 0-19-921986-9

15.14 References

15.14.1 Mathematical references

- Ahlfors, Lars (1979), *Complex analysis* (3rd ed.), McGraw-Hill, ISBN 978-0-07-000657-7

- Conway, John B. (1986), *Functions of One Complex Variable I*, Springer, ISBN 0-387-90328-3

- Joshi, Kapil D. (1989), *Foundations of Discrete Mathematics*, New York: John Wiley & Sons, ISBN 978-0-470-21152-6

- Pedoe, Dan (1988), *Geometry: A comprehensive course*, Dover, ISBN 0-486-65812-0

- Press, WH; Teukolsky, SA; Vetterling, WT; Flannery, BP (2007), "Section 5.5 Complex Arithmetic", *Numerical Recipes: The Art of Scientific Computing* (3rd ed.), New York: Cambridge University Press, ISBN 978-0-521-88068-8

- Solomentsev, E.D. (2001), "Complex number", in Hazewinkel, Michiel, *Encyclopedia of Mathematics*, Springer, ISBN 978-1-55608-010-4

15.14.2 Historical references

- Burton, David M. (1995), *The History of Mathematics* (3rd ed.), New York: McGraw-Hill, ISBN 978-0-07-009465-9

- Katz, Victor J. (2004), *A History of Mathematics, Brief Version*, Addison-Wesley, ISBN 978-0-321-16193-2

- Nahin, Paul J. (1998), *An Imaginary Tale: The Story of* $\sqrt{-1}$ (hardcover edition ed.), Princeton University Press, ISBN 0-691-02795-1

    A gentle introduction to the history of complex numbers and the beginnings of complex analysis.

- H.D. Ebbinghaus; H. Hermes; F. Hirzebruch; M. Koecher; K. Mainzer; J. Neukirch; A. Prestel; R. Remmert (1991), *Numbers* (hardcover ed.), Springer, ISBN 0-387-97497-0

    An advanced perspective on the historical development of the concept of number.

15.15 Further reading

- *The Road to Reality: A Complete Guide to the Laws of the Universe*, by Roger Penrose; Alfred A. Knopf, 2005; ISBN 0-679-45443-8. Chapters 4–7 in particular deal extensively (and enthusiastically) with complex numbers.

- *Unknown Quantity: A Real and Imaginary History of Algebra*, by John Derbyshire; Joseph Henry Press; ISBN 0-309-09657-X (hardcover 2006). A very readable history with emphasis on solving polynomial equations and the structures of modern algebra.

- *Visual Complex Analysis*, by Tristan Needham; Clarendon Press; ISBN 0-19-853447-7 (hardcover, 1997). History of complex numbers and complex analysis with compelling and useful visual interpretations.

- Conway, John B., *Functions of One Complex Variable I* (Graduate Texts in Mathematics), Springer; 2 edition (12 September 2005). ISBN 0-387-90328-3.

15.16 External links

- Hazewinkel, Michiel, ed. (2001), "Complex number", *Encyclopedia of Mathematics*, Springer, ISBN 978-1-55608-010-4

- Introduction to Complex Numbers from Khan Academy

-

- Imaginary Numbers on *In Our Time* at the BBC.

- Euler's work on Complex Roots of Polynomials at Convergence. MAA Mathematical Sciences Digital Library.

- John and Betty's Journey Through Complex Numbers

- The Origin of Complex Numbers by John H. Mathews and Russell W. Howell

- Dimensions: a math film. Chapter 5 presents an introduction to complex arithmetic and stereographic projection. Chapter 6 discusses transformations of the complex plane, Julia sets, and the Mandelbrot set.

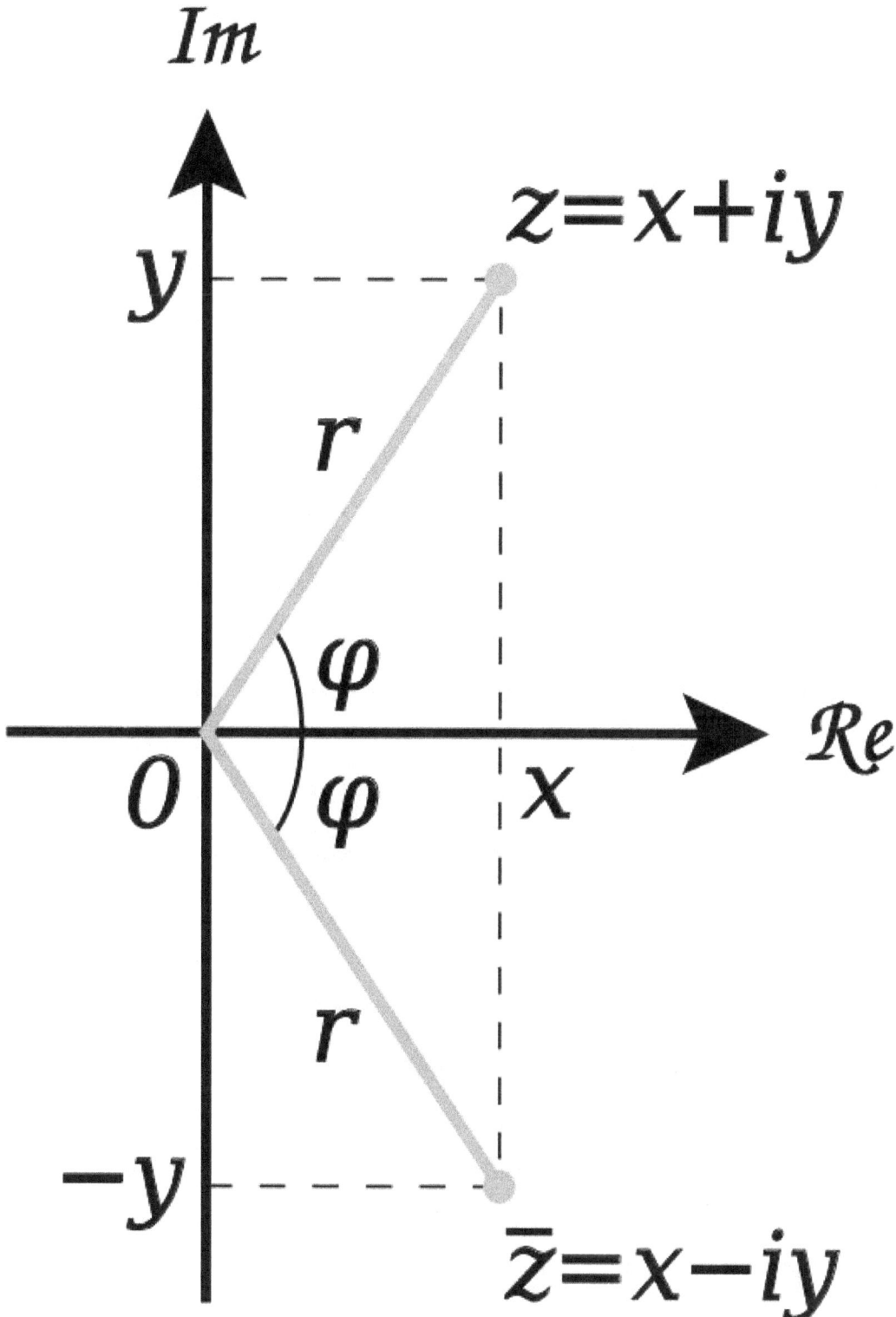

An illustration of the complex plane. The real part of a complex number z = x + iy is x, and its imaginary part is y.

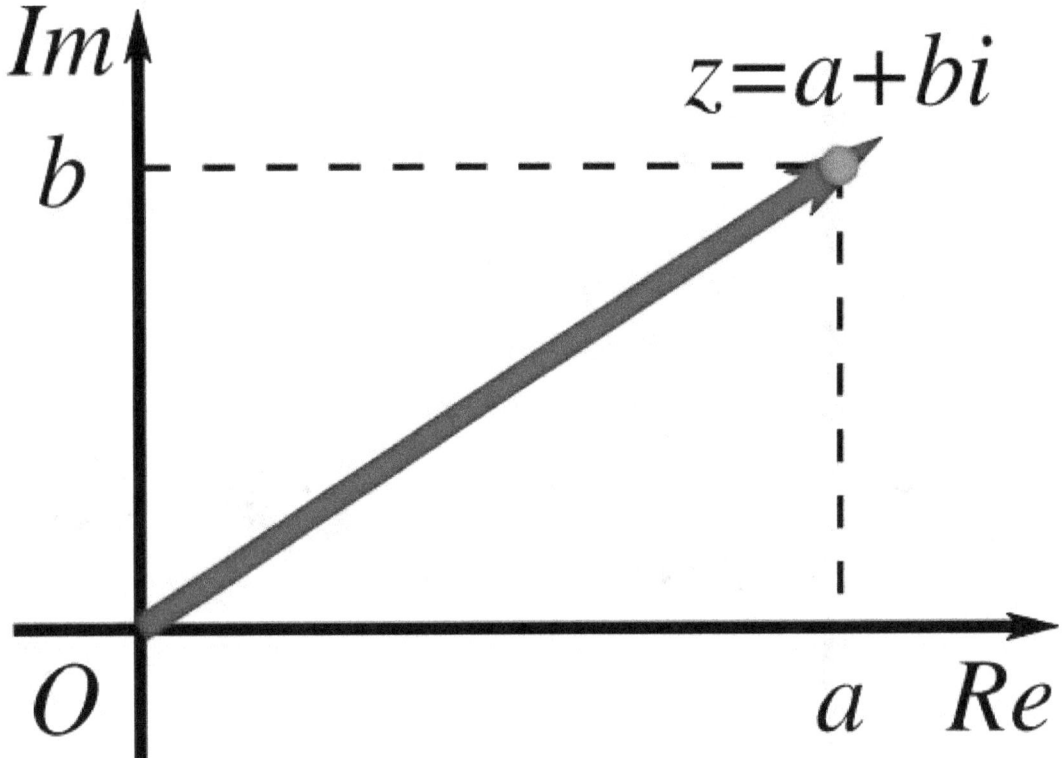

Figure 1: A complex number plotted as a point (red) and position vector (blue) on an Argand diagram; a+bi is the rectangular expression of the point.

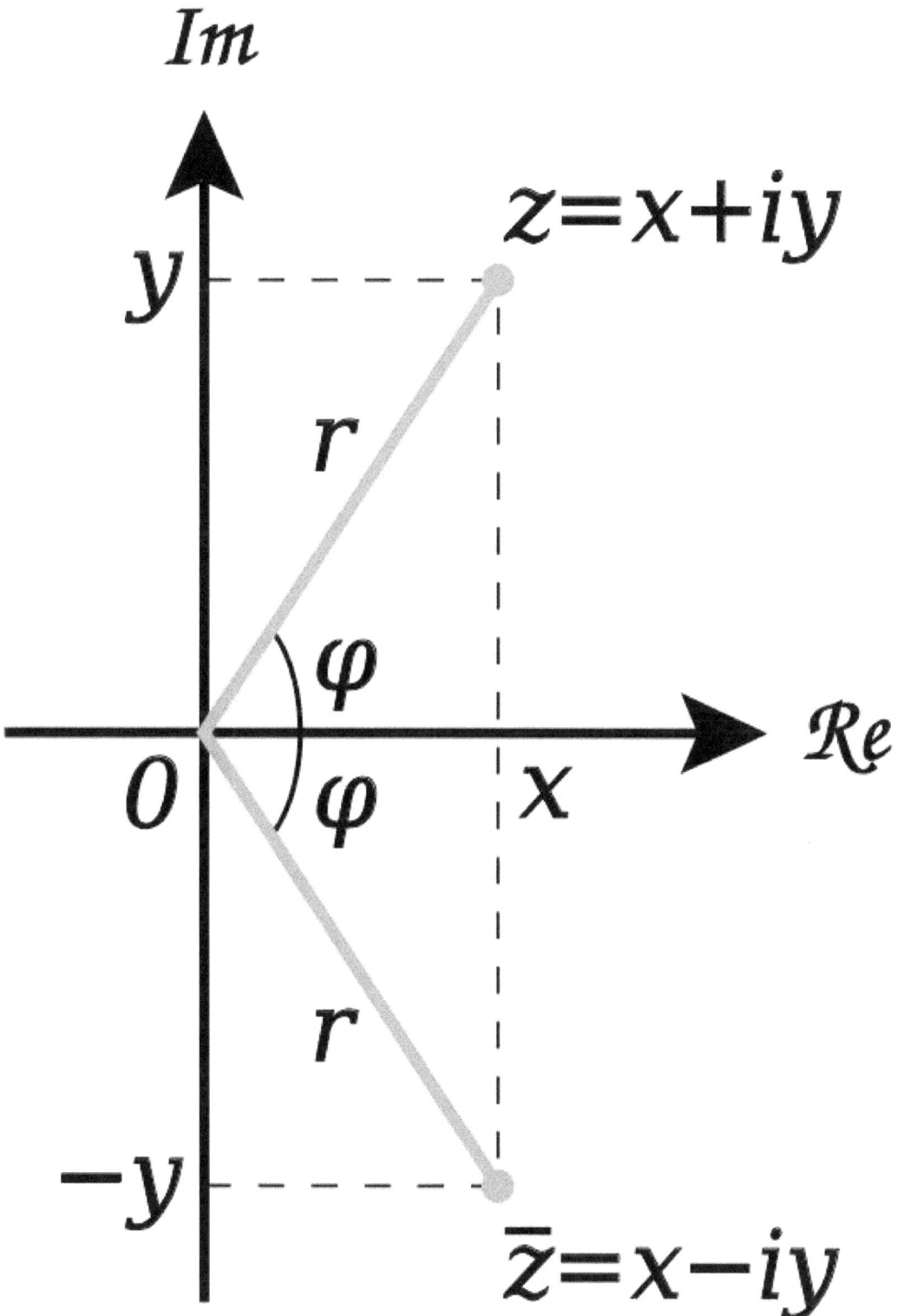

Geometric representation of z and its conjugate z̄ in the complex plane

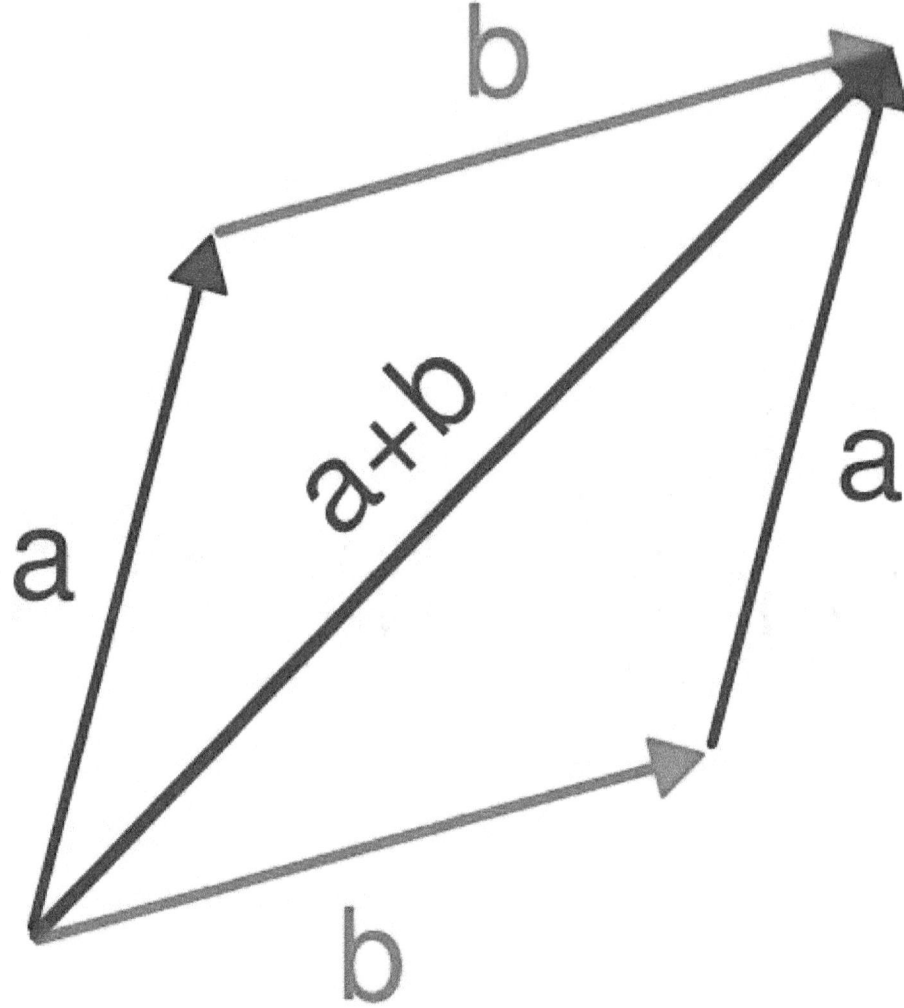

Addition of two complex numbers can be done geometrically by constructing a parallelogram.

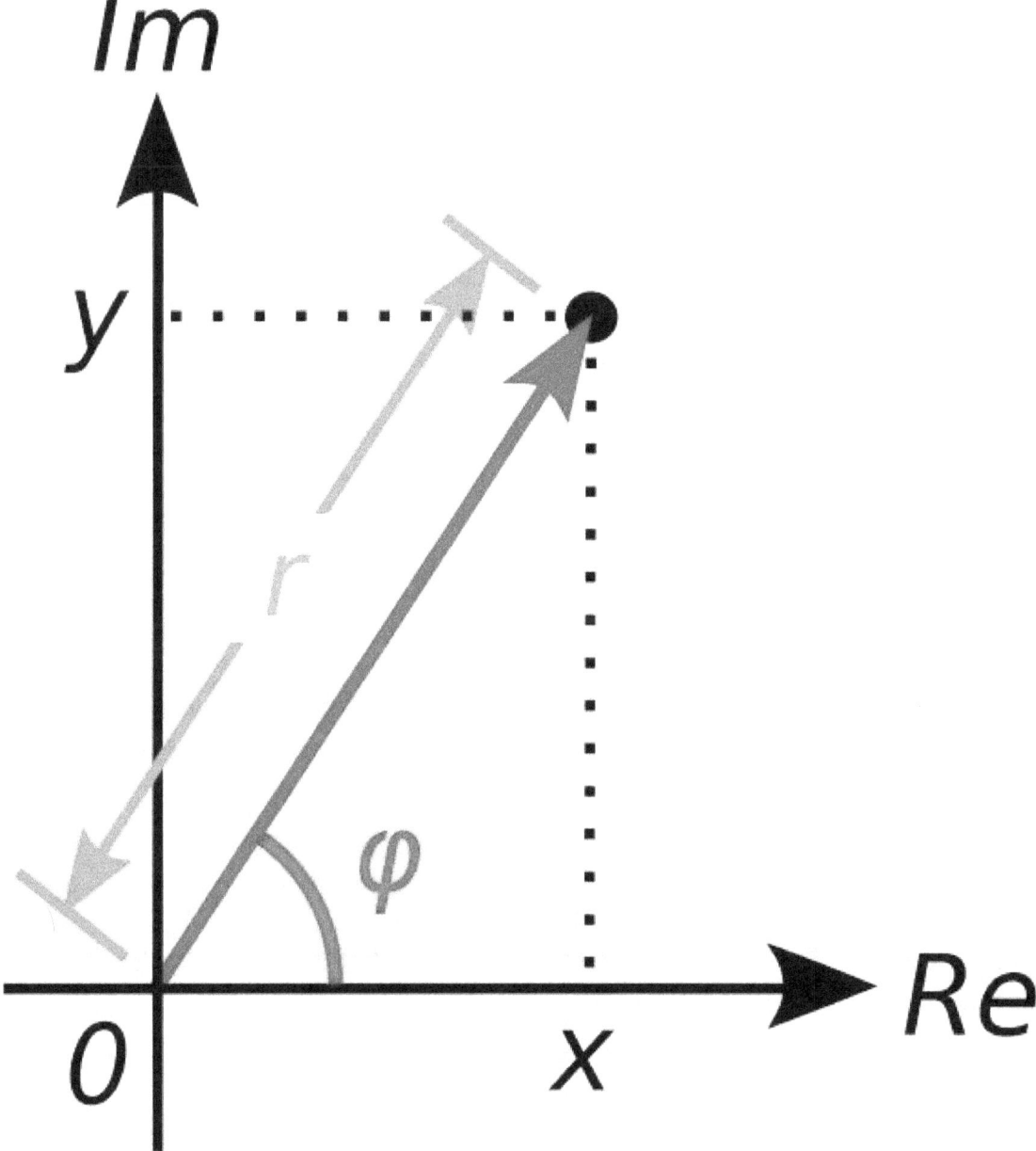

Figure 2: The argument φ and modulus r locate a point on an Argand diagram; $r(\cos\varphi + i\sin\varphi)$ or $re^{i\varphi}$ are polar expressions of the point.

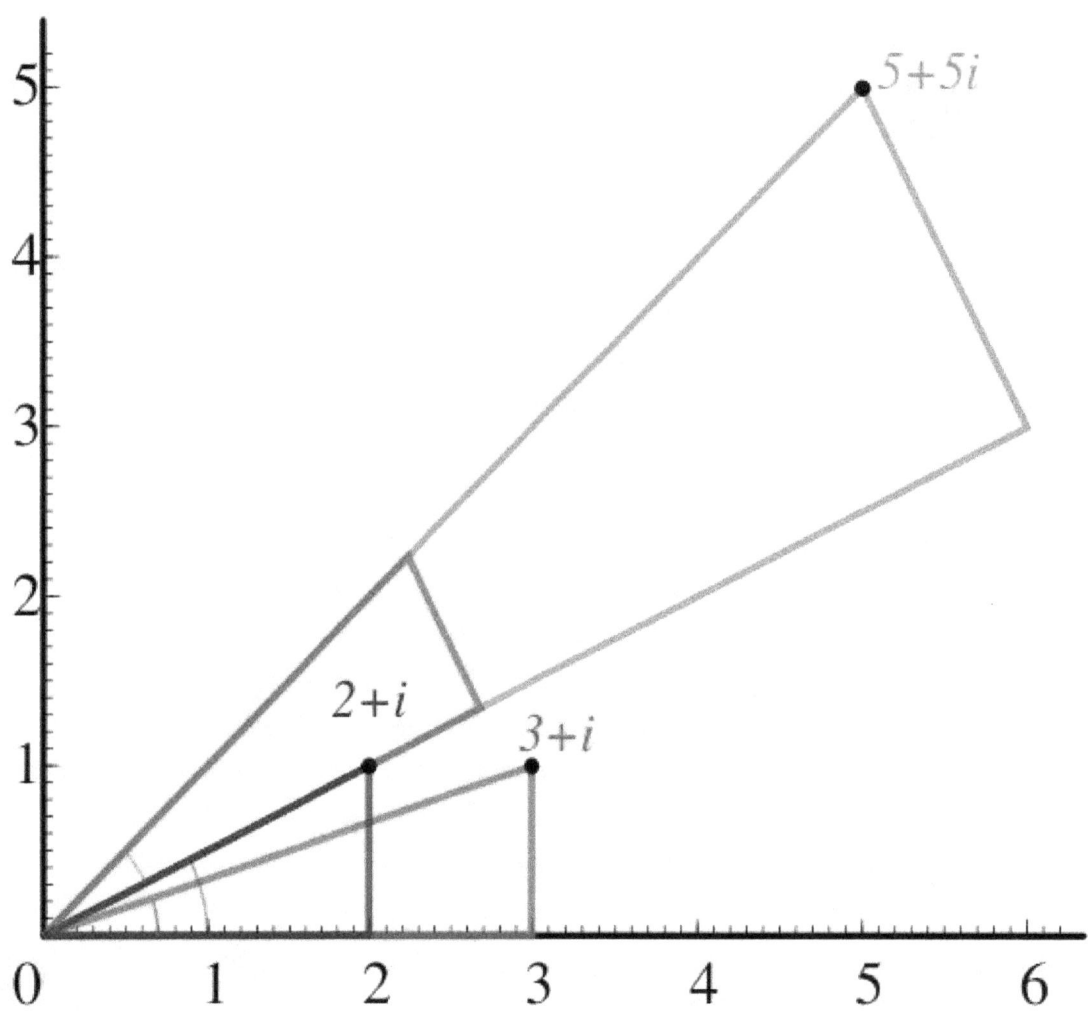

Multiplication of 2 + i (blue triangle) and 3 + i (red triangle). The red triangle is rotated to match the vertex of the blue one and stretched by √5, the length of the hypotenuse of the blue triangle.

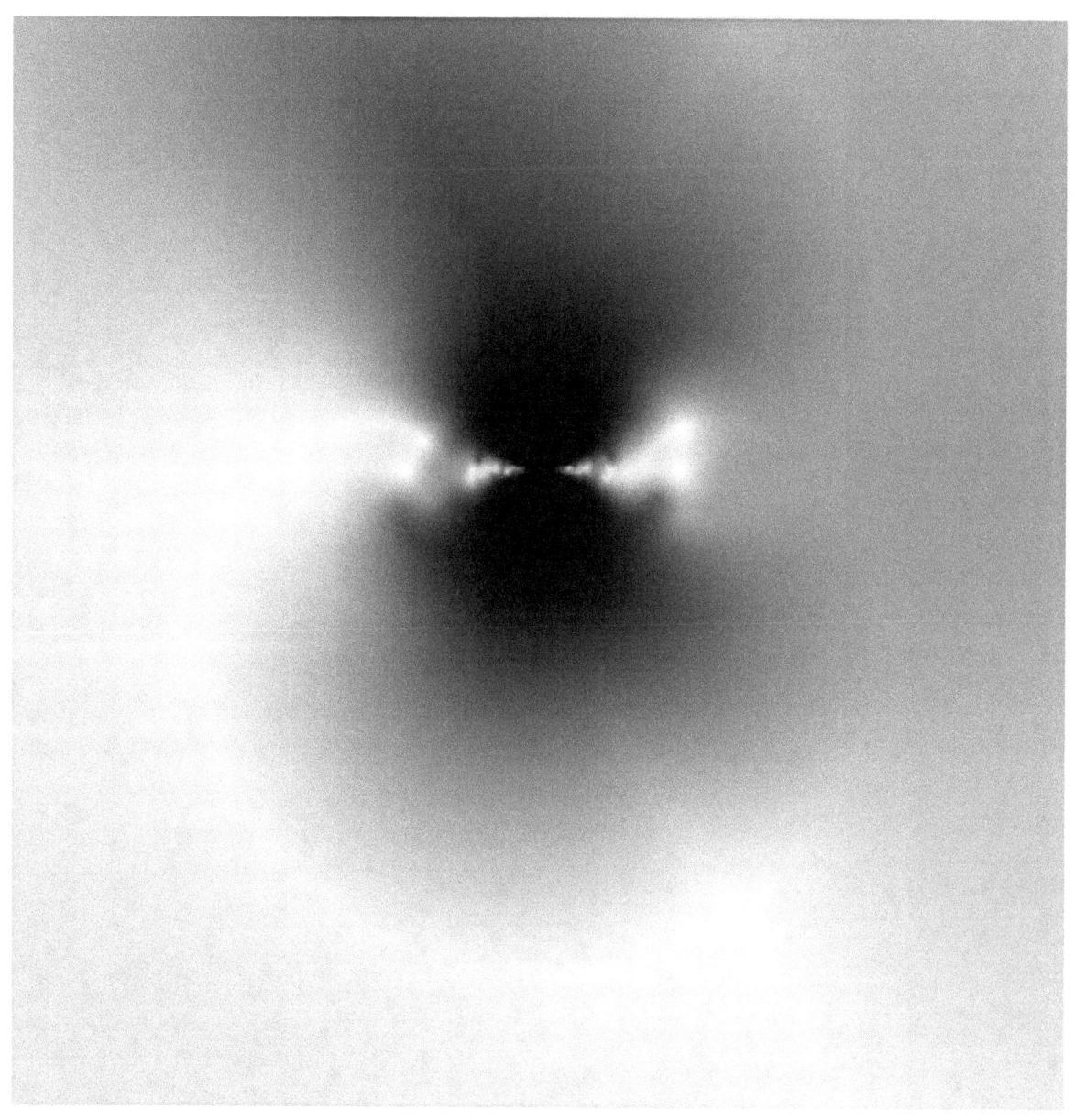

Color wheel graph of sin(1/z). Black parts inside refer to numbers having large absolute values.

Construction of a regular pentagon using straightedge and compass.

Chapter 16

Hypercomplex number

Not to be confused with surcomplex number.

In mathematics, a **hypercomplex number** is a traditional term for an element of a unital algebra over the field of real numbers. In the nineteenth century number systems called quaternions, tessarines, coquaternions, biquaternions, and octonions became established concepts in mathematical literature, added to the real and complex numbers. The concept of a hypercomplex number covered them all, and called for a discipline to explain and classify them.

The cataloguing project began in 1872 when Benjamin Peirce first published his *Linear Associative Algebra*, and was carried forward by his son Charles Sanders Peirce.[1] Most significantly, they identified the nilpotent and the idempotent elements as useful hypercomplex numbers for classifications. The Cayley–Dickson construction used involutions to generate complex numbers, quaternions, and octonions out of the real number system. Hurwitz and Frobenius proved theorems that put limits on hypercomplexity: Hurwitz's theorem (normed division algebras), and Frobenius theorem (associative division algebras). Finally in 1958 J. Frank Adams used topological methods to prove that there exist only four finite-dimensional real division algebras: the reals $\mathbb{R}$, the complexes $\mathbb{C}$, the quaternions $\mathbb{H}$, and the octonions $\mathbb{O}$.[2]

It was matrix algebra that harnessed the hypercomplex systems. First, matrices contributed new hypercomplex numbers like 2×2 real matrices. Soon the matrix paradigm began to explain the others as they became represented by matrices and their operations. In 1907 Joseph Wedderburn showed that associative hypercomplex systems could be represented by matrices, or direct sums of systems of matrices. From that date the preferred term for a hypercomplex system became associative algebra as seen in the title of Wedderburn's thesis at University of Edinburgh. Note however, that non-associative systems like octonions and hyperbolic quaternions represent another type of hypercomplex number.

As Hawkins (1972) explains, the hypercomplex numbers are stepping stones to learning about Lie groups and group representation theory. For instance, in 1929 Emmy Noether wrote on "hypercomplex quantities and representation theory". Review of the historic particulars gives body to the generalities of modern theory. In 1973 Kantor and Solodovnikov published a textbook on hypercomplex numbers which was translated in 1989; a reviewer says it has a "highly classical flavour". See Karen Parshall (1985) for a detailed exposition of the heyday of hypercomplex numbers, including the role of such luminaries as Theodor Molien and Eduard Study. For the transition to modern algebra, Bartel van der Waerden devotes thirty pages to hypercomplex numbers in his *History of Algebra* (1985).

16.1 Definition

A definition of a **hypercomplex number** is given by Kantor & Solodovnikov (1989) as an element of a finite-dimensional algebra over the real numbers that is unital and distributive (but not necessarily associative). Elements are generated with real number coefficients $(a_0, \ldots, a_n)$ for a basis $\{1, i_1, \ldots, i_n\}$. Where possible, it is conventional to choose the basis so that $i_k^2 \in \{-1, 0, +1\}$. A technical approach to hypercomplex numbers directs attention first to those of dimension two. Higher dimensions are configured as Cliffordian or algebraic sums of other algebras.

126

16.2 Two-dimensional real algebras

Theorem:[3][4][5] Up to isomorphism, there are exactly three 2-dimensional unital algebras over the reals: the ordinary complex numbers, the split-complex numbers, and the dual numbers.

> Proof: Since the algebra is closed under squaring, and it has but two dimensions, the non-real basis element
> u squares to an arbitrary linear combination of 1 and u:

$$u^2 = a_0 + a_1 u$$

with arbitrary real numbers a_0 and a_1. Using the common method of completing the square by subtracting $a_1 u$ and adding the quadratic complement $a_1^2/4$ to both sides yields

$$u^2 - a_1 u + \frac{a_1^2}{4} = a_0 + \frac{a_1^2}{4}.$$

$$u^2 - a_1 u + \frac{a_1^2}{4} = \left(u - \frac{a_1}{2}\right)^2 = \tilde{u}^2 \text{ so that}$$

$$\tilde{u}^2 = a_0 + \frac{a_1^2}{4}.$$

The three cases depend on this real value:

- If $4a_0 = -a_1^2$, the above formula yields $\tilde{u}^2 = 0$. Hence, $\tilde{u}$ can directly be identified with the nilpotent element ϵ of the basis $\{1, \epsilon\}$ of the dual numbers.

- If $4a_0 > -a_1^2$, the above formula yields $\tilde{u}^2 > 0$. This leads to the split-complex numbers which have normalized basis $\{1, j\}$ with $j^2 = +1$. To obtain j from $\tilde{u}$, the latter must be divided by the positive real number $a := \sqrt{a_0 + \frac{a_1^2}{4}}$ which has the same square as $\tilde{u}$ has.

- If $4a_0 < -a_1^2$, the above formula yields $\tilde{u}^2 < 0$. This leads to the complex numbers which have normalized basis $\{1, i\}$ with $i^2 = -1$. To yield i from $\tilde{u}$, the latter has to be divided by a positive real number $a := \sqrt{\frac{a_1^2}{4} - a_0}$ which squares to the negative of $\tilde{u}^2$.

The complex numbers are the only two-dimensional hypercomplex algebra that is a field. Algebras such as the split-complex numbers that include non-real roots of 1 also contain idempotents $\frac{1}{2}(1 \pm j)$ and zero divisors $(1+j)(1-j) = 0$, so such algebras cannot be division algebras. However, these properties can turn out to be very meaningful, for instance in describing the Lorentz transformations of special relativity.

In a 2004 edition of Mathematics Magazine the two-dimensional real algebras have been styled the "generalized complex numbers".[6] The idea of cross-ratio of four complex numbers can be extended to the two-dimensional real algebras.[7]

16.3 Higher-dimensional examples (more than one non-real axis)

16.3.1 Clifford algebras

A Clifford algebra is the unital associative algebra generated over an underlying vector space equipped with a quadratic form. Over the real numbers this is equivalent to being able to define a symmetric scalar product, $u \cdot v = \frac{1}{2}(uv + vu)$ that can be used to orthogonalise the quadratic form, to give a set of bases $\{e_1, ..., ek\}$ such that:

$$\frac{1}{2}(e_i e_j + e_j e_i) = \begin{cases} -1, 0, +1 & i = j, \\ 0 & i \neq j. \end{cases}$$

Imposing closure under multiplication generates a multivector space spanned by a basis of 2^k elements, $\{1, e_1, e_2, e_3, ..., e_1e_2, ..., e_1e_2e_3, ...\}$. These can be interpreted as the basis of a hypercomplex number system. Unlike the basis $\{e_1, ..., ek\}$, the remaining basis elements may or may not anti-commute, depending on how many simple exchanges must be carried out to swap the two factors. So $e_1e_2 = -e_2e_1$, but $e_1(e_2e_3) = +(e_2e_3)e_1$.

Putting aside the bases for which $ei^2 = 0$ (i.e. directions in the original space over which the quadratic form was degenerate), the remaining Clifford algebras can be identified by the label $C\ell p,q(\mathbf{R})$, indicating that the algebra is constructed from p simple basis elements with $ei^2 = +1$, q with $ei^2 = -1$, and where $\mathbf{R}$ indicates that this is to be a Clifford algebra over the reals—i.e. coefficients of elements of the algebra are to be real numbers.

These algebras, called geometric algebras, form a systematic set, which turn out to be very useful in physics problems which involve rotations, phases, or spins, notably in classical and quantum mechanics, electromagnetic theory and relativity.

Examples include: the complex numbers $C\ell_{0,1}(\mathbf{R})$, split-complex numbers $C\ell_{1,0}(\mathbf{R})$, quaternions $C\ell_{0,2}(\mathbf{R})$, split-biquaternions $C\ell_{0,3}(\mathbf{R})$, coquaternions $C\ell_{1,1}(\mathbf{R}) \approx C\ell_{2,0}(\mathbf{R})$ (the natural algebra of two-dimensional space); $C\ell_{3,0}(\mathbf{R})$ (the natural algebra of three-dimensional space, and the algebra of the Pauli matrices); and the spacetime algebra $C\ell_{1,3}(\mathbf{R})$.

The elements of the algebra $C\ell p,q(\mathbf{R})$ form an even subalgebra $C\ell^0_{q+1,p}(\mathbf{R})$ of the algebra $C\ell_{q+1,p}(\mathbf{R})$, which can be used to parametrise rotations in the larger algebra. There is thus a close connection between complex numbers and rotations in two-dimentional space; between quaternions and rotations in three-dimensional space; between split-complex numbers and (hyperbolic) rotations (Lorentz transformations) in 1+1-dimensional space, and so on.

Whereas Cayley–Dickson and split-complex constructs with eight or more dimensions are not associative with respect to multiplication, Clifford algebras retain associativity at any number of dimensions.

In 1995 Ian R. Porteous wrote on "The recognition of subalgebras" in his book on Clifford algebras. His Proposition 11.4 summarizes the hypercomplex cases:[8]

> Let A be a real associative algebra with unit element 1. Then

- 1 generates $\mathbf{R}$ (algebra of real numbers),

- any two-dimensional subalgebra generated by an element e_0 of A such that $e_0{}^2 = -1$ is isomorphic to $\mathbf{C}$ (algebra of complex numbers),

- any two-dimensional subalgebra generated by an element e_0 of A such that $e_0{}^2 = 1$ is isomorphic to $^2\mathbf{R}$ (algebra of split-complex numbers),

- any four-dimensional subalgebra generated by a set $\{e_0, e_1\}$ of mutually anti-commuting elements of A such that $e_0^2 = e_1^2 = -1$ is isomorphic to $\mathbf{H}$ (algebra of quaternions),

- any four-dimensional subalgebra generated by a set $\{e_0, e_1\}$ of mutually anti-commuting elements of A such that $e_0^2 = e_1^2 = 1$ is isomorphic to $M_2(\mathbf{R})$ (2×2 real matrices, coquaternions),

- any eight-dimensional subalgebra generated by a set $\{e_0, e_1, e_2\}$ of mutually anti-commuting elements of A such that $e_0^2 = e_1^2 = e_2^2 = -1$ is isomorphic to $^2\mathbf{H}$ (split-biquaternions),

- any eight-dimensional subalgebra generated by a set $\{e_0, e_1, e_2\}$ of mutually anti-commuting elements of A such that $e_0^2 = e_1^2 = e_2^2 = 1$ is isomorphic to $M_2(\mathbf{C})$ (biquaternions, Pauli algebra, 2×2 complex matrices).

For extension beyond the classical algebras, see Classification of Clifford algebras.

16.3.2 Cayley–Dickson construction

For more details on this topic, see Cayley–Dickson construction.

All of the Clifford algebras $C\ell p,q(\mathbf{R})$ apart from the real numbers, complex numbers and the quaternions contain non-real elements that square to +1; and so cannot be division algebras. A different approach to extending the complex numbers is taken by the Cayley–Dickson construction. This generates number systems of dimension 2^n, n in $\{2, 3, 4, ...\}$, with bases $\{1, i_1, ..., i_{2^n-1}\}$, where all the non-real basis elements anti-commute and satisfy $i_m^2 = -1$. In 8 or more dimensions ($n \geq 3$) these algebras are non-associative. In 16 or more dimensions ($n \geq 4$) these algebras also have zero-divisors.

The first algebras in this sequence are the four-dimensional quaternions, eight-dimensional octonions, and 16-dimensional sedenions. An algebraic symmetry is lost with each increase in dimensionality: quaternion multiplication is not commutative, octonion multiplication is non-associative, and the norm of sedenions is not multiplicative.

The Cayley–Dickson construction can be modified by inserting an extra sign at some stages. It then generates two of the "split algebras" in the collection of composition algebras:

split-quaternions with basis $\{1, i_1, i_2, i_3\}$ satisfying $i_1^2 = -1, i_2^2 = i_3^2 = +1$,) and

split-octonions with basis $\{1, i_1, ..., i_7\}$ satisfying $i_1^2 = i_2^2 = i_3^2 = -1$, $i_4^2 = \cdots = i_7^2 = +1$.

As with quaternions, split-quaternions are not commutative, but further contain nilpotents; they are isomorphic to the 2 × 2 real matrices. Split-octonions are non-associative and contain nilpotents.

16.3.3 Tensor products

The tensor product of any two algebras is another algebra, which can be used to produce many more examples of hypercomplex number systems.

In particular taking tensor products with the complex numbers (considered as algebras over the reals) leads to four-dimensional tessarines $\mathbb{C} \otimes_{\mathbb{R}} \mathbb{C}$, eight-dimensional biquaternions $\mathbb{C} \otimes_{\mathbb{R}} \mathbb{H}$, and 16-dimensional complex octonions $\mathbb{C} \otimes_{\mathbb{R}} \mathbb{O}$.

16.3.4 Further examples

- bicomplex numbers: a 4D vector space over the reals, or 2D over the complex numbers
- multicomplex numbers: 2^{n-1}-dimensional vector spaces over the complex numbers
- composition algebra: algebras with a quadratic form that composes with the product

16.4 See also

- Thomas Kirkman
- Georg Scheffers
- Richard Brauer
- Hypercomplex analysis

16.5 Notes and references

[1] Linear Associative Algebra (1881) American Journal of Mathematics 4(1):221–6

[2] Adams, J. F. (July 1960). "On the Non-Existence of Elements of Hopf Invariant One". *Annals of Mathematics* **72** (1): 20–104.

[3] Isaak Yaglom (1968) *Complex Numbers in Geometry*, pages 10 to 14

[4] John H. Ewing editor (1991) *Numbers*, page 237, Springer, ISBN 3-540-97497-0

[5] Kantor & Solodovnikov (1978) 14,15

[6] Anthony A. Harkin & Joseph B. Harkin (2004) Geometry of Generalized Complex Numbers, Mathematics Magazine 77(2):118–29

[7] Sky Brewer (2013) "Projective Cross-ratio on Hypercomplex Numbers", Advances in Applied Clifford Algebras 23(1):1–14

[8] Ian R. Porteous (1995) *Clifford Algebras and the Classical Groups*, pages 88–89, Cambridge University Press ISBN 0-521-55177-3

- Daniel Alfsmann (2006) On families of 2^N dimensional hypercomplex algebras suitable for digital signal processing, 14th European Signal Processing Conference, Florence, Italy.

- Emil Artin (1928) "Zur Theorie der hyperkomplexen Zahlen" and "Zur Arithmetik hyperkomplexer Zahlen", in *The Collected Papers of* Emil Artin, Serge Lang and John T. Tate editors, pp 301–45, Addison-Wesley, 1965.

- Baez, John (2002), "The Octonions", *Bulletin of the American Mathematical Society* **39**: 145–205, doi:10.1090/S0273-0979-01-00934-X, ISSN 0002-9904

- Elie Cartan (1908) "Les systems de nombres complex et les groupes de transformations", *Encyclopédie des sciences mathématiques pures et appliquées* I 1. and *Ouvres Completes* T.2 pt. 1, pp 107–246.

- Thomas Hawkins (1972) "Hypercomplex numbers, Lie groups, and the creation of group representation theory", *Archive for History of Exact Sciences* 8:243–87.

- Kantor, I.L., Solodownikow (1978), Hyperkomplexe Zahlen, BSB B.G. Teubner Verlagsgesellschaft, Leipzig.

- Kantor, I. L.; Solodovnikov, A. S. (1989), *Hypercomplex numbers*, Berlin, New York: Springer-Verlag, ISBN 978-0-387-96980-0, MR 996029

- Jeanne La Duke (1983) "The study of linear associative algebras in the United States, 1870–1927", see pp. 147–159 of *Emmy Noether in Bryn Mawr* Bhama Srinivasan & Judith Sally editors, Springer Verlag.

- Theodor Molien (1893) "Über Systeme höher complexen Zahlen", *Mathematische Annalen* 41:83–156.

- Silviu Olariu (2002) *Complex Numbers in N Dimensions*, North-Holland Mathematics Studies #190, Elsevier ISBN 0-444-51123-7 .

- Karen Parshall (1985) "Wedderburn and the Structure of Algebras" *Archive for History of Exact Sciences* 32:223

- Irene Sabadini, Michael Shapiro & Frank Sommen, editors (2009) *Hypercomplex Analysis and Applications* Birkhauser   ISBN 978-3-7643-9892-7 .

- Eduard Study (1898) "Theorie der gemeinen und höhern komplexen Grössen", *Encyclopädie der mathematischen Wissenschaften* I A 4 147–83.

- Henry Taber (1904) "On Hypercomplex Number Systems", Transactions of the American Mathematical Society 5:509.

- B.L. van der Waerden (1985) *A History of Algebra*, Chapter 10: The discovery of algebras, Chapter 11: Structure of algebras, Springer, ISBN 3-540-13610X .

- Joseph Wedderburn (1908) "On Hypercomplex Numbers", *Proceedings of the London Mathematical Society* 6:77–118.

16.6 External links

- Hazewinkel, Michiel, ed. (2001), "Hypercomplex number", *Encyclopedia of Mathematics*, Springer, ISBN 978-1-55608-010-4

- History of the Hypercomplexes on hyperjeff.com

- Hypercomplex.info

- Weisstein, Eric W., "Hypercomplex number", *MathWorld*.

- E. Study, "On systems of complex numbers and their application to the theory of transformation groups" (English translation)

- G. Frobenius, "Theory of hypercomplex quantities" (English translation)

Chapter 17

Infinity

For other uses of "Infinity" and "Infinite", see Infinity (disambiguation).

Infinity (symbol: ∞) is an abstract concept describing something *without any limit* and is relevant in a number of fields,

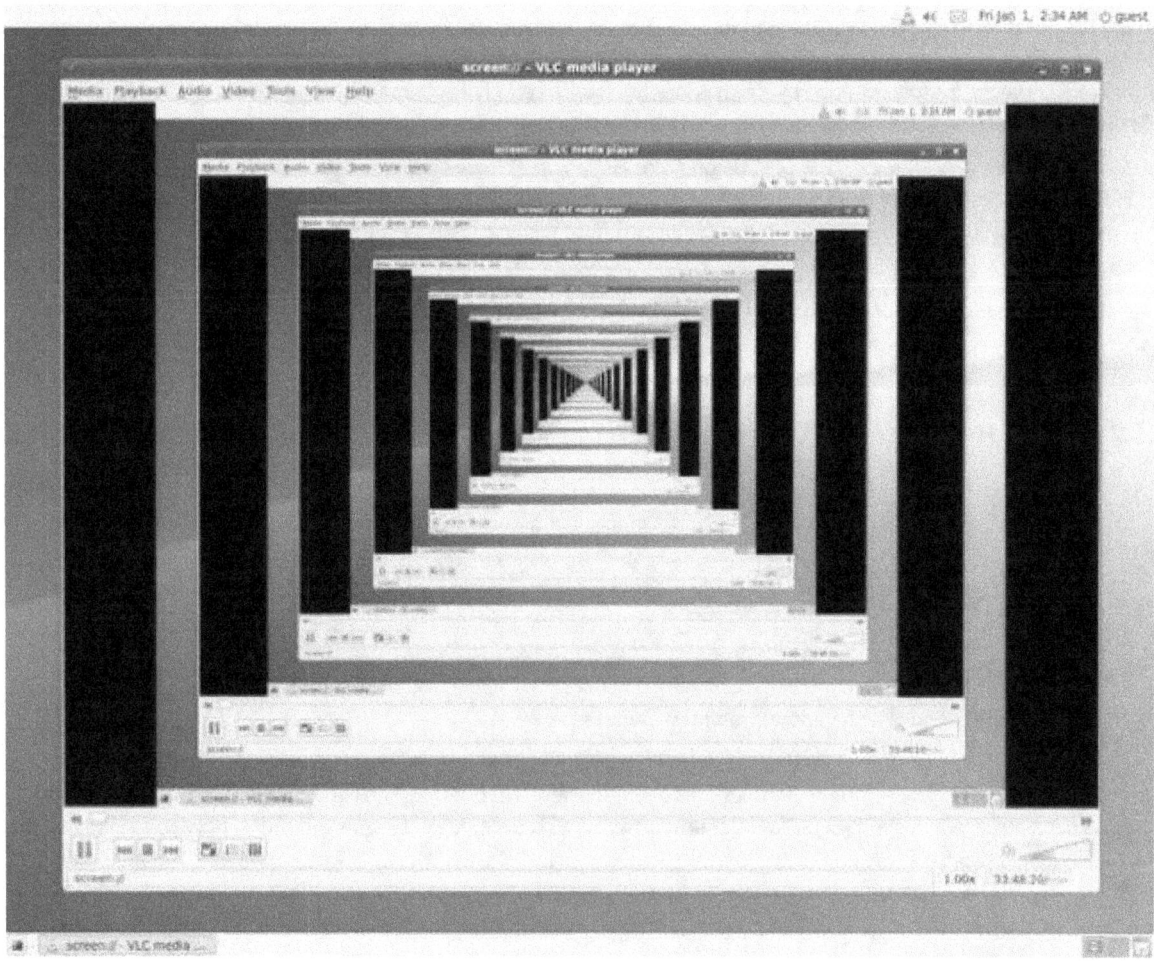

Infinity represented in screenshot form

predominantly mathematics and physics. In mathematics, "infinity" is often treated as if it were a number (i.e., it counts or measures things: "an infinite number of terms") but it is not the same sort of number as natural or real numbers.

In number systems incorporating infinitesimals, the reciprocal of an infinitesimal is an infinite number, i.e., a number greater than any real number; see $1/\infty$.

Georg Cantor formalized many ideas related to infinity and infinite sets during the late 19th and early 20th centuries. In the theory he developed, there are infinite sets of different sizes (called cardinalities).[1] For example, the set of integers is countably infinite, while the infinite set of real numbers is uncountable.[2]

17.1 History

Main article: Infinity (philosophy)

Ancient cultures had various ideas about the nature of infinity. The ancient Indians and Greeks did not define infinity in precise formalism as does modern mathematics, and instead approached infinity as a philosophical concept.

17.1.1 Early Greek

The earliest recorded idea of infinity comes from Anaximander, a pre-Socratic Greek philosopher who lived in Miletus. He used the word apeiron which means infinite or limitless.[3] However, the earliest attestable accounts of mathematical infinity come from Zeno of Elea (c. 490 BCE? – c. 430 BCE?), a pre-Socratic Greek philosopher of southern Italy and member of the Eleatic School founded by Parmenides. Aristotle called him the inventor of the dialectic. He is best known for his paradoxes, described by Bertrand Russell as "immeasurably subtle and profound".

In accordance with the traditional view of Aristotle, the Hellenistic Greeks generally preferred to distinguish the potential infinity from the actual infinity; for example, instead of saying that there are an infinity of primes, Euclid prefers instead to say that there are more prime numbers than contained in any given collection of prime numbers (Elements, Book IX, Proposition 20).

However, recent readings of the Archimedes Palimpsest have hinted that Archimedes at least had an intuition about actual infinite quantities.

17.1.2 Early Indian

The Indian mathematical text Surya Prajnapti (c. 4th–3rd century BCE) classifies all numbers into three sets: enumerable, innumerable, and infinite. Each of these was further subdivided into three orders:

- Enumerable: lowest, intermediate, and highest
- Innumerable: nearly innumerable, truly innumerable, and innumerably innumerable
- Infinite: nearly infinite, truly infinite, infinitely infinite

In this work, two basic types of infinite numbers are distinguished. On both physical and ontological grounds, a distinction was made between *asaṃkhyāta* ("countless, innumerable") and *ananta* ("endless, unlimited"), between rigidly bounded and loosely bounded infinities.

17.1.3 17th century

European mathematicians started using infinite numbers in a systematic fashion in the 17th century. John Wallis first used the notation ∞ for such a number, and exploited it in area calculations by dividing the region into infinitesimal strips of width on the order of $\frac{1}{\infty}$. Euler used the notation i for an infinite number, and exploited it by applying the binomial formula to the i 'th power, and infinite products of i factors.

17.2   Mathematics

17.2.1   Infinity symbol

Main article: Infinity symbol

The infinity symbol ∞ (sometimes called the lemniscate) is a mathematical symbol representing the concept of infinity. The symbol is encoded in Unicode at U+221E ∞ infinity (HTML ∞ · ∞) and in LaTeX as \infty.

It was introduced in 1655 by John Wallis,[4][5] and, since its introduction, has also been used outside mathematics in modern mysticism[6] and literary symbology.[7]

17.2.2   Calculus

Leibniz, one of the co-inventors of infinitesimal calculus, speculated widely about infinite numbers and their use in mathematics. To Leibniz, both infinitesimals and infinite quantities were ideal entities, not of the same nature as appreciable quantities, but enjoying the same properties in accordance with the Law of Continuity.[8][9]

Real analysis

In real analysis, the symbol ∞ , called "infinity", is used to denote an unbounded limit.[10] $x \to \infty$ means that x grows without bound, and $x \to -\infty$ means the value of x is decreasing without bound. If $f(t) \geq 0$ for every t, then[11]

- $\int_a^b f(t)\, dt = \infty$ means that $f(t)$ does not bound a finite area from a to b

- $\int_{-\infty}^{\infty} f(t)\, dt = \infty$ means that the area under $f(t)$ is infinite.

- $\int_{-\infty}^{\infty} f(t)\, dt = a$ means that the total area under $f(t)$ is finite, and equals a

Infinity is also used to describe infinite series:

- $\sum_{i=0}^{\infty} f(i) = a$ means that the sum of the infinite series converges to some real value a .

- $\sum_{i=0}^{\infty} f(i) = \infty$ means that the sum of the infinite series diverges in the specific sense that the partial sums grow without bound.

Infinity can be used not only to define a limit but as a value in the extended real number system. Points labeled $+\infty$ and $-\infty$ can be added to the topological space of the real numbers, producing the two-point compactification of the real numbers. Adding algebraic properties to this gives us the extended real numbers.[12] We can also treat $+\infty$ and $-\infty$ as the same, leading to the one-point compactification of the real numbers, which is the real projective line.[13] Projective geometry also refers to a line at infinity in plane geometry, a plane at infinity in three-dimensional space, and so forth for higher dimensions.

Complex analysis

In complex analysis the symbol ∞ , called "infinity", denotes an unsigned infinite limit. $x \to \infty$ means that the magnitude $|x|$ of x grows beyond any assigned value. A point labeled ∞ can be added to the complex plane as a topological space giving the one-point compactification of the complex plane. When this is done, the resulting space is a one-dimensional complex manifold, or Riemann surface, called the extended complex plane or the Riemann sphere. Arithmetic operations similar to those given above for the extended real numbers can also be defined, though there is no distinction in the signs (therefore one exception is that infinity cannot be added to itself). On the other hand, this kind of infinity enables division by zero, namely $z/0 = \infty$ for any nonzero complex number z. In this context it is often useful to consider meromorphic

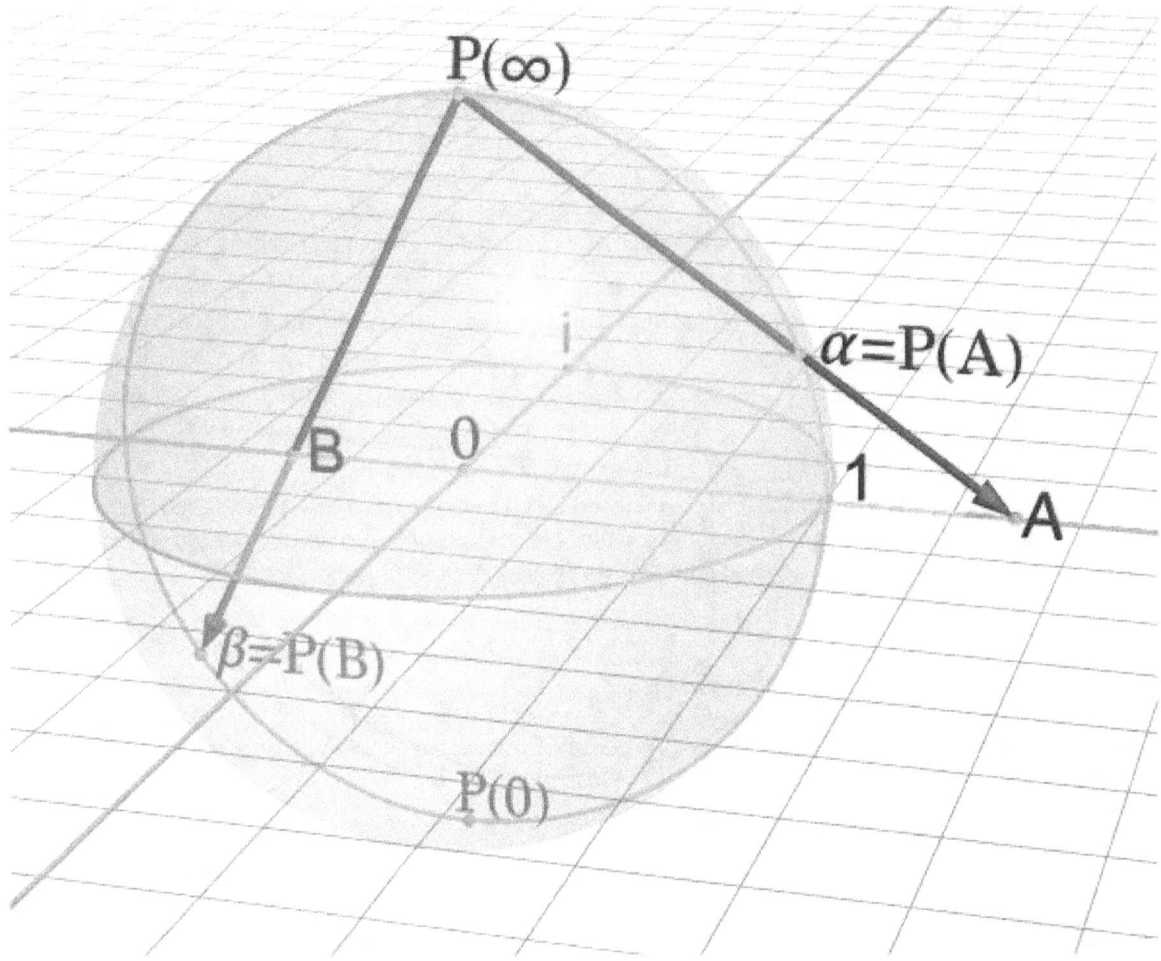

By stereographic projection, the complex plane can be "wrapped" onto a sphere, with the top point of the sphere corresponding to infinity. This is the called the Riemann sphere.

functions as maps into the Riemann sphere taking the value of ∞ at the poles. The domain of a complex-valued function may be extended to include the point at infinity as well. One important example of such functions is the group of Möbius transformations.

17.2.3 Nonstandard analysis

The original formulation of infinitesimal calculus by Isaac Newton and Gottfried Leibniz used infinitesimal quantities. In the twentieth century, it was shown that this treatment could be put on a rigorous footing through various logical systems, including smooth infinitesimal analysis and nonstandard analysis. In the latter, infinitesimals are invertible, and their inverses are infinite numbers. The infinities in this sense are part of a hyperreal field; there is no equivalence between them as with the Cantorian transfinites. For example, if H is an infinite number, then H + H = 2H and H + 1 are distinct infinite numbers. This approach to non-standard calculus is fully developed in Keisler (1986).

17.2.4 Set theory

Main articles: Cardinality and Ordinal number
 A different form of "infinity" are the ordinal and cardinal infinities of set theory. Georg Cantor developed a system of transfinite numbers, in which the first transfinite cardinal is aleph-null ($\aleph_0$), the cardinality of the set of natural numbers.

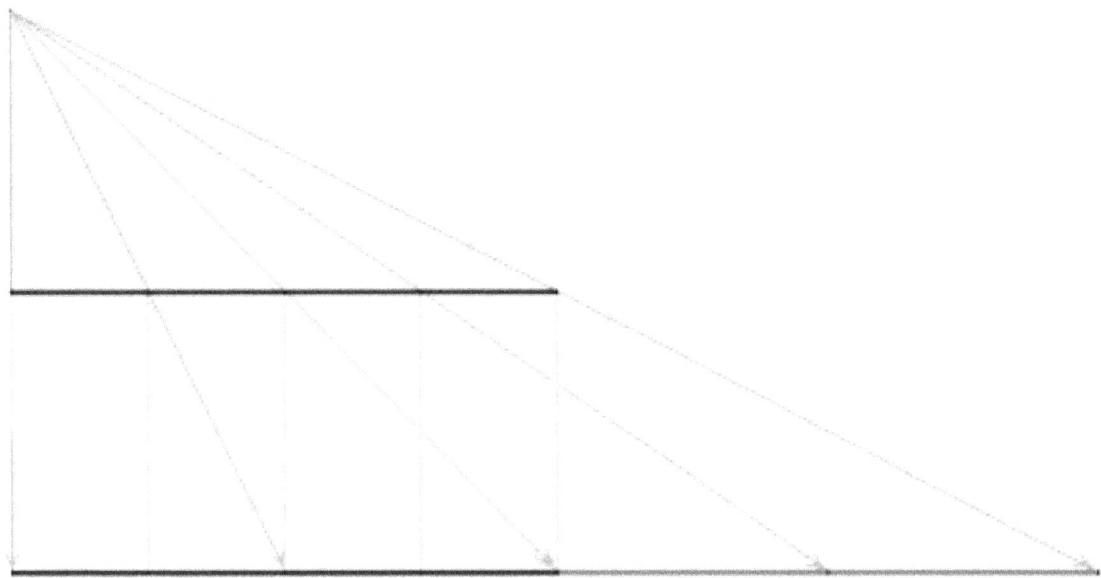

One-to-one correspondence between infinite set and proper subset

This modern mathematical conception of the quantitative infinite developed in the late nineteenth century from work by Cantor, Gottlob Frege, Richard Dedekind and others, using the idea of collections, or sets.

Dedekind's approach was essentially to adopt the idea of one-to-one correspondence as a standard for comparing the size of sets, and to reject the view of Galileo (which derived from Euclid) that the whole cannot be the same size as the part (however, see Galileo's paradox where he concludes that positive integers which are squares and all positive integers are the same size). An infinite set can simply be defined as one having the same size as at least one of its proper parts; this notion of infinity is called Dedekind infinite. The diagram gives an example: viewing lines as infinite sets of points, the left half of the lower blue line can be mapped in a one-to-one manner (green correspondences) to the higher blue line, and, in turn, to the whole lower blue line (red correspondences); therefore the whole lower blue line and its left half have the same cardinality, i.e. "size".

Cantor defined two kinds of infinite numbers: ordinal numbers and cardinal numbers. Ordinal numbers may be identified with well-ordered sets, or counting carried on to any stopping point, including points after an infinite number have already been counted. Generalizing finite and the ordinary infinite sequences which are maps from the positive integers leads to mappings from ordinal numbers, and transfinite sequences. Cardinal numbers define the size of sets, meaning how many members they contain, and can be standardized by choosing the first ordinal number of a certain size to represent the cardinal number of that size. The smallest ordinal infinity is that of the positive integers, and any set which has the cardinality of the integers is countably infinite. If a set is too large to be put in one to one correspondence with the positive integers, it is called *uncountable*. Cantor's views prevailed and modern mathematics accepts actual infinity.[14] Certain extended number systems, such as the hyperreal numbers, incorporate the ordinary (finite) numbers and infinite numbers of different sizes.

Cardinality of the continuum

Main article: Cardinality of the continuum

One of Cantor's most important results was that the cardinality of the continuum c is greater than that of the natural numbers $\aleph_0$; that is, there are more real numbers **R** than natural numbers **N**. Namely, Cantor showed that $c = 2^{\aleph_0} > \aleph_0$ (see Cantor's diagonal argument or Cantor's first uncountability proof).

The continuum hypothesis states that there is no cardinal number between the cardinality of the reals and the cardinality of the natural numbers, that is, $\mathfrak{c} = \aleph_1 = \beth_1$ (see Beth one). However, this hypothesis can neither be proved nor disproved within the widely accepted Zermelo–Fraenkel set theory, even assuming the Axiom of Choice.

Cardinal arithmetic can be used to show not only that the number of points in a real number line is equal to the number of points in any segment of that line, but that this is equal to the number of points on a plane and, indeed, in any finite-dimensional space.

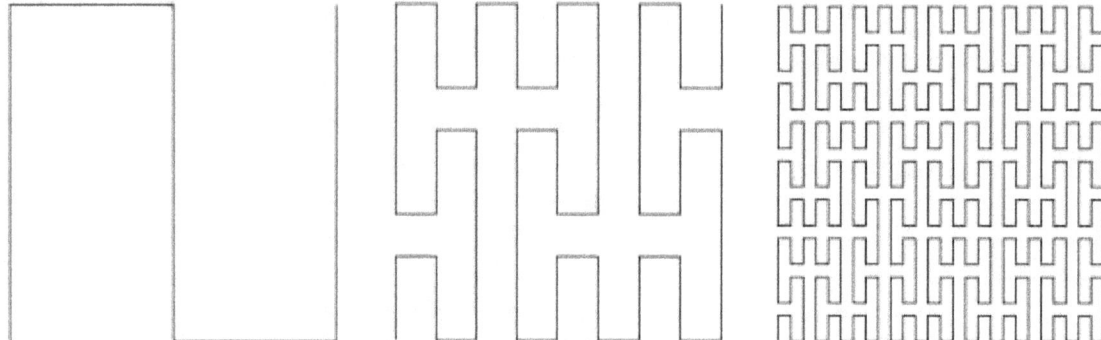

The first three steps of a fractal construction whose limit is a space-filling curve, showing that there are as many points in a one-dimensional line as in a two-dimensional square.

The first of these results is apparent by considering, for instance, the tangent function, which provides a one-to-one correspondence between the interval $(-\pi/2, \pi/2)$ and **R** (see also Hilbert's paradox of the Grand Hotel). The second result was proved by Cantor in 1878, but only became intuitively apparent in 1890, when Giuseppe Peano introduced the space-filling curves, curved lines that twist and turn enough to fill the whole of any square, or cube, or hypercube, or finite-dimensional space. These curves can be used to define a one-to-one correspondence between the points in the side of a square and those in the square.

17.2.5 Geometry and topology

Main article: Dimension (vector space)

Infinite-dimensional spaces are widely used in geometry and topology, particularly as classifying spaces, notably Eilenberg–MacLane spaces. Common examples are the infinite-dimensional complex projective space K(Z,2) and the infinite-dimensional real projective space K(Z/2Z,1).

17.2.6 Fractals

The structure of a fractal object is reiterated in its magnifications. Fractals can be magnified indefinitely without losing their structure and becoming "smooth"; they have infinite perimeters—some with infinite, and others with finite surface areas. One such fractal curve with an infinite perimeter and finite surface area is the Koch snowflake.

17.2.7 Mathematics without infinity

Leopold Kronecker was skeptical of the notion of infinity and how his fellow mathematicians were using it in the 1870s and 1880s. This skepticism was developed in the philosophy of mathematics called finitism, an extreme form of the philosophical and mathematical schools of constructivism and intuitionism.[15]

17.3 Physics

In physics, approximations of real numbers are used for continuous measurements and natural numbers are used for discrete measurements (i.e. counting). It is therefore assumed by physicists that no measurable quantity could have an infinite value, for instance by taking an infinite value in an extended real number system, or by requiring the counting of an infinite number of events. It is, for example, presumed impossible for any type of body to have infinite mass or infinite energy. Concepts of infinite things such as an infinite plane wave exist, but there are no experimental means to generate them.[16]

17.3.1 Theoretical applications of physical infinity

The practice of refusing infinite values for measurable quantities does not come from *a priori* or ideological motivations, but rather from more methodological and pragmatic motivations. One of the needs of any physical and scientific theory is to give usable formulas that correspond to or at least approximate reality. As an example if any object of infinite gravitational mass were to exist, any usage of the formula to calculate the gravitational force would lead to an infinite result, which would be of no benefit since the result would be always the same regardless of the position and the mass of the other object. The formula would be useful neither to compute the force between two objects of finite mass nor to compute their motions. If an infinite mass object were to exist, any object of finite mass would be attracted with infinite force (and hence acceleration) by the infinite mass object, which is not what we can observe in reality. Sometimes infinite result of a physical quantity may mean that the theory being used to compute the result may be approaching the point where it fails. This may help to indicate the limitations of a theory.

This point of view does not mean that infinity cannot be used in physics. For convenience's sake, calculations, equations, theories and approximations often use infinite series, unbounded functions, etc., and may involve infinite quantities. Physicists however require that the end result be physically meaningful. In quantum field theory infinities arise which need to be interpreted in such a way as to lead to a physically meaningful result, a process called renormalization.

However, there are some theoretical circumstances where the end result is infinity. One example is the singularity in the description of black holes. Some solutions of the equations of the general theory of relativity allow for finite mass distributions of zero size, and thus infinite density. This is an example of what is called a mathematical singularity, or a point where a physical theory breaks down. This does not necessarily mean that physical infinities exist; it may mean simply that the theory is incapable of describing the situation properly. Two other examples occur in inverse-square force laws of the gravitational force equation of Newtonian gravity and Coulomb's law of electrostatics. At r=0 these equations evaluate to infinities.

17.3.2 Cosmology

The first published proposal that the universe is infinite came from Thomas Digges in 1576.[17] Eight years later, in 1584, the Italian philosopher and astronomer Giordano Bruno proposed an unbounded universe in *On the Infinite Universe and Worlds*: "Innumerable suns exist; innumerable earths revolve around these suns in a manner similar to the way the seven planets revolve around our sun. Living beings inhabit these worlds."

Cosmologists have long sought to discover whether infinity exists in our physical universe: Are there an infinite number of stars? Does the universe have infinite volume? Does space "go on forever"? This is an open question of cosmology. Note that the question of being infinite is logically separate from the question of having boundaries. The two-dimensional surface of the Earth, for example, is finite, yet has no edge. By travelling in a straight line with respect to the Earth's curvature one will eventually return to the exact spot one started from. The universe, at least in principle, might have a similar topology. If so, one might eventually return to one's starting point after travelling in a straight line through the universe for long enough.

If, on the other hand, the universe were not curved like a sphere but had a flat topology, it could be both unbounded and infinite. The curvature of the universe can be measured through multipole moments in the spectrum of the cosmic background radiation. As to date, analysis of the radiation patterns recorded by the WMAP spacecraft hints that the universe has a flat topology. This would be consistent with an infinite physical universe.

However, the universe could also be finite, even if its curvature is flat. An easy way to understand this is to consider two-dimensional examples, such as video games where items that leave one edge of the screen reappear on the other. The topology of such games is toroidal and the geometry is flat. Many possible bounded, flat possibilities also exist for three-dimensional space.[18]

The concept of infinity also extends to the multiverse hypothesis, which, when explained by astrophysicists such as Michio Kaku, posits that there are an infinite number and variety of universes.[19]

17.4 Logic

In logic an infinite regress argument is "a distinctively philosophical kind of argument purporting to show that a thesis is defective because it generates an infinite series when either (form A) no such series exists or (form B) were it to exist, the thesis would lack the role (e.g., of justification) that it is supposed to play."[20]

17.5 Computing

The IEEE floating-point standard (IEEE 754) specifies the positive and negative infinity values. These are defined as the result of arithmetic overflow, division by zero, and other exceptional operations.

Some programming languages, such as Java[21] and J,[22] allow the programmer an explicit access to the positive and negative infinity values as language constants. These can be used as greatest and least elements, as they compare (respectively) greater than or less than all other values. They have uses as sentinel values in algorithms involving sorting, searching, or windowing.

In languages that do not have greatest and least elements, but do allow overloading of relational operators, it is possible for a programmer to *create* the greatest and least elements. In languages that do not provide explicit access to such values from the initial state of the program, but do implement the floating-point data type, the infinity values may still be accessible and usable as the result of certain operations.

17.6 Arts and cognitive sciences

Perspective artwork utilizes the concept of imaginary vanishing points, or points at infinity, located at an infinite distance from the observer. This allows artists to create paintings that realistically render space, distances, and forms.[23] Artist M. C. Escher is specifically known for employing the concept of infinity in his work in this and other ways.

Cognitive scientist George Lakoff considers the concept of infinity in mathematics and the sciences as a metaphor. This perspective is based on the basic metaphor of infinity (BMI), defined as the ever-increasing sequence <1,2,3,...>.

The symbol is often used romantically to represent eternal love. Several types of jewelry are fashioned into the infinity shape for this purpose.

17.7 See also

- 0.999...

- Aleph number

- Exponentiation

- Indeterminate form

- Infinite monkey theorem

- Infinite set

- Paradoxes of infinity

- Surreal number

17.8   Notes

[1] Gowers, Timothy; Barrow-Green, June; Leader, Imre (2008). *The Princeton Companion to Mathematics*. Princeton University Press. p. 616. ISBN 0-691-11880-9., Extract of page 616

[2] Maddox 2002, pp. 113 –117

[3] Wallace 2004, pg. 44

[4] Scott, Joseph Frederick (1981), *The mathematical work of John Wallis, D.D., F.R.S., (1616–1703)* (2 ed.), American Mathematical Society, p. 24, ISBN 0-8284-0314-7.

[5] Martin-Löf, Per (1990), "Mathematics of infinity", *COLOG-88 (Tallinn, 1988)*, Lecture Notes in Computer Science **417**, Berlin: Springer, pp. 146–197, doi:10.1007/3-540-52335-9_54, MR 1064143.

[6] O'Flaherty, Wendy Doniger (1986), *Dreams, Illusion, and Other Realities*, University of Chicago Press, p. 243, ISBN 9780226618555.

[7] Toker, Leona (1989), *Nabokov: The Mystery of Literary Structures*, Cornell University Press, p. 159, ISBN 9780801422119.

[8] Continuity and Infinitesimals entry by John Lane Bell in the *Stanford Encyclopedia of Philosophy*

[9] Jesseph, Douglas Michael (1998). "Leibniz on the Foundations of the Calculus: The Question of the Reality of Infinitesimal Magnitudes". *Perspectives on Science* **6** (1&2): 6–40. ISSN 1063-6145. OCLC 42413222. Archived from the original on 16 February 2010. Retrieved 16 February 2010.

[10] Taylor 1955, p. 63

[11] These uses of infinity for integrals and series can be found in any standard calculus text, such as, Swokoski 1983, pp. 468-510

[12] Aliprantis, Charalambos D.; Burkinshaw, Owen (1998), *Principles of Real Analysis* (3rd ed.), San Diego, CA: Academic Press, Inc., p. 29, ISBN 0-12-050257-7, MR 1669668.

[13] Gemignani 1990, p. 177

[14] Moore, A. W. (1991). *The Infinite*. Routledge.

[15] Kline, Morris (1972). *Mathematical Thought from Ancient to Modern Times*. New York: Oxford University Press. pp. 1197–1198. ISBN 0-19-506135-7.

[16] Doric Lenses - Application Note - Axicons - 2. Intensity Distribution. Retrieved 7 April 2014.

[17] John Gribbin (2009), *In Search of the Multiverse: Parallel Worlds, Hidden Dimensions, and the Ultimate Quest for the Frontiers of Reality*, ISBN 9780470613528. p. 88

[18] Weeks, Jeffrey (December 12, 2001). *The Shape of Space*. CRC Press. ISBN 978-0824707095.

[19] Kaku, M. (2006). Parallel worlds. Knopf Doubleday Publishing Group.

[20] *Cambridge Dictionary of Philosophy*, Second Edition, p. 429

[21] Gosling, James; et. al. (27 July 2012). "4.2.3.". *The Java™ Language Specification* (Java SE 7 ed.). California, U.S.A.: Oracle America, Inc. Retrieved 6 September 2012.

[22] Stokes, Roger (July 2012). "19.2.1". *Learning J*. Retrieved 6 September 2012.

[23] Kline, Morris (1985). *Mathematics for the nonmathematician*. Courier Dover Publications. p. 229. ISBN 0-486-24823-2., Section 10-7, p. 229

17.9 References

- Gemignani, Michael C. (1990), *Elementary Topology* (2nd ed.), Dover, ISBN 0-486-66522-4

- Keisler, H. Jerome (1986), *Elementary Calculus: An Approach Using Infinitesimals* (2nd ed.)

- Maddox, Randall B. (2002), *Mathematical Thinking and Writing: A Transition to Abstract Mathematics*, Academic Press, ISBN 0-12-464976-9

- Swokowski, Earl W. (1983), *Calculus with Analytic Geometry* (Alternate ed.), Prindle, Weber & Schmidt, ISBN 0-87150-341-7

- Taylor, Angus E. (1955), *Advanced Calculus*, Blaisdell Publishing Company

- David Foster Wallace (2004). *Everything and More: A Compact History of Infinity*. Norton, W. W. & Company, Inc. ISBN 0-393-32629-2.

17.10 Further reading

- Amir D. Aczel (2001). *The Mystery of the Aleph: Mathematics, the Kabbalah, and the Search for Infinity*. New York: Pocket Books. ISBN 0-7434-2299-6.

- D. P. Agrawal (2000). *Ancient Jaina Mathematics: an Introduction*, Infinity Foundation.

- Bell, J. L.: Continuity and infinitesimals. Stanford Encyclopedia of philosophy. Revised 2009.

- L. C. Jain (1982). *Exact Sciences from Jaina Sources*.

- L. C. Jain (1973). "Set theory in the Jaina school of mathematics", *Indian Journal of History of Science*.

- George G. Joseph (2000). *The Crest of the Peacock: Non-European Roots of Mathematics* (2nd ed.). Penguin Books. ISBN 0-14-027778-1.

- Eli Maor (1991). *To Infinity and Beyond*. Princeton University Press. ISBN 0-691-02511-8.

- Rudy Rucker (1995). *Infinity and the Mind: The Science and Philosophy of the Infinite*. Princeton University Press. ISBN 0-691-00172-3.

- Navjyoti Singh (1988). *Jaina Theory of Actual Infinity and Transfinite Numbers. Journal of Asiatic Society* **30**.

17.11 External links

- The Infinite entry in the *Internet Encyclopedia of Philosophy*

-

- Infinity on *In Our Time* at the BBC. (listen now)

- *A Crash Course in the Mathematics of Infinite Sets*, by Peter Suber. From the St. John's Review, XLIV, 2 (1998) 1–59. The stand-alone appendix to *Infinite Reflections*, below. A concise introduction to Cantor's mathematics of infinite sets.

- *Infinite Reflections*, by Peter Suber. How Cantor's mathematics of the infinite solves a handful of ancient philosophical problems of the infinite. From the St. John's Review, XLIV, 2 (1998) 1–59.

- Grime, James. "Infinity is bigger than you think". *Numberphile*. Brady Haran.

- *Infinity*, Principia Cybernetica

- Hotel Infinity

- John J. O'Connor and Edmund F. Robertson (1998). 'Georg Ferdinand Ludwig Philipp Cantor', *MacTutor History of Mathematics archive*.

- John J. O'Connor and Edmund F. Robertson (2000). 'Jaina mathematics', *MacTutor History of Mathematics archive*.

- Ian Pearce (2002). 'Jainism', *MacTutor History of Mathematics archive*.

- Source page on medieval and modern writing on Infinity

- The Mystery Of The Aleph: Mathematics, the Kabbalah, and the Search for Infinity

- Dictionary of the Infinite (compilation of articles about infinity in physics, mathematics, and philosophy)

Chapter 18

Abstract algebra

This article is about the branch of mathematics. For the Swedish band, see Abstrakt Algebra.

"Modern algebra" redirects here. For van der Waerden's book, see Moderne Algebra.

In algebra, which is a broad division of mathematics, **abstract algebra** (occasionally called **modern algebra**) is the study of algebraic structures. Algebraic structures include groups, rings, fields, modules, vector spaces, lattices, and algebra over a field. The term *abstract algebra* was coined in the early 20th century to distinguish this area of study from the other parts of algebra.

Algebraic structures, with their associated homomorphisms, form mathematical categories. Category theory is a powerful formalism for analyzing and comparing different algebraic structures.

Universal algebra is a related subject that studies the nature and theories of various types of algebraic structures as a whole. For example, universal algebra studies the overall theory of groups, as distinguished from studying particular groups.

18.1 History

As in other parts of mathematics, concrete problems and examples have played important roles in the development of abstract algebra. Through the end of the nineteenth century, many -- perhaps most -- of these problems were in some way related to the theory of algebraic equations. Major themes include:

- Solving of systems of linear equations, which led to linear algebra

- Attempts to find formulae for solutions of general polynomial equations of higher degree that resulted in discovery of groups as abstract manifestations of symmetry

- Arithmetical investigations of quadratic and higher degree forms and diophantine equations, that directly produced the notions of a ring and ideal.

Numerous textbooks in abstract algebra start with axiomatic definitions of various algebraic structures and then proceed to establish their properties. This creates a false impression that in algebra axioms had come first and then served as a motivation and as a basis of further study. The true order of historical development was almost exactly the opposite. For example, the hypercomplex numbers of the nineteenth century had kinematic and physical motivations but challenged comprehension. Most theories that are now recognized as parts of algebra started as collections of disparate facts from various branches of mathematics, acquired a common theme that served as a core around which various results were grouped, and finally became unified on a basis of a common set of concepts. An archetypical example of this progressive synthesis can be seen in the history of group theory.

The permutations of Rubik's Cube have a group structure; the group is a fundamental concept within abstract algebra.

18.1.1   Early group theory

There were several threads in the early development of group theory, in modern language loosely corresponding to *number theory*, *theory of equations*, and *geometry*.

Leonhard Euler considered algebraic operations on numbers modulo an integer, modular arithmetic, in his generalization of Fermat's little theorem. These investigations were taken much further by Carl Friedrich Gauss, who considered the structure of multiplicative groups of residues mod n and established many properties of cyclic and more general abelian groups that arise in this way. In his investigations of composition of binary quadratic forms, Gauss explicitly stated the associative law for the composition of forms, but like Euler before him, he seems to have been more interested in concrete results than in general theory. In 1870, Leopold Kronecker gave a definition of an abelian group in the context of ideal class groups of a number field, generalizing Gauss's work; but it appears he did not tie his definition with previous work on groups, particularly permutation groups. In 1882, considering the same question, Heinrich M. Weber realized the connection and gave a similar definition that involved the cancellation property but omitted the existence of the inverse element, which was sufficient in his context (finite groups).

Permutations were studied by Joseph-Louis Lagrange in his 1770 paper *Réflexions sur la résolution algébrique des équations (Thoughts on the algebraic solution of equations)* devoted to solutions of algebraic equations, in which he introduced Lagrange resolvents. Lagrange's goal was to understand why equations of third and fourth degree admit formulae for solutions, and he identified as key objects permutations of the roots. An important novel step taken by Lagrange in this paper was the abstract view of the roots, i.e. as symbols and not as numbers. However, he did not consider composition of permutations. Serendipitously, the first edition of Edward Waring's *Meditationes Algebraicae (Meditations on Algebra)* appeared in the same year, with an expanded version published in 1782. Waring proved the main theorem on symmetric functions, and specially considered the relation between the roots of a quartic equation and its resolvent cubic. *Mémoire sur la résolution des équations (Memoire on the Solving of Equations)* of Alexandre Vandermonde (1771) developed the theory of symmetric functions from a slightly different angle, but like Lagrange, with the goal of understanding solvability of algebraic equations.

> *Kronecker claimed in 1888 that the study of modern algebra began with this first paper of Vandermonde. Cauchy states quite clearly that Vandermonde had priority over Lagrange for this remarkable idea, which eventually led to the study of group theory.*[1]

Paolo Ruffini was the first person to develop the theory of permutation groups, and like his predecessors, also in the context of solving algebraic equations. His goal was to establish the impossibility of an algebraic solution to a general algebraic equation of degree greater than four. En route to this goal he introduced the notion of the order of an element of a group, conjugacy, the cycle decomposition of elements of permutation groups and the notions of primitive and imprimitive and proved some important theorems relating these concepts, such as

> *if G is a subgroup of S_5 whose order is divisible by 5 then G contains an element of order 5.*

Note, however, that he got by without formalizing the concept of a group, or even of a permutation group. The next step was taken by Évariste Galois in 1832, although his work remained unpublished until 1846, when he considered for the first time what is now called the *closure property* of a group of permutations, which he expressed as

> ... if in such a group one has the substitutions S and T then one has the substitution ST.

The theory of permutation groups received further far-reaching development in the hands of Augustin Cauchy and Camille Jordan, both through introduction of new concepts and, primarily, a great wealth of results about special classes of permutation groups and even some general theorems. Among other things, Jordan defined a notion of isomorphism, still in the context of permutation groups and, incidentally, it was he who put the term *group* in wide use.

The abstract notion of a group appeared for the first time in Arthur Cayley's papers in 1854. Cayley realized that a group need not be a permutation group (or even *finite*), and may instead consist of matrices, whose algebraic properties, such as multiplication and inverses, he systematically investigated in succeeding years. Much later Cayley would revisit the question whether abstract groups were more general than permutation groups, and establish that, in fact, any group is isomorphic to a group of permutations.

18.1.2 Modern algebra

The end of the 19th and the beginning of the 20th century saw a tremendous shift in the methodology of mathematics. Abstract algebra emerged around the start of the 20th century, under the name *modern algebra*. Its study was part of the drive for more intellectual rigor in mathematics. Initially, the assumptions in classical algebra, on which the whole of mathematics (and major parts of the natural sciences) depend, took the form of axiomatic systems. No longer satisfied with establishing properties of concrete objects, mathematicians started to turn their attention to general theory. Formal definitions of certain algebraic structures began to emerge in the 19th century. For example, results about various groups of permutations came to be seen as instances of general theorems that concern a general notion of an *abstract group*. Questions of structure and classification of various mathematical objects came to forefront.

These processes were occurring throughout all of mathematics, but became especially pronounced in algebra. Formal definition through primitive operations and axioms were proposed for many basic algebraic structures, such as groups, rings, and fields. Hence such things as group theory and ring theory took their places in pure mathematics. The algebraic investigations of general fields by Ernst Steinitz and of commutative and then general rings by David Hilbert, Emil Artin and Emmy Noether, building up on the work of Ernst Kummer, Leopold Kronecker and Richard Dedekind, who had considered ideals in commutative rings, and of Georg Frobenius and Issai Schur, concerning representation theory of groups, came to define abstract algebra. These developments of the last quarter of the 19th century and the first quarter of 20th century were systematically exposed in Bartel van der Waerden's *Moderne algebra*, the two-volume monograph published in 1930–1931 that forever changed for the mathematical world the meaning of the word *algebra* from *the theory of equations* to the *theory of algebraic structures*.

18.2   Basic concepts

Main article: Algebraic structures

By abstracting away various amounts of detail, mathematicians have created theories of various algebraic structures that apply to many objects. For instance, almost all systems studied are sets, to which the theorems of set theory apply. Those sets that have a certain binary operation defined on them form magmas, to which the concepts concerning magmas, as well those concerning sets, apply. We can add additional constraints on the algebraic structure, such as associativity (to form semigroups); identity, and inverses (to form groups); and other more complex structures. With additional structure, more theorems could be proved, but the generality is reduced. The "hierarchy" of algebraic objects (in terms of generality) creates a hierarchy of the corresponding theories: for instance, the theorems of group theory apply to rings (algebraic objects that have two binary operations with certain axioms) since a ring is a group over one of its operations. Mathematicians choose a balance between the amount of generality and the richness of the theory.

Examples of algebraic structures with a single binary operation are:

- Magmas
- Quasigroups
- Monoids
- Semigroups
- Groups

More complicated examples include:

- Rings
- Fields
- Modules
- Vector spaces
- Algebras over fields
- Associative algebras
- Lie algebras
- Lattices
- Boolean algebras

18.3 Applications

Because of its generality, abstract algebra is used in many fields of mathematics and science. For instance, algebraic topology uses algebraic objects to study topologies. The recently (As of 2006) proved Poincaré conjecture asserts that the fundamental group of a manifold, which encodes information about connectedness, can be used to determine whether a manifold is a sphere or not. Algebraic number theory studies various number rings that generalize the set of integers. Using tools of algebraic number theory, Andrew Wiles proved Fermat's Last Theorem.

In physics, groups are used to represent symmetry operations, and the usage of group theory could simplify differential equations. In gauge theory, the requirement of local symmetry can be used to deduce the equations describing a system. The groups that describe those symmetries are Lie groups, and the study of Lie groups and Lie algebras reveals much about the physical system; for instance, the number of force carriers in a theory is equal to dimension of the Lie algebra, and these bosons interact with the force they mediate if the Lie algebra is nonabelian.[2]

18.4 See also

Main article: Outline of abstract algebra

* Coding theory

* Publications in abstract algebra

18.5 References

[1] Vandermonde biography in Mac Tutor History of Mathematics Archive.

[2] Schumm, Bruce (2004), *Deep Down Things*, Baltimore: Johns Hopkins University Press, ISBN 0-8018-7971-X

18.6 Sources

* Allenby, R.B.J.T. (1991), *Rings, Fields and Groups*, Butterworth-Heinemann, ISBN 978-0-340-54440-2

* Artin, Michael (1991), *Algebra*, Prentice Hall, ISBN 978-0-89871-510-1

* Burris, Stanley N.; Sankappanavar, H. P. (1999) [1981], *A Course in Universal Algebra*

* Gilbert, Jimmie; Gilbert, Linda (2005), *Elements of Modern Algebra*, Thomson Brooks/Cole, ISBN 978-0-534-40264-8

* Lang, Serge (2002), *Algebra*, Graduate Texts in Mathematics **211** (Revised third ed.), New York: Springer-Verlag, ISBN 978-0-387-95385-4, MR 1878556

* Sethuraman, B. A. (1996), *Rings, Fields, Vector Spaces, and Group Theory: An Introduction to Abstract Algebra via Geometric Constructibility*, Berlin, New York: Springer-Verlag, ISBN 978-0-387-94848-5

* Whitehead, C. (2002), *Guide to Abstract Algebra* (2nd ed.), Houndmills: Palgrave, ISBN 978-0-333-79447-0

* W. Keith Nicholson (2012) *Introduction to Abstract Algebra*, 4th edition, John Wiley & Sons ISBN 978-1-118-13535-8 .

* John R. Durbin (1992) *Modern Algebra : an introduction*, John Wiley & Sons

18.7   External links

- John Beachy: *Abstract Algebra On Line*, Comprehensive list of definitions and theorems.

- Edwin Connell "Elements of Abstract and Linear Algebra ", Free online textbook.

- Fredrick M. Goodman: *Algebra: Abstract and Concrete*.

- Judson, Thomas W. (1997), *Abstract Algebra: Theory and Applications* An introductory undergraduate text in the spirit of texts by Gallian or Herstein, covering groups, rings, integral domains, fields and Galois theory. Free downloadable PDF with open-source GFDL license.

- Sethuraman, B.A.. (2015), *A Gentle Introduction to Abstract Algebra* A very gentle introduction that introduces rings and fields, vector spaces, and groups, with a focus on examples. Free downloadable PDF, under GNU-FDL license. Sized for both tablets and as a regular book. Contains many short video tutorials.

- Zeidler, A. Bernhard (2014), *Abstract Algebra* (PDF) A web-book on algebra and commutative algebra. Warning: work in progress! Free downloadable PDF under Open Publication License.

Chapter 19

Linear algebra

Not to be confused with Elementary algebra.

Linear algebra is the branch of mathematics concerning vector spaces and linear mappings between such spaces. It

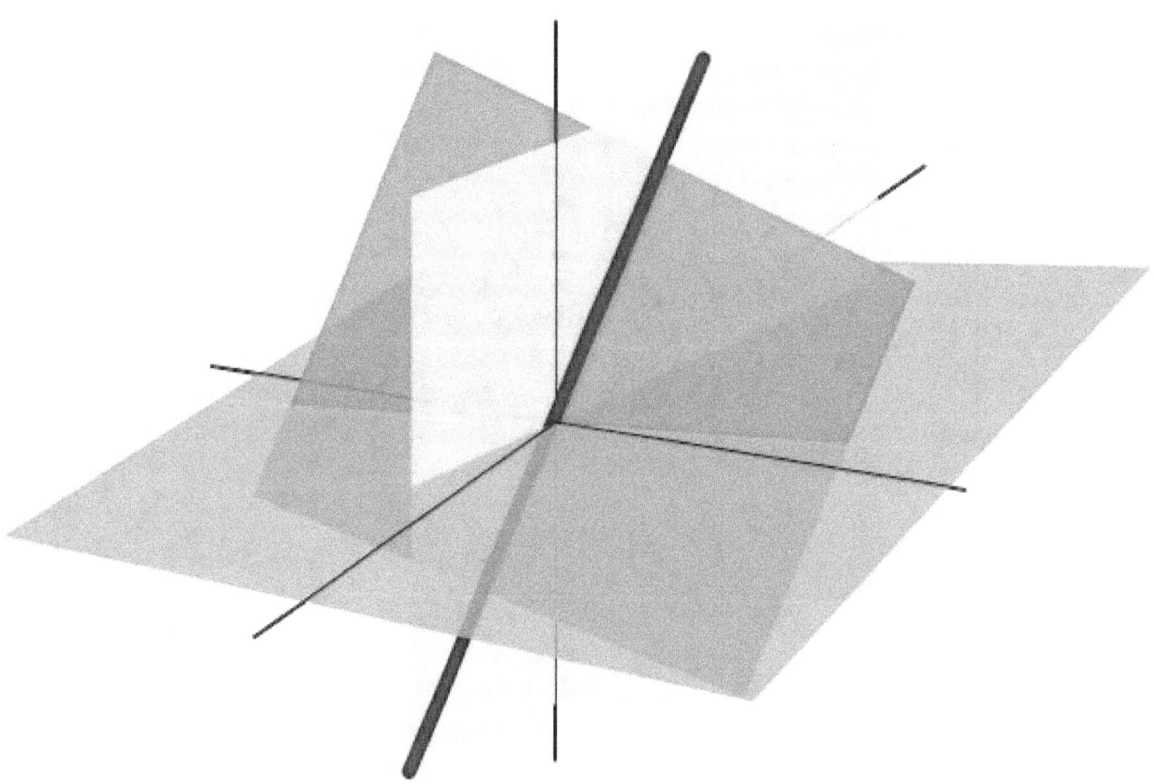

The three-dimensional Euclidean space R^3 is a vector space, and lines and planes passing through the origin are vector subspaces in R^3.

includes the study of lines, planes, and subspaces, but is also concerned with properties common to all vector spaces.

The set of points with coordinates that satisfy a linear equation forms a hyperplane in an n-dimensional space. The conditions under which a set of n hyperplanes intersect in a single point is an important focus of study in linear algebra. Such an investigation is initially motivated by a system of linear equations containing several unknowns. Such equations are naturally represented using the formalism of matrices and vectors.[1][2]

Linear algebra is central to both pure and applied mathematics. For instance, abstract algebra arises by relaxing the axioms of a vector space, leading to a number of generalizations. Functional analysis studies the infinite-dimensional version of the theory of vector spaces. Combined with calculus, linear algebra facilitates the solution of linear systems of differential equations.

Techniques from linear algebra are also used in analytic geometry, engineering, physics, natural sciences, computer science, computer animation, and the social sciences (particularly in economics). Because linear algebra is such a well-developed theory, nonlinear mathematical models are sometimes approximated by linear models.

19.1   History

The study of linear algebra first emerged from the study of determinants, which were used to solve systems of linear equations. Determinants were used by Leibniz in 1693, and subsequently, Gabriel Cramer devised Cramer's Rule for solving linear systems in 1750. Later, Gauss further developed the theory of solving linear systems by using Gaussian elimination, which was initially listed as an advancement in geodesy.[3]

The study of matrix algebra first emerged in England in the mid-1800s. In 1844 Hermann Grassmann published his "Theory of Extension" which included foundational new topics of what is today called linear algebra. In 1848, James Joseph Sylvester introduced the term matrix, which is Latin for "womb". While studying compositions of linear transformations, Arthur Cayley was led to define matrix multiplication and inverses. Crucially, Cayley used a single letter to denote a matrix, thus treating a matrix as an aggregate object. He also realized the connection between matrices and determinants, and wrote "There would be many things to say about this theory of matrices which should, it seems to me, precede the theory of determinants".[3]

In 1882, Hüseyin Tevfik Pasha wrote the book titled "Linear Algebra".[4][5] The first modern and more precise definition of a vector space was introduced by Peano in 1888;[3] by 1900, a theory of linear transformations of finite-dimensional vector spaces had emerged. Linear algebra first took its modern form in the first half of the twentieth century, when many ideas and methods of previous centuries were generalized as abstract algebra. The use of matrices in quantum mechanics, special relativity, and statistics helped spread the subject of linear algebra beyond pure mathematics. The development of computers led to increased research in efficient algorithms for Gaussian elimination and matrix decompositions, and linear algebra became an essential tool for modelling and simulations.[3]

The origin of many of these ideas is discussed in the articles on determinants and Gaussian elimination.

19.1.1   Educational history

Linear algebra first appeared in graduate textbooks in the 1940s and in undergraduate textbooks in the 1950s.[6] Following work by the School Mathematics Study Group, U.S. high schools asked 12th grade students to do "matrix algebra, formerly reserved for college" in the 1960s.[7] In France during the 1960s, educators attempted to teach linear algebra through affine dimensional vector spaces in the first year of secondary school. This was met with a backlash in the 1980s that removed linear algebra from the curriculum.[8] In 1993, the U.S.-based Linear Algebra Curriculum Study Group recommended that undergraduate linear algebra courses be given an application-based "matrix orientation" as opposed to a theoretical orientation.[9]

19.2   Scope of study

19.2.1   Vector spaces

The main structures of linear algebra are vector spaces. A vector space over a field F is a set V together with two binary operations. Elements of V are called *vectors* and elements of F are called *scalars*. The first operation, *vector addition*, takes any two vectors v and w and outputs a third vector $v + w$. The second operation, *scalar multiplication*, takes any scalar a and any vector v and outputs a new vector av. The operations of addition and multiplication in a vector space must satisfy the following axioms.[10] In the list below, let u, v and w be arbitrary vectors in V, and a and b scalars in F.

The first four axioms are those of V being an abelian group under vector addition. Vector spaces may be diverse in nature, for example, containing functions, polynomials or matrices. Linear algebra is concerned with properties common to all vector spaces.

19.2.2 Linear transformations

Similarly as in the theory of other algebraic structures, linear algebra studies mappings between vector spaces that preserve the vector-space structure. Given two vector spaces V and W over a field **F**, a linear transformation (also called linear map, linear mapping or linear operator) is a map

$$T : V \to W$$

that is compatible with addition and scalar multiplication:

$$T(u + v) = T(u) + T(v), \quad T(av) = aT(v)$$

for any vectors $u, v \in V$ and a scalar $a \in$ **F**.

Additionally for any vectors $u, v \in V$ and scalars $a, b \in$ **F**:

$$T(au + bv) = T(au) + T(bv) = aT(u) + bT(v)$$

When a bijective linear mapping exists between two vector spaces (that is, every vector from the second space is associated with exactly one in the first), we say that the two spaces are isomorphic. Because an isomorphism preserves linear structure, two isomorphic vector spaces are "essentially the same" from the linear algebra point of view. One essential question in linear algebra is whether a mapping is an isomorphism or not, and this question can be answered by checking if the determinant is nonzero. If a mapping is not an isomorphism, linear algebra is interested in finding its range (or image) and the set of elements that get mapped to zero, called the kernel of the mapping.

Linear transformations have geometric significance. For example, 2×2 real matrices denote standard planar mappings that preserve the origin.

19.2.3 Subspaces, span, and basis

Again, in analogue with theories of other algebraic objects, linear algebra is interested in subsets of vector spaces that are themselves vector spaces; these subsets are called linear subspaces. For example, both the range and kernel of a linear mapping are subspaces, and are thus often called the range space and the nullspace; these are important examples of subspaces. Another important way of forming a subspace is to take a linear combination of a set of vectors v_1, v_2, ..., vk:

$$a_1 v_1 + a_2 v_2 + \cdots + a_k v_k,$$

where a_1, a_2, ..., ak are scalars. The set of all linear combinations of vectors v_1, v_2, ..., vk is called their span, which forms a subspace.

A linear combination of any system of vectors with all zero coefficients is the zero vector of V. If this is the only way to express the zero vector as a linear combination of v_1, v_2, ..., vk then these vectors are linearly independent. Given a set of vectors that span a space, if any vector w is a linear combination of other vectors (and so the set is not linearly independent), then the span would remain the same if we remove w from the set. Thus, a set of linearly dependent vectors is redundant in the sense that there will be a linearly independent subset which will span the same subspace. Therefore, we are mostly interested in a linearly independent set of vectors that spans a vector space V, which we call a basis of V. Any set of vectors that spans V contains a basis, and any linearly independent set of vectors in V can be extended to a basis.[11] It turns out that if we accept the axiom of choice, every vector space has a basis;[12] nevertheless, this basis may be unnatural, and indeed, may not even be constructible. For instance, there exists a basis for the real numbers considered as a vector space over the rationals, but no explicit basis has been constructed.

Any two bases of a vector space V have the same cardinality, which is called the dimension of V. The dimension of a vector space is well-defined by the dimension theorem for vector spaces. If a basis of V has finite number of elements, V is called a finite-dimensional vector space. If V is finite-dimensional and U is a subspace of V, then $\dim U \leq \dim V$. If U_1 and U_2 are subspaces of V, then

$$\dim(U_1 + U_2) = \dim U_1 + \dim U_2 - \dim(U_1 \cap U_2) .[13]$$

One often restricts consideration to finite-dimensional vector spaces. A fundamental theorem of linear algebra states that all vector spaces of the same dimension are isomorphic,[14] giving an easy way of characterizing isomorphism.

19.2.4 Matrix theory

Main article: Matrix (mathematics)

A particular basis $\{v_1, v_2, ..., vn\}$ of V allows one to construct a coordinate system in V: the vector with coordinates $(a_1, a_2, ..., an)$ is the linear combination

$$a_1 v_1 + a_2 v_2 + \cdots + a_n v_n.$$

The condition that $v_1, v_2, ..., vn$ span V guarantees that each vector v can be assigned coordinates, whereas the linear independence of $v_1, v_2, ..., vn$ assures that these coordinates are unique (i.e. there is only one linear combination of the basis vectors that is equal to v). In this way, once a basis of a vector space V over $\mathbf{F}$ has been chosen, V may be identified with the coordinate n-space $\mathbf{F}^n$. Under this identification, addition and scalar multiplication of vectors in V correspond to addition and scalar multiplication of their coordinate vectors in $\mathbf{F}^n$. Furthermore, if V and W are an n-dimensional and m-dimensional vector space over $\mathbf{F}$, and a basis of V and a basis of W have been fixed, then any linear transformation $T: V \rightarrow W$ may be encoded by an $m \times n$ matrix A with entries in the field $\mathbf{F}$, called the matrix of T with respect to these bases. Two matrices that encode the same linear transformation in different bases are called similar. Matrix theory replaces the study of linear transformations, which were defined axiomatically, by the study of matrices, which are concrete objects. This major technique distinguishes linear algebra from theories of other algebraic structures, which usually cannot be parameterized so concretely.

There is an important distinction between the coordinate n-space $\mathbf{R}^n$ and a general finite-dimensional vector space V. While $\mathbf{R}^n$ has a standard basis $\{e_1, e_2, ..., en\}$, a vector space V typically does not come equipped with such a basis and many different bases exist (although they all consist of the same number of elements equal to the dimension of V).

One major application of the matrix theory is calculation of determinants, a central concept in linear algebra. While determinants could be defined in a basis-free manner, they are usually introduced via a specific representation of the mapping; the value of the determinant does not depend on the specific basis. It turns out that a mapping has an inverse if and only if the determinant has an inverse (every non-zero real or complex number has an inverse[15]). If the determinant is zero, then the nullspace is nontrivial. Determinants have other applications, including a systematic way of seeing if a set of vectors is linearly independent (we write the vectors as the columns of a matrix, and if the determinant of that matrix is zero, the vectors are linearly dependent). Determinants could also be used to solve systems of linear equations (see Cramer's rule), but in real applications, Gaussian elimination is a faster method.

19.2.5 Eigenvalues and eigenvectors

In general, the action of a linear transformation may be quite complex. Attention to low-dimensional examples gives an indication of the variety of their types. One strategy for a general n-dimensional transformation T is to find "characteristic lines" that are invariant sets under T. If v is a non-zero vector such that Tv is a scalar multiple of v, then the line through 0 and v is an invariant set under T and v is called a characteristic vector or **eigenvector**. The scalar λ such that $Tv = \lambda v$ is called a characteristic value or **eigenvalue** of T.

To find an eigenvector or an eigenvalue, we note that

$$Tv - \lambda v = (T - \lambda I)v = 0,$$

where I is the identity matrix. For there to be nontrivial solutions to that equation, $\det(T - \lambda I) = 0$. The determinant is a polynomial, and so the eigenvalues are not guaranteed to exist if the field is **R**. Thus, we often work with an algebraically closed field such as the complex numbers when dealing with eigenvectors and eigenvalues so that an eigenvalue will always exist. It would be particularly nice if given a transformation T taking a vector space V into itself we can find a basis for V consisting of eigenvectors. If such a basis exists, we can easily compute the action of the transformation on any vector: if $v_1, v_2, ..., vn$ are linearly independent eigenvectors of a mapping of n-dimensional spaces T with (not necessarily distinct) eigenvalues $\lambda_1, \lambda_2, ..., \lambda n$, and if $v = a_1 v_1 + ... + an\, vn$, then,

$$T(v) = T(a_1 v_1) + \cdots + T(a_n v_n) = a_1 T(v_1) + \cdots + a_n T(v_n) = a_1 \lambda_1 v_1 + \cdots + a_n \lambda_n v_n.$$

Such a transformation is called a diagonalizable matrix since in the eigenbasis, the transformation is represented by a diagonal matrix. Because operations like matrix multiplication, matrix inversion, and determinant calculation are simple on diagonal matrices, computations involving matrices are much simpler if we can bring the matrix to a diagonal form. Not all matrices are diagonalizable (even over an algebraically closed field).

19.2.6 Inner-product spaces

Besides these basic concepts, linear algebra also studies vector spaces with additional structure, such as an inner product. The inner product is an example of a bilinear form, and it gives the vector space a geometric structure by allowing for the definition of length and angles. Formally, an *inner product* is a map

$$\langle \cdot, \cdot \rangle : V \times V \to F$$

that satisfies the following three axioms for all vectors u, v, w in V and all scalars a in F:[16][17]

- Conjugate symmetry:

$$\langle u, v \rangle = \overline{\langle v, u \rangle}.$$

Note that in **R**, it is symmetric.

- Linearity in the first argument:

$$\langle au, v \rangle = a \langle u, v \rangle.$$

$$\langle u + v, w \rangle = \langle u, w \rangle + \langle v, w \rangle.$$

- Positive-definiteness:

$\langle v, v \rangle \geq 0$ with equality only for $v = 0$.

We can define the length of a vector v in V by

$$\|v\|^2 = \langle v, v \rangle.$$

and we can prove the Cauchy–Schwarz inequality:

$$|\langle u, v \rangle| \leq \|u\| \cdot \|v\|.$$

In particular, the quantity

$$\frac{|\langle u, v \rangle|}{\|u\| \cdot \|v\|} \leq 1.$$

and so we can call this quantity the cosine of the angle between the two vectors.

Two vectors are orthogonal if $\langle u, v \rangle = 0$. An orthonormal basis is a basis where all basis vectors have length 1 and are orthogonal to each other. Given any finite-dimensional vector space, an orthonormal basis could be found by the Gram–Schmidt procedure. Orthonormal bases are particularly nice to deal with, since if $v = a_1 \, v_1 + ... + an \, vn$, then $a_i = \langle v, v_i \rangle$.

The inner product facilitates the construction of many useful concepts. For instance, given a transform T, we can define its Hermitian conjugate $T*$ as the linear transform satisfying

$$\langle Tu, v \rangle = \langle u, T^*v \rangle.$$

If T satisfies $TT* = T*T$, we call T normal. It turns out that normal matrices are precisely the matrices that have an orthonormal system of eigenvectors that span V.

19.3   Some main useful theorems

Main article: Theorems and definitions in linear algebra

- A matrix is invertible, or non-singular, if and only if the linear map represented by the matrix is an isomorphism.

- Any vector space over a field **F** of dimension n is isomorphic to $\mathbf{F}^n$ as a vector space over **F**.

- Corollary: Any two vector spaces over **F** of the same finite dimension are isomorphic to each other.

- A linear map is an isomorphism if and only if the determinant is nonzero.

19.4   Applications

Because of the ubiquity of vector spaces, linear algebra is used in many fields of mathematics, natural sciences, computer science, and social science. Below are just some examples of applications of linear algebra.

19.4.1 Solution of linear systems

Main article: System of linear equations

Linear algebra provides the formal setting for the linear combination of equations used in the Gaussian method. Suppose the goal is to find and describe the solution(s), if any, of the following system of linear equations:

$$
\begin{aligned}
2x + y - z &= 8 \quad (L_1) \\
-3x - y + 2z &= -11 \quad (L_2) \\
-2x + y + 2z &= -3 \quad (L_3)
\end{aligned}
$$

The Gaussian-elimination algorithm is as follows: eliminate x from all equations below L_1, and then eliminate y from all equations below L_2. This will put the system into triangular form. Then, using back-substitution, each unknown can be solved for.

In the example, x is eliminated from L_2 by adding $(3/2)L_1$ to L_2. x is then eliminated from L_3 by adding L_1 to L_3. Formally:

$$L_2 + \tfrac{3}{2}L_1 \to L_2$$

$$L_3 + L_1 \to L_3$$

The result is:

$$
\begin{aligned}
2x + y - z &= 8 \\
\tfrac{1}{2}y + \tfrac{1}{2}z &= 1 \\
2y + z &= 5
\end{aligned}
$$

Now y is eliminated from L_3 by adding $-4L_2$ to L_3:

$$L_3 + -4L_2 \to L_3$$

The result is:

$$
\begin{aligned}
2x + y - z &= 8 \\
\tfrac{1}{2}y + \tfrac{1}{2}z &= 1 \\
-z &= 1
\end{aligned}
$$

This result is a system of linear equations in triangular form, and so the first part of the algorithm is complete.

The last part, back-substitution, consists of solving for the known in reverse order. It can thus be seen that

$$z = -1 \quad (L_3)$$

Then, z can be substituted into L_2, which can then be solved to obtain

$$y = 3 \quad (L_2)$$

Next, z and y can be substituted into L_1, which can be solved to obtain

$$x = 2 \quad (L_1)$$

The system is solved.

We can, in general, write any system of linear equations as a matrix equation:

$$Ax = b.$$

The solution of this system is characterized as follows: first, we find a particular solution x_0 of this equation using Gaussian elimination. Then, we compute the solutions of $Ax = 0$; that is, we find the null space N of A. The solution set of this equation is given by $x_0 + N = \{x_0 + n : n \in N\}$. If the number of variables is equal to the number of equations, then we can characterize when the system has a unique solution: since N is trivial if and only if $\det A \neq 0$, the equation has a unique solution if and only if $\det A \neq 0$.[18]

19.4.2   Least-squares best fit line

The least squares method is used to determine the best fit line for a set of data.[19] This line will minimize the sum of the squares of the residuals.

19.4.3   Fourier series expansion

Fourier series are a representation of a function $f: [-\pi, \pi] \to \mathbf{R}$ as a trigonometric series:

$$f(x) = \frac{a_0}{2} + \sum_{n=1}^{\infty} [a_n \cos(nx) + b_n \sin(nx)].$$

This series expansion is extremely useful in solving partial differential equations. In this article, we will not be concerned with convergence issues; it is nice to note that all Lipschitz-continuous functions have a converging Fourier series expansion, and nice enough discontinuous functions have a Fourier series that converges to the function value at most points.

The space of all functions that can be represented by a Fourier series form a vector space (technically speaking, we call functions that have the same Fourier series expansion the "same" function, since two different discontinuous functions might have the same Fourier series). Moreover, this space is also an inner product space with the inner product

$$\langle f, g \rangle = \frac{1}{\pi} \int_{-\pi}^{\pi} f(x)g(x)\,dx.$$

The functions $g_n(x) = \sin(nx)$ for $n > 0$ and $h_n(x) = \cos(nx)$ for $n \geq 0$ are an orthonormal basis for the space of Fourier-expandable functions. We can thus use the tools of linear algebra to find the expansion of any function in this space in terms of these basis functions. For instance, to find the coefficient a_k, we take the inner product with h_k:

$$\langle f, h_k \rangle = \frac{a_0}{2} \langle h_0, h_k \rangle + \sum_{n=1}^{\infty} [a_n \langle h_n, h_k \rangle + b_n \langle g_n, h_k \rangle].$$

and by orthonormality, $\langle f, h_k \rangle = a_k$; that is,

$$a_k = \frac{1}{\pi} \int_{-\pi}^{\pi} f(x) \cos(kx)\,dx.$$

19.4.4   Quantum mechanics

Quantum mechanics is highly inspired by notions in linear algebra. In quantum mechanics, the physical state of a particle is represented by a vector, and observables (such as momentum, energy, and angular momentum) are represented by linear operators on the underlying vector space. More concretely, the wave function of a particle describes its physical state and lies in the vector space L^2 (the functions $\varphi: \mathbf{R}^3 \to \mathbf{C}$ such that $\int_{-\infty}^{\infty} \int_{-\infty}^{\infty} \int_{-\infty}^{\infty} |\phi|^2 dx dy dz$ is finite), and it evolves according to the Schrödinger equation. Energy is represented as the operator $H = -\frac{\hbar^2}{2m} \nabla^2 + V(x, y, z)$, where V is the potential energy. H is also known as the Hamiltonian operator. The eigenvalues of H represents the possible energies that can be observed. Given a particle in some state φ, we can expand φ into a linear combination of eigenstates of H. The component of H in each eigenstate determines the probability of measuring the corresponding eigenvalue, and the measurement forces the particle to assume that eigenstate (wave function collapse).

19.5   Geometric introduction

Many of the principles and techniques of linear algebra can be seen in the geometry of lines in a real two dimensional plane E. When formulated using vectors and matrices the geometry of points and lines in the plane can be extended to the geometry of points and hyperplanes in high-dimensional spaces.

Point coordinates in the plane E are ordered pairs of real numbers, (x,y), and a line is defined as the set of points (x,y) that satisfy the linear equation[20]

$$\lambda : ax + by + c = 0,$$

where a, b and c are not all zero. Then,

$$\lambda : \begin{bmatrix} a & b & c \end{bmatrix} \begin{Bmatrix} x \\ y \\ 1 \end{Bmatrix} = 0.$$

or

$$A\mathbf{x} = 0,$$

where $\mathbf{x} = (x, y, 1)$ is the 3×1 set of homogeneous coordinates associated with the point (x, y).[21]

Homogeneous coordinates identify the plane E with the $z = 1$ plane in three dimensional space. The x–y coordinates in E are obtained from homogeneous coordinates $\mathbf{y} = (y_1, y_2, y_3)$ by dividing by the third component (if it is nonzero) to obtain $\mathbf{y} = (y_1/y_3, y_2/y_3, 1)$.

The linear equation, λ, has the important property, that if $\mathbf{x}_1$ and $\mathbf{x}_2$ are homogeneous coordinates of points on the line, then the point $\alpha\mathbf{x}_1 + \beta\mathbf{x}_2$ is also on the line, for any real α and β.

Now consider the equations of the two lines λ_1 and λ_2,

$$\lambda_1 : a_1 x + b_1 y + c_1 = 0, \quad \lambda_2 : a_2 x + b_2 y + c_2 = 0,$$

which forms a system of linear equations. The intersection of these two lines is defined by $\mathbf{x} = (x, y, 1)$ that satisfy the matrix equation,

$$\lambda_{1,2} : \begin{bmatrix} a_1 & b_1 & c_1 \\ a_2 & b_2 & c_2 \end{bmatrix} \begin{Bmatrix} x \\ y \\ 1 \end{Bmatrix} = \begin{Bmatrix} 0 \\ 0 \end{Bmatrix}.$$

or using homogeneous coordinates,

$$B\mathbf{x} = 0.$$

The point of intersection of these two lines is the unique non-zero solution of these equations. In homogeneous coordinates, the solutions are multiples of the following solution:[21]

$$x_1 = \begin{vmatrix} b_1 & c_1 \\ b_2 & c_2 \end{vmatrix}, \ x_2 = -\begin{vmatrix} a_1 & c_1 \\ a_2 & c_2 \end{vmatrix}, \ x_3 = \begin{vmatrix} a_1 & b_1 \\ a_2 & b_2 \end{vmatrix}$$

if the rows of **B** are linearly independent (i.e., λ_1 and λ_2 represent distinct lines). Divide through by x_3 to get Cramer's rule for the solution of a set of two linear equations in two unknowns.[22] Notice that this yields a point in the $z = 1$ plane only when the 2×2 submatrix associated with x_3 has a non-zero determinant.

It is interesting to consider the case of three lines, λ_1, λ_2 and λ_3, which yield the matrix equation,

$$\lambda_{1,2,3}: \begin{bmatrix} a_1 & b_1 & c_1 \\ a_2 & b_2 & c_2 \\ a_3 & b_3 & c_3 \end{bmatrix} \begin{Bmatrix} x \\ y \\ 1 \end{Bmatrix} = \begin{Bmatrix} 0 \\ 0 \\ 0 \end{Bmatrix}.$$

which in homogeneous form yields,

$$C\mathbf{x} = 0.$$

Clearly, this equation has the solution $\mathbf{x} = (0,0,0)$, which is not a point on the $z = 1$ plane E. For a solution to exist in the plane E, the coefficient matrix C must have rank 2, which means its determinant must be zero. Another way to say this is that the columns of the matrix must be linearly dependent.

19.6   Introduction to linear transformations

Another way to approach linear algebra is to consider linear functions on the two dimensional real plane $E=\mathbf{R}^2$. Here **R** denotes the set of real numbers. Let $\mathbf{x}=(x, y)$ be an arbitrary vector in E and consider the linear function $\lambda: E{\rightarrow}\mathbf{R}$, given by

$$\lambda : \begin{bmatrix} a & b \end{bmatrix} \begin{Bmatrix} x \\ y \end{Bmatrix} = c.$$

or

$$A\mathbf{x} = c.$$

This transformation has the important property that if $A\mathbf{y}=d$, then

$$A(\alpha\mathbf{x} + \beta\mathbf{y}) = \alpha A\mathbf{x} + \beta A\mathbf{y} = \alpha c + \beta d.$$

This shows that the sum of vectors in E map to the sum of their images in **R**. This is the defining characteristic of a linear map, or linear transformation.[20] For this case, where the image space is a real number the map is called a linear functional.[22]

Consider the linear functional a little more carefully. Let $\mathbf{i}=(1,0)$ and $\mathbf{j}=(0,1)$ be the natural basis vectors on E, so that $\mathbf{x}=x\mathbf{i}+y\mathbf{j}$. It is now possible to see that

$$A\mathbf{x} = A(x\mathbf{i}+y\mathbf{j}) = xA\mathbf{i} + yA\mathbf{j} = \begin{bmatrix} A\mathbf{i} & A\mathbf{j} \end{bmatrix} \begin{Bmatrix} x \\ y \end{Bmatrix} = \begin{bmatrix} a & b \end{bmatrix} \begin{Bmatrix} x \\ y \end{Bmatrix} = c.$$

Thus, the columns of the matrix A are the image of the basis vectors of E in $\mathbf{R}$.

This is true for any pair of vectors used to define coordinates in E. Suppose we select a non-orthogonal non-unit vector basis $\mathbf{v}$ and $\mathbf{w}$ to define coordinates of vectors in E. This means a vector $\mathbf{x}$ has coordinates (α,β), such that $\mathbf{x}=\alpha\mathbf{v}+\beta\mathbf{w}$. Then, we have the linear functional

$$\lambda : A\mathbf{x} = \begin{bmatrix} A\mathbf{v} & A\mathbf{w} \end{bmatrix} \begin{Bmatrix} \alpha \\ \beta \end{Bmatrix} = \begin{bmatrix} d & e \end{bmatrix} \begin{Bmatrix} \alpha \\ \beta \end{Bmatrix} = c.$$

where $A\mathbf{v}=d$ and $A\mathbf{w}=e$ are the images of the basis vectors $\mathbf{v}$ and $\mathbf{w}$. This is written in matrix form as

$$\begin{bmatrix} a & b \end{bmatrix} \begin{bmatrix} v_1 & w_1 \\ v_2 & w_2 \end{bmatrix} = \begin{bmatrix} d & e \end{bmatrix}.$$

19.6.1 Coordinates relative to a basis

This leads to the question of how to determine the coordinates of a vector $\mathbf{x}$ relative to a general basis $\mathbf{v}$ and $\mathbf{w}$ in E. Assume that we know the coordinates of the vectors, $\mathbf{x}$, $\mathbf{v}$ and $\mathbf{w}$ in the natural basis $\mathbf{i}=(1,0)$ and $\mathbf{j}=(0,1)$. Our goal is two find the real numbers α, β, so that $\mathbf{x}=\alpha\mathbf{v}+\beta\mathbf{w}$, that is

$$\begin{Bmatrix} x \\ y \end{Bmatrix} = \begin{bmatrix} v_1 & w_1 \\ v_2 & w_2 \end{bmatrix} \begin{Bmatrix} \alpha \\ \beta \end{Bmatrix}.$$

To solve this equation for α, β, we compute the linear coordinate functionals σ and τ for the basis $\mathbf{v}$, $\mathbf{w}$, which are given by,[21]

$$\sigma = \begin{bmatrix} \sigma_1 & \sigma_2 \end{bmatrix} = \frac{1}{v_1 w_2 - v_2 w_1} \begin{bmatrix} w_2 & -w_1 \end{bmatrix}, \tau = \begin{bmatrix} \tau_1 & \tau_2 \end{bmatrix} = \frac{1}{v_1 w_2 - v_2 w_1} \begin{bmatrix} -v_2 & v_1 \end{bmatrix}.$$

The functionals σ and τ compute the components of $\mathbf{x}$ along the basis vectors $\mathbf{v}$ and $\mathbf{w}$, respectively, that is,

$$\sigma\mathbf{x} = \alpha, \tau\mathbf{x} = \beta,$$

which can be written in matrix form as

$$\begin{bmatrix} \sigma_1 & \sigma_2 \\ \tau_1 & \tau_2 \end{bmatrix} \begin{Bmatrix} x \\ y \end{Bmatrix} = \begin{Bmatrix} \alpha \\ \beta \end{Bmatrix}.$$

These coordinate functionals have the properties,

$$\sigma\mathbf{v} = 1, \sigma\mathbf{w} = 0, \tau\mathbf{w} = 1, \tau\mathbf{v} = 0.$$

These equations can be assembled into the single matrix equation,

$$\begin{bmatrix} \sigma_1 & \sigma_2 \\ \tau_1 & \tau_2 \end{bmatrix} \begin{bmatrix} v_1 & w_1 \\ v_2 & w_2 \end{bmatrix} = \begin{bmatrix} 1 & 0 \\ 0 & 1 \end{bmatrix}.$$

Thus, the matrix formed by the coordinate linear functionals is the inverse of the matrix formed by the basis vectors.[20][22]

19.6.2   Inverse image

The set of points in the plane E that map to the same image in $\mathbf{R}$ under the linear functional λ define a line in E. This line is the image of the inverse map, $\lambda^{-1}: \mathbf{R} \rightarrow E$. This inverse image is the set of the points $\mathbf{x}=(x, y)$ that solve the equation,

$$A\mathbf{x} = \begin{bmatrix} a & b \end{bmatrix} \begin{Bmatrix} x \\ y \end{Bmatrix} = c.$$

Notice that a linear functional operates on known values for $\mathbf{x}=(x, y)$ to compute a value c in $\mathbf{R}$, while the inverse image seeks the values for $\mathbf{x}=(x, y)$ that yield a specific value c.

In order to solve the equation, we first recognize that only one of the two unknowns (x,y) can be determined, so we select y to be determined, and rearrange the equation

$$by = c - ax.$$

Solve for y and obtain the inverse image as the set of points,

$$\mathbf{x}(t) = \begin{Bmatrix} 0 \\ c/b \end{Bmatrix} + t \begin{Bmatrix} 1 \\ -a/b \end{Bmatrix} = \mathbf{p} + t\mathbf{h}.$$

For convenience the free parameter x has been relabeled t.

The vector $\mathbf{p}$ defines the intersection of the line with the y-axis, known as the y-intercept. The vector $\mathbf{h}$ satisfies the homogeneous equation,

$$A\mathbf{h} = \begin{bmatrix} a & b \end{bmatrix} \begin{Bmatrix} 1 \\ -a/b \end{Bmatrix} = 0.$$

Notice that if $\mathbf{h}$ is a solution to this homogeneous equation, then $t\,\mathbf{h}$ is also a solution.

The set of points of a linear functional that map to zero define the *kernel* of the linear functional. The line can be considered to be the set of points $\mathbf{h}$ in the kernel translated by the vector $\mathbf{p}$.[20][22]

19.7   Generalizations and related topics

Since linear algebra is a successful theory, its methods have been developed and generalized in other parts of mathematics. In module theory, one replaces the field of scalars by a ring. The concepts of linear independence, span, basis, and dimension (which is called rank in module theory) still make sense. Nevertheless, many theorems from linear algebra become false in module theory. For instance, not all modules have a basis (those that do are called free modules), the rank of a free module is not necessarily unique, not every linearly independent subset of a module can be extended to form a basis, and not every subset of a module that spans the space contains a basis.

In multilinear algebra, one considers multivariable linear transformations, that is, mappings that are linear in each of a number of different variables. This line of inquiry naturally leads to the idea of the dual space, the vector space V^* consisting of linear maps $f\colon V \to F$ where F is the field of scalars. Multilinear maps $T\colon V^n \to F$ can be described via tensor products of elements of V^*.

If, in addition to vector addition and scalar multiplication, there is a bilinear vector product $V \times V \to V$, the vector space is called an algebra; for instance, associative algebras are algebras with an associate vector product (like the algebra of square matrices, or the algebra of polynomials).

Functional analysis mixes the methods of linear algebra with those of mathematical analysis and studies various function spaces, such as L^p spaces.

Representation theory studies the actions of algebraic objects on vector spaces by representing these objects as matrices. It is interested in all the ways that this is possible, and it does so by finding subspaces invariant under all transformations of the algebra. The concept of eigenvalues and eigenvectors is especially important.

Algebraic geometry considers the solutions of systems of polynomial equations.

There are several related topics in the field of Computer Programming that utilizes much of the techniques and theorems Linear Algebra encompasses and refers to.

19.8 See also

- Linear equation

- Linear equation over a ring

- System of linear equations

- Gaussian elimination

- Eigenvectors

- Fundamental matrix in computer vision

- Linear regression, a statistical estimation method

- List of linear algebra topics

- Numerical linear algebra

- Simplex method, a solution technique for linear programs

- Transformation matrix

19.9 Notes

[1] Strang, Gilbert (July 19, 2005), *Linear Algebra and Its Applications* (4th ed.), Brooks Cole, ISBN 978-0-03-010567-8

[2] Weisstein, Eric. "Linear Algebra". *From MathWorld--A Wolfram Web Resource*. Wolfram. Retrieved 16 April 2012.

[3] Vitulli, Marie. "A Brief History of Linear Algebra and Matrix Theory". *Department of Mathematics*. University of Oregon. Archived from the original on 2012-09-10. Retrieved 2014-07-08.

[4] http://www.journals.istanbul.edu.tr/tr/index.php/oba/article/download/9103/8452

[5] http://archive.org/details/linearalgebra00tevfgoog

[6] Tucker, Alan (1993). "The Growing Importance of Linear Algebra in Undergraduate Mathematics". *College Mathematics Journal* **24** (1): 3–9. doi:10.2307/2686426.

[7] Goodlad, John I.; von stoephasius, Reneta; Klein, M. Frances (1966). "The changing school curriculum". U.S. Department of Health, Education, and Welfare; Office of Education. Retrieved 9 July 2014.

[8] Dorier, Jean-Luc; Robert, Aline; Robinet, Jacqueline; Rogalsiu, Marc (2000). Dorier, Jean-Luc. ed. *The Obstacle of Formalism in Linear Algebra*. Springer. pp. 85–124. ISBN 978-0-7923-6539-6. Retrieved 9 July 2014.

[9] Carlson, David; Johnson, Charles R.; Lay, David C.; Porter, A. Duane (1993). "The Linear Algebra Curriculum Study Group Recommendations for the First Course in Linear Algebra". *The College Mathematics Journal* **24** (1): 41–46. doi:10.2307/2686430.

[10] Roman 2005, ch. 1, p. 27

[11] Axler (2004), pp. 28–29

[12] The existence of a basis is straightforward for countably generated vector spaces, and for well-ordered vector spaces, but in full generality it is logically equivalent to the axiom of choice.

[13] Axler (2204), p. 33

[14] Axler (2004), p. 55

[15] If we restrict to integers, then only 1 and −1 have an inverse. Consequently, the inverse of an integer matrix is an integer matrix if and only if the determinant is 1 or −1.

[16] P. K. Jain, Khalil Ahmad (1995). "5.1 Definitions and basic properties of inner product spaces and Hilbert spaces". *Functional analysis* (2nd ed.). New Age International. p. 203. ISBN 81-224-0801-X.

[17] Eduard Prugovečki (1981). "Definition 2.1". *Quantum mechanics in Hilbert space* (2nd ed.). Academic Press. pp. 18 *ff*. ISBN 0-12-566060-X.

[18] Gunawardena, Jeremy. "Matrix algebra for beginners, Part I" (PDF). *Harvard Medical School*. Retrieved 2 May 2012.

[19] Miller, Steven. "The Method of Least Squares" (PDF). *Brown University*. Retrieved 1 May 2013.

[20] Strang, Gilbert (July 19, 2005), *Linear Algebra and Its Applications* (4th ed.), Brooks Cole, ISBN 978-0-03-010567-8.

[21] J. G. Semple and G. T. Kneebone, *Algebraic Projective Geometry*, Clarendon Press, London, 1952.

[22] E. D. Nering, *Linear Algebra and Matrix Theory*, John-Wiley, New York, NY, 1963

[1] This axiom is not asserting the associativity of an operation, since there are two operations in question, scalar multiplication: *bv*; and field multiplication: *ab*.

19.10   Further reading

19.10.1   History

- Fearnley-Sander, Desmond, "Hermann Grassmann and the Creation of Linear Algebra" (), American Mathematical Monthly **86** (1979), pp. 809–817.

- Grassmann, Hermann, *Die lineale Ausdehnungslehre ein neuer Zweig der Mathematik: dargestellt und durch Anwendungen auf die übrigen Zweige der Mathematik, wie auch auf die Statik, Mechanik, die Lehre vom Magnetismus und die Krystallonomie erläutert*, O. Wigand, Leipzig, 1844.

19.10.2 Introductory textbooks

- Strang, Gilbert (February 2009), *Introduction to Linear Algebra* (4th ed.), Wellesley-Cambridge Press, ISBN 978-0-98-02327-14

- Murty, Katta G. (2014) *Computational and Algorithmic Linear Algebra and n-Dimensional Geometry*, World Scientific Publishing, ISBN 978-981-4366-62-5. *Chapter 1: Systems of Simultaneous Linear Equations*

- Bretscher, Otto (June 28, 2004), *Linear Algebra with Applications* (3rd ed.), Prentice Hall, ISBN 978-0-13-145334-0

- Farin, Gerald; Hansford, Dianne (December 15, 2004), *Practical Linear Algebra: A Geometry Toolbox*, AK Peters, ISBN 978-1-56881-234-2

- Hefferon, Jim (2008), *Linear Algebra*

- Anton, Howard (2005), *Elementary Linear Algebra (Applications Version)* (9th ed.), Wiley International

- Lay, David C. (August 22, 2005), *Linear Algebra and Its Applications* (3rd ed.), Addison Wesley, ISBN 978-0-321-28713-7

- Kolman, Bernard; Hill, David R. (May 3, 2007), *Elementary Linear Algebra with Applications* (9th ed.), Prentice Hall, ISBN 978-0-13-229654-0

- Leon, Steven J. (2006), *Linear Algebra With Applications* (7th ed.), Pearson Prentice Hall, ISBN 978-0-13-185785-8

- Poole, David (2010), *Linear Algebra: A Modern Introduction* (3rd ed.), Cengage – Brooks/Cole, ISBN 978-0-538-73545-2

- Ricardo, Henry (2010), *A Modern Introduction To Linear Algebra* (1st ed.), CRC Press, ISBN 978-1-4398-0040-9

- Sadun, Lorenzo (2008), *Applied Linear Algebra: the decoupling principle* (2nd ed.), AMS, ISBN 978-0-8218-4441-0

19.10.3 Advanced textbooks

- Axler, Sheldon (February 26, 2004), *Linear Algebra Done Right* (2nd ed.), Springer, ISBN 978-0-387-98258-8

- Bhatia, Rajendra (November 15, 1996), *Matrix Analysis*, Graduate Texts in Mathematics, Springer, ISBN 978-0-387-94846-1

- Demmel, James W. (August 1, 1997), *Applied Numerical Linear Algebra*, SIAM, ISBN 978-0-89871-389-3

- Dym, Harry (2007), *Linear Algebra in Action*, AMS, ISBN 978-0-8218-3813-6

- Gantmacher, F.R. (2005, 1959 edition), *Applications of the Theory of Matrices*, Dover Publications, ISBN 978-0-486-44554-0 Check date values in: |date= (help)

- Gantmacher, Felix R. (1990), *Matrix Theory Vol. 1* (2nd ed.), American Mathematical Society, ISBN 978-0-8218-1376-8

- Gantmacher, Felix R. (2000), *Matrix Theory Vol. 2* (2nd ed.), American Mathematical Society, ISBN 978-0-8218-2664-5

- Gelfand, I. M. (1989), *Lectures on Linear Algebra*, Dover Publications, ISBN 978-0-486-66082-0

- Glazman, I. M.; Ljubic, Ju. I. (2006), *Finite-Dimensional Linear Analysis*, Dover Publications, ISBN 978-0-486-45332-3

- Golan, Johnathan S. (January 2007), *The Linear Algebra a Beginning Graduate Student Ought to Know* (2nd ed.), Springer, ISBN 978-1-4020-5494-5

- Golan, Johnathan S. (August 1995), *Foundations of Linear Algebra*, Kluwer, ISBN 0-7923-3614-3

- Golub, Gene H.; Van Loan, Charles F. (October 15, 1996), *Matrix Computations*, Johns Hopkins Studies in Mathematical Sciences (3rd ed.), The Johns Hopkins University Press, ISBN 978-0-8018-5414-9

- Greub, Werner H. (October 16, 1981), *Linear Algebra*, Graduate Texts in Mathematics (4th ed.), Springer, ISBN 978-0-8018-5414-9

- Hoffman, Kenneth; Kunze, Ray (1971), *Linear algebra* (2nd ed.), Englewood Cliffs, N.J.: Prentice-Hall, Inc., MR 0276251

- Halmos, Paul R. (August 20, 1993), *Finite-Dimensional Vector Spaces*, Undergraduate Texts in Mathematics, Springer, ISBN 978-0-387-90093-3

- Friedberg, Stephen H.; Insel, Arnold J.; Spence, Lawrence E. (November 11, 2002), *Linear Algebra* (4th ed.), Prentice Hall, ISBN 978-0-13-008451-4

- Horn, Roger A.; Johnson, Charles R. (February 23, 1990), *Matrix Analysis*, Cambridge University Press, ISBN 978-0-521-38632-6

- Horn, Roger A.; Johnson, Charles R. (June 24, 1994), *Topics in Matrix Analysis*, Cambridge University Press, ISBN 978-0-521-46713-1

- Lang, Serge (March 9, 2004), *Linear Algebra*, Undergraduate Texts in Mathematics (3rd ed.), Springer, ISBN 978-0-387-96412-6

- Marcus, Marvin; Minc, Henryk (2010), *A Survey of Matrix Theory and Matrix Inequalities*, Dover Publications, ISBN 978-0-486-67102-4

- Meyer, Carl D. (February 15, 2001), *Matrix Analysis and Applied Linear Algebra*, Society for Industrial and Applied Mathematics (SIAM), ISBN 978-0-89871-454-8

- Mirsky, L. (1990), *An Introduction to Linear Algebra*, Dover Publications, ISBN 978-0-486-66434-7

- Roman, Steven (March 22, 2005), *Advanced Linear Algebra*, Graduate Texts in Mathematics (2nd ed.), Springer, ISBN 978-0-387-24766-3

- Shafarevich, I. R.; A. O. Remizov (2012), *Linear Algebra and Geometry*, Springer, ISBN 978-3-642-30993-9

- Shilov, Georgi E. (June 1, 1977), *Linear algebra*, Dover Publications, ISBN 978-0-486-63518-7

- Shores, Thomas S. (December 6, 2006), *Applied Linear Algebra and Matrix Analysis*, Undergraduate Texts in Mathematics, Springer, ISBN 978-0-387-33194-2

- Smith, Larry (May 28, 1998), *Linear Algebra*, Undergraduate Texts in Mathematics, Springer, ISBN 978-0-387-98455-1

- Trefethen, Lloyd N.; Bau, David (1997), *Numerical Linear Algebra*, SIAM, ISBN 978-0-898-71361-9

19.10.4   Study guides and outlines

- Leduc, Steven A. (May 1, 1996), *Linear Algebra (Cliffs Quick Review)*, Cliffs Notes, ISBN 978-0-8220-5331-6

- Lipschutz, Seymour; Lipson, Marc (December 6, 2000), *Schaum's Outline of Linear Algebra* (3rd ed.), McGraw-Hill, ISBN 978-0-07-136200-9

- Lipschutz, Seymour (January 1, 1989), *3,000 Solved Problems in Linear Algebra*, McGraw–Hill, ISBN 978-0-07-038023-3

- McMahon, David (October 28, 2005), *Linear Algebra Demystified*, McGraw–Hill Professional, ISBN 978-0-07-146579-3

- Zhang, Fuzhen (April 7, 2009), *Linear Algebra: Challenging Problems for Students*, The Johns Hopkins University Press, ISBN 978-0-8018-9125-0

19.11 External links

- International Linear Algebra Society

- MIT Professor Gilbert Strang's Linear Algebra Course Homepage : MIT Course Website

- MIT Linear Algebra Lectures: free videos from MIT OpenCourseWare

- Linear Algebra - Foundations to Frontiers Free MOOC launched by edX

- Linear Algebra Toolkit.

- Hazewinkel, Michiel, ed. (2001), "Linear algebra", *Encyclopedia of Mathematics*, Springer, ISBN 978-1-55608-010-4

- Linear Algebra on MathWorld.

- Linear Algebra tutorial with online interactive programs.

- Matrix and Linear Algebra Terms on Earliest Known Uses of Some of the Words of Mathematics

- Earliest Uses of Symbols for Matrices and Vectors on Earliest Uses of Various Mathematical Symbols

- Linear Algebra by Elmer G. Wiens. Interactive web pages for vectors, matrices, linear equations, etc.

- Linear Algebra Solved Problems: Interactive forums for discussion of linear algebra problems, from the lowest up to the hardest level (*Putnam*).

- Linear Algebra for Informatics. José Figueroa-O'Farrill, University of Edinburgh

- Online Notes / Linear Algebra Paul Dawkins, Lamar University

- Elementary Linear Algebra textbook with solutions

- Linear Algebra Wiki

- Linear algebra (math 21b) homework and exercises

- Textbook and solutions manual, Saylor Foundation.

- An Intuitive Guide to Linear Algebra on BetterExplained

19.11.1 Online books

- Beezer, Rob, *A First Course in Linear Algebra*

- Connell, Edwin H., *Elements of Abstract and Linear Algebra*

- Hefferon, Jim, *Linear Algebra*

- Matthews, Keith, *Elementary Linear Algebra*

- Sharipov, Ruslan, *Course of linear algebra and multidimensional geometry*

- Treil, Sergei, *Linear Algebra Done Wrong*

Chapter 20

Number theory

Not to be confused with Numerology.

Number theory (or **arithmetic**[note 1]) is a branch of pure mathematics devoted primarily to the study of the integers. It is sometimes called "The Queen of Mathematics" because of its foundational place in the discipline.[1] Number theorists study prime numbers as well as the properties of objects made out of integers (e.g., rational numbers) or defined as generalizations of the integers (e.g., algebraic integers).

Integers can be considered either in themselves or as solutions to equations (Diophantine geometry). Questions in number theory are often best understood through the study of analytical objects (e.g., the Riemann zeta function) that encode properties of the integers, primes or other number-theoretic objects in some fashion (analytic number theory). One may also study real numbers in relation to rational numbers, e.g., as approximated by the latter (Diophantine approximation).

The older term for number theory is *arithmetic*. By the early twentieth century, it had been superseded by "number theory".[note 2] (The word "arithmetic" is used by the general public to mean "elementary calculations"; it has also acquired other meanings in mathematical logic, as in *Peano arithmetic*, and computer science, as in *floating point arithmetic*.) The use of the term *arithmetic* for *number theory* regained some ground in the second half of the 20th century, arguably in part due to French influence.[note 3] In particular, *arithmetical* is preferred as an adjective to *number-theoretic*.

20.1 History

20.1.1 Origins

Dawn of arithmetic

The first historical find of an arithmetical nature is a fragment of a table: the broken clay tablet Plimpton 322 (Larsa, Mesopotamia, ca. 1800 BCE) contains a list of "Pythagorean triples", i.e., integers (a, b, c) such that $a^2 + b^2 = c^2$. The triples are too many and too large to have been obtained by brute force. The heading over the first column reads: "The *takiltum* of the diagonal which has been subtracted such that the width..."[2]

The table's layout suggests[3] that it was constructed by means of what amounts, in modern language, to the identity

$$\left(\tfrac{1}{2} \left(x - \tfrac{1}{x} \right) \right)^2 + 1 = \left(\tfrac{1}{2} \left(x + \tfrac{1}{x} \right) \right)^2.$$

which is implicit in routine Old Babylonian exercises.[4] If some other method was used,[5] the triples were first constructed and then reordered by c/a, presumably for actual use as a "table", i.e., with a view to applications.

It is not known what these applications may have been, or whether there could have been any; Babylonian astronomy, for example, truly flowered only later. It has been suggested instead that the table was a source of numerical examples for school problems.[6][note 4]

While Babylonian number theory—or what survives of Babylonian mathematics that can be called thus—consists of this single, striking fragment, Babylonian algebra (in the secondary-school sense of "algebra") was exceptionally well developed.[7] Late Neoplatonic sources[8] state that Pythagoras learned mathematics from the Babylonians. Much earlier sources[9] state that Thales and Pythagoras traveled and studied in Egypt.

Euclid IX 21—34 is very probably Pythagorean;[10] it is very simple material ("odd times even is even", "if an odd number measures [= divides] an even number, then it also measures [= divides] half of it"), but it is all that is needed to prove that $\sqrt{2}$ is irrational.[11] Pythagorean mystics gave great importance to the odd and the even.[12] The discovery that $\sqrt{2}$ is irrational is credited to the early Pythagoreans (pre-Theodorus).[13] By revealing (in modern terms) that numbers could be irrational, this discovery seems to have provoked the first foundational crisis in mathematical history; its proof or its divulgation are sometimes credited to Hippasus, who was expelled or split from the Pythagorean sect.[14] This forced a distinction between *numbers* (integers and the rationals—the subjects of arithmetic), on the one hand, and *lengths* and *proportions* (which we would identify with real numbers, whether rational or not), on the other hand.

The Pythagorean tradition spoke also of so-called polygonal or figurate numbers.[15] While square numbers, cubic numbers, etc., are seen now as more natural than triangular numbers, pentagonal numbers, etc., the study of the sums of triangular and pentagonal numbers would prove fruitful in the early modern period (17th to early 19th century).

We know of no clearly arithmetical material in ancient Egyptian or Vedic sources, though there is some algebra in both. The Chinese remainder theorem appears as an exercise [16] in Sun Zi's *Suan Ching*, also known as *The Mathematical Classic of Sun Zi* (3rd, 4th or 5th century CE.)[17] (There is one important step glossed over in Sun Zi's solution:[note 5] it is the problem that was later solved by Āryabhaṭa's kuṭṭaka – see below.)

There is also some numerical mysticism in Chinese mathematics,[note 6] but, unlike that of the Pythagoreans, it seems to have led nowhere. Like the Pythagoreans' perfect numbers, magic squares have passed from superstition into recreation.

Classical Greece and the early Hellenistic period

Aside from a few fragments, the mathematics of Classical Greece is known to us either through the reports of contemporary non-mathematicians or through mathematical works from the early Hellenistic period.[18] In the case of number theory, this means, by and large, *Plato* and *Euclid*, respectively.

Plato had a keen interest in mathematics, and distinguished clearly between arithmetic and calculation. (By *arithmetic* he meant, in part, theorising on number, rather than what *arithmetic* or *number theory* have come to mean.) It is through one of Plato's dialogues—namely, *Theaetetus*—that we know that Theodorus had proven that $\sqrt{3}, \sqrt{5}, \ldots, \sqrt{17}$ are irrational. Theaetetus was, like Plato, a disciple of Theodorus's; he worked on distinguishing different kinds of incommensurables, and was thus arguably a pioneer in the study of number systems. (Book X of Euclid's Elements is described by Pappus as being largely based on Theaetetus's work.)

Euclid devoted part of his *Elements* to prime numbers and divisibility, topics that belong unambiguously to number theory and are basic to it (Books VII to IX of Euclid's Elements). In particular, he gave an algorithm for computing the greatest common divisor of two numbers (the Euclidean algorithm; *Elements*, Prop. VII.2) and the first known proof of the infinitude of primes (*Elements*, Prop. IX.20).

In 1773, Lessing published an epigram he had found in a manuscript during his work as a librarian; it claimed to be a letter sent by Archimedes to Eratosthenes.[19][20] The epigram proposed what has become known as Archimedes' cattle problem; its solution (absent from the manuscript) requires solving an indeterminate quadratic equation (which reduces to what would later be misnamed Pell's equation). As far as we know, such equations were first successfully treated by the Indian school. It is not known whether Archimedes himself had a method of solution.

Diophantus

Very little is known about Diophantus of Alexandria; he probably lived in the third century CE, that is, about five hundred years after Euclid. Six out of the thirteen books of Diophantus's *Arithmetica* survive in the original Greek; four more books survive in an Arabic translation. The *Arithmetica* is a collection of worked-out problems where the task is invariably to find rational solutions to a system of polynomial equations, usually of the form $f(x, y) = z^2$ or $f(x, y, z) = w^2$. Thus, nowadays, we speak of *Diophantine equations* when we speak of polynomial equations to which rational or integer solutions must be found.

One may say that Diophantus was studying rational points — i.e., points whose coordinates are rational — on curves and algebraic varieties; however, unlike the Greeks of the Classical period, who did what we would now call basic algebra in geometrical terms, Diophantus did what we would now call basic algebraic geometry in purely algebraic terms. In modern language, what Diophantus did was to find rational parametrizations of varieties; that is, given an equation of the form (say) $f(x_1, x_2, x_3) = 0$, his aim was to find (in essence) three rational functions g_1, g_2, g_3 such that, for all values of r and s, setting $x_i = g_i(r, s)$ for $i = 1, 2, 3$ gives a solution to $f(x_1, x_2, x_3) = 0$.

Diophantus also studied the equations of some non-rational curves, for which no rational parametrisation is possible. He managed to find some rational points on these curves (elliptic curves, as it happens, in what seems to be their first known occurrence) by means of what amounts to a tangent construction: translated into coordinate geometry (which did not exist in Diophantus's time), his method would be visualised as drawing a tangent to a curve at a known rational point, and then finding the other point of intersection of the tangent with the curve; that other point is a new rational point. (Diophantus also resorted to what could be called a special case of a secant construction.)

While Diophantus was concerned largely with rational solutions, he assumed some results on integer numbers, in particular that every integer is the sum of four squares (though he never stated as much explicitly).

Āryabhaṭa, Brahmagupta, Bhāskara

While Greek astronomy probably influenced Indian learning, to the point of introducing trigonometry,[21] it seems to be the case that Indian mathematics is otherwise an indigenous tradition;[22] in particular, there is no evidence that Euclid's Elements reached India before the 18th century.[23]

Āryabhaṭa (476–550 CE) showed that pairs of simultaneous congruences $n \equiv a_1 \pmod{m}_1$, $n \equiv a_2 \pmod{m}_2$ could be solved by a method he called *kuṭṭaka*, or *pulveriser*;[24] this is a procedure close to (a generalisation of) the Euclidean algorithm, which was probably discovered independently in India.[25] Āryabhaṭa seems to have had in mind applications to astronomical calculations.[21]

Brahmagupta (628 CE) started the systematic study of indefinite quadratic equations—in particular, the misnamed Pell equation, in which Archimedes may have first been interested, and which did not start to be solved in the West until the time of Fermat and Euler. Later Sanskrit authors would follow, using Brahmagupta's technical terminology. A general procedure (the chakravala, or "cyclic method") for solving Pell's equation was finally found by Jayadeva (cited in the eleventh century; his work is otherwise lost); the earliest surviving exposition appears in Bhāskara II's Bīja-gaṇita (twelfth century).[26]

Unfortunately, Indian mathematics remained largely unknown in the West until the late eighteenth century;[27] Brahmagupta and Bhāskara's work was translated into English in 1817 by Henry Colebrooke.[28]

Arithmetic in the Islamic golden age

In the early ninth century, the caliph Al-Ma'mun ordered translations of many Greek mathematical works and at least one Sanskrit work (the *Sindhind*, which may [29] or may not[30] be Brahmagupta's *Brāhmasphuṭasiddhānta*). Diophantus's main work, the *Arithmetica*, was translated into Arabic by Qusta ibn Luqa (820–912). Part of the treatise *al-Fakhrī* (by al-Karajī, 953 – ca. 1029) builds on it to some extent. According to Rashed Roshdi, Al-Karajī's contemporary Ibn al-Haytham knew[31] what would later be called Wilson's theorem.

Western Europe in the Middle Ages

Other than a treatise on squares in arithmetic progression by Fibonacci — who lived and studied in north Africa and Constantinople during his formative years, ca. 1175–1200 — no number theory to speak of was done in western Europe during the Middle Ages. Matters started to change in Europe in the late Renaissance, thanks to a renewed study of the works of Greek antiquity. A catalyst was the textual emendation and translation into Latin of Diophantus's *Arithmetica* (Bachet, 1621, following a first attempt by Xylander, 1575).

20.1.2 Early modern number theory

Fermat

Pierre de Fermat (1601–1665) never published his writings; in particular, his work on number theory is contained almost entirely in letters to mathematicians and in private marginal notes.[32] He wrote down nearly no proofs in number theory; he had no models in the area.[33] He did make repeated use of mathematical induction, introducing the method of infinite descent.

One of Fermat's first interests was perfect numbers (which appear in Euclid, *Elements* IX) and amicable numbers;[note 7] this led him to work on integer divisors, which were from the beginning among the subjects of the correspondence (1636 onwards) that put him in touch with the mathematical community of the day.[34] He had already studied Bachet's edition of Diophantus carefully;[35] by 1643, his interests had shifted largely to Diophantine problems and sums of squares[36] (also treated by Diophantus).

Fermat's achievements in arithmetic include:

- Fermat's little theorem (1640),[37] stating that, if a is not divisible by a prime p, then $a^{p-1} \equiv 1 \pmod{p}$. [note 8]

- If a and b are coprime, then $a^2 + b^2$ is not divisible by any prime congruent to −1 modulo 4;[38] *and* Every prime congruent to 1 modulo 4 can be written in the form $a^2 + b^2$.[39] These two statements also date from 1640; in 1659, Fermat stated to Huygens that he had proven the latter statement by the method of infinite descent.[40] Fermat and Frenicle also did some work (some of it erroneous)[41] on other quadratic forms.

- Fermat posed the problem of solving $x^2 - Ny^2 = 1$ as a challenge to English mathematicians (1657). The problem was solved in a few months by Wallis and Brouncker.[42] Fermat considered their solution valid, but pointed out they had provided an algorithm without a proof (as had Jayadeva and Bhaskara, though Fermat would never know this.) He states that a proof can be found by descent.

- Fermat developed methods for (doing what in our terms amounts to) finding points on curves of genus 0 and 1. As in Diophantus, there are many special procedures and what amounts to a tangent construction, but no use of a secant construction.[43]

- Fermat states and proves (by descent) in the appendix to *Observations on Diophantus* (Obs. XLV)[44] that $x^4 + y^4 = z^4$ has no non-trivial solutions in the integers. Fermat also mentioned to his correspondents that $x^3 + y^3 = z^3$ has no non-trivial solutions, and that this could be proven by descent.[45] The first known proof is due to Euler (1753; indeed by descent).[46]

Fermat's claim ("Fermat's last theorem") to have shown there are no solutions to $x^n + y^n = z^n$ for all $n \geq 3$ (a fact the only known proof of which is beyond his methods) appears only in his annotations on the margin of his copy of Diophantus; he never claimed this to others[47] and thus would have had no need to retract it if he found any mistake in his supposed proof.

Euler

The interest of Leonhard Euler (1707–1783) in number theory was first spurred in 1729, when a friend of his, the amateur[note 9] Goldbach, pointed him towards some of Fermat's work on the subject.[48][49] This has been called the "rebirth" of modern number theory,[35] after Fermat's relative lack of success in getting his contemporaries' attention for the subject.[50] Euler's work on number theory includes the following:[51]

- *Proofs for Fermat's statements.* This includes Fermat's little theorem (generalised by Euler to non-prime moduli); the fact that $p = x^2 + y^2$ if and only if $p \equiv 1 \bmod 4$; initial work towards a proof that every integer is the sum of four squares (the first complete proof is by Joseph-Louis Lagrange (1770), soon improved by Euler himself[52]); the lack of non-zero integer solutions to $x^4 + y^4 = z^2$ (implying the case *n=4* of Fermat's last theorem, the case *n=3* of which Euler also proved by a related method).

- *Pell's equation*, first misnamed by Euler.[53] He wrote on the link between continued fractions and Pell's equation.[54]

- *First steps towards analytic number theory.* In his work of sums of four squares, partitions, pentagonal numbers, and the distribution of prime numbers, Euler pioneered the use of what can be seen as analysis (in particular, infinite series) in number theory. Since he lived before the development of complex analysis, most of his work is restricted to the formal manipulation of power series. He did, however, do some very notable (though not fully rigorous) early work on what would later be called the Riemann zeta function.[55]

- *Quadratic forms.* Following Fermat's lead, Euler did further research on the question of which primes can be expressed in the form $x^2 + Ny^2$, some of it prefiguring quadratic reciprocity.[56] [57][58]

- *Diophantine equations.* Euler worked on some Diophantine equations of genus 0 and 1.[59][60] In particular, he studied Diophantus's work; he tried to systematise it, but the time was not yet ripe for such an endeavour – algebraic geometry was still in its infancy.[61] He did notice there was a connection between Diophantine problems and elliptic integrals,[61] whose study he had himself initiated.

Lagrange, Legendre and Gauss

Joseph-Louis Lagrange (1736–1813) was the first to give full proofs of some of Fermat's and Euler's work and observations - for instance, the four-square theorem and the basic theory of the misnamed "Pell's equation" (for which an algorithmic solution was found by Fermat and his contemporaries, and also by Jayadeva and Bhaskara II before them.) He also studied quadratic forms in full generality (as opposed to $mX^2 + nY^2$) — defining their equivalence relation, showing how to put them in reduced form, etc.

Adrien-Marie Legendre (1752–1833) was the first to state the law of quadratic reciprocity. He also conjectured what amounts to the prime number theorem and Dirichlet's theorem on arithmetic progressions. He gave a full treatment of the equation $ax^2 + by^2 + cz^2 = 0$ [62] and worked on quadratic forms along the lines later developed fully by Gauss.[63] In his old age, he was the first to prove "Fermat's last theorem" for $n = 5$ (completing work by Peter Gustav Lejeune Dirichlet, and crediting both him and Sophie Germain).[64]

In his *Disquisitiones Arithmeticae* (1798), Carl Friedrich Gauss (1777–1855) proved the law of quadratic reciprocity and developed the theory of quadratic forms (in particular, defining their composition). He also introduced some basic notation (congruences) and devoted a section to computational matters, including primality tests.[65] The last section of the *Disquisitiones* established a link between roots of unity and number theory:

> The theory of the division of the circle...which is treated in sec. 7 does not belong by itself to arithmetic, but its principles can only be drawn from higher arithmetic.[66]

In this way, Gauss arguably made a first foray towards both Évariste Galois's work and algebraic number theory.

20.1.3   Maturity and division into subfields

Starting early in the nineteenth century, the following developments gradually took place:

- The rise to self-consciousness of number theory (or *higher arithmetic*) as a field of study.[67]

- The development of much of modern mathematics necessary for basic modern number theory: complex analysis, group theory, Galois theory—accompanied by greater rigor in analysis and abstraction in algebra.

- The rough subdivision of number theory into its modern subfields—in particular, analytic and algebraic number theory.

Algebraic number theory may be said to start with the study of reciprocity and cyclotomy, but truly came into its own with the development of abstract algebra and early ideal theory and valuation theory; see below. A conventional starting point for analytic number theory is Dirichlet's theorem on arithmetic progressions (1837),[68] [69] whose proof introduced

L-functions and involved some asymptotic analysis and a limiting process on a real variable.[70] The first use of analytic ideas in number theory actually goes back to Euler (1730s),[71] [72] who used formal power series and non-rigorous (or implicit) limiting arguments. The use of *complex* analysis in number theory comes later: the work of Bernhard Riemann (1859) on the zeta function is the canonical starting point;[73] Jacobi's four-square theorem (1839), which predates it, belongs to an initially different strand that has by now taken a leading role in analytic number theory (modular forms).[74]

The history of each subfield is briefly addressed in its own section below; see the main article of each subfield for fuller treatments. Many of the most interesting questions in each area remain open and are being actively worked on.

20.2 Main subdivisions

20.2.1 Elementary tools

The term *elementary* generally denotes a method that does not use complex analysis. For example, the prime number theorem was first proven using complex analysis in 1896, but an elementary proof was found only in 1949 by Erdős and Selberg.[75] The term is somewhat ambiguous: for example, proofs based on complex Tauberian theorems (e.g. Wiener–Ikehara) are often seen as quite enlightening but not elementary, in spite of using Fourier analysis, rather than complex analysis as such. Here as elsewhere, an *elementary* proof may be longer and more difficult for most readers than a non-elementary one.

Number theory has the reputation of being a field many of whose results can be stated to the layperson. At the same time, the proofs of these results are not particularly accessible, in part because the range of tools they use is, if anything, unusually broad within mathematics.[76]

20.2.2 Analytic number theory

Main article: Analytic number theory
Analytic number theory may be defined

- in terms of its tools, as the study of the integers by means of tools from real and complex analysis;[68] or

- in terms of its concerns, as the study within number theory of estimates on size and density, as opposed to identities.[77]

Some subjects generally considered to be part of analytic number theory, e.g., sieve theory,[note 10] are better covered by the second rather than the first definition: some of sieve theory, for instance, uses little analysis,[note 11] yet it does belong to analytic number theory.

The following are examples of problems in analytic number theory: the prime number theorem, the Goldbach conjecture (or the twin prime conjecture, or the Hardy–Littlewood conjectures), the Waring problem and the Riemann Hypothesis. Some of the most important tools of analytic number theory are the circle method, sieve methods and L-functions (or, rather, the study of their properties). The theory of modular forms (and, more generally, automorphic forms) also occupies an increasingly central place in the toolbox of analytic number theory.[78]

One may ask analytic questions about algebraic numbers, and use analytic means to answer such questions; it is thus that algebraic and analytic number theory intersect. For example, one may define prime ideals (generalizations of prime numbers in the field of algebraic numbers) and ask how many prime ideals there are up to a certain size. This question can be answered by means of an examination of Dedekind zeta functions, which are generalizations of the Riemann zeta function, a key analytic object at the roots of the subject.[79] This is an example of a general procedure in analytic number theory: deriving information about the distribution of a sequence (here, prime ideals or prime numbers) from the analytic behavior of an appropriately constructed complex-valued function.[80]

20.2.3   Algebraic number theory

Main article: Algebraic number theory

An *algebraic number* is any complex number that is a solution to some polynomial equation $f(x) = 0$ with rational coefficients; for example, every solution x of $x^5 + (11/2)x^3 - 7x^2 + 9 = 0$ (say) is an algebraic number. Fields of algebraic numbers are also called *algebraic number fields*, or shortly *number fields*. Algebraic number theory studies algebraic number fields.[81] Thus, analytic and algebraic number theory can and do overlap: the former is defined by its methods, the latter by its objects of study.

It could be argued that the simplest kind of number fields (viz., quadratic fields) were already studied by Gauss, as the discussion of quadratic forms in *Disquisitiones arithmeticae* can be restated in terms of ideals and norms in quadratic fields. (A *quadratic field* consists of all numbers of the form $a + b\sqrt{d}$, where a and b are rational numbers and d is a fixed rational number whose square root is not rational.) For that matter, the 11th-century chakravala method amounts—in modern terms—to an algorithm for finding the units of a real quadratic number field. However, neither Bhāskara nor Gauss knew of number fields as such.

The grounds of the subject as we know it were set in the late nineteenth century, when *ideal numbers*, the *theory of ideals* and *valuation theory* were developed; these are three complementary ways of dealing with the lack of unique factorisation in algebraic number fields. (For example, in the field generated by the rationals and $\sqrt{-5}$, the number 6 can be factorised both as $6 = 2 \cdot 3$ and $6 = (1 + \sqrt{-5})(1 - \sqrt{-5})$; all of 2, 3, $1 + \sqrt{-5}$ and $1 - \sqrt{-5}$ are irreducible, and thus, in a naïve sense, analogous to primes among the integers.) The initial impetus for the development of ideal numbers (by Kummer) seems to have come from the study of higher reciprocity laws,[82] i.e., generalisations of quadratic reciprocity.

Number fields are often studied as extensions of smaller number fields: a field L is said to be an *extension* of a field K if L contains K. (For example, the complex numbers C are an extension of the reals R, and the reals R are an extension of the rationals Q.) Classifying the possible extensions of a given number field is a difficult and partially open problem. Abelian extensions—that is, extensions L of K such that the Galois group[note 12] Gal(L/K) of L over K is an abelian group—are relatively well understood. Their classification was the object of the programme of class field theory, which was initiated in the late 19th century (partly by Kronecker and Eisenstein) and carried out largely in 1900—1950.

An example of an active area of research in algebraic number theory is Iwasawa theory. The Langlands program, one of the main current large-scale research plans in mathematics, is sometimes described as an attempt to generalise class field theory to non-abelian extensions of number fields.

20.2.4   Diophantine geometry

Main articles: Diophantine geometry and Glossary of arithmetic and Diophantine geometry

The central problem of *Diophantine geometry* is to determine when a Diophantine equation has solutions, and if it does, how many. The approach taken is to think of the solutions of an equation as a geometric object.

For example, an equation in two variables defines a curve in the plane. More generally, an equation, or system of equations, in two or more variables defines a curve, a surface or some other such object in n-dimensional space. In Diophantine geometry, one asks whether there are any *rational points* (points all of whose coordinates are rationals) or *integral points* (points all of whose coordinates are integers) on the curve or surface. If there are any such points, the next step is to ask how many there are and how they are distributed. A basic question in this direction is: are there finitely or infinitely many rational points on a given curve (or surface)? What about integer points?

An example here may be helpful. Consider the Pythagorean equation $x^2 + y^2 = 1$; we would like to study its rational solutions, i.e., its solutions (x, y) such that x and y are both rational. This is the same as asking for all integer solutions to $a^2 + b^2 = c^2$; any solution to the latter equation gives us a solution $x = a/c$, $y = b/c$ to the former. It is also the same as asking for all points with rational coordinates on the curve described by $x^2 + y^2 = 1$. (This curve happens to be a circle of radius 1 around the origin.)

The rephrasing of questions on equations in terms of points on curves turns out to be felicitous. The finiteness or not of the number of rational or integer points on an algebraic curve—that is, rational or integer solutions to an equation $f(x, y) = 0$, where f is a polynomial in two variables—turns out to depend crucially on the *genus* of the curve. The *genus* can be defined as follows:[note 13] allow the variables in $f(x, y) = 0$ to be complex numbers; then $f(x, y) = 0$ defines a 2-dimensional surface in (projective) 4-dimensional space (since two complex variables can be decomposed into four real variables, i.e., four dimensions). Count the number of (doughnut) holes in the surface; call this number the *genus* of $f(x, y) = 0$. Other geometrical notions turn out to be just as crucial.

There is also the closely linked area of Diophantine approximations: given a number x, how well can it be approximated by rationals? (We are looking for approximations that are good relative to the amount of space that it takes to write the rational: call a/q (with $\gcd(a, q) = 1$) a good approximation to x if $|x - a/q| < \frac{1}{q^c}$, where c is large.) This question is of special interest if x is an algebraic number. If x cannot be well approximated, then some equations do not have integer or rational solutions. Moreover, several concepts (especially that of height) turn out to be crucial both in Diophantine geometry and in the study of Diophantine approximations. This question is also of special interest in transcendence theory: if a number can be better approximated than any algebraic number, then it is a transcendental number. It is by this argument that π and e have been shown to be transcendental.

Diophantine geometry should not be confused with the geometry of numbers, which is a collection of graphical methods for answering certain questions in algebraic number theory. *Arithmetic geometry*, on the other hand, is a contemporary term for much the same domain as that covered by the term *Diophantine geometry*. The term *arithmetic geometry* is arguably used most often when one wishes to emphasise the connections to modern algebraic geometry (as in, for instance, Faltings' theorem) rather than to techniques in Diophantine approximations.

20.3 Recent approaches and subfields

The areas below date as such from no earlier than the mid-twentieth century, even if they are based on older material. For example, as is explained below, the matter of algorithms in number theory is very old, in some sense older than the concept of proof; at the same time, the modern study of computability dates only from the 1930s and 1940s, and computational complexity theory from the 1970s.

20.3.1 Probabilistic number theory

Main article: Probabilistic number theory

Take a number at random between one and a million. How likely is it to be prime? This is just another way of asking how many primes there are between one and a million. Further: how many prime divisors will it have, on average? How many divisors will it have altogether, and with what likelihood? What is the probability that it have many more or many fewer divisors or prime divisors than the average?

Much of probabilistic number theory can be seen as an important special case of the study of variables that are almost, but not quite, mutually independent. For example, the event that a random integer between one and a million be divisible by two and the event that it be divisible by three are almost independent, but not quite.

It is sometimes said that probabilistic combinatorics uses the fact that whatever happens with probability greater than 0 must happen sometimes; one may say with equal justice that many applications of probabilistic number theory hinge on the fact that whatever is unusual must be rare. If certain algebraic objects (say, rational or integer solutions to certain equations) can be shown to be in the tail of certain sensibly defined distributions, it follows that there must be few of them; this is a very concrete non-probabilistic statement following from a probabilistic one.

At times, a non-rigorous, probabilistic approach leads to a number of heuristic algorithms and open problems, notably Cramér's conjecture.

20.3.2   Arithmetic combinatorics

Main articles: Arithmetic combinatorics and Additive number theory

Let A be a set of N integers. Consider the set $A + A = \{\ m + n \mid m, n \in A\ \}$ consisting of all sums of two elements of A. Is $A + A$ much larger than A? Barely larger? If $A + A$ is barely larger than A, must A have plenty of arithmetic structure, for example, does A resemble an arithmetic progression?

If we begin from a fairly "thick" infinite set A, does it contain many elements in arithmetic progression: a, $a + b$, $a + 2b$, $a + 3b$, ..., $a + 10b$, say? Should it be possible to write large integers as sums of elements of A?

These questions are characteristic of *arithmetic combinatorics*. This is a presently coalescing field; it subsumes *additive number theory* (which concerns itself with certain very specific sets A of arithmetic significance, such as the primes or the squares) and, arguably, some of the *geometry of numbers*, together with some rapidly developing new material. Its focus on issues of growth and distribution accounts in part for its developing links with ergodic theory, finite group theory, model theory, and other fields. The term *additive combinatorics* is also used; however, the sets A being studied need not be sets of integers, but rather subsets of non-commutative groups, for which the multiplication symbol, not the addition symbol, is traditionally used; they can also be subsets of rings, in which case the growth of $A + A$ and $A \cdot A$ may be compared.

20.3.3   Computations in number theory

Main article: Computational number theory

While the word *algorithm* goes back only to certain readers of al-Khwārizmī, careful descriptions of methods of solution are older than proofs: such methods (that is, algorithms) are as old as any recognisable mathematics—ancient Egyptian, Babylonian, Vedic, Chinese—whereas proofs appeared only with the Greeks of the classical period. An interesting early case is that of what we now call the Euclidean algorithm. In its basic form (namely, as an algorithm for computing the greatest common divisor) it appears as Proposition 2 of Book VII in *Elements*, together with a proof of correctness. However, in the form that is often used in number theory (namely, as an algorithm for finding integer solutions to an equation $ax + by = c$, or, what is the same, for finding the quantities whose existence is assured by the Chinese remainder theorem) it first appears in the works of Āryabhaṭa (5th–6th century CE) as an algorithm called *kuṭṭaka* ("pulveriser"), without a proof of correctness.

There are two main questions: "can we compute this?" and "can we compute it rapidly?". Anybody can test whether a number is prime or, if it is not, split it into prime factors; doing so rapidly is another matter. We now know fast algorithms for testing primality, but, in spite of much work (both theoretical and practical), no truly fast algorithm for factoring.

The difficulty of a computation can be useful: modern protocols for encrypting messages (e.g., RSA) depend on functions that are known to all, but whose inverses (a) are known only to a chosen few, and (b) would take one too long a time to figure out on one's own. For example, these functions can be such that their inverses can be computed only if certain large integers are factorized. While many difficult computational problems outside number theory are known, most working encryption protocols nowadays are based on the difficulty of a few number-theoretical problems.

On a different note — some things may not be computable at all; in fact, this can be proven in some instances. For instance, in 1970, it was proven, as a solution to Hilbert's 10th problem, that there is no Turing machine which can solve all Diophantine equations.[83] In particular, this means that, given a computably enumerable set of axioms, there are Diophantine equations for which there is no proof, starting from the axioms, of whether the set of equations has or does not have integer solutions. (We would necessarily be speaking of Diophantine equations for which there are no integer solutions, since, given a Diophantine equation with at least one solution, the solution itself provides a proof of the fact that a solution exists. We cannot prove, of course, that a particular Diophantine equation is of this kind, since this would imply that it has no solutions.)

20.4 Applications

The number-theorist Leonard Dickson (1874-1954) said "Thank God that number theory is unsullied by any application". Such a view is no longer applicable to number theory.[84] In 1974, Donald Knuth said "...virtually every theorem in elementary number theory arises in a natural, motivated way in connection with the problem of making computers do high-speed numerical calculations".[85] Elementary number theory is taught in discrete mathematics courses for computer scientists; and, on the other hand, number theory also has applications to the continuous in numerical analysis.[86] As well as the well-known applications to cryptography, there are also applications to many other areas of mathematics.[87][88]

20.5 Literature

Two of the most popular introductions to the subject are:

- G. H. Hardy; E. M. Wright (2008) [1938]. *An introduction to the theory of numbers* (rev. by D. R. Heath-Brown and J. H. Silverman, 6th ed.). Oxford University Press. ISBN 978-0-19-921986-5.

- Vinogradov, I. M. (2003) [1954]. *Elements of Number Theory* (reprint of the 1954 ed.). Mineola, NY: Dover Publications.

Hardy and Wright's book is a comprehensive classic, though its clarity sometimes suffers due to the authors' insistence on elementary methods.[89] Vinogradov's main attraction consists in its set of problems, which quickly lead to Vinogradov's own research interests; the text itself is very basic and close to minimal. Other popular first introductions are:

- Ivan M. Niven; Herbert S. Zuckerman; Hugh L. Montgomery (2008) [1960]. *An introduction to the theory of numbers* (reprint of the 5th edition 1991 ed.). John Wiley & Sons. ISBN 978-8-12-651811-1.

- Kenneth H. Rosen (2010). *Elementary Number Theory* (6th ed.). Pearson Education. ISBN 978-0-32-171775-7.

Popular choices for a second textbook include:

- Borevich, A. I.; Shafarevich, Igor R. (1966). *Number theory*. Pure and Applied Mathematics **20**. Boston, MA: Academic Press. ISBN 978-0-12-117850-5. MR 0195803.

- Serre, Jean-Pierre (1996) [1973]. *A course in arithmetic*. Graduate texts in mathematics **7**. Springer. ISBN 978-0-387-90040-7.

20.6 Prizes

The American Mathematical Society awards the *Cole Prize in Number Theory*. Moreover number theory is one of the three mathematical subdisciplines rewarded by the *Fermat Prize*.

20.7 See also

- Algebraic function field
- Finite field
- p-adic number

20.8   Notes

[1] Especially in older sources; see two following notes.

[2] Already in 1921, T. L. Heath had to explain: "By arithmetic, Plato meant, not arithmetic in our sense, but the science which considers numbers in themselves, in other words, what we mean by the Theory of Numbers." (Heath 1921, p. 13)

[3] Take, e.g. Serre 1973. In 1952, Davenport still had to specify that he meant *The Higher Arithmetic*. Hardy and Wright wrote in the introduction to *An Introduction to the Theory of Numbers* (1938): "We proposed at one time to change [the title] to *An introduction to arithmetic*, a more novel and in some ways a more appropriate title; but it was pointed out that this might lead to misunderstandings about the content of the book." (Hardy & Wright 2008)

[4] Robson 2001, p. 201. This is controversial. See Plimpton 322. Robson's article is written polemically (Robson 2001, p. 202) with a view to "perhaps [...] knocking [Plimpton 322] off its pedestal" (Robson 2001, p. 167); at the same time, it settles to the conclusion that

> [...] the question "how was the tablet calculated?" does not have to have the same answer as the question "what problems does the tablet set?" The first can be answered most satisfactorily by reciprocal pairs, as first suggested half a century ago, and the second by some sort of right-triangle problems (Robson 2001, p. 202).

Robson takes issue with the notion that the scribe who produced Plimpton 322 (who had to "work for a living", and would not have belonged to a "leisured middle class") could have been motivated by his own "idle curiosity" in the absence of a "market for new mathematics".(Robson 2001, pp. 199–200)

[5] Sun Zi, *Suan Ching*, Ch. 3, Problem 26, in Lam & Ang 2004, pp. 219–220:

> [26] Now there are an unknown number of things. If we count by threes, there is a remainder 2; if we count by fives, there is a remainder 3; if we count by sevens, there is a remainder 2. Find the number of things. *Answer*: 23.
> *Method*: If we count by threes and there is a remainder 2, put down 140. If we count by fives and there is a remainder 3, put down 63. If we count by sevens and there is a remainder 2, put down 30. Add them to obtain 233 and subtract 210 to get the answer. If we count by threes and there is a remainder 1, put down 70. If we count by fives and there is a remainder 1, put down 21. If we count by sevens and there is a remainder 1, put down 15. When [a number] exceeds 106, the result is obtained by subtracting 105.

[6] See, e.g., Sun Zi, *Suan Ching*, Ch. 3, Problem 36, in Lam & Ang 2004, pp. 223–224:

> [36] Now there is a pregnant woman whose age is 29. If the gestation period is 9 months, determine the sex of the unborn child. *Answer*: Male.
> *Method*: Put down 49, add the gestation period and subtract the age. From the remainder take away 1 representing the heaven, 2 the earth, 3 the man, 4 the four seasons, 5 the five phases, 6 the six pitch-pipes, 7 the seven stars [of the Dipper], 8 the eight winds, and 9 the nine divisions [of China under Yu the Great]. If the remainder is odd, [the sex] is male and if the remainder is even, [the sex] is female.

This is the last problem in Sun Zi's otherwise matter-of-fact treatise.

[7] Perfect and especially amicable numbers are of little or no interest nowadays. The same was not true in medieval times – whether in the West or the Arab-speaking world – due in part to the importance given to them by the Neopythagorean (and hence mystical) Nicomachus (ca. 100 CE), who wrote a primitive but influential "Introduction to Arithmetic". See van der Waerden 1961, Ch. IV.

[8] Here, as usual, given two integers a and b and a non-zero integer m, we write $a \equiv b \pmod{m}$ (read "a is congruent to b modulo m") to mean that m divides $a - b$, or, what is the same, a and b leave the same residue when divided by m. This notation is actually much later than Fermat's; it first appears in section 1 of Gauss's Disquisitiones Arithmeticae. Fermat's little theorem is a consequence of the fact that the order of an element of a group divides the order of the group. The modern proof would have been within Fermat's means (and was indeed given later by Euler), even though the modern concept of a group came long after Fermat or Euler. (It helps to know that inverses exist modulo p (i.e., given a not divisible by a prime p, there is an integer x such that $xa \equiv 1 \pmod{p}$); this fact (which, in modern language, makes the residues mod p into a group, and which was already known to Āryabhaṭa; see above) was familiar to Fermat thanks to its rediscovery by Bachet (Weil 1984, p. 7). Weil goes on to say that Fermat would have recognised that Bachet's argument is essentially Euclid's algorithm.

[9] Up to the second half of the seventeenth century, academic positions were very rare, and most mathematicians and scientists earned their living in some other way (Weil 1984, pp. 159, 161). (There were already some recognisable features of professional *practice*, viz., seeking correspondents, visiting foreign colleagues, building private libraries (Weil 1984, pp. 160–161). Matters started to shift in the late 17th century (Weil 1984, p. 161); scientific academies were founded in England (the Royal Society, 1662) and France (the Académie des sciences, 1666) and Russia (1724). Euler was offered a position at this last one in 1726; he accepted, arriving in St. Petersburg in 1727 (Weil 1984, p. 163 and Varadarajan 2006, p. 7). In this context, the term *amateur* usually applied to Goldbach is well-defined and makes some sense: he has been described as a man of letters who earned a living as a spy (Truesdell 1984, p. xv); cited in Varadarajan 2006, p. 9). Notice, however, that Goldbach published some works on mathematics and sometimes held academic positions.

[10] Sieve theory figures as one of the main subareas of analytic number theory in many standard treatments; see, for instance, Iwaniec & Kowalski 2004 or Montgomery & Vaughan 2007

[11] This is the case for small sieves (in particular, some combinatorial sieves such as the Brun sieve) rather than for large sieves; the study of the latter now includes ideas from harmonic and functional analysis.

[12] The Galois group of an extension K/L consists of the operations (isomorphisms) that send elements of L to other elements of L while leaving all elements of K fixed. Thus, for instance, $Gal(C/R)$ consists of two elements: the identity element (taking every element $x + iy$ of C to itself) and complex conjugation (the map taking each element $x + iy$ to $x - iy$). The Galois group of an extension tells us many of its crucial properties. The study of Galois groups started with Évariste Galois; in modern language, the main outcome of his work is that an equation $f(x) = 0$ can be solved by radicals (that is, x can be expressed in terms of the four basic operations together with square roots, cubic roots, etc.) if and only if the extension of the rationals by the roots of the equation $f(x) = 0$ has a Galois group that is solvable in the sense of group theory. ("Solvable", in the sense of group theory, is a simple property that can be checked easily for finite groups.)

[13] It may be useful to look at an example here. Say we want to study the curve $y^2 = x^3 + 7$. We allow x and y to be complex numbers: $(a + bi)^2 = (c + di)^3 + 7$. This is, in effect, a set of two equations on four variables, since both the real and the imaginary part on each side must match. As a result, we get a surface (two-dimensional) in four-dimensional space. After we choose a convenient hyperplane on which to project the surface (meaning that, say, we choose to ignore the coordinate a), we can plot the resulting projection, which is a surface in ordinary three-dimensional space. It then becomes clear that the result is a torus, i.e., the surface of a doughnut (somewhat stretched). A doughnut has one hole; hence the genus is 1.

20.9   References

[1] Long 1972, p. 1.

[2] Neugebauer & Sachs 1945, p. 40. The term *takiltum* is problematic. Robson prefers the rendering "The holding-square of the diagonal from which 1 is torn out, so that the short side comes up...".Robson 2001, p. 192

[3] Robson 2001, p. 189. Other sources give the modern formula $(p^2 - q^2, 2pq, p^2 + q^2)$. Van der Waerden gives both the modern formula and what amounts to the form preferred by Robson.(van der Waerden 1961, p. 79)

[4] van der Waerden 1961, p. 184.

[5] Neugebauer (Neugebauer 1969, pp. 36–40) discusses the table in detail and mentions in passing Euclid's method in modern notation (Neugebauer 1969, p. 39).

[6] Friberg 1981, p. 302.

[7] van der Waerden 1961, p. 43.

[8] Iamblichus, *Life of Pythagoras*,(trans. e.g. Guthrie 1987) cited in van der Waerden 1961, p. 108. See also Porphyry, *Life of Pythagoras*, paragraph 6, in Guthrie 1987 Van der Waerden (van der Waerden 1961, pp. 87–90) sustains the view that Thales knew Babylonian mathematics.

[9] Herodotus (II. 81) and Isocrates (*Busiris* 28), cited in: Huffman 2011. On Thales, see Eudemus ap. Proclus, 65.7, (e.g. Morrow 1992, p. 52) cited in: O'Grady 2004, p. 1. Proclus was using a work by Eudemus of Rhodes (now lost), the *Catalogue of Geometers*. See also introduction, Morrow 1992, p. xxx on Proclus' reliability.

[10] Becker 1936, p. 533, cited in: van der Waerden 1961, p. 108.

[11] Becker 1936.

[12] van der Waerden 1961, p. 109.

[13] Plato, *Theaetetus*, p. 147 B, (e.g. Jowett 1871), cited in von Fritz 2004, p. 212: "Theodorus was writing out for us something about roots, such as the roots of three or five, showing that they are incommensurable by the unit:..." *See also* Spiral of Theodorus.

[14] von Fritz 2004.

[15] Heath 1921, p. 76.

[16] Sun Zi, *Suan Ching*, Chapter 3, Problem 26. This can be found in Lam & Ang 2004, pp. 219–220, which contains a full translation of the *Suan Ching* (based on Qian 1963). See also the discussion in Lam & Ang 2004, pp. 138–140.

[17] The date of the text has been narrowed down to 220–420 AD (Yan Dunjie) or 280–473 AD (Wang Ling) through internal evidence (= taxation systems assumed in the text). See Lam & Ang 2004, pp. 27–28.

[18] Boyer & Merzbach 1991, p. 82.

[19] Vardi 1998, p. 305-319.

[20] Weil 1984, pp. 17–24.

[21] Plofker 2008, p. 119.

[22] Any early contact between Babylonian and Indian mathematics remains conjectural (Plofker 2008, p. 42).

[23] Mumford 2010, p. 387.

[24] Āryabhata, Āryabhatīya, Chapter 2, verses 32–33, cited in: Plofker 2008, pp. 134–140. See also Clark 1930, pp. 42–50. A slightly more explicit description of the kuṭṭaka was later given in Brahmagupta, *Brāhmasphuṭasiddhānta*, XVIII, 3–5 (in Colebrooke 1817, p. 325, cited in Clark 1930, p. 42).

[25] Mumford 2010, p. 388.

[26] Plofker 2008, p. 194.

[27] Plofker 2008, p. 283.

[28] Colebrooke 1817.

[29] Colebrooke 1817, p. lxv, cited in Hopkins 1990, p. 302. See also the preface in Sachau 1888 cited in Smith 1958, pp. 168

[30] Pingree 1968, pp. 97–125, and Pingree 1970, pp. 103–123, cited in Plofker 2008, p. 256.

[31] Rashed 1980, p. 305–321.

[32] Weil 1984, pp. 45–46.

[33] Weil 1984, p. 118. This was more so in number theory than in other areas (remark in Mahoney 1994, p. 284). Bachet's own proofs were "ludicrously clumsy" (Weil 1984, p. 33).

[34] Mahoney 1994, pp. 48, 53–54. The initial subjects of Fermat's correspondence included divisors ("aliquot parts") and many subjects outside number theory; see the list in the letter from Fermat to Roberval, 22.IX.1636, Tannery & Henry 1891, Vol. II, pp. 72, 74, cited in Mahoney 1994, p. 54.

[35] Weil 1984, pp. 1–2.

[36] Weil 1984, p. 53.

[37] Tannery & Henry 1891, Vol. II, p. 209, Letter XLVI from Fermat to Frenicle, 1640, cited in Weil 1984, p. 56

[38] Tannery & Henry 1891, Vol. II, p. 204, cited in Weil 1984, p. 63. All of the following citations from Fermat's *Varia Opera* are taken from Weil 1984, Chap. II. The standard Tannery & Henry work includes a revision of Fermat's posthumous *Varia Opera Mathematica* originally prepared by his son (Fermat 1679).

[39] Tannery & Henry 1891, Vol. II, p. 213.

[40] Tannery & Henry 1891, Vol. II, p. 423.

[41] Weil 1984, pp. 80, 91–92.

[42] Weil 1984, p. 92.

[43] Weil 1984, Ch. II, sect. XV and XVI.

[44] Tannery & Henry 1891, Vol. I, pp. 340–341.

[45] Weil 1984, p. 115.

[46] Weil 1984, pp. 115–116.

[47] Weil 1984, p. 104.

[48] Weil 1984, pp. 2, 172.

[49] Varadarajan 2006, p. 9.

[50] Weil 1984, p. 2 and Varadarajan 2006, p. 37

[51] Varadarajan 2006, p. 39 and Weil 1984, pp. 176–189

[52] Weil 1984, pp. 178–179.

[53] Weil 1984, p. 174. Euler was generous in giving credit to others (Varadarajan 2006, p. 14), not always correctly.

[54] Weil 1984, p. 183.

[55] Varadarajan 2006, pp. 45–55; see also chapter III.

[56] Varadarajan 2006, pp. 44–47.

[57] Weil 1984, pp. 177–179.

[58] Edwards 1983, pp. 285–291.

[59] Varadarajan 2006, pp. 55–56.

[60] Weil 1984, pp. 179–181.

[61] Weil 1984, p. 181.

[62] Weil 1984, pp. 327–328.

[63] Weil 1984, pp. 332–334.

[64] Weil 1984, pp. 337–338.

[65] Goldstein & Schappacher 2007, p. 14.

[66] From the preface of *Disquisitiones Arithmeticae*; the translation is taken from Goldstein & Schappacher 2007, p. 16

[67] See the discussion in section 5 of Goldstein & Schappacher 2007. Early signs of self-consciousness are present already in letters by Fermat: thus his remarks on what number theory is, and how "Diophantus's work [...] does not really belong to [it]" (quoted in Weil 1984, p. 25).

[68] Apostol 1976, p. 7.

[69] Davenport & Montgomery 2000, p. 1.

[70] See the proof in Davenport & Montgomery 2000, section 1

[71] Iwaniec & Kowalski 2004, p. 1.

[72] Varadarajan 2006, sections 2.5, 3.1 and 6.1.

[73] Granville 2008, pp. 322–348.

[74] See the comment on the importance of modularity in Iwaniec & Kowalski 2004, p. 1

[75] Goldfeld 2003.

[76] See, e.g., the initial comment in Iwaniec & Kowalski 2004, p. 1.

[77] Granville 2008, section 1: "The main difference is that in algebraic number theory [...] one typically considers questions with answers that are given by exact formulas, whereas in analytic number theory [...] one looks for *good approximations*."

[78] See the remarks in the introduction to Iwaniec & Kowalski 2004, p. 1: "However much stronger...".

[79] Granville 2008, section 3: "[Riemann] defined what we now call the Riemann zeta function [...] Riemann's deep work gave birth to our subject [...]"

[80] See, e.g., Montgomery & Vaughan 2007, p. 1.

[81] CITEREFMilne2014, p. 2.

[82] Edwards 2000, p. 79.

[83] Davis, Martin; Matiyasevich, Yuri; Robinson, Julia (1976). "Hilbert's Tenth Problem: Diophantine Equations: Positive Aspects of a Negative Solution". In Felix E. Browder. *Mathematical Developments Arising from Hilbert Problems*. Proceedings of Symposia in Pure Mathematics. XXVIII.2. American Mathematical Society. pp. 323–378. ISBN 0-8218-1428-1. Zbl 0346.02026. Reprinted in *The Collected Works of Julia Robinson*, Solomon Feferman, editor, pp.269–378, American Mathematical Society 1996.

[84] "The Unreasonable Effectiveness of Number Theory", Stefan Andrus Burr, George E. Andrews, American Mathematical Soc., 1992, ISBN 9780821855010

[85] Computer science and its relation to mathematics" DE Knuth - The American Mathematical Monthly, 1974

[86] "Applications of number theory to numerical analysis", Lo-keng Hua, Luogeng Hua, Yuan Wang, Springer-Verlag, 1981, ISBN 978-3-540-10382-0

[87] "Practical applications of algebraic number theory". Mathoverflow.net. Retrieved 2012-05-18.

[88] "Where is number theory used in the rest of mathematics?". Mathoverflow.net. 2008-09-23. Retrieved 2012-05-18.

[89] Apostol n.d..

20.10 Sources

- Apostol, Tom M. (1976). *Introduction to analytic number theory*. Undergraduate Texts in Mathematics. Springer. ISBN 978-0-387-90163-3.

- Apostol, Tom M. (n.d.). "An Introduction to the Theory of Numbers". (Review of Hardy & Wright.) Mathematical Reviews (MathSciNet) MR0568909. American Mathematical Society. (Subscription needed)

- Becker, Oskar (1936). "Die Lehre von Geraden und Ungeraden im neunten Buch der euklidischen Elemente". *Quellen und Studien zur Geschichte der Mathematik, Astronomie und Physik*. Abteilung B:Studien (in German) (Berlin: J. Springer Verlag) **3**: 533–53.

- Boyer, Carl Benjamin; Merzbach, Uta C. (1991) [1968]. *A History of Mathematics* (2nd ed.). New York: Wiley. ISBN 978-0-471-54397-8. 1968 edition at archive.org

- Clark, Walter Eugene (trans.) (1930). *The Āryabhaṭīya of Āryabhaṭa: An ancient Indian work on Mathematics and Astronomy*. University of Chicago Press.

- Colebrooke, Henry Thomas (1817). *Algebra, with Arithmetic and Mensuration, from the Sanscrit of Brahmegupta and Bháscara*. London: J. Murray.

- Davenport, Harold; Montgomery, Hugh L. (2000). *Multiplicative Number Theory*. Graduate texts in mathematics **74** (revised 3rd ed.). Springer. ISBN 978-0-387-95097-6.

- Edwards, Harold M. (November 1983). "Euler and Quadratic Reciprocity". *Mathematics Magazine* (Mathematical Association of America) **56** (5): 285–291. doi:10.2307/2690368. JSTOR 2690368.

- Edwards, Harold M. (2000) [1977]. *Fermat's Last Theorem: a Genetic Introduction to Algebraic Number Theory*. Graduate Texts in Mathematics **50** (reprint of 1977 ed.). Springer Verlag. ISBN 978-0-387-95002-0.

- Fermat, Pierre de (1679). *Varia Opera Mathematica* (in French and Latin). Toulouse: Joannis Pech.

- Friberg, Jöran (August 1981). "Methods and Traditions of Babylonian Mathematics: Plimpton 322, Pythagorean Triples and the Babylonian Triangle Parameter Equations". *Historia Mathematica* (Elsevier) **8** (3): 277–318. doi:10.1016/0315-0860(81)90069-0.

- von Fritz, Kurt (2004). "The Discovery of Incommensurability by Hippasus of Metapontum". In Christianidis, J. *Classics in the History of Greek Mathematics*. Berlin: Kluwer (Springer). ISBN 978-1-4020-0081-2.

- Gauss, Carl Friedrich; Waterhouse, William C. (trans.) (1966) [1801]. *Disquisitiones Arithmeticae*. Springer. ISBN 978-0-387-96254-2.

- Goldfeld, Dorian M. (2003). "Elementary Proof of the Prime Number Theorem: a Historical Perspective" (PDF).

- Goldstein, Catherine; Schappacher, Norbert (2007). "A book in search of a discipline". In Goldstein, C.; Schappacher, N.; Schwermer, Joachim. *The Shaping of Arithmetic after Gauss' "Disquisitiones Arithmeticae"*. Berlin & Heidelberg: Springer. pp. 3–66. ISBN 978-3-540-20441-1.

- Granville, Andrew (2008). "Analytic number theory". In Gowers, Timothy; Barrow-Green, June; Leader, Imre. *The Princeton Companion to Mathematics*. Princeton University Press. ISBN 978-0-691-11880-2.

- Porphyry; Guthrie, K. S. (trans.) (1920). *Life of Pythagoras*. Alpine, New Jersey: Platonist Press.

- Guthrie, Kenneth Sylvan (1987). *The Pythagorean Sourcebook and Library*. Grand Rapids, Michigan: Phanes Press. ISBN 978-0-933999-51-0.

- Hardy, Godfrey Harold; Wright, E. M. (2008) [1938]. *An Introduction to the Theory of Numbers* (Sixth ed.). Oxford University Press. ISBN 978-0-19-921986-5. MR 2445243.

- Heath, Thomas L. (1921). *A History of Greek Mathematics, Volume 1: From Thales to Euclid*. Oxford: Clarendon Press.

- Hopkins, J. F. P. (1990). "Geographical and Navigational Literature". In Young, M. J. L.; Latham, J. D.; Serjeant, R. B. *Religion, Learning and Science in the 'Abbasid Period*. The Cambridge history of Arabic literature. Cambridge University Press. ISBN 978-0-521-32763-3.

- Huffman, Carl A. (8 August 2011). Zalta, Edward N., ed. "Pythagoras". *Stanford Encyclopaedia of Philosophy* (Fall 2011 ed.). Retrieved 7 February 2012.

- Iwaniec, Henryk; Kowalski, Emmanuel (2004). *Analytic Number Theory*. American Mathematical Society Colloquium Publications **53**. Providence, RI.: American Mathematical Society. ISBN 0-8218-3633-1.

- Plato; Jowett, Benjamin (trans.) (1871). *Theaetetus*.

- Lam, Lay Yong; Ang, Tian Se (2004). *Fleeting Footsteps: Tracing the Conception of Arithmetic and Algebra in Ancient China* (revised ed.). Singapore: World Scientific. ISBN 978-981-238-696-0.

- Long, Calvin T. (1972). *Elementary Introduction to Number Theory* (2nd ed.). Lexington, VA: D. C. Heath and Company. LCCN 77171950.

- Mahoney, M. S. (1994). *The Mathematical Career of Pierre de Fermat, 1601–1665* (Reprint, 2nd ed.). Princeton University Press. ISBN 978-0-691-03666-3.

- Milne, J. S. (2014). "Algebraic Number Theory". Available at www.jmilne.org/math.

- Montgomery, Hugh L.; Vaughan, Robert C. (2007). *Multiplicative Number Theory: I. Classical Theory,*. Cambridge University Press. ISBN 978-0-521-84903-6.

- Morrow, Glenn Raymond (trans., ed.); Proclus (1992). *A Commentary on Book 1 of Euclid's Elements*. Princeton University Press. ISBN 978-0-691-02090-7.

- Mumford, David (March 2010). "Mathematics in India: reviewed by David Mumford" (PDF). *Notices of the American Mathematical Society* **57** (3): 387. ISSN 1088-9477.

- Neugebauer, Otto E. (1969). *The Exact Sciences in Antiquity* (corrected reprint of the 1957 ed.). New York: Dover Publications. ISBN 978-0-486-22332-2.

- Neugebauer, Otto E.; Sachs, Abraham Joseph; Götze, Albrecht (1945). *Mathematical Cuneiform Texts*. American Oriental Series **29**. American Oriental Society etc.

- O'Grady, Patricia (September 2004). "Thales of Miletus". The Internet Encyclopaedia of Philosophy. Retrieved 7 February 2012.

- Pingree, David; Ya'qub, ibn Tariq (1968). "The Fragments of the Works of Ya'qub ibn Tariq". *Journal of Near Eastern Studies* (University of Chicago Press) **26**.

- Pingree, D.; al-Fazari (1970). "The Fragments of the Works of al-Fazari". *Journal of Near Eastern Studies* (University of Chicago Press) **28**.

- Plofker, Kim (2008). *Mathematics in India*. Princeton University Press. ISBN 978-0-691-12067-6.

- Qian, Baocong, ed. (1963). *Suanjing shi shu (Ten Mathematical Classics)* (in Chinese). Beijing: Zhonghua shuju.

- Rashed, Roshdi (1980). "Ibn al-Haytham et le théorème de Wilson". *Archive for History of Exact Sciences* **22** (4): 305–321. doi:10.1007/BF00717654.

- Robson, Eleanor (2001). "Neither Sherlock Holmes nor Babylon: a Reassessment of Plimpton 322" (PDF). *Historia Mathematica* (Elsevier) **28** (28): 167–206. doi:10.1006/hmat.2001.2317.

- Sachau, Eduard; Bīrūni, Muḥammad ibn Aḥmad (1888). *Alberuni's India: An Account of the Religion, Philosophy, Literature, Geography, Chronology, Astronomy and Astrology of India, Vol. 1*. London: Kegan, Paul, Trench, Trübner & Co.

- Serre, Jean-Pierre (1996) [1973]. *A Course in Arithmetic*. Graduate texts in mathematics **7**. Springer. ISBN 978-0-387-90040-7.

- Smith, D. E. (1958). *History of Mathematics, Vol I*. New York: Dover Publications.

- Tannery, Paul; Henry, Charles (eds.); Fermat, Pierre de (1891). *Oeuvres de Fermat*. (4 Vols.) (in French and Latin). Paris: Imprimerie Gauthier-Villars et Fils. Volume 1 Volume 2 Volume 3 Volume 4 (1912)

- Iamblichus; Taylor, Thomas (trans.) (1818). *Life of Pythagoras or, Pythagoric Life*. London: J. M. Watkins. For other editions, see Iamblichus#List of editions and translations

- Truesdell, C. A. (1984). "Leonard Euler, Supreme Geometer". In Hewlett, John (trans.). *Leonard Euler, Elements of Algebra* (reprint of 1840 5th ed.). New York: Springer-Verlag. ISBN 978-0-387-96014-2. This Google books preview of *Elements of algebra* lacks Truesdell's intro, which is reprinted (slightly abridged) in the following book:

- Truesdell, C. A. (2007). "Leonard Euler, Supreme Geometer". In Dunham, William. *The Genius of Euler: reflections on his life and work*. Volume 2 of MAA tercentenary Euler celebration. New York: Mathematical Association of America. ISBN 978-0-88385-558-4.

- Varadarajan, V. S. (2006). *Euler Through Time: A New Look at Old Themes*. American Mathematical Society. ISBN 978-0-8218-3580-7.

- Vardi, Ilan (April 1998). "Archimedes' Cattle Problem" (PDF). *American Mathematical Monthly* **105** (4): 305–319. doi:10.2307/2589706.

- van der Waerden. Bartel L.; Dresden, Arnold (trans) (1961). *Science Awakening*. Vol. 1 or Vol 2. New York: Oxford University Press.

- Weil, André (1984). *Number Theory: an Approach Through History — from Hammurapi to Legendre*. Boston: Birkhäuser. ISBN 978-0-8176-3141-3.

This article incorporates material from the Citizendium article "Number theory", which is licensed under the Creative Commons Attribution-ShareAlike 3.0 Unported License but not under the GFDL.

20.11 External links

- Hazewinkel, Michiel, ed. (2001), "Number theory", *Encyclopedia of Mathematics*, Springer, ISBN 978-1-55608-010-4

- Quotations related to Number theory at Wikiquote

- Number Theory Web

A Lehmer sieve, which is a primitive digital computer once used for finding primes and solving simple Diophantine equations.

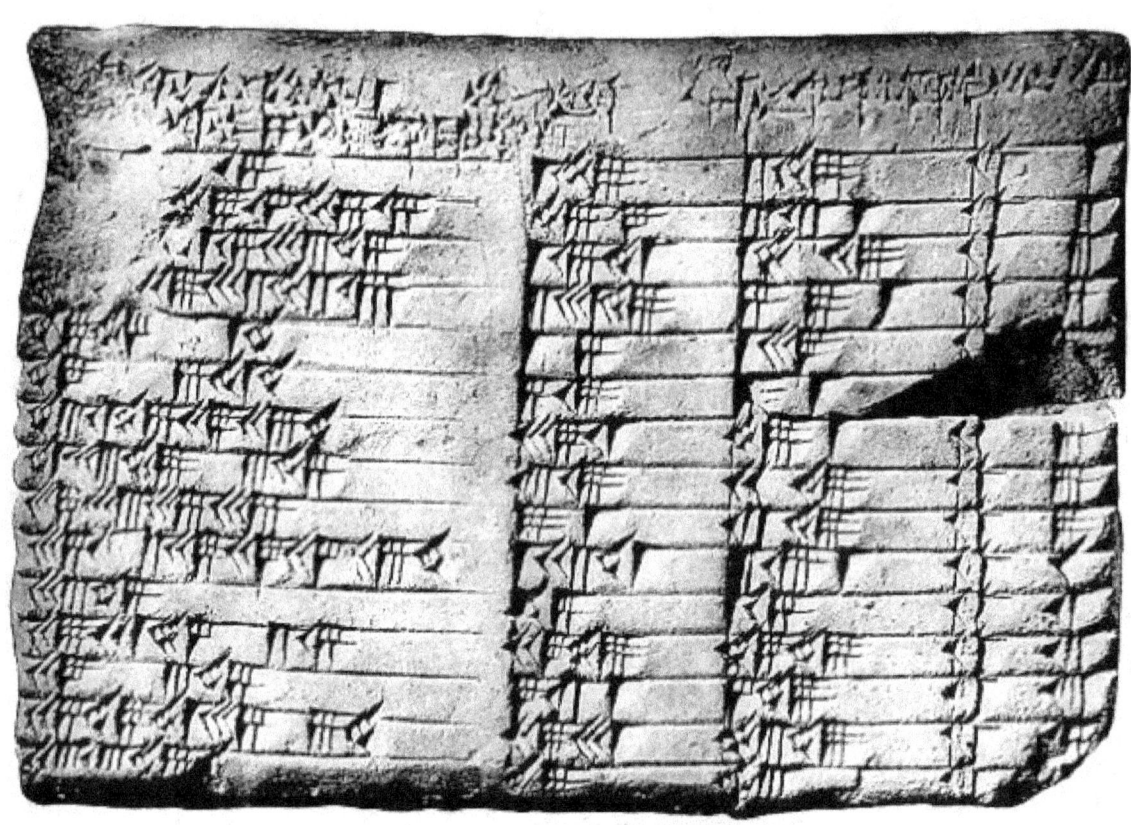

The Plimpton 322 tablet

DIOPHANTI
ALEXANDRINI
ARITHMETICORVM
LIBRI SEX.

ET DE NVMERIS MVLTANGVLIS
LIBER VNVS.

Nunc primùm Græcè & Latinè editi, atque absolutissimis
Commentariis illustrati.

AVCTORE CLAVDIO GASPARE BACHETO
M. ZIRIACO SEBVSIANO, V. C.

LVTETIAE PARISIORVM,

Sumptibus SEBASTIANI CRAMOISY, via
Iacobæa, sub Ciconiis.

M. DC. XXI.

CVM PRIVILEGIO REGIS.

Leonhard Euler

(library stamp, illegible)

DISQVISITIONES

ARITHMETICAE

AVCTORE

D. CAROLO FRIDERICO GAVSS

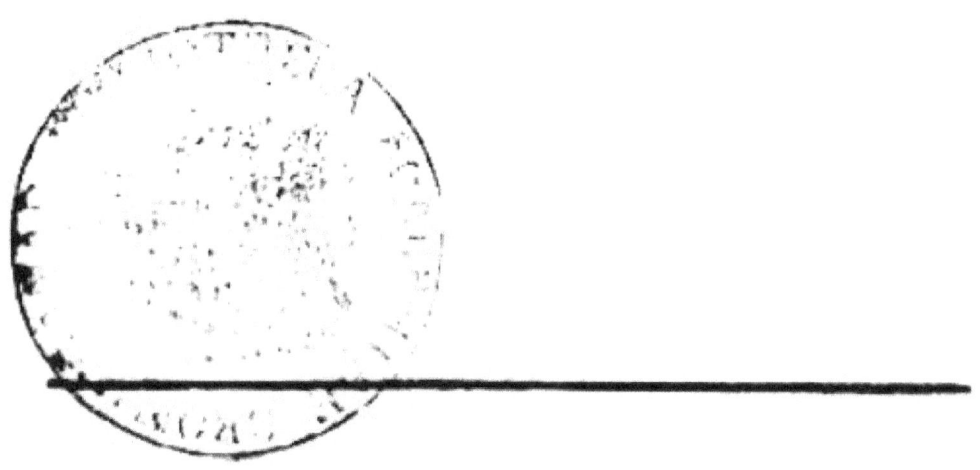

LIPSIAE

IN COMMISSIS AFVD GERH. FLEISCHER, JVN.

1801.

Carl Friedrich Gauss

Ernst Kummer

Peter Gustav Lejeune Dirichlet

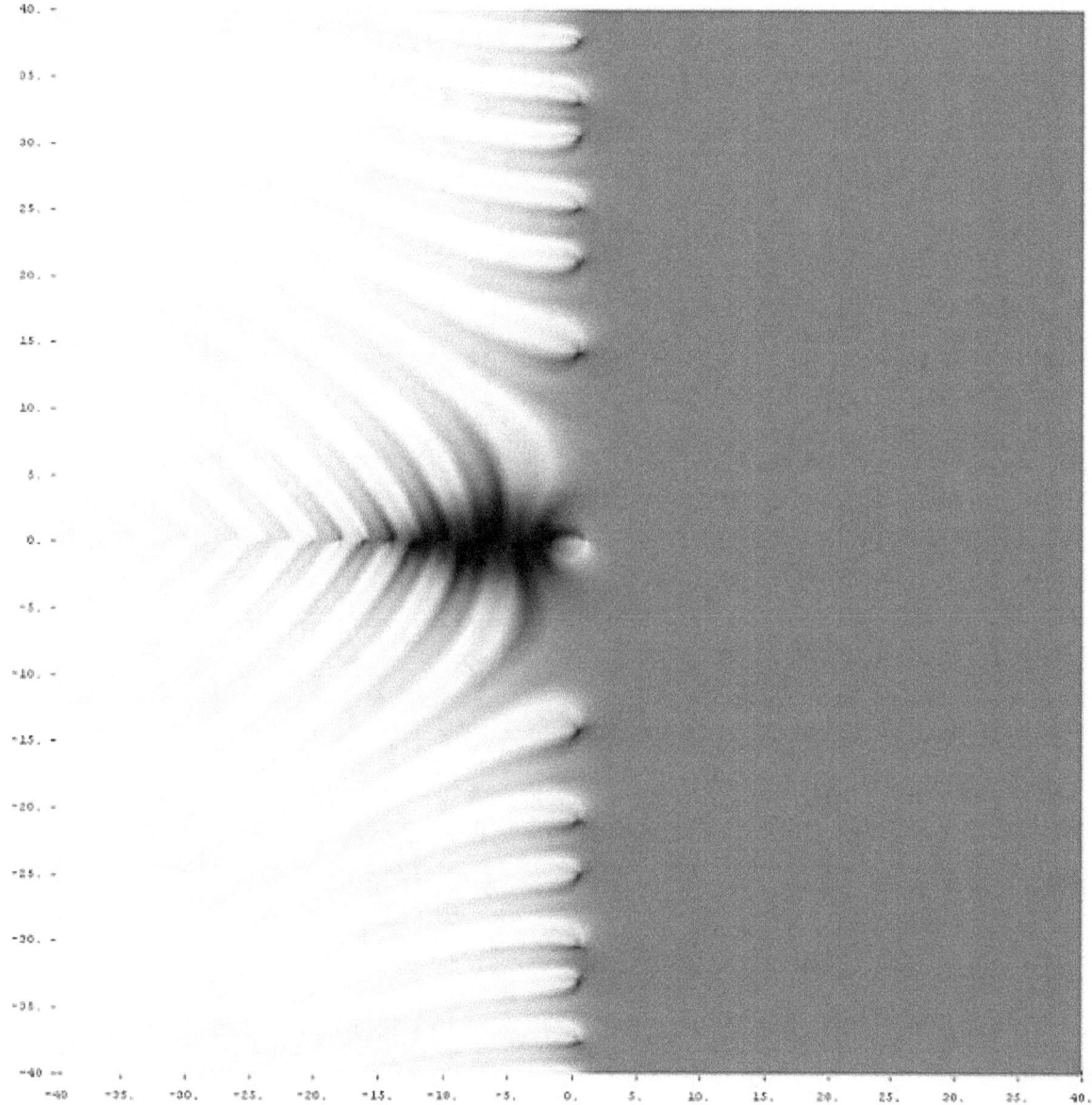

Riemann zeta function ζ(s) in the complex plane. The color of a point s gives the value of ζ(s): dark colors denote values close to zero and hue gives the value's argument.

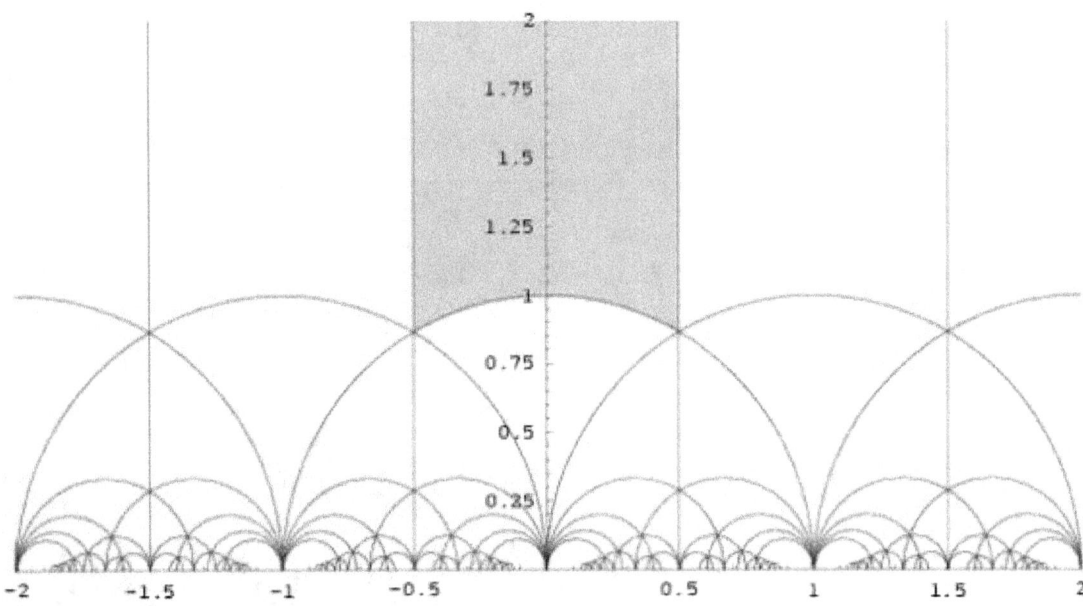

The action of the modular group on the upper half plane. The region in grey is the standard fundamental domain.

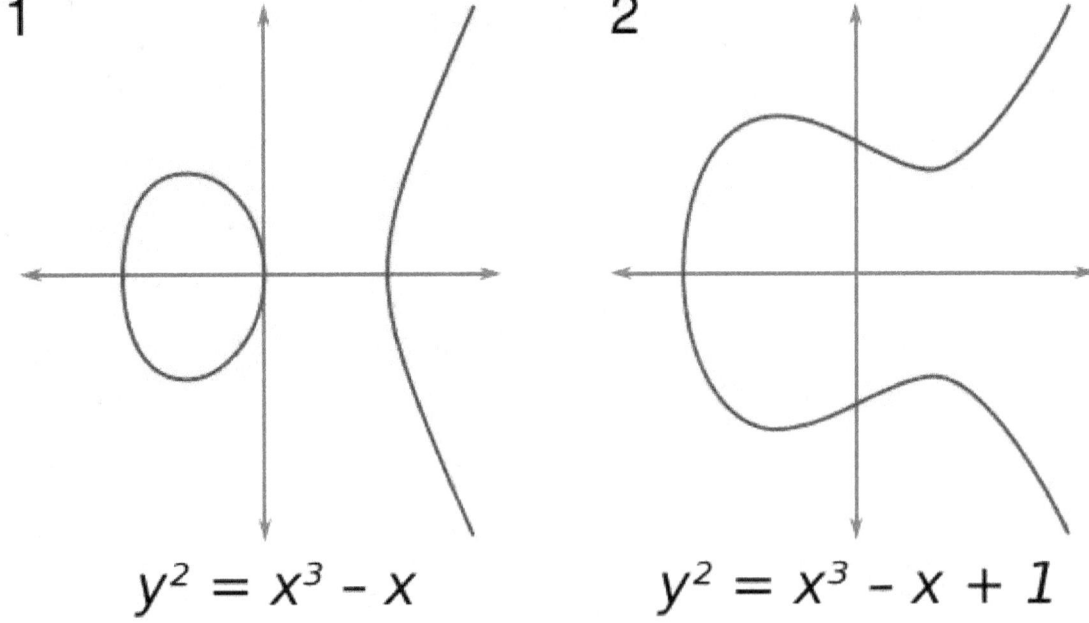

Two examples of an elliptic curve, i.e., a curve of genus 1 having at least one rational point. (Either graph can be seen as a slice of a torus in four-dimensional space.)

Chapter 21

Order theory

For a topical guide to this subject, see Outline of order theory.

Order theory is a branch of mathematics which investigates our intuitive notion of order using binary relations. It provides a formal framework for describing statements such as "this is less than that" or "this precedes that". This article introduces the field and provides basic definitions. A list of order-theoretic terms can be found in the order theory glossary.

21.1 Background and motivation

Orders are everywhere in mathematics and related fields like computer science. The first order often discussed in primary school is the standard order on the natural numbers e.g. "2 is less than 3", "10 is greater than 5", or "Does Tom have fewer cookies than Sally?". This intuitive concept can be extended to orders on other sets of numbers, such as the integers and the reals. The idea of being greater than or less than another number is one of the basic intuitions of number systems (compare with numeral systems) in general (although one usually is also interested in the actual difference of two numbers, which is not given by the order). Another familiar example of an ordering is the lexicographic order of words in a dictionary.

The above types of orders have a special property: each element can be *compared* to any other element, i.e. it is greater, smaller, or equal. However, this is not always a desired requirement. For example, consider the subset ordering of sets. If a set A contains all the elements of a set B, then B is said to be smaller than or equal to A. Yet there are some sets that cannot be related in this fashion. Whenever both contain some elements that are not in the other, the two sets are not related by subset-inclusion. Hence, subset-inclusion is only a *partial order*, as opposed to the *total orders* given before.

Order theory captures the intuition of orders that arises from such examples in a general setting. This is achieved by specifying properties that a relation $\leq$ must have to be a mathematical order. This more abstract approach makes much sense, because one can derive numerous theorems in the general setting, without focusing on the details of any particular order. These insights can then be readily transferred to many less abstract applications.

Driven by the wide practical usage of orders, numerous special kinds of ordered sets have been defined, some of which have grown into mathematical fields of their own. In addition, order theory does not restrict itself to the various classes of ordering relations, but also considers appropriate functions between them. A simple example of an order theoretic property for functions comes from analysis where monotone functions are frequently found.

21.2 Basic definitions

This section introduces ordered sets by building upon the concepts of set theory, arithmetic, and binary relations.

21.2.1 Partially ordered sets

Orders are special binary relations. Suppose that P is a set and that $\leq$ is a relation on P. Then $\leq$ is a **partial order** if it is reflexive, antisymmetric, and transitive, i.e., for all a, b and c in P, we have that:

$a \leq a$ (reflexivity)

if $a \leq b$ and $b \leq a$ then $a = b$ (antisymmetry)

if $a \leq b$ and $b \leq c$ then $a \leq c$ (transitivity).

A set with a partial order on it is called a **partially ordered set**, **poset**, or just an **ordered set** if the intended meaning is clear. By checking these properties, one immediately sees that the well-known orders on natural numbers, integers, rational numbers and reals are all orders in the above sense. However, they have the additional property of being **total**, i.e., for all a and b in P, we have that:

$a \leq b$ or $b \leq a$ (totality).

These orders can also be called **linear orders** or **chains**. While many classical orders are linear, the subset order on sets provides an example where this is not the case. Another example is given by the divisibility relation "|". For two natural numbers n and m, we write $n|m$ if n divides m without remainder. One easily sees that this yields a partial order. The identity relation $=$ on any set is also a partial order in which every two distinct elements are incomparable. It is also the only relation that is both a partial order and an equivalence relation. Many advanced properties of posets are interesting mainly for non-linear orders.

21.2.2 Visualizing a poset

Hasse diagrams can visually represent the elements and relations of a partial ordering. These are graph drawings where the vertices are the elements of the poset and the ordering relation is indicated by both the edges and the relative positioning of the vertices. Orders are drawn bottom-up: if an element x is smaller than (precedes) y then there exists a path from x to y that is directed upwards. It is often necessary for the edges connecting elements to cross each other, but elements must never be located within an edge. An instructive exercise is to draw the Hasse diagram for the set of natural numbers that are smaller than or equal to 13, ordered by | (the *divides* relation).

Even some infinite sets can be diagrammed by superimposing an ellipsis (...) on a finite sub-order. This works well for the natural numbers, but it fails for the reals, where there is no immediate successor above 0; however, quite often one can obtain an intuition related to diagrams of a similar kind.

21.2.3 Special elements within an order

In a partially ordered set there may be some elements that play a special role. The most basic example is given by the **least element** of a poset. For example, 1 is the least element of the positive integers and the empty set is the least set under the subset order. Formally, an element m is a least element if:

$m \leq a$, for all elements a of the order.

The notation 0 is frequently found for the least element, even when no numbers are concerned. However, in orders on sets of numbers, this notation might be inappropriate or ambiguous, since the number 0 is not always least. An example is given by the above divisibility order |, where 1 is the least element since it divides all other numbers. In contrast, 0 is the number that is divided by all other numbers. Hence it is the **greatest element** of the order. Other frequent terms for the least and greatest elements is **bottom** and **top** or **zero** and **unit**.

Least and greatest elements may fail to exist, as the example of the real numbers shows. But if they exist, they are always unique. In contrast, consider the divisibility relation | on the set $\{2,3,4,5,6\}$. Although this set has neither top nor bottom, the elements 2, 3, and 5 have no elements below them, while 4, 5, and 6 have none above. Such elements are called **minimal** and **maximal**, respectively. Formally, an element m is minimal if:

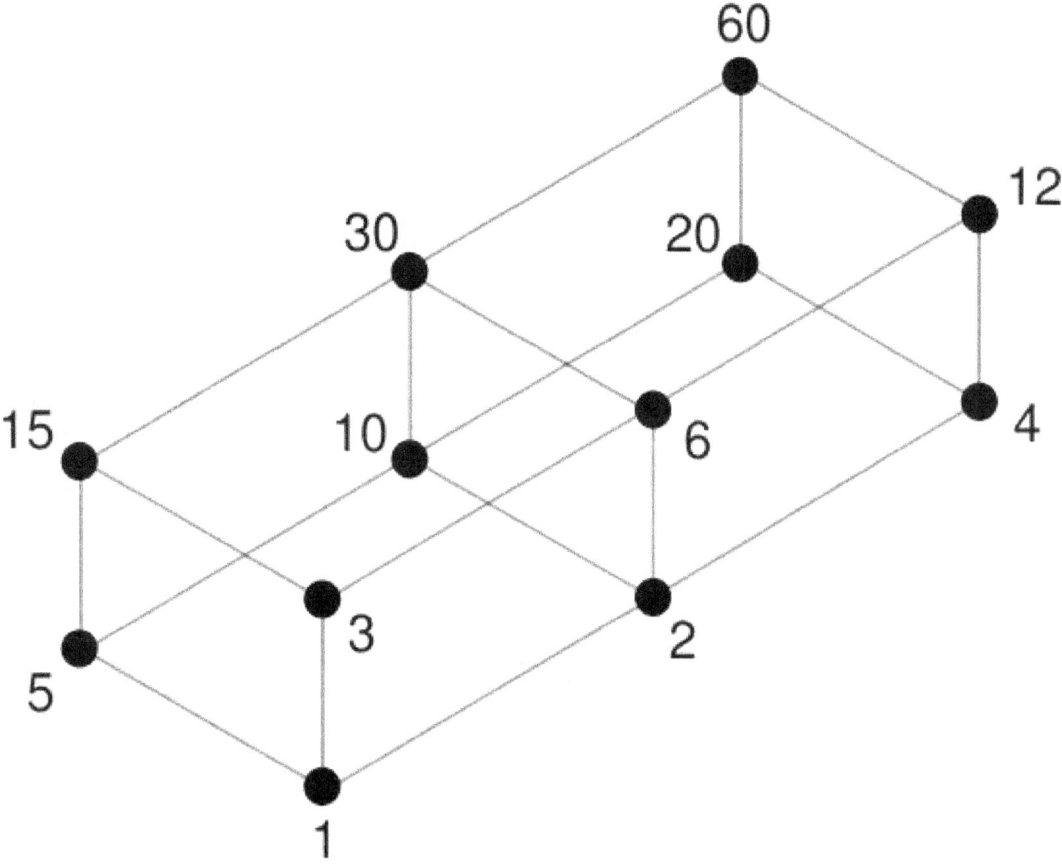

Hasse diagram of the set of all divisors of 60, partially ordered by divisibility

$a \leq m$ implies $a = m$, for all elements a of the order.

Exchanging $\leq$ with $\geq$ yields the definition of maximality. As the example shows, there can be many maximal elements and some elements may be both maximal and minimal (e.g. 5 above). However, if there is a least element, then it is the only minimal element of the order. Again, in infinite posets maximal elements do not always exist - the set of all *finite* subsets of a given infinite set, ordered by subset inclusion, provides one of many counterexamples. An important tool to ensure the existence of maximal elements under certain conditions is **Zorn's Lemma**.

Subsets of partially ordered sets inherit the order. We already applied this by considering the subset {2,3,4,5,6} of the natural numbers with the induced divisibility ordering. Now there are also elements of a poset that are special with respect to some subset of the order. This leads to the definition of **upper bounds**. Given a subset S of some poset P, an upper bound of S is an element b of P that is above all elements of S. Formally, this means that

$s \leq b$, for all s in S.

Lower bounds again are defined by inverting the order. For example, -5 is a lower bound of the natural numbers as a subset of the integers. Given a set of sets, an upper bound for these sets under the subset ordering is given by their union. In fact, this upper bound is quite special: it is the smallest set that contains all of the sets. Hence, we have found the **least upper bound** of a set of sets. This concept is also called **supremum** or **join**, and for a set S one writes sup(S) or $\vee S$ for its least upper bound. Conversely, **the greatest lower bound** is known as **infimum** or **meet** and denoted inf(S) or $\wedge S$. These concepts play an important role in many applications of order theory. For two elements x and y, one also writes $x \vee y$ and $x \wedge y$ for sup({x,y}) and inf({x,y}), respectively.

For another example, consider again the relation | on natural numbers. The least upper bound of two numbers is the smallest number that is divided by both of them, i.e. the least common multiple of the numbers. Greatest lower bounds in turn are given by the greatest common divisor.

21.2.4 Duality

In the previous definitions, we often noted that a concept can be defined by just inverting the ordering in a former definition. This is the case for "least" and "greatest", for "minimal" and "maximal", for "upper bound" and "lower bound", and so on. This is a general situation in order theory: A given order can be inverted by just exchanging its direction, pictorially flipping the Hasse diagram top-down. This yields the so-called **dual**, **inverse**, or **opposite order**.

Every order theoretic definition has its dual: it is the notion one obtains by applying the definition to the inverse order. Since all concepts are symmetric, this operation preserves the theorems of partial orders. For a given mathematical result, one can just invert the order and replace all definitions by their duals and one obtains another valid theorem. This is important and useful, since one obtains two theorems for the price of one. Some more details and examples can be found in the article on duality in order theory.

21.2.5 Constructing new orders

There are many ways to construct orders out of given orders. The dual order is one example. Another important construction is the cartesian product of two partially ordered sets, taken together with the product order on pairs of elements. The ordering is defined by $(a, x) \leq (b, y)$ if (and only if) $a \leq b$ and $x \leq y$. (Notice carefully that there are three distinct meanings for the relation symbol $\leq$ in this definition.) The disjoint union of two posets is another typical example of order construction, where the order is just the (disjoint) union of the original orders.

Every partial order $\leq$ gives rise to a so-called strict order $<$, by defining $a < b$ if $a \leq b$ and not $b \leq a$. This transformation can be inverted by setting $a \leq b$ if $a < b$ or $a = b$. The two concepts are equivalent although in some circumstances one can be more convenient to work with than the other.

21.3 Functions between orders

It is reasonable to consider functions between partially ordered sets having certain additional properties that are related to the ordering relations of the two sets. The most fundamental condition that occurs in this context is monotonicity. A function f from a poset P to a poset Q is **monotone**, or **order-preserving**, if $a \leq b$ in P implies $f(a) \leq f(b)$ in Q (Noting that, strictly, the two relations here are different since they apply to different sets.). The converse of this implication leads to functions that are **order-reflecting**, i.e. functions f as above for which $f(a) \leq f(b)$ implies $a \leq b$. On the other hand, a function may also be **order-reversing** or **antitone**, if $a \leq b$ implies $f(b) \leq f(a)$.

An **order-embedding** is a function f between orders that is both order-preserving and order-reflecting. Examples for these definitions are found easily. For instance, the function that maps a natural number to its successor is clearly monotone with respect to the natural order. Any function from a discrete order, i.e. from a set ordered by the identity order "=", is also monotone. Mapping each natural number to the corresponding real number gives an example for an order embedding. The set complement on a powerset is an example of an antitone function.

An important question is when two orders are "essentially equal", i.e. when they are the same up to renaming of elements. **Order isomorphisms** are functions that define such a renaming. An order-isomorphism is a monotone bijective function that has a monotone inverse. This is equivalent to being a surjective order-embedding. Hence, the image $f(P)$ of an order-embedding is always isomorphic to P, which justifies the term "embedding".

A more elaborate type of functions is given by so-called **Galois connections**. Monotone Galois connections can be viewed as a generalization of order-isomorphisms, since they constitute of a pair of two functions in converse directions, which are "not quite" inverse to each other, but that still have close relationships.

Another special type of self-maps on a poset are **closure operators**, which are not only monotonic, but also idempotent, i.e. $f(x) = f(f(x))$, and **extensive** (or *inflationary*), i.e. $x \leq f(x)$. These have many applications in all kinds of "closures" that appear in mathematics.

Besides being compatible with the mere order relations, functions between posets may also behave well with respect to special elements and constructions. For example, when talking about posets with least element, it may seem reasonable to consider only monotonic functions that preserve this element, i.e. which map least elements to least elements. If binary infima $\wedge$ exist, then a reasonable property might be to require that $f(x \wedge y) = f(x) \wedge f(y)$, for all x and y. All of these properties, and indeed many more, may be compiled under the label of limit-preserving functions.

Finally, one can invert the view, switching from *functions of orders* to *orders of functions*. Indeed, the functions between two posets P and Q can be ordered via the pointwise order. For two functions f and g, we have $f \leq g$ if $f(x) \leq g(x)$ for all elements x of P. This occurs for example in domain theory, where function spaces play an important role.

21.4   Special types of orders

Many of the structures that are studied in order theory employ order relations with further properties. In fact, even some relations that are not partial orders are of special interest. Mainly the concept of a preorder has to be mentioned. A preorder is a relation that is reflexive and transitive, but not necessarily antisymmetric. Each preorder induces an equivalence relation between elements, where a is equivalent to b, if $a \leq b$ and $b \leq a$. Preorders can be turned into orders by identifying all elements that are equivalent with respect to this relation.

Several types of orders can be defined from numerical data on the items of the order: a total order results from attaching distinct real numbers to each item and using the numerical comparisons to order the items; instead, if distinct items are allowed to have equal numerical scores, one obtains a strict weak ordering. Requiring two scores to be separated by a fixed threshold before they may be compared leads to the concept of a semiorder, while allowing the threshold to vary on a per-item basis produces an interval order.

An additional simple but useful property leads to so-called **well-orders**, for which all non-empty subsets have a minimal element. Generalizing well-orders from linear to partial orders, a set is **well partially ordered** if all its non-empty subsets have a finite number of minimal elements.

Many other types of orders arise when the existence of infima and suprema of certain sets is guaranteed. Focusing on this aspect, usually referred to as completeness of orders, one obtains:

- Bounded posets, i.e. posets with a least and greatest element (which are just the supremum and infimum of the empty subset),

- Lattices, in which every non-empty finite set has a supremum and infimum,

- Complete lattices, where every set has a supremum and infimum, and

- Directed complete partial orders (dcpos), that guarantee the existence of suprema of all directed subsets and that are studied in domain theory.

- Partial orders with complements, or *poc sets*,[1] are posets S having a unique bottom element $0 \in S$, along with an order-reversing involution, such that $a \leq a^* \Rightarrow a = 0$.

However, one can go even further: if all finite non-empty infima exist, then $\wedge$ can be viewed as a total binary operation in the sense of universal algebra. Hence, in a lattice, two operations $\wedge$ and $\vee$ are available, and one can define new properties by giving identities, such as

$$x \wedge (y \vee z) = (x \wedge y) \vee (x \wedge z), \text{ for all } x, y, \text{ and } z.$$

This condition is called **distributivity** and gives rise to distributive lattices. There are some other important distributivity laws which are discussed in the article on distributivity in order theory. Some additional order structures that are often specified via algebraic operations and defining identities are

- Heyting algebras and

- Boolean algebras,

which both introduce a new operation ~ called **negation**. Both structures play a role in mathematical logic and especially Boolean algebras have major applications in computer science. Finally, various structures in mathematics combine orders with even more algebraic operations, as in the case of quantales, that allow for the definition of an addition operation.

Many other important properties of posets exist. For example, a poset is **locally finite** if every closed interval $[a, b]$ in it is finite. Locally finite posets give rise to incidence algebras which in turn can be used to define the Euler characteristic of finite bounded posets.

21.5 Subsets of ordered sets

In an ordered set, one can define many types of special subsets based on the given order. A simple example are **upper sets**; i.e. sets that contain all elements that are above them in the order. Formally, the **upper closure** of a set S in a poset P is given by the set $\{x$ in $P \mid$ there is some y in S with $y \leq x\}$. A set that is equal to its upper closure is called an upper set. **Lower sets** are defined dually.

More complicated lower subsets are ideals, which have the additional property that each two of their elements have an upper bound within the ideal. Their duals are given by filters. A related concept is that of a directed subset, which like an ideal contains upper bounds of finite subsets, but does not have to be a lower set. Furthermore it is often generalized to preordered sets.

A subset which is - as a sub-poset - linearly ordered, is called a chain. The opposite notion, the antichain, is a subset that contains no two comparable elements; i.e. that is a discrete order.

21.6 Related mathematical areas

Although most mathematical areas *use* orders in one or the other way, there are also a few theories that have relationships which go far beyond mere application. Together with their major points of contact with order theory, some of these are to be presented below.

21.6.1 Universal algebra

As already mentioned, the methods and formalisms of universal algebra are an important tool for many order theoretic considerations. Beside formalizing orders in terms of algebraic structures that satisfy certain identities, one can also establish other connections to algebra. An example is given by the correspondence between Boolean algebras and Boolean rings. Other issues are concerned with the existence of free constructions, such as *free lattices* based on a given set of generators. Furthermore, closure operators are important in the study of universal algebra.

21.6.2 Topology

In topology orders play a very prominent role. In fact, the set of open sets provides a classical example of a complete lattice, more precisely a complete Heyting algebra (or "**frame**" or "**locale**"). Filters and nets are notions closely related to order theory and the closure operator of sets can be used to define topology. Beyond these relations, topology can be looked at solely in terms of the open set lattices, which leads to the study of pointless topology. Furthermore, a natural preorder of elements of the underlying set of a topology is given by the so-called specialization order, that is actually a partial order if the topology is T_0.

Conversely, in order theory, one often makes use of topological results. There are various ways to define subsets of an order which can be considered as open sets of a topology. Especially, it is interesting to consider topologies on a poset

$(X, \leq)$ that in turn induce $\leq$ as their specialization order. The *finest* such topology is the Alexandrov topology, given by taking all upper sets as opens. Conversely, the *coarsest* topology that induces the specialization order is the upper topology, having the complements of principal ideals (i.e. sets of the form $\{y \text{ in } X \mid y \leq x\}$ for some x) as a subbase. Additionally, a topology with specialization order $\leq$ may be order consistent, meaning that their open sets are "inaccessible by directed suprema" (with respect to $\leq$). The finest order consistent topology is the Scott topology, which is coarser than the Alexandrov topology. A third important topology in this spirit is the Lawson topology. There are close connections between these topologies and the concepts of order theory. For example, a function preserves directed suprema iff it is continuous with respect to the Scott topology (for this reason this order theoretic property is also called Scott-continuity).

21.6.3 Category theory

The visualization of orders with Hasse diagrams has a straightforward generalization: instead of displaying lesser elements *below* greater ones, the direction of the order can also be depicted by giving directions to the edges of a graph. In this way, each order is seen to be equivalent to a directed acyclic graph, where the nodes are the elements of the poset and there is a directed path from a to b if and only if $a \leq b$. Dropping the requirement of being acyclic, one can also obtain all preorders.

When equipped with all transitive edges, these graphs in turn are just special categories, where elements are objects and each set of morphisms between two elements is at most singleton. Functions between orders become functors between categories. Interestingly, many ideas of order theory are just concepts of category theory in small. For example, an infimum is just a categorical product. More generally, one can capture infima and suprema under the abstract notion of a categorical limit (or *colimit*, respectively). Another place where categorical ideas occur is the concept of a (monotone) Galois connection, which is just the same as a pair of adjoint functors.

But category theory also has its impact on order theory on a larger scale. Classes of posets with appropriate functions as discussed above form interesting categories. Often one can also state constructions of orders, like the product order, in terms of categories. Further insights result when categories of orders are found categorically equivalent to other categories, for example of topological spaces. This line of research leads to various *representation theorems*, often collected under the label of Stone duality.

21.7 History

As explained before, orders are ubiquitous in mathematics. However, earliest explicit mentionings of partial orders are probably to be found not before the 19th century. In this context the works of George Boole are of great importance. Moreover, works of Charles Sanders Peirce, Richard Dedekind, and Ernst Schröder also consider concepts of order theory. Certainly, there are others to be named in this context and surely there exists more detailed material on the history of order theory.

The term *poset* as an abbreviation for partially ordered set was coined by Garrett Birkhoff in the second edition of his influential book *Lattice Theory*.[2][3]

21.8 See also

- Cyclic order

- Hierarchy

- Incidence algebra

- Important publications in order theory

- Causal Sets

21.9 Notes

[1] Roller, Martin A. (1998), *Poc sets, median algebras and group actions. An extended study of Dunwoody's construction and Sageev's theorem* (PDF), Southampton Preprint Archive

[2] Birkhoff 1948, p.1

[3] Earliest Known Uses of Some of the Words of Mathematics

21.10 References

• Birkhoff, Garrett (1940). *Lattice Theory* **25** (3rd Revised ed.). American Mathematical Society. ISBN 978-0-8218-1025-5.

• Burris, S. N.; Sankappanavar, H. P. (1981). *A Course in Universal Algebra*. Springer. ISBN 978-0-387-90578-5.

• Davey, B. A.; Priestley, H. A. (2002). *Introduction to Lattices and Order* (2nd ed.). Cambridge University Press. ISBN 0-521-78451-4.

• Gierz, G.; Hofmann, K. H.; Keimel, K.; Mislove, M.; Scott, D. S. (2003). *Continuous Lattices and Domains*. Encyclopedia of Mathematics and its Applications **93**. Cambridge University Press. ISBN 978-0-521-80338-0.

21.11 External links

• Orders at ProvenMath partial order, linear order, well order, initial segment; formal definitions and proofs within the axioms of set theory.

• Nagel, Felix (2013). Set Theory and Topology. An Introduction to the Foundations of Analysis

Chapter 22

Function (mathematics)

In mathematics, a **function**[1] is a relation between a set of inputs and a set of permissible outputs with the property that each input is related to exactly one output. An example is the function that relates each real number x to its square x^2. The output of a function f corresponding to an input x is denoted by $f(x)$ (read "f of x"). In this example, if the input is -3, then the output is 9, and we may write $f(-3) = 9$. Likewise, if the input is 3, then the output is also 9, and we may write $f(3) = 9$. (The same output may be produced by more than one input, but each input gives only one output.) The input variable(s) are sometimes referred to as the argument(s) of the function.

Functions of various kinds are "the central objects of investigation"[2] in most fields of modern mathematics. There are many ways to describe or represent a function. Some functions may be defined by a formula or algorithm that tells how to compute the output for a given input. Others are given by a picture, called the graph of the function. In science, functions are sometimes defined by a table that gives the outputs for selected inputs. A function could be described implicitly, for example as the inverse to another function or as a solution of a differential equation.

The input and output of a function can be expressed as an ordered pair, ordered so that the first element is the input (or tuple of inputs, if the function takes more than one input), and the second is the output. In the example above, $f(x) = x^2$, we have the ordered pair $(-3, 9)$. If both input and output are real numbers, this ordered pair can be viewed as the Cartesian coordinates of a point on the graph of the function.

In modern mathematics,[3] a function is defined by its set of inputs, called the *domain*; a set containing the set of outputs, and possibly additional elements, as members, called its *codomain*; and the set of all input-output pairs, called its *graph*. Sometimes the codomain is called the function's "range", but more commonly the word "range" is used to mean, instead, specifically the set of outputs (this is also called the *image* of the function). For example, we could define a function using the rule $f(x) = x^2$ by saying that the domain and codomain are the real numbers, and that the graph consists of all pairs of real numbers (x, x^2). The image of this function is the set of non-negative real numbers. Collections of functions with the same domain and the same codomain are called function spaces, the properties of which are studied in such mathematical disciplines as real analysis, complex analysis, and functional analysis.

In analogy with arithmetic, it is possible to define addition, subtraction, multiplication, and division of functions, in those cases where the output is a number. Another important operation defined on functions is function composition, where the output from one function becomes the input to another function.

22.1 Introduction and examples

For an example of a function, let X be the set consisting of four shapes: a red triangle, a yellow rectangle, a green hexagon, and a red square; and let Y be the set consisting of five colors: red, blue, green, pink, and yellow. Linking each shape to its color is a function from X to Y: each shape is linked to a color (i.e., an element in Y), and each shape is "linked", or "mapped", to exactly one color. There is no shape that lacks a color and no shape that has two or more colors. This function will be referred to as the "color-of-the-shape function".

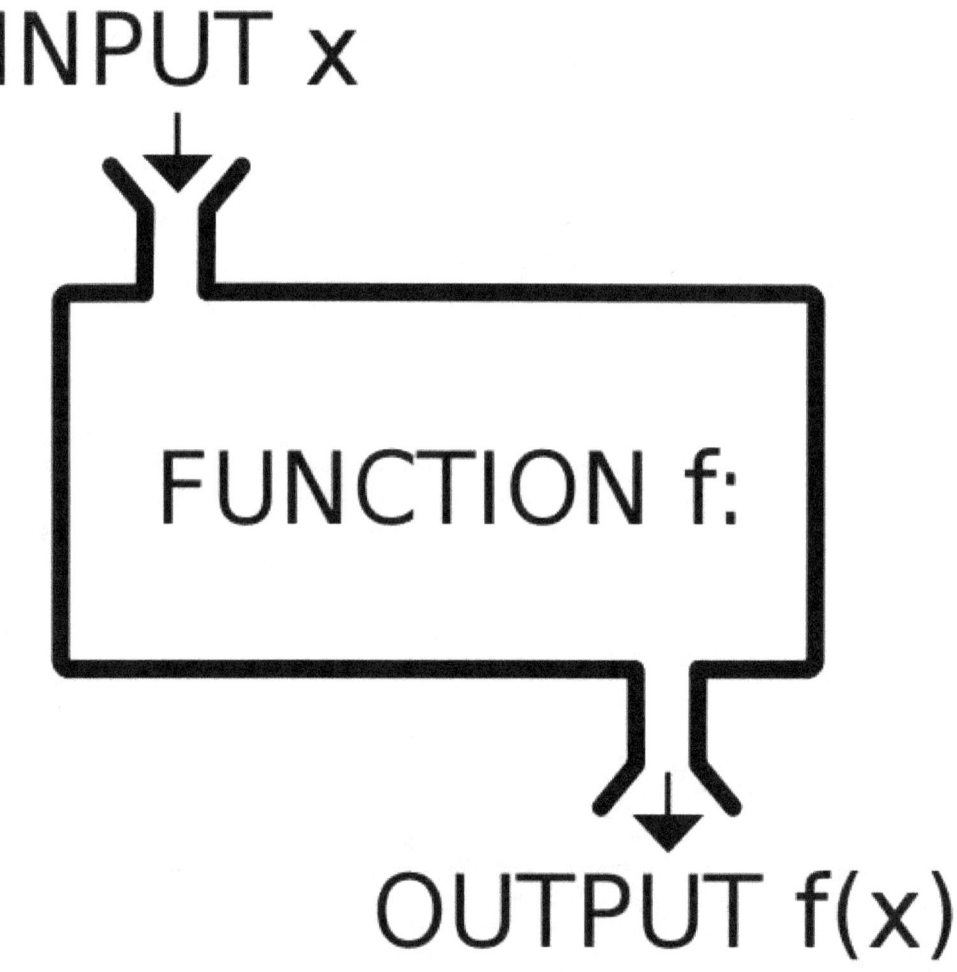

A function f takes an input x, and returns a single output f(x). One metaphor describes the function as a "machine" or "black box" that for each input returns a corresponding output.

The input to a function is called the argument and the output is called the value. The set of all permitted inputs to a given function is called the domain of the function, while the set of permissible outputs is called the codomain. Thus, the domain of the "color-of-the-shape function" is the set of the four shapes, and the codomain consists of the five colors. The concept of a function does *not* require that every possible output is the value of some argument, e.g. the color blue is not the color of any of the four shapes in X.

A second example of a function is the following: the domain is chosen to be the set of natural numbers (1, 2, 3, 4, ...), and the codomain is the set of integers (..., −3, −2, −1, 0, 1, 2, 3, ...). The function associates to any natural number n the number $4−n$. For example, to 1 it associates 3 and to 10 it associates −6.

A third example of a function has the set of polygons as domain and the set of natural numbers as codomain. The function associates a polygon with its number of vertices. For example, a triangle is associated with the number 3, a square with the number 4, and so on.

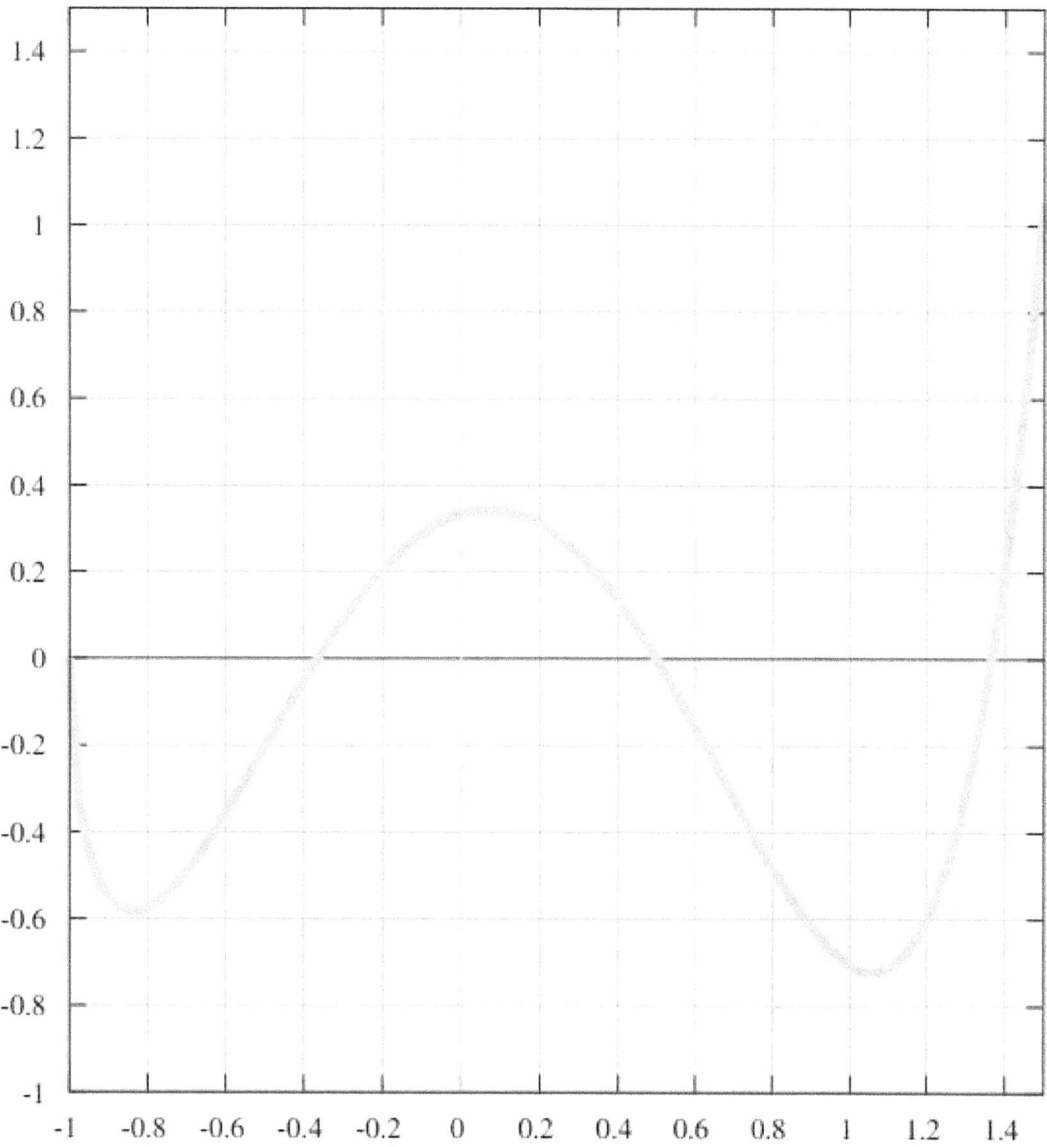

The red curve is the graph of a function f in the Cartesian plane, consisting of all points with coordinates of the form (x,f(x)). The property of having one output for each input is represented geometrically by the fact that each vertical line (such as the yellow line through the origin) has exactly one crossing point with the curve.

The term range is sometimes used either for the codomain or for the set of all the actual values a function has.

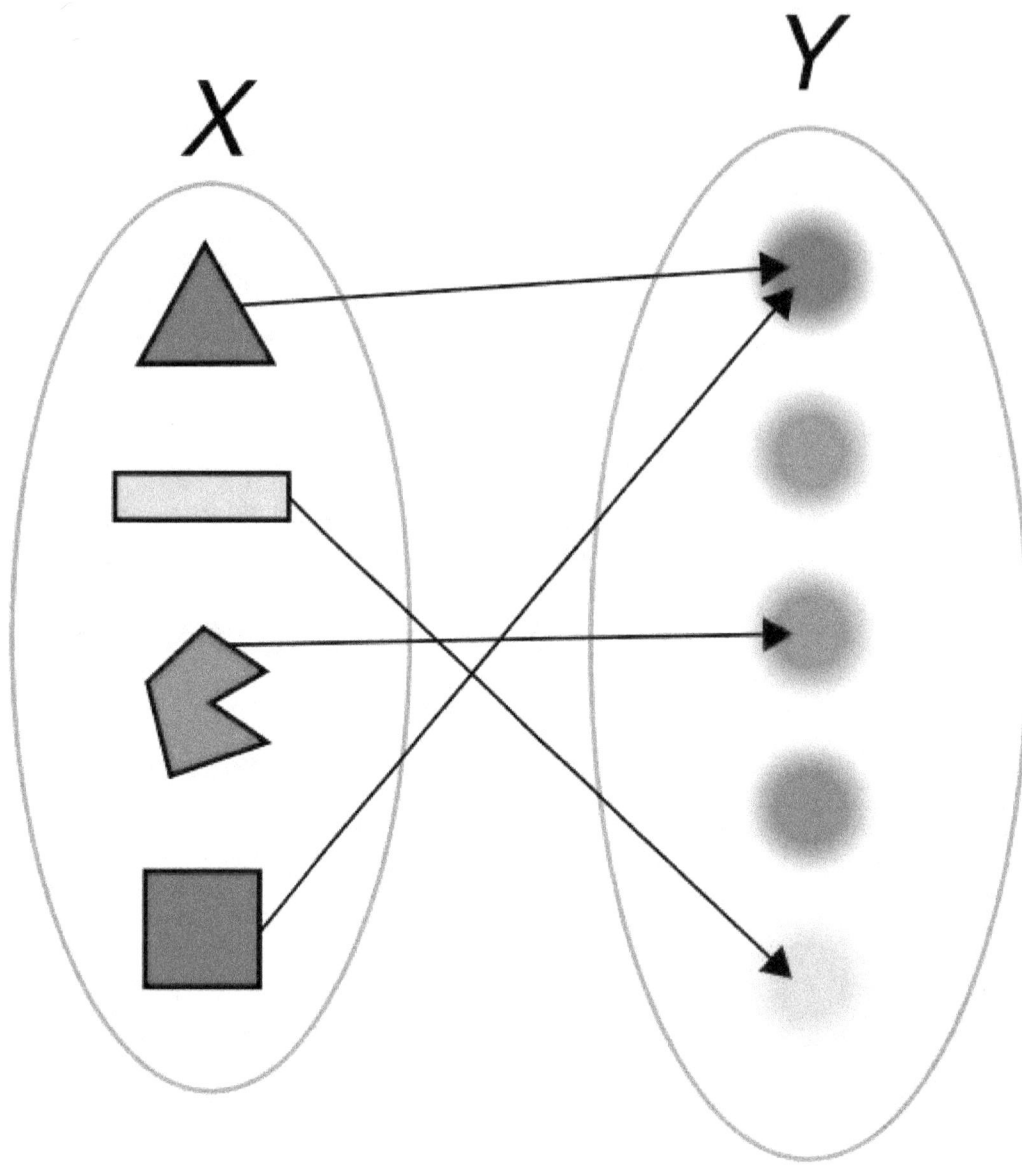

A function that associates to any of the four colored shapes its color.

22.2   Definition

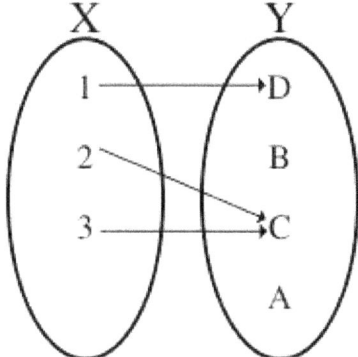

The above diagram represents a function with domain {1, 2, 3}, codomain {A, B, C, D} and set of ordered pairs {(1,D), (2,C), (3,C)}. The image is {C,D}.

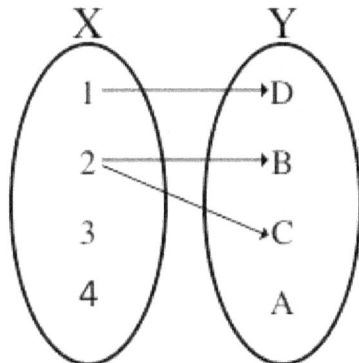

However, this second diagram does *not* represent a function. One reason is that 2 is the first element in more than one ordered pair. In particular, (2, B) and (2, C) are both elements of the set of ordered pairs. Another reason, sufficient by itself, is that 3 is not the first element (input) for any ordered pair. A third reason, likewise, is that 4 is not the first element of any ordered pair.

In order to avoid the use of the informally defined concepts of "rules" and "associates", the above intuitive explanation of functions is completed with a formal definition. This definition relies on the notion of the Cartesian product. The Cartesian product of two sets X and Y is the set of all ordered pairs, written (x, y), where x is an element of X and y is an element of Y. The x and the y are called the components of the ordered pair. The Cartesian product of X and Y is denoted by $X \times Y$.

A function f from X to Y is a subset of the Cartesian product $X \times Y$ subject to the following condition: every element of X is the first component of one and only one ordered pair in the subset.[4] In other words, for every x in X there is exactly one element y such that the ordered pair (x, y) is contained in the subset defining the function f. This formal definition is a precise rendition of the idea that to each x is associated an element y of Y, namely the uniquely specified element y with the property just mentioned.

Considering the "color-of-the-shape" function above, the set X is the domain consisting of the four shapes, while Y is the codomain consisting of five colors. There are twenty possible ordered pairs (four shapes times five colors), one of which is

("yellow rectangle", "red").

The "color-of-the-shape" function described above consists of the set of those ordered pairs.

(shape, color)

where the color is the actual color of the given shape. Thus, the pair ("red triangle", "red") is in the function, but the pair ("yellow rectangle", "red") is not.

22.3 Notation

For more details on this topic, see functional notation.

A function f with domain X and codomain Y is commonly denoted by

$$f : X \to Y$$

or

$$X \xrightarrow{f} Y.$$

In this context, the elements of X are called arguments of f. For each argument x, the corresponding unique y in the codomain is called the function value at x or the *image* of x under f. It is written as $f(x)$. One says that f associates y with x or maps x to y. This is abbreviated by

$$y = f(x).$$

A general function is often denoted by f. Special functions have names, for example, the signum function is denoted by sgn. Given a real number x, its image under the signum function is then written as sgn(x). Here, the argument is denoted by the symbol x, but different symbols may be used in other contexts. For example, in physics, the velocity of some body, depending on the time, is denoted $v(t)$. The parentheses around the argument may be omitted when there is little chance of confusion, thus: sin x; this is known as prefix notation.

In order to denote a specific function, the notation $\mapsto$ (an arrow with a bar at its tail) is used. For example, the above function reads

$$f : \mathbb{N} \to \mathbb{Z}$$
$$x \mapsto 4 - x.$$

The first part can be read as:

- "f is a function from $\mathbb{N}$ (the set of natural numbers) to $\mathbb{Z}$ (the set of integers)" or

- "f is a $\mathbb{Z}$-valued function of an $\mathbb{N}$-valued variable".

The second part is read:

- "x maps to $4-x$."

In other words, this function has the natural numbers as domain, the integers as codomain. Strictly speaking, a function is properly defined only when the domain and codomain are specified. For example, the formula $f(x) = 4 - x$ alone (without specifying the codomain and domain) is not a properly defined function. Moreover, the function

$$g : \mathbb{Z} \to \mathbb{Z}$$
$$x \mapsto 4 - x.$$

(with different domain) is not considered the same function, even though the formulas defining f and g agree, and similarly with a different codomain. Despite that, many authors drop the specification of the domain and codomain, especially if these are clear from the context. So in this example many just write $f(x) = 4 - x$. Sometimes, the maximal possible domain is also understood implicitly: a formula such as $f(x) = \sqrt{x^2 - 5x + 6}$ may mean that the domain of f is the set of real numbers x where the square root is defined (in this case $x \le 2$ or $x \ge 3$).[5]

To define a function, sometimes a dot notation is used in order to emphasize the functional nature of an expression without assigning a special symbol to the variable. For instance, $a(\cdot)^2$ stands for the function $x \mapsto ax^2$, $\int_a^\cdot f(u)du$ stands for the integral function $x \mapsto \int_a^x f(u)du$, and so on.

22.4   Specifying a function

A function can be defined by any mathematical condition relating each argument (input value) to the corresponding output value. If the domain is finite, a function f may be defined by simply tabulating all the arguments x and their corresponding function values $f(x)$. More commonly, a function is defined by a formula, or (more generally) an algorithm — a recipe that tells how to compute the value of $f(x)$ given any x in the domain.

There are many other ways of defining functions. Examples include piecewise definitions, induction or recursion, algebraic or analytic closure, limits, analytic continuation, infinite series, and as solutions to integral and differential equations. The lambda calculus provides a powerful and flexible syntax for defining and combining functions of several variables. In advanced mathematics, some functions exist because of an axiom, such as the Axiom of Choice.

22.4.1   Graph

Main article: Graph of a function

The *graph* of a function is its set of ordered pairs F. This is an abstraction of the idea of a graph as a picture showing the function plotted on a pair of coordinate axes; for example, (3, 9), the point above 3 on the horizontal axis and to the right of 9 on the vertical axis, lies on the graph of $y=x^2$.

22.4.2   Formulas and algorithms

Different formulas or algorithms may describe the same function. For instance $f(x) = (x + 1)(x - 1)$ is exactly the same function as $f(x) = x^2 - 1$.[6] Furthermore, a function need not be described by a formula, expression, or algorithm, nor need it deal with numbers at all: the domain and codomain of a function may be arbitrary sets. One example of a function that acts on non-numeric inputs takes English words as inputs and returns the first letter of the input word as output.

As an example, the factorial function is defined on the nonnegative integers and produces a nonnegative integer. It is defined by the following inductive algorithm: 0! is defined to be 1, and $n!$ is defined to be $n(n - 1)!$ for all positive integers n. The factorial function is denoted with the exclamation mark (serving as the symbol of the function) after the variable (postfix notation).

22.4.3   Computability

Main article: computable function

Functions that send integers to integers, or finite strings to finite strings, can sometimes be defined by an algorithm, which gives a precise description of a set of steps for computing the output of the function from its input. Functions definable by an algorithm are called *computable functions*. For example, the Euclidean algorithm gives a precise process to compute the greatest common divisor of two positive integers. Many of the functions studied in the context of number theory are computable.

Fundamental results of computability theory show that there are functions that can be precisely defined but are not computable. Moreover, in the sense of cardinality, almost all functions from the integers to integers are not computable. The number of computable functions from integers to integers is countable, because the number of possible algorithms is. The number of all functions from integers to integers is higher: the same as the cardinality of the real numbers. Thus most functions from integers to integers are not computable. Specific examples of uncomputable functions are known, including the busy beaver function and functions related to the halting problem and other undecidable problems.

22.5 Basic properties

There are a number of general basic properties and notions. In this section, f is a function with domain X and codomain Y.

22.5.1 Image and preimage

Main article: Image (mathematics)

If A is any subset of the domain X, then $f(A)$ is the subset of the codomain Y consisting of all images of elements of

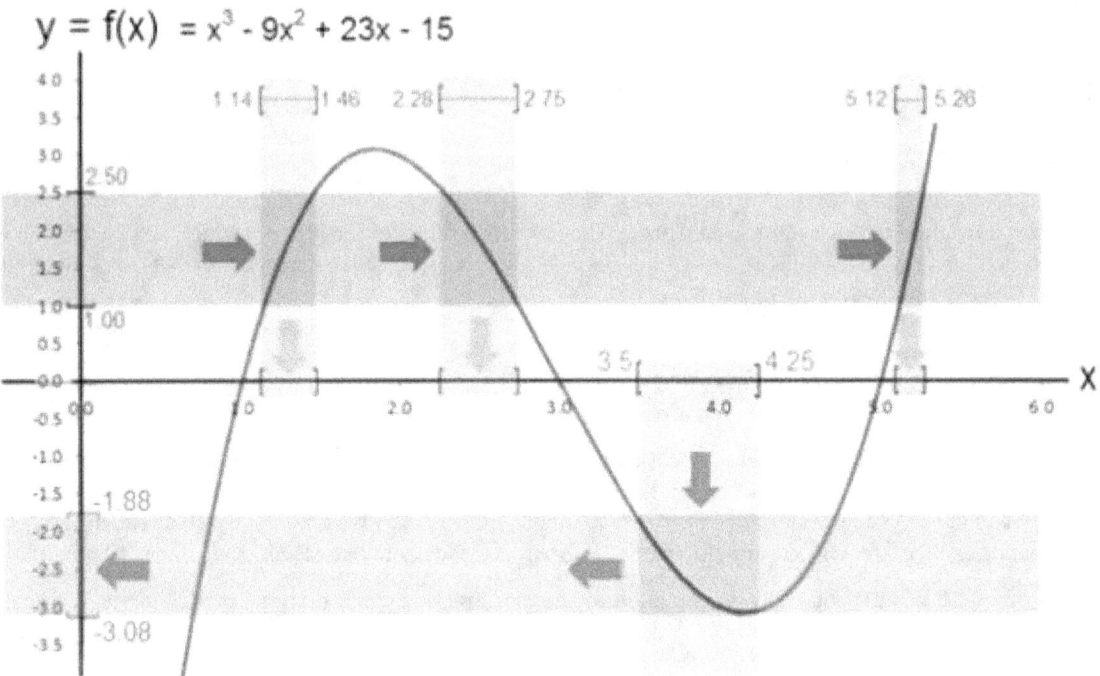

The graph of the function $f(x) = x^3 - 9x^2 + 23x - 15$. *The interval* $A = [3.5, 4.25]$ *is a subset of the domain, thus it is shown as part of the x-axis (green). The image of A is (approximately) the interval* $[-3.08, -1.88]$. *It is obtained by projecting to the y-axis (along the blue arrows) the intersection of the graph with the light green area consisting of all points whose x-coordinate is between 3.5 and 4.25, the part of the (vertical) y-axis shown in blue. The preimage of* $B = [1, 2.5]$ *consists of three intervals. They are obtained by projecting the intersection of the light red area with the graph to the x-axis.*

A. We say the $f(A)$ is the *image* of A under f. The *image* of f is given by $f(X)$. On the other hand, the *inverse image* (or *preimage, complete inverse image*) of a subset B of the codomain Y under a function f is the subset of the domain X defined by

$$f^{-1}(B) = \{x \in X : f(x) \in B\}.$$

So, for example, the preimage of $\{4, 9\}$ under the squaring function is the set $\{-3,-2,2,3\}$. The term range usually refers to the image,[7] but sometimes it refers to the codomain.

By definition of a function, the image of an element x of the domain is always a single element y of the codomain. Conversely, though, the preimage of a singleton set (a set with exactly one element) may in general contain any number of elements. For example, if $f(x) = 7$ (the constant function taking value 7), then the preimage of $\{5\}$ is the empty set but the preimage of $\{7\}$ is the entire domain. It is customary to write $f^{-1}(b)$ instead of $f^{-1}(\{b\})$, i.e.

$$f^{-1}(b) = \{x \in X : f(x) = b\}.$$

This set is sometimes called the fiber of b under f.

Use of $f(A)$ to denote the image of a subset $A \subseteq X$ is consistent so long as no subset of the domain is also an element of the domain. In some fields (e.g., in set theory, where ordinals are also sets of ordinals) it is convenient or even necessary to distinguish the two concepts; the customary notation is $f[A]$ for the set $\{ f(x) : x \in A \}$. Likewise, some authors use square brackets to avoid confusion between the inverse image and the inverse function. Thus they would write $f^{-1}[B]$ and $f^{-1}[b]$ for the preimage of a set and a singleton.

22.5.2   Injective and surjective functions

A function is called *injective* (or *one-to-one*, or an injection) if $f(a) \neq f(b)$ for any two *different* elements a and b of the domain. It is called surjective (or *onto*) if $f(X) = Y$. That is, it is surjective if for every element y in the codomain there is an x in the domain such that $f(x) = y$. Finally f is called *bijective* if it is both injective and surjective. This nomenclature was introduced by the Bourbaki group.

The above "color-of-the-shape" function is not injective, since two distinct shapes (the red triangle and the red rectangle) are assigned the same value. Moreover, it is not surjective, since the image of the function contains only three, but not all five colors in the codomain.

22.5.3   Function composition

Main article: Function composition

The *function composition* of two functions takes the output of one function as the input of a second one. More specifically, the composition of f with a function g: $Y \to Z$ is the function $g \circ f : X \to Z$ defined by

$$(g \circ f)(x) = g(f(x)).$$

That is, the value of x is obtained by first applying f to x to obtain $y = f(x)$ and then applying g to y to obtain $z = g(y)$. In the notation $g \circ f$, the function on the right, f, acts first and the function on the left, g acts second, reversing English reading order. The notation can be memorized by reading the notation as "g of f" or "g after f". The composition $g \circ f$ is only defined when the codomain of f is the domain of g. Assuming that, the composition in the opposite order $f \circ g$ need not be defined. Even if it is, i.e., if the codomain of f is the codomain of g, it is *not* in general true that

$$g \circ f = f \circ g.$$

That is, the order of the composition is important. For example, suppose $f(x) = x^2$ and $g(x) = x+1$. Then $g(f(x)) = x^2+1$, while $f(g(x)) = (x+1)^2$, which is x^2+2x+1, a different function.

22.5.4   Identity function

Main article: Identity function

The unique function over a set X that maps each element to itself is called the *identity function* for X, and typically denoted by $\mathrm{id}X$. Each set has its own identity function, so the subscript cannot be omitted unless the set can be inferred from context. Under composition, an identity function is "neutral": if f is any function from X to Y, then

$$f \circ \mathrm{id}_X = f,$$
$$\mathrm{id}_Y \circ f = f.$$

22.5.5   Restrictions and extensions

Main article: Restriction (mathematics)

Informally, a *restriction* of a function f is the result of trimming its domain. More precisely, if S is any subset of X, the restriction of f to S is the function $f|S$ from S to Y such that $f|S(s) = f(s)$ for all s in S. If g is a restriction of f, then it is said that f is an *extension* of g.

The *overriding* of $f\colon X \to Y$ by $g\colon W \to Y$ (also called *overriding union*) is an extension of g denoted as $(f \oplus g)\colon (X \cup W) \to Y$. Its graph is the set-theoretical union of the graphs of g and $f|X \setminus W$. Thus, it relates any element of the domain of g to its image under g, and any other element of the domain of f to its image under f. Overriding is an associative operation; it has the empty function as an identity element. If $f|X \cap W$ and $g|X \cap W$ are pointwise equal (e.g., the domains of f and g are disjoint), then the union of f and g is defined and is equal to their overriding union. This definition agrees with the definition of union for binary relations.

22.5.6   Inverse function

Main article: Inverse function

An *inverse function* for f, denoted by f^{-1}, is a function in the opposite direction, from Y to X, satisfying

$$f \circ f^{-1} = \mathrm{id}_Y, \, f^{-1} \circ f = \mathrm{id}_X \,.$$

That is, the two possible compositions of f and f^{-1} need to be the respective identity maps of X and Y.

As a simple example, if f converts a temperature in degrees Celsius C to degrees Fahrenheit F, the function converting degrees Fahrenheit to degrees Celsius would be a suitable f^{-1}.

$$f(C) = \frac{9}{5}C + 32$$
$$f^{-1}(F) = \frac{5}{9}(F - 32)$$

Such an inverse function exists if and only if f is bijective. In this case, f is called invertible. The notation $g \circ f$ (or, in some texts, just gf) and f^{-1} are akin to multiplication and reciprocal notation. With this analogy, identity functions are like the multiplicative identity, 1, and inverse functions are like reciprocals (hence the notation).

22.6   Types of functions

For a more extensive list, see list of types of functions.

22.6.1   Real-valued functions

A real-valued function f is one whose codomain is the set of real numbers or a subset thereof. If, in addition, the domain is also a subset of the reals, f is a real valued function of a real variable. The study of such functions is called real analysis.

Real-valued functions enjoy so-called pointwise operations. That is, given two functions

$$f, g: X \to Y$$

where Y is a subset of the reals (and X is an arbitrary set), their (pointwise) sum $f+g$ and product $f \cdot g$ are functions with the same domain and codomain. They are defined by the formulas:

$$(f + g)(x) = f(x) + g(x),$$
$$(f \cdot g)(x) = f(x) \cdot g(x).$$

In a similar vein, complex analysis studies functions whose domain and codomain are both the set of complex numbers. In most situations, the domain and codomain are understood from context, and only the relationship between the input and output is given, but if $f(x) = \sqrt{x}$, then in real variables the domain is limited to non-negative numbers.

The following table contains a few particularly important types of real-valued functions:

22.6.2   Further types of functions

Further information: List of mathematical functions

There are many other special classes of functions that are important to particular branches of mathematics, or particular applications. Here is a partial list:

- differentiable, integrable

- polynomial, rational

- algebraic, transcendental

- odd or even

- convex, monotonic

- holomorphic, meromorphic, entire

- vector-valued

- computable

22.7 Function spaces

Main article: Function space

The set of all functions from a set X to a set Y is denoted by $X \to Y$, by $[X \to Y]$, or by Y^X. The latter notation is motivated by the fact that, when X and Y are finite and of size $|X|$ and $|Y|$, then the number of functions $X \to Y$ is $|Y^X| = |Y|^{|X|}$. This is an example of the convention from enumerative combinatorics that provides notations for sets based on their cardinalities. If X is infinite and there is more than one element in Y then there are uncountably many functions from X to Y, though only countably many of them can be expressed with a formula or algorithm.

22.7.1 Currying

Main article: Currying

An alternative approach to handling functions with multiple arguments is to transform them into a chain of functions that each takes a single argument. For instance, one can interpret Add(3,5) to mean "first produce a function that adds 3 to its argument, and then apply the 'Add 3' function to 5". This transformation is called currying: Add 3 is curry(Add) applied to 3. There is a bijection between the function spaces $C^{A \times B}$ and $(C^B)^A$.

When working with curried functions it is customary to use prefix notation with function application considered left-associative, since juxtaposition of multiple arguments—as in $(f \; x \; y)$—naturally maps to evaluation of a curried function. Conversely, the $\to$ and $\mapsto$ symbols are considered to be right-associative, so that curried functions may be defined by a notation such as $f \colon \mathbf{Z} \to \mathbf{Z} \to \mathbf{Z} = x \mapsto y \mapsto x \cdot y$.

22.8 Variants and generalizations

22.8.1 Alternative definition of a function

The above definition of "a function from X to Y" is generally agreed on, however there are two different ways a "function" is normally defined where the domain X and codomain Y are not explicitly or implicitly specified. Usually this is not a problem as the domain and codomain normally will be known. With one definition saying the function defined by $f(x) = x^2$ on the reals does not completely specify a function as the codomain is not specified, and in the other it is a valid definition.

In the other definition a function is defined as a set of ordered pairs where each first element only occurs once. The domain is the set of all the first elements of a pair and there is no explicit codomain separate from the image.[8][9] Concepts like surjective have to be refined for such functions, more specifically by saying that a (given) function is *surjective on a (given) set* if its image equals that set. For example, we might say a function f is surjective on the set of real numbers.

If a function is defined as a set of ordered pairs with no specific codomain, then $f \colon X \to Y$ indicates that f is a function whose domain is X and whose image is a subset of Y. This is the case in the ISO standard.[7] Y may be referred to as the codomain but then any set including the image of f is a valid codomain of f. This is also referred to by saying that "f maps X into Y".[7] In some usages X and Y may subset the ordered pairs, e.g. the function f on the real numbers such that $y = x^2$ when used as in $f \colon [0,4] \to [0,4]$ means the function defined only on the interval $[0,2]$.[10] With the definition of a function as an ordered triple this would always be considered a partial function.

An alternative definition of the composite function $g(f(x))$ defines it for the set of all x in the domain of f such that $f(x)$ is in the domain of g.[11] Thus the real square root of $-x^2$ is a function only defined at 0 where it has the value 0.

Functions are commonly defined as a type of relation. A relation from X to Y is a set of ordered pairs (x, y) with $x \in X$ and $y \in Y$. A function from X to Y can be described as a relation from X to Y that is left-total and right-unique. However when X and Y are not specified there is a disagreement about the definition of a relation that parallels that for functions. Normally a relation is just defined as a set of ordered pairs and a correspondence is defined as a triple (X, Y, F), however the distinction between the two is often blurred or a relation is never referred to without specifying the two sets. The definition of a function as a triple defines a function as a type of correspondence, whereas the definition of a function as a set of ordered pairs defines a function as a type of relation.

Many operations in set theory, such as the power set, have the class of all sets as their domain, and therefore, although they are informally described as functions, they do not fit the set-theoretical definition outlined above, because a class is not necessarily a set. However some definitions of relations and functions define them as classes of pairs rather than sets of pairs and therefore do include the power set as a function.[12]

22.8.2   Partial and multi-valued functions

In some parts of mathematics, including recursion theory and functional analysis, it is convenient to study *partial functions* in which some values of the domain have no association in the graph; i.e., single-valued relations. For example, the function f such that $f(x) = 1/x$ does not define a value for $x = 0$, since division by zero is not defined. Hence f is only a partial function from the real line to the real line. The term total function can be used to stress the fact that every element of the domain does appear as the first element of an ordered pair in the graph. In other parts of mathematics, non-single-valued relations are similarly conflated with functions: these are called *multivalued functions*, with the corresponding term single-valued function for ordinary functions.

22.8.3   Functions with multiple inputs and outputs

The concept of function can be extended to an object that takes a combination of two (or more) argument values to a single result. This intuitive concept is formalized by a function whose domain is the Cartesian product of two or more sets.

For example, consider the function that associates two integers to their product: $f(x, y) = x \cdot y$. This function can be defined formally as having domain $\mathbf{Z} \times \mathbf{Z}$, the set of all integer pairs; codomain $\mathbf{Z}$; and, for graph, the set of all pairs $((x,y), x \cdot y)$. Note that the first component of any such pair is itself a pair (of integers), while the second component is a single integer.

The function value of the pair (x,y) is $f((x,y))$. However, it is customary to drop one set of parentheses and consider $f(x,y)$ a function of two variables, x and y. Functions of two variables may be plotted on the three-dimensional Cartesian as ordered triples of the form $(x,y,f(x,y))$.

The concept can still further be extended by considering a function that also produces output that is expressed as several variables. For example, consider the integer divide function, with domain $\mathbf{Z} \times \mathbf{N}$ and codomain $\mathbf{Z} \times \mathbf{N}$. The resultant (quotient, remainder) pair is a single value in the codomain seen as a Cartesian product.

Binary operations

The familiar binary operations of arithmetic, addition and multiplication, can be viewed as functions from $\mathbf{R} \times \mathbf{R}$ to $\mathbf{R}$. This view is generalized in abstract algebra, where n-ary functions are used to model the operations of arbitrary algebraic structures. For example, an abstract group is defined as a set X and a function f from $X \times X$ to X that satisfies certain properties.

Traditionally, addition and multiplication are written in the infix notation: $x+y$ and $x \times y$ instead of $+(x, y)$ and $\times (x, y)$.

22.8.4   Functors

The idea of structure-preserving functions, or homomorphisms, led to the abstract notion of morphism, the key concept of category theory. In fact, functions $f : X \rightarrow Y$ are the morphisms in the category of sets, including the empty set: if the domain X is the empty set, then the subset of $X \times Y$ describing the function is necessarily empty, too. However, this is still a well-defined function. Such a function is called an empty function. In particular, the identity function of the empty set is defined, a requirement for sets to form a category.

The concept of categorification is an attempt to replace set-theoretic notions by category-theoretic ones. In particular, according to this idea, sets are replaced by categories, while functions between sets are replaced by functors.[13]

22.9 History

Main article: History of the function concept

22.10 See also

- Associative array
- Functional
- Functional decomposition
- Function fitting
- Functional predicate
- Functional programming
- Generalized function
- Implicit function
- List of functions
- Multivalued function
- Parametric equation

22.11 Notes

[1] The words **map** or **mapping, transformation, correspondence**, and **operator** are often used synonymously. Halmos 1970, p. 30.

[2] Spivak 2008, p. 39.

[3] MacLane, Saunders; Birkhoff, Garrett (1967). *Algebra* (First ed.). New York: Macmillan. pp. 1–13.

[4] Hamilton, A. G. *Numbers, sets, and axioms: the apparatus of mathematics*. Cambridge University Press. p. 83. ISBN 0-521-24509-5.

[5] Bloch 2011, p. 133.

[6] Hartley Rogers, Jr (1987). *Theory of Recursive Functions and Effective Computation*. MIT Press. pp. 1–2. ISBN 0-262-68052-1.

[7] *Quantities and Units - Part 2: Mathematical signs and symbols to be used in the natural sciences and technology*, page 15. ISO 80000-2 (ISO/IEC 2009-12-01)

[8] Apostol, Tom (1967). *Calculus vol 1*. John Wiley. p. 53. ISBN 0-471-00005-1.

[9] Heins, Maurice (1968). *Complex function theory*. Academic Press. p. 4.

[10] Bartle 1967, p. 13.

[11] Bartle 1967, p. 21.

[12] Tarski, Alfred; Givant, Steven (1987). *A formalization of set theory without variables*. American Mathematical Society. p. 3. ISBN 0-8218-1041-3.

[13] John C. Baez; James Dolan (1998). "Categorification". arXiv:math/9802029.

22.12   References

- Bartle, Robert (1967). *The Elements of Real Analysis*. John Wiley & Sons.

- Bloch, Ethan D. (2011). *Proofs and Fundamentals: A First Course in Abstract Mathematics*. Springer. ISBN 978-1-4419-7126-5.

- Halmos, Paul R. (1970). *Naive Set Theory*. Springer-Verlag. ISBN 0-387-90092-6.

- Spivak, Michael (2008). *Calculus* (4th ed.). Publish or Perish. ISBN 978-0-914098-91-1.

22.13   Further reading

- Anton, Howard (1980). *Calculus with Analytical Geometry*. Wiley. ISBN 978-0-471-03248-9.

- Bartle, Robert G. (1976). *The Elements of Real Analysis* (2nd ed.). Wiley. ISBN 978-0-471-05464-1.

- Dubinsky, Ed; Harel, Guershon (1992). *The Concept of Function: Aspects of Epistemology and Pedagogy*. Mathematical Association of America. ISBN 0-88385-081-8.

- Hammack, Richard (2009). "12. Functions" (PDF). *Book of Proof*. Virginia Commonwealth University. Retrieved 2012-08-01.

- Husch, Lawrence S. (2001). *Visual Calculus*. University of Tennessee. Retrieved 2007-09-27.

- Katz, Robert (1964). *Axiomatic Analysis*. D. C. Heath and Company.

- Kleiner, Israel (1989). *Evolution of the Function Concept: A Brief Survey*. The College Mathematics Journal **20** (4) (Mathematical Association of America). pp. 282–300. doi:10.2307/2686848. JSTOR 2686848.

- Lützen, Jesper (2003). "Between rigor and applications: Developments in the concept of function in mathematical analysis". In Roy Porter, ed. *The Cambridge History of Science: The modern physical and mathematical sciences*. Cambridge University Press. ISBN 0521571995. An approachable and diverting historical presentation.

- Malik, M. A. (1980). *Historical and pedagogical aspects of the definition of function*. International Journal of Mathematical Education in Science and Technology **11** (4). pp. 489–492. doi:10.1080/0020739800110404.

- Reichenbach, Hans (1947) *Elements of Symbolic Logic*, Dover Publishing Inc., New York NY, ISBN 0-486-24004-5.

- Ruthing, D. (1984). *Some definitions of the concept of function from Bernoulli, Joh. to Bourbaki, N. Mathematical Intelligencer* **6** (4). pp. 72–77.

- Thomas, George B.; Finney, Ross L. (1995). *Calculus and Analytic Geometry* (9th ed.). Addison-Wesley. ISBN 978-0-201-53174-9.

22.14   External links

- Khan Academy: Functions, free online micro lectures

- Hazewinkel, Michiel, ed. (2001), "Function", *Encyclopedia of Mathematics*, Springer, ISBN 978-1-55608-010-4

- Weisstein, Eric W., "Function", *MathWorld*.

- The Wolfram Functions Site gives formulae and visualizations of many mathematical functions.

- Shodor: Function Flyer, interactive Java applet for graphing and exploring functions.

- xFunctions, a Java applet for exploring functions graphically.

- Draw Function Graphs, online drawing program for mathematical functions.

- Functions from cut-the-knot.

- Function at ProvenMath.

- Comprehensive web-based function graphing & evaluation tool.

- Abstractmath.org articles on functions

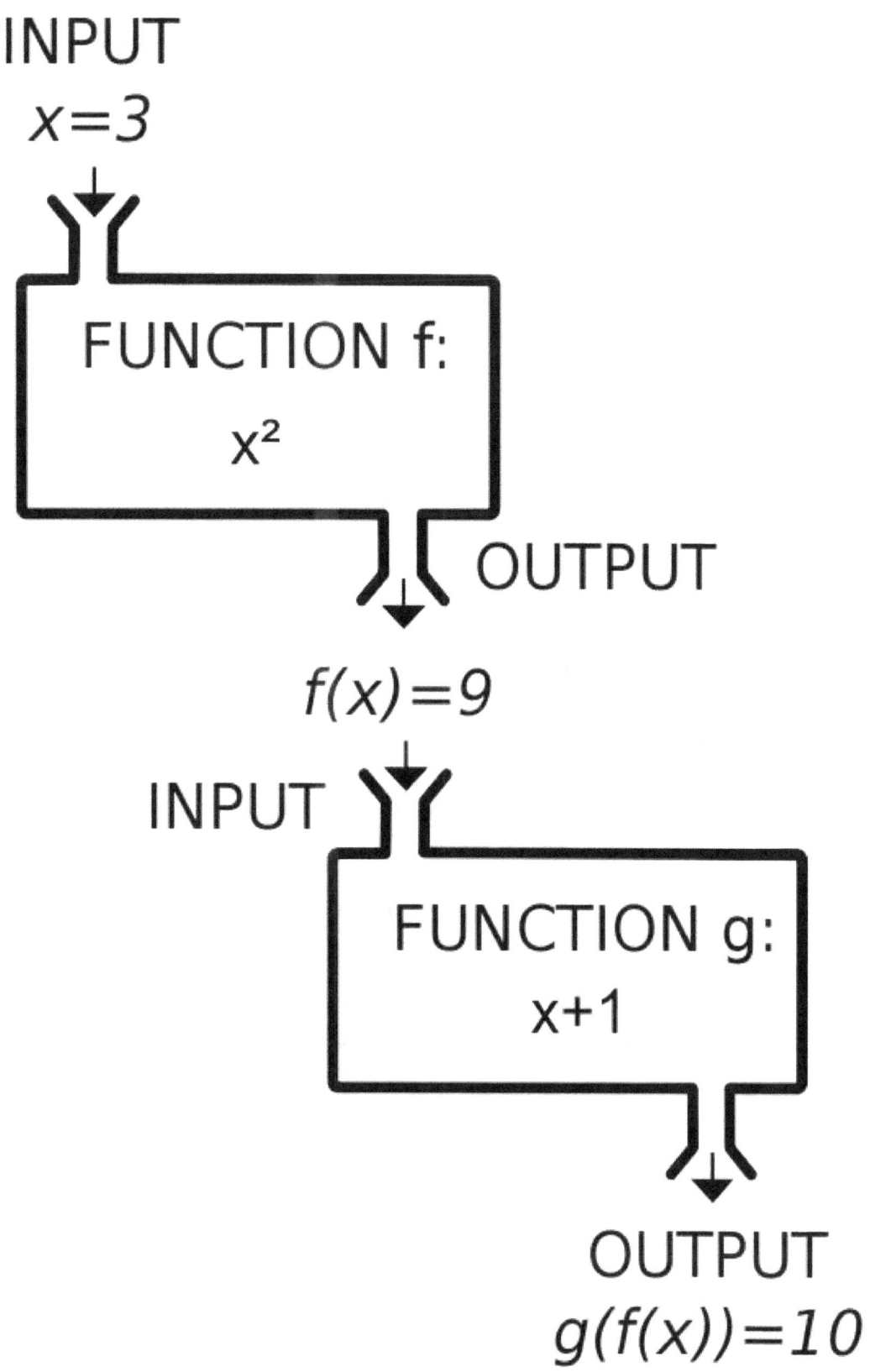

A composite function g(f(x)) can be visualized as the combination of two "machines". The first takes input x and outputs f(x). The second takes f(x) and outputs g(f(x)).

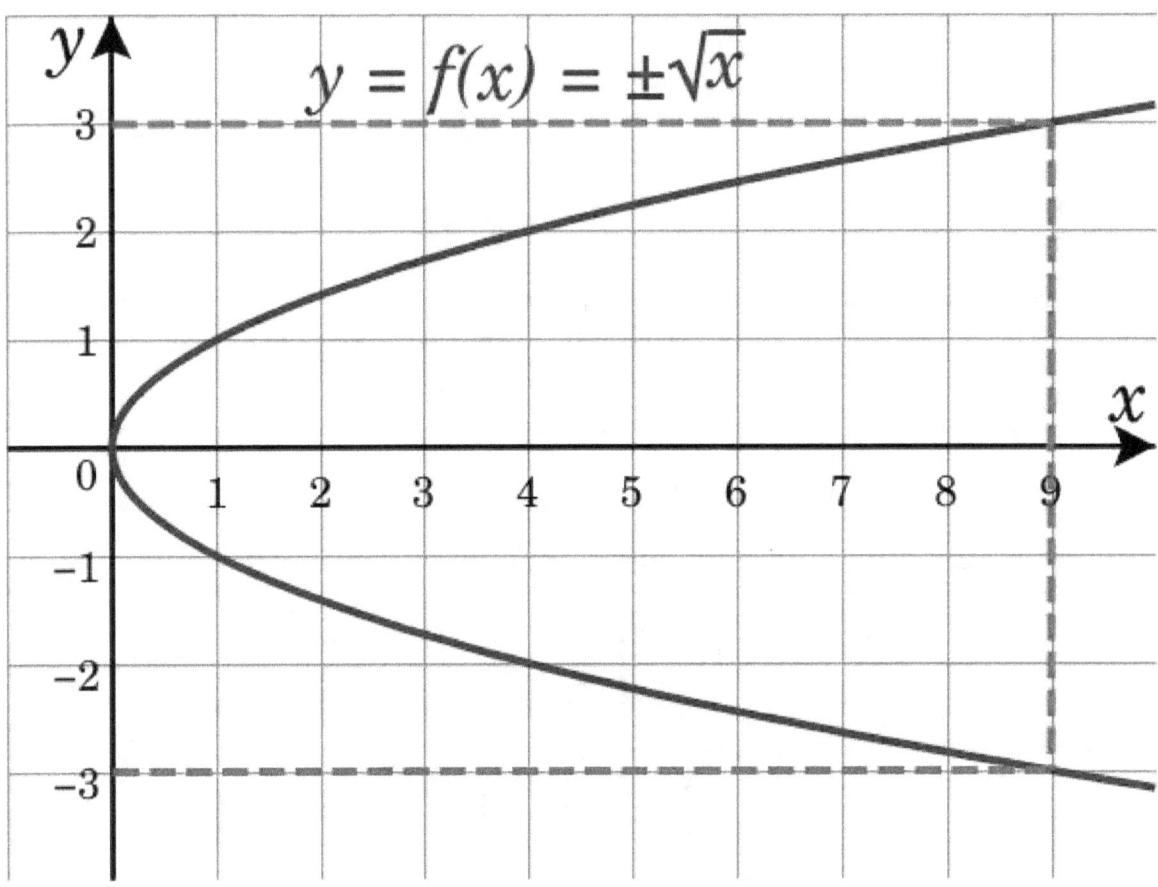

$$y = f(x) = \pm\sqrt{x}$$

$f(x) = \pm\sqrt{x}$ is not a function in the proper sense, but a multi-valued function: it assigns to each positive real number x two values: the (positive) square root of x, and $-\sqrt{x}$.

Chapter 23

Geometry

For other uses, see Geometry (disambiguation).

Geometry (from the Ancient Greek: γεωμετρία; *geo-* "earth", *-metron* "measurement") is a branch of mathematics

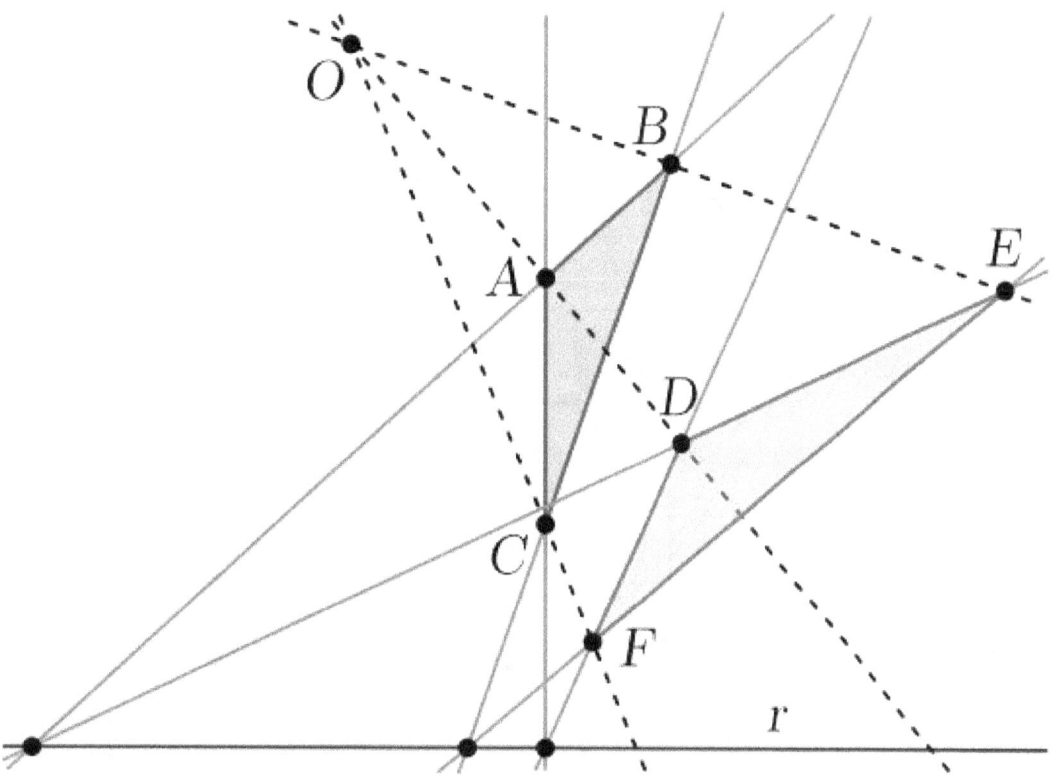

An illustration of Desargues' theorem, an important result in Euclidean and projective geometry

concerned with questions of shape, size, relative position of figures, and the properties of space. A mathematician who works in the field of geometry is called a geometer. Geometry arose independently in a number of early cultures as a body of practical knowledge concerning lengths, areas, and volumes, with elements of formal mathematical science emerging in the West as early as Thales (6th century BC). By the 3rd century BC, geometry was put into an axiomatic form by Euclid, whose treatment—Euclidean geometry—set a standard for many centuries to follow.[1] Archimedes developed ingenious techniques for calculating areas and volumes, in many ways anticipating modern integral calculus. The field of astronomy, especially as it relates to mapping the positions of stars and planets on the celestial sphere and describing the relationship between movements of celestial bodies, served as an important source of geometric problems during the next one and a half millennia. In the classical world, both geometry and astronomy were considered to be part of the Quadrivium, a subset of the seven liberal arts considered essential for a free citizen to master.

The introduction of coordinates by René Descartes and the concurrent developments of algebra marked a new stage for geometry, since geometric figures such as plane curves could now be represented analytically in the form of functions and equations. This played a key role in the emergence of infinitesimal calculus in the 17th century. Furthermore, the theory of perspective showed that there is more to geometry than just the metric properties of figures: perspective is the origin of projective geometry. The subject of geometry was further enriched by the study of the intrinsic structure of geometric objects that originated with Euler and Gauss and led to the creation of topology and differential geometry.

In Euclid's time, there was no clear distinction between physical and geometrical space. Since the 19th-century discovery of non-Euclidean geometry, the concept of space has undergone a radical transformation and raised the question of which geometrical space best fits physical space. With the rise of formal mathematics in the 20th century, 'space' (whether 'point', 'line', or 'plane') lost its intuitive contents, so today one has to distinguish between physical space, geometrical spaces (in which 'space', 'point' etc. still have their intuitive meanings) and abstract spaces. Contemporary geometry considers manifolds, spaces that are considerably more abstract than the familiar Euclidean space, which they only approximately resemble at small scales. These spaces may be endowed with additional structure which allow one to speak about length. Modern geometry has many ties to physics as is exemplified by the links between pseudo-Riemannian geometry and general relativity. One of the youngest physical theories, string theory, is also very geometric in flavour.

While the visual nature of geometry makes it initially more accessible than other mathematical areas such as algebra or number theory, geometric language is also used in contexts far removed from its traditional, Euclidean provenance (for example, in fractal geometry and algebraic geometry).[2]

23.1 Overview

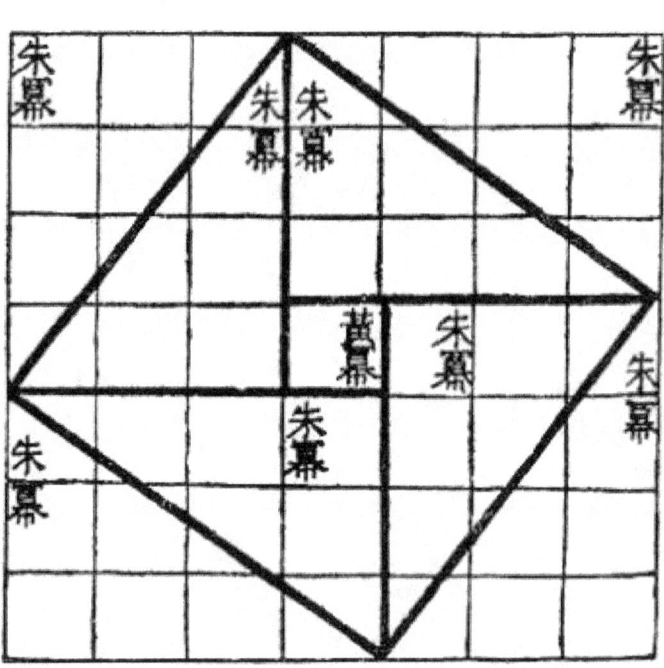

Visual checking of the Pythagorean theorem for the (3, 4, 5) triangle as in the Chou Pei Suan Ching 500–200 BC.

Because the recorded development of geometry spans more than two millennia, perceptions of what constitutes geometry have evolved throughout the ages:

23.1.1   Practical geometry

Geometry originated as a practical science concerned with surveys, measurements, areas, and volumes. Among other highlights, notable accomplishments include formulas for lengths, areas and volumes, such as the Pythagorean theorem, circumference and area of a circle, area of a triangle, volume of a cylinder, sphere, and a pyramid. A method of computing certain inaccessible distances or heights based on similarity of geometric figures is attributed to Thales. The development of astronomy led to the emergence of trigonometry and spherical trigonometry, together with the attendant computational techniques.

23.1.2   Axiomatic geometry

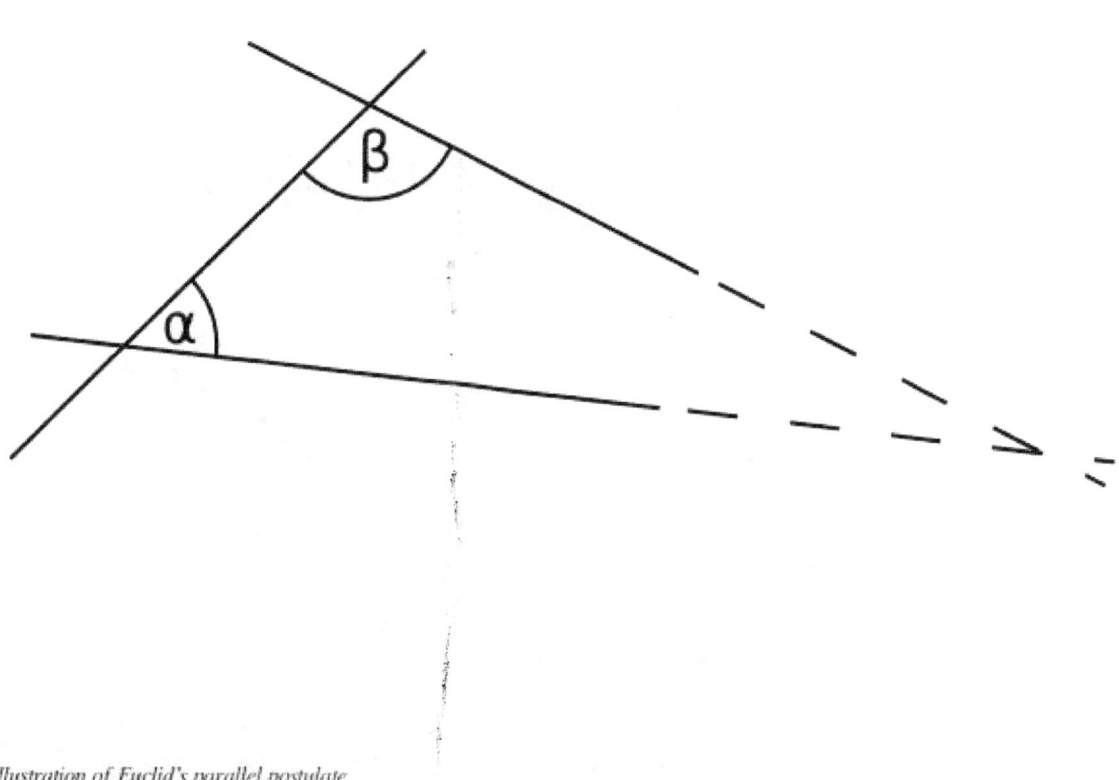

An illustration of Euclid's parallel postulate

See also: Euclidean geometry

Euclid took a more abstract approach in his Elements, one of the most influential books ever written. Euclid introduced certain axioms, or postulates, expressing primary or self-evident properties of points, lines, and planes. He proceeded to rigorously deduce other properties by mathematical reasoning. The characteristic feature of Euclid's approach to geometry was its rigor, and it has come to be known as *axiomatic* or *synthetic* geometry. At the start of the 19th century, the discovery of non-Euclidean geometries by Nikolai Ivanovich Lobachevsky (1792–1856), János Bolyai (1802–1860) and Carl Friedrich Gauss (1777–1855) and others led to a revival of interest in this discipline, and in the 20th century, David Hilbert (1862–1943) employed axiomatic reasoning in an attempt to provide a modern foundation of geometry.

Geometry lessons in the 20th century

23.1.3   Geometric constructions

Main article: Compass and straightedge constructions

Classical geometers paid special attention to constructing geometric objects that had been described in some other way. Classically, the only instruments allowed in geometric constructions are the compass and straightedge. Also, every construction had to be complete in a finite number of steps. However, some problems turned out to be difficult or impossible to solve by these means alone, and ingenious constructions using parabolas and other curves, as well as mechanical devices, were found.

23.1.4   Numbers in geometry

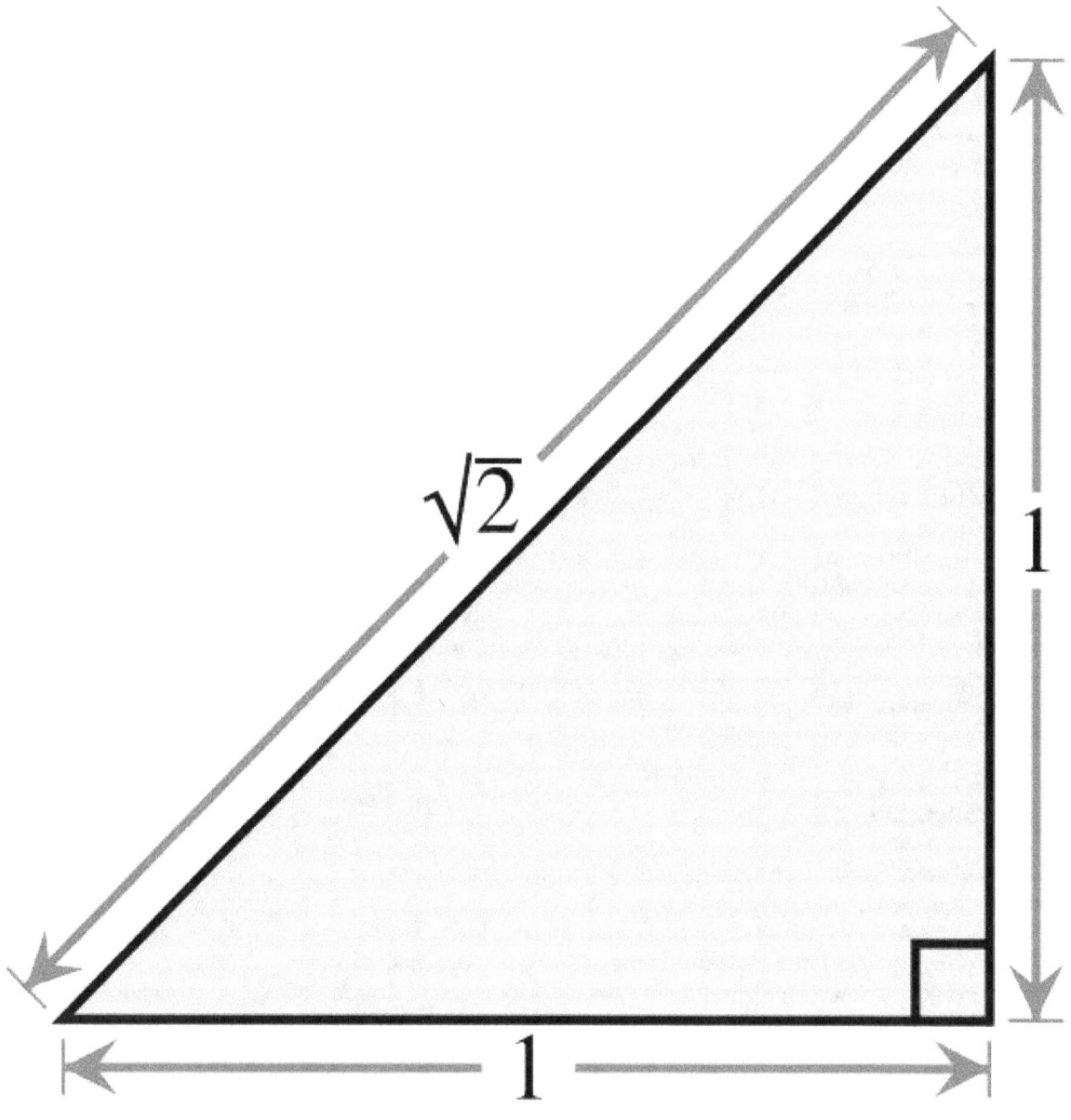

The Pythagoreans discovered that the sides of a triangle could have incommensurable lengths.

In ancient Greece the Pythagoreans considered the role of numbers in geometry. However, the discovery of incommensurable lengths, which contradicted their philosophical views, made them abandon abstract numbers in favor of concrete geometric quantities, such as length and area of figures. Numbers were reintroduced into geometry in the form of coordinates by Descartes, who realized that the study of geometric shapes can be facilitated by their algebraic representation, and for whom the Cartesian plane is named. Analytic geometry applies methods of algebra to geometric questions, typically by relating geometric curves to algebraic equations. These ideas played a key role in the development of calculus in the 17th century and led to the discovery of many new properties of plane curves. Modern algebraic geometry considers similar questions on a vastly more abstract level.

23.1.5 Geometry of position

Main articles: Projective geometry and Topology

Even in ancient times, geometers considered questions of relative position or spatial relationship of geometric figures and shapes. Some examples are given by inscribed and circumscribed circles of polygons, lines intersecting and tangent to conic sections, the Pappus and Menelaus configurations of points and lines. In the Middle Ages, new and more complicated questions of this type were considered: What is the maximum number of spheres simultaneously touching a given sphere of the same radius (kissing number problem)? What is the densest packing of spheres of equal size in space (Kepler conjecture)? Most of these questions involved 'rigid' geometrical shapes, such as lines or spheres. Projective, convex, and discrete geometry are three sub-disciplines within present day geometry that deal with these types of questions.

Leonhard Euler, in studying problems like the Seven Bridges of Königsberg, considered the most fundamental properties of geometric figures based solely on shape, independent of their metric properties. Euler called this new branch of geometry *geometria situs* (geometry of place), but it is now known as topology. Topology grew out of geometry, but turned into a large independent discipline. It does not differentiate between objects that can be continuously deformed into each other. The objects may nevertheless retain some geometry, as in the case of hyperbolic knots.

23.1.6 Geometry beyond Euclid

In the nearly two thousand years since Euclid, while the range of geometrical questions asked and answered inevitably expanded, the basic understanding of space remained essentially the same. Immanuel Kant argued that there is only one, *absolute*, geometry, which is known to be true *a priori* by an inner faculty of mind: Euclidean geometry was synthetic a priori.[3] This dominant view was overturned by the revolutionary discovery of non-Euclidean geometry in the works of Bolyai, Lobachevsky, and Gauss (who never published his theory). They demonstrated that ordinary Euclidean space is only one possibility for development of geometry. A broad vision of the subject of geometry was then expressed by Riemann in his 1867 inauguration lecture *Über die Hypothesen, welche der Geometrie zu Grunde liegen* (*On the hypotheses on which geometry is based*),[4] published only after his death. Riemann's new idea of space proved crucial in Einstein's general relativity theory, and Riemannian geometry, that considers very general spaces in which the notion of length is defined, is a mainstay of modern geometry.

23.1.7 Dimension

Where the traditional geometry allowed dimensions 1 (a line), 2 (a plane) and 3 (our ambient world conceived of as three-dimensional space), mathematicians have used higher dimensions for nearly two centuries. Dimension has gone through stages of being any natural number n, possibly infinite with the introduction of Hilbert space, and any positive real number in fractal geometry. Dimension theory is a technical area, initially within general topology, that discusses *definitions*; in common with most mathematical ideas, dimension is now defined rather than an intuition. Connected topological manifolds have a well-defined dimension; this is a theorem (invariance of domain) rather than anything *a priori*.

The issue of dimension still matters to geometry, in the absence of complete answers to classic questions. Dimensions 3 of space and 4 of space-time are special cases in geometric topology. Dimension 10 or 11 is a key number in string theory. Research may bring a satisfactory *geometric* reason for the significance of 10 and 11 dimensions.

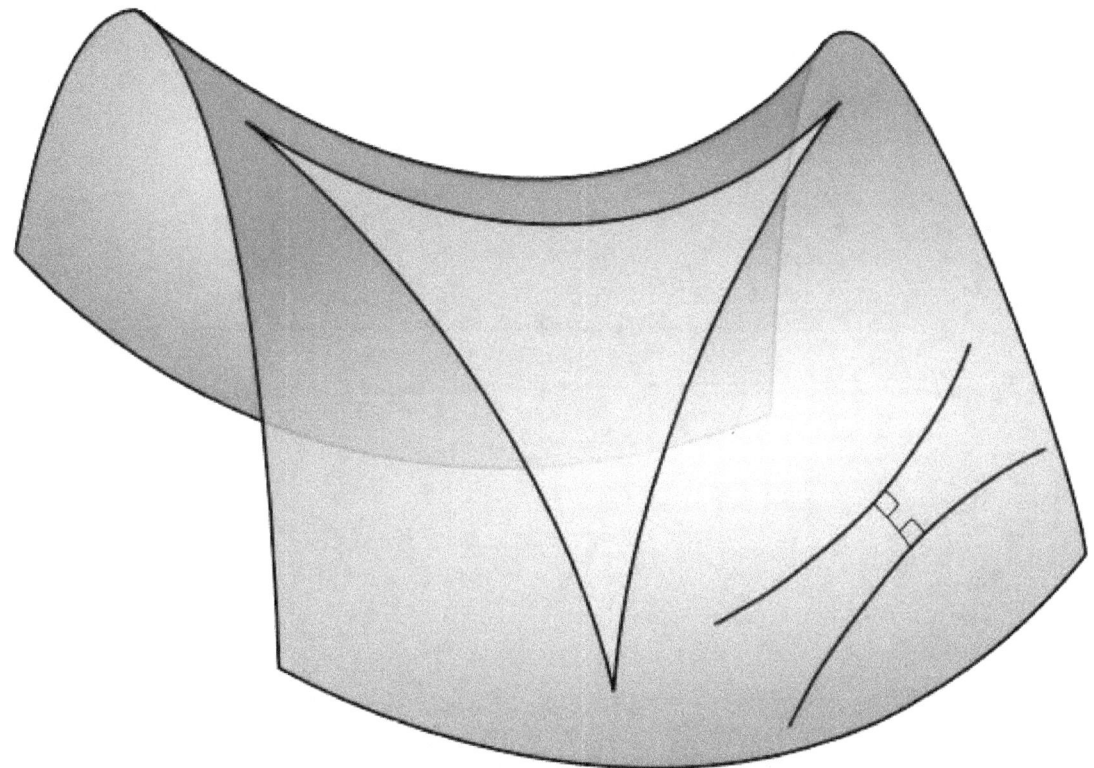

Differential geometry uses tools from calculus to study problems involving curvature.

23.1.8 Symmetry

The theme of symmetry in geometry is nearly as old as the science of geometry itself. Symmetric shapes such as the circle, regular polygons and platonic solids held deep significance for many ancient philosophers and were investigated in detail before the time of Euclid. Symmetric patterns occur in nature and were artistically rendered in a multitude of forms, including the graphics of M. C. Escher. Nonetheless, it was not until the second half of 19th century that the unifying role of symmetry in foundations of geometry was recognized. Felix Klein's Erlangen program proclaimed that, in a very precise sense, symmetry, expressed via the notion of a transformation group, determines what geometry *is*. Symmetry in classical Euclidean geometry is represented by congruences and rigid motions, whereas in projective geometry an analogous role is played by collineations, geometric transformations that take straight lines into straight lines. However it was in the new geometries of Bolyai and Lobachevsky, Riemann, Clifford and Klein, and Sophus Lie that Klein's idea to 'define a geometry via its symmetry group' proved most influential. Both discrete and continuous symmetries play prominent roles in geometry, the former in topology and geometric group theory, the latter in Lie theory and Riemannian geometry.

A different type of symmetry is the principle of duality in projective geometry (see Duality (projective geometry)) among other fields. This meta-phenomenon can roughly be described as follows: in any theorem, exchange *point* with *plane*, *join* with *meet*, *lies in* with *contains*, and you will get an equally true theorem. A similar and closely related form of duality exists between a vector space and its dual space.

23.2 History

Main article: History of geometry

 The earliest recorded beginnings of geometry can be traced to ancient Mesopotamia and Egypt in the 2nd millennium BC.[5][6] Early geometry was a collection of empirically discovered principles concerning lengths, angles, areas, and

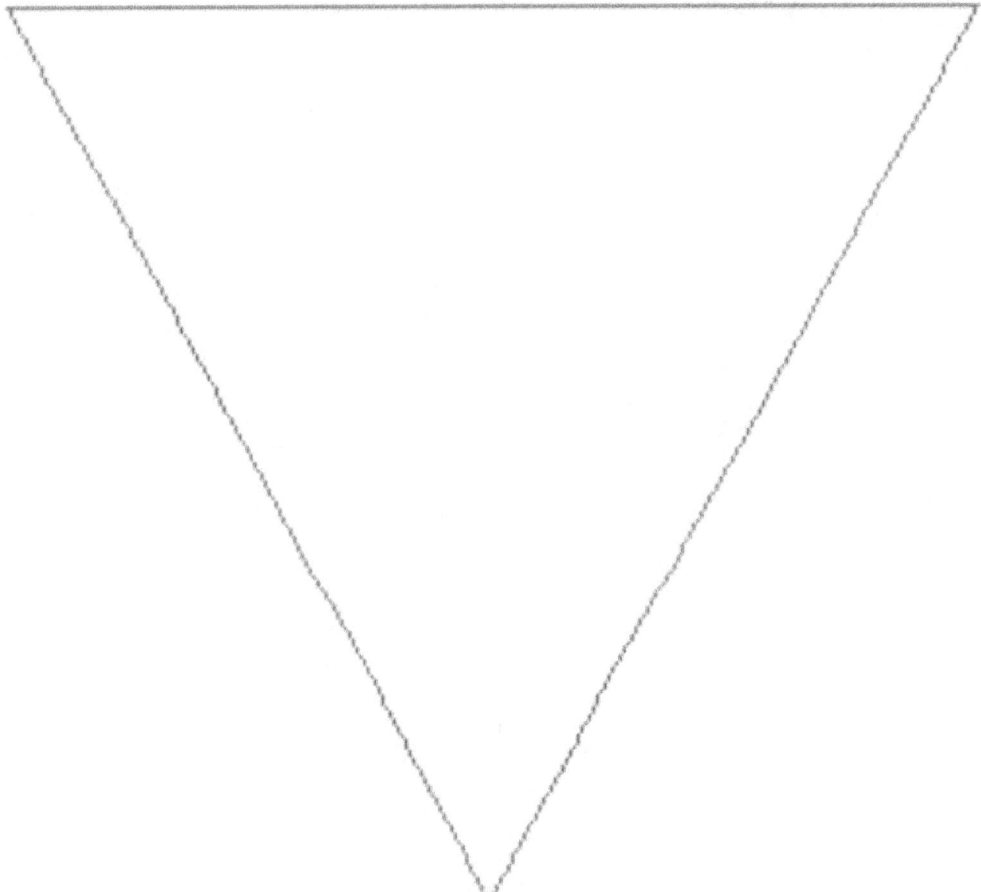

The Koch snowflake, with fractal dimension=log4/log3 and topological dimension=1

volumes, which were developed to meet some practical need in surveying, construction, astronomy, and various crafts. The earliest known texts on geometry are the Egyptian *Rhind Papyrus* (2000–1800 BC) and *Moscow Papyrus* (c. 1890 BC), the Babylonian clay tablets such as Plimpton 322 (1900 BC). For example, the Moscow Papyrus gives a formula for calculating the volume of a truncated pyramid, or frustum.[7] South of Egypt the ancient Nubians established a system of geometry including early versions of sun clocks.[8][9]

In the 7th century BC, the Greek mathematician Thales of Miletus used geometry to solve problems such as calculating the height of pyramids and the distance of ships from the shore. He is credited with the first use of deductive reasoning applied to geometry, by deriving four corollaries to Thales' Theorem.[10] Pythagoras established the Pythagorean School, which is credited with the first proof of the Pythagorean theorem,[11] though the statement of the theorem has a long history[12][13] Eudoxus (408–c. 355 BC) developed the method of exhaustion, which allowed the calculation of areas and volumes of curvilinear figures,[14] as well as a theory of ratios that avoided the problem of incommensurable magnitudes, which

A tiling of the hyperbolic plane

enabled subsequent geometers to make significant advances. Around 300 BC, geometry was revolutionized by Euclid, whose *Elements*, widely considered the most successful and influential textbook of all time,[15] introduced mathematical rigor through the axiomatic method and is the earliest example of the format still used in mathematics today, that of definition, axiom, theorem, and proof. Although most of the contents of the *Elements* were already known, Euclid arranged them into a single, coherent logical framework.[16] The *Elements* was known to all educated people in the West until the middle of the 20th century and its contents are still taught in geometry classes today.[17] Archimedes (c. 287–212 BC) of Syracuse used the method of exhaustion to calculate the area under the arc of a parabola with the summation of an infinite series, and gave remarkably accurate approximations of Pi.[18] He also studied the spiral bearing his name and obtained formulas for the volumes of surfaces of revolution.

Indian mathematicians also made many important contributions in geometry. The *Satapatha Brahmana* (ninth century BC) contains rules for ritual geometric constructions that are similar to the *Sulba Sutras*.[19] According to (Hayashi 2005, p. 363), the *Śulba Sūtras* contain "the earliest extant verbal expression of the Pythagorean Theorem in the world, although it had already been known to the Old Babylonians. They contain lists of Pythagorean triples,[20] which are particular cases

A European and an Arab practicing geometry in the 15th century.

of Diophantine equations.[21] In the Bakhshali manuscript, there is a handful of geometric problems (including problems about volumes of irregular solids). The Bakhshali manuscript also "employs a decimal place value system with a dot for zero."[22] Aryabhata's *Aryabhatiya* (499) includes the computation of areas and volumes. Brahmagupta wrote his astronomical work *Brāhma Sphuṭa Siddhānta* in 628. Chapter 12, containing 66 Sanskrit verses, was divided into two sections: "basic operations" (including cube roots, fractions, ratio and proportion, and barter) and "practical mathematics" (including mixture, mathematical series, plane figures, stacking bricks, sawing of timber, and piling of grain).[23] In the latter section, he stated his famous theorem on the diagonals of a cyclic quadrilateral. Chapter 12 also included a formula for the area of a cyclic quadrilateral (a generalization of Heron's formula), as well as a complete description of rational triangles (*i.e.* triangles with rational sides and rational areas).[23]

In the Middle Ages, mathematics in medieval Islam contributed to the development of geometry, especially algebraic geometry[24] and geometric algebra.[25] Al-Mahani (b. 853) conceived the idea of reducing geometrical problems such as duplicating the cube to problems in algebra.[26] Thābit ibn Qurra (known as Thebit in Latin) (836–901) dealt with arithmetic operations applied to ratios of geometrical quantities, and contributed to the development of analytic geome-

Woman teaching geometry. *Illustration at the beginning of a medieval translation of Euclid's Elements, (c. 1310)*

try.[27] Omar Khayyám (1048–1131) found geometric solutions to cubic equations.[28] The theorems of Ibn al-Haytham (Alhazen), Omar Khayyam and Nasir al-Din al-Tusi on quadrilaterals, including the Lambert quadrilateral and Saccheri quadrilateral, were early results in hyperbolic geometry, and along with their alternative postulates, such as Playfair's axiom, these works had a considerable influence on the development of non-Euclidean geometry among later European geometers, including Witelo (c. 1230–c. 1314), Gersonides (1288–1344), Alfonso, John Wallis, and Giovanni Girolamo Saccheri.[29]

In the early 17th century, there were two important developments in geometry. The first was the creation of analytic geometry, or geometry with coordinates and equations, by René Descartes (1596–1650) and Pierre de Fermat (1601–1665). This was a necessary precursor to the development of calculus and a precise quantitative science of physics.

The second geometric development of this period was the systematic study of projective geometry by Girard Desargues (1591–1661). Projective geometry is a geometry without measurement or parallel lines, just the study of how points are related to each other.

Two developments in geometry in the 19th century changed the way it had been studied previously. These were the discovery of non-Euclidean geometries by Nikolai Ivanovich Lobachevsky, János Bolyai and Carl Friedrich Gauss and of the formulation of symmetry as the central consideration in the Erlangen Programme of Felix Klein (which generalized the Euclidean and non-Euclidean geometries). Two of the master geometers of the time were Bernhard Riemann (1826–1866), working primarily with tools from mathematical analysis, and introducing the Riemann surface, and Henri Poincaré, the founder of algebraic topology and the geometric theory of dynamical systems. As a consequence of these major changes in the conception of geometry, the concept of "space" became something rich and varied, and the natural background for theories as different as complex analysis and classical mechanics.

23.3 Contemporary geometry

23.3.1 Euclidean geometry

Euclidean geometry has become closely connected with computational geometry, computer graphics, convex geometry, incidence geometry, finite geometry, discrete geometry, and some areas of combinatorics. Attention was given to further work on Euclidean geometry and the Euclidean groups by crystallography and the work of H. S. M. Coxeter, and can be seen in theories of Coxeter groups and polytopes. Geometric group theory is an expanding area of the theory of more general discrete groups, drawing on geometric models and algebraic techniques.

23.3.2 Differential geometry

Differential geometry has been of increasing importance to mathematical physics due to Einstein's general relativity postulation that the universe is curved. Contemporary differential geometry is *intrinsic*, meaning that the spaces it considers are smooth manifolds whose geometric structure is governed by a Riemannian metric, which determines how distances are measured near each point, and not *a priori* parts of some ambient flat Euclidean space.

23.3.3 Topology and geometry

The field of topology, which saw massive development in the 20th century, is in a technical sense a type of transformation geometry, in which transformations are homeomorphisms. This has often been expressed in the form of the dictum 'topology is rubber-sheet geometry'. Contemporary geometric topology and differential topology, and particular subfields such as Morse theory, would be counted by most mathematicians as part of geometry. Algebraic topology and general topology have gone their own ways.

23.3.4 Algebraic geometry

The field of algebraic geometry is the modern incarnation of the Cartesian geometry of co-ordinates. From late 1950s through mid-1970s it had undergone major foundational development, largely due to work of Jean-Pierre Serre and Alexander Grothendieck. This led to the introduction of schemes and greater emphasis on topological methods, including various cohomology theories. One of seven Millennium Prize problems, the Hodge conjecture, is a question in algebraic geometry.

The study of low-dimensional algebraic varieties, algebraic curves, algebraic surfaces and algebraic varieties of dimension 3 ("algebraic threefolds"), has been far advanced. Gröbner basis theory and real algebraic geometry are among more applied subfields of modern algebraic geometry. Arithmetic geometry is an active field combining algebraic geometry and number theory. Other directions of research involve moduli spaces and complex geometry. Algebro-geometric methods are commonly applied in string and brane theory.

The 4₂₁polytope, orthogonally projected into the Eₙ Lie group Coxeter plane

23.4   See also

23.4.1   Lists

- List of geometers
    - Category:Algebraic geometers
    - Category:Differential geometers
    - Category:Geometers
    - Category:Topologists
- List of formulas in elementary geometry

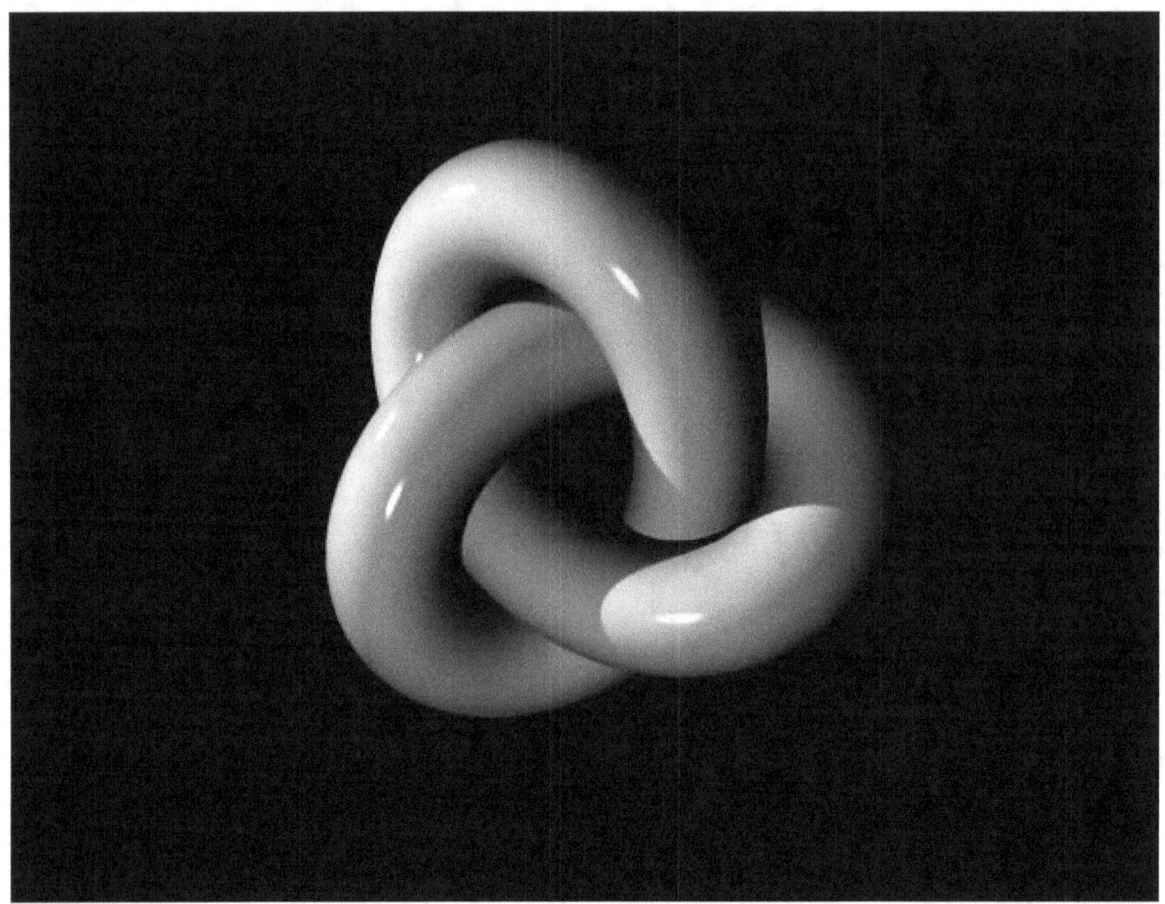

A thickening of the trefoil knot

- List of geometry topics

- List of important publications in geometry

- List of mathematics articles

23.4.2   Related topics

- Descriptive geometry

- *Flatland*, a book written by Edwin Abbott Abbott about two- and three-dimensional space, to understand the concept of four dimensions

- Interactive geometry software

23.4.3   Other fields

- Molecular geometry

Quintic Calabi–Yau threefold

23.5   Notes

[1] Martin J. Turner,Jonathan M. Blackledge,Patrick R. Andrews (1998). *Fractal geometry in digital imaging*. Academic Press. p. 1. ISBN 0-12-703970-8

[2] It is quite common in algebraic geometry to speak about *geometry of algebraic varieties over finite fields*, possibly singular. From a naïve perspective, these objects are just finite sets of points, but by invoking powerful geometric imagery and using well developed geometric techniques, it is possible to find structure and establish properties that make them somewhat analogous to the ordinary spheres or cones.

[3] Kline (1972) "Mathematical thought from ancient to modern times", Oxford University Press, p. 1032. Kant did not reject the logical (analytic a priori) *possibility* of non-Euclidean geometry, see Jeremy Gray, "Ideas of Space Euclidean, Non-Euclidean, and Relativistic", Oxford, 1989; p. 85. Some have implied that, in light of this, Kant had in fact *predicted* the development of

non-Euclidean geometry, cf. Leonard Nelson, "Philosophy and Axiomatics," Socratic Method and Critical Philosophy, Dover, 1965, p. 164.

[4] "Ueber die Hypothesen, welche der Geometrie zu Grunde liegen.".

[5] J. Friberg, "Methods and traditions of Babylonian mathematics. Plimpton 322, Pythagorean triples, and the Babylonian triangle parameter equations", Historia Mathematica, 8, 1981, pp. 277—318.

[6] Neugebauer, Otto (1969) [1957]. *The Exact Sciences in Antiquity* (2 ed.). Dover Publications. ISBN 978-0-486-22332-2. Chap. IV "Egyptian Mathematics and Astronomy", pp. 71–96.

[7] (Boyer 1991, "Egypt" p. 19)

[8] The Journal of Egyptian Archaeology. Vol. 84, 1998 Gnomons at Meroë and Early Trigonometry. pg. 171

[9] Slayman, Andrew (May 27, 1998). "Neolithic Skywatchers". *Archaeology Magazine Archive*.

[10] (Boyer 1991, "Ionia and the Pythagoreans" p. 43)

[11] Eves, Howard, An Introduction to the History of Mathematics, Saunders, 1990, ISBN 0-03-029558-0.

[12] Kurt Von Fritz (1945). "The Discovery of Incommensurability by Hippasus of Metapontum". *The Annals of Mathematics*.

[13] James R. Choike (1980). "The Pentagram and the Discovery of an Irrational Number". *The Two-Year College Mathematics Journal*.

[14] (Boyer 1991, "The Age of Plato and Aristotle" p. 92)

[15] (Boyer 1991, "Euclid of Alexandria" p. 119)

[16] (Boyer 1991, "Euclid of Alexandria" p. 104)

[17] Howard Eves, *An Introduction to the History of Mathematics*, Saunders, 1990, ISBN 0-03-029558-0 p. 141: "No work, except The Bible, has been more widely used...."

[18] O'Connor, J.J. and Robertson, E.F. (February 1996). "A history of calculus". University of St Andrews. Retrieved 2007-08-07.

[19] (Staal 1999)

[20] Pythagorean triples are triples of integers (a, b, c) with the property: $a^2 + b^2 = c^2$. Thus, $3^2 + 4^2 = 5^2$, $8^2 + 15^2 = 17^2$, $12^2 + 35^2 = 37^2$ etc.

[21] (Cooke 2005, p. 198): "The arithmetic content of the *Śulva Sūtras* consists of rules for finding Pythagorean triples such as as (3, 4, 5), (5, 12, 13), (8, 15, 17), and (12, 35, 37). It is not certain what practical use these arithmetic rules had. The best conjecture is that they were part of religious ritual. A Hindu home was required to have three fires burning at three different altars. The three altars were to be of different shapes, but all three were to have the same area. These conditions led to certain "Diophantine" problems, a particular case of which is the generation of Pythagorean triples, so as to make one square integer equal to the sum of two others."

[22] (Hayashi 2005, p. 371)

[23] (Hayashi 2003, pp. 121–122)

[24] R. Rashed (1994), *The development of Arabic mathematics: between arithmetic and algebra*, p. 35 London

[25] Boyer (1991). "The Arabic Hegemony". *A History of Mathematics*. pp. 241–242. Omar Khayyam (ca. 1050–1123), the "tent-maker," wrote an *Algebra* that went beyond that of al-Khwarizmi to include equations of third degree. Like his Arab predecessors, Omar Khayyam provided for quadratic equations both arithmetic and geometric solutions; for general cubic equations, he believed (mistakenly, as the 16th century later showed), arithmetic solutions were impossible; hence he gave only geometric solutions. The scheme of using intersecting conics to solve cubics had been used earlier by Menaechmus, Archimedes, and Alhazan, but Omar Khayyam took the praiseworthy step of generalizing the method to cover all third-degree equations (having positive roots). .. For equations of higher degree than three, Omar Khayyam evidently did not envision similar geometric methods, for space does not contain more than three dimensions, ... One of the most fruitful contributions of Arabic eclecticism was the tendency to close the gap between numerical and geometric algebra. The decisive step in this direction came much later with Descartes, but Omar Khayyam was moving in this direction when he wrote, "Whoever thinks algebra is a trick in obtaining unknowns has thought it in vain. No attention should be paid to the fact that algebra and geometry are different in appearance. Algebras are geometric facts which are proved."

[26] O'Connor, John J.; Robertson, Edmund F., "Al-Mahani", *MacTutor History of Mathematics archive*, University of St Andrews.

[27] O'Connor, John J.; Robertson, Edmund F., "Al-Sabi Thabit ibn Qurra al-Harrani", *MacTutor History of Mathematics archive*, University of St Andrews.

[28] O'Connor, John J.; Robertson, Edmund F., "Omar Khayyam", *MacTutor History of Mathematics archive*, University of St Andrews.

[29] Boris A. Rosenfeld and Adolf P. Youschkevitch (1996), "Geometry", in Roshdi Rashed, ed., *Encyclopedia of the History of Arabic Science*, Vol. 2, p. 447–494 [470], Routledge, London and New York:

> "Three scientists, Ibn al-Haytham, Khayyam, and al-Tusi, had made the most considerable contribution to this branch of geometry whose importance came to be completely recognized only in the 19th century. In essence, their propositions concerning the properties of quadrangles which they considered, assuming that some of the angles of these figures were acute of obtuse, embodied the first few theorems of the hyperbolic and the elliptic geometries. Their other proposals showed that various geometric statements were equivalent to the Euclidean postulate V. It is extremely important that these scholars established the mutual connection between this postulate and the sum of the angles of a triangle and a quadrangle. By their works on the theory of parallel lines Arab mathematicians directly influenced the relevant investigations of their European counterparts. The first European attempt to prove the postulate on parallel lines – made by Witelo, the Polish scientists of the 13th century, while revising Ibn al-Haytham's *Book of Optics* (*Kitab al-Manazir*) – was undoubtedly prompted by Arabic sources. The proofs put forward in the 14th century by the Jewish scholar Levi ben Gerson, who lived in southern France, and by the above-mentioned Alfonso from Spain directly border on Ibn al-Haytham's demonstration. Above, we have demonstrated that *Pseudo-Tusi's Exposition of Euclid* had stimulated both J. Wallis's and G. Saccheri's studies of the theory of parallel lines."

23.6   Sources

- Boyer, C. B. (1991) [1989]. *A History of Mathematics* (Second edition, revised by Uta C. Merzbach ed.). New York: Wiley. ISBN 0-471-54397-7.

- Nikolai I. Lobachevsky, *Pangeometry*, translator and editor: A. Papadopoulos, Heritage of European Mathematics Series, Vol. 4, European Mathematical Society, 2010.

23.7   Further reading

- Jay Kappraff, *A Participatory Approach to Modern Geometry*, 2014, World Scientific Publishing, ISBN 978-981-4556-70-5.

- Leonard Mlodinow, *Euclid's Window – The Story of Geometry from Parallel Lines to Hyperspace*, UK edn. Allen Lane, 1992.

23.8   External links

- A geometry course from Wikiversity

- *Unusual Geometry Problems*

- *The Math Forum* — Geometry

  - *The Math Forum* — K–12 Geometry

  - *The Math Forum* — College Geometry

  - *The Math Forum* — Advanced Geometry

- Nature Precedings — *Pegs and Ropes Geometry at Stonehenge*

- *The Mathematical Atlas* — Geometric Areas of Mathematics

- "4000 Years of Geometry", lecture by Robin Wilson given at Gresham College, 3 October 2007 (available for MP3 and MP4 download as well as a text file)

    - Finitism in Geometry at the Stanford Encyclopedia of Philosophy

- The Geometry Junkyard

- Interactive Geometry Applications (Java and Cabri 3D)

- Interactive geometry reference with hundreds of applets

- Dynamic Geometry Sketches (with some Student Explorations)

- Geometry classes at Khan Academy

Chapter 24

Algebraic geometry

Not to be confused with Geometric algebra, an application of Clifford algebra to geometry.
For the book by Robin Hartshorne, see Algebraic Geometry (book).

Algebraic geometry is a branch of mathematics, classically studying zeros of multivariate polynomials. Modern algebraic geometry is based on the use of abstract algebraic techniques, mainly from commutative algebra, for solving geometrical problems about these sets of zeros.

The fundamental objects of study in algebraic geometry are algebraic varieties, which are geometric manifestations of solutions of systems of polynomial equations. Examples of the most studied classes of algebraic varieties are: plane algebraic curves, which include lines, circles, parabolas, ellipses, hyperbolas, cubic curves like elliptic curves and quartic curves like lemniscates, and Cassini ovals. A point of the plane belongs to an algebraic curve if its coordinates satisfy a given polynomial equation. Basic questions involve the study of the points of special interest like the singular points, the inflection points and the points at infinity. More advanced questions involve the topology of the curve and relations between the curves given by different equations.

Algebraic geometry occupies a central place in modern mathematics and has multiple conceptual connections with such diverse fields as complex analysis, topology and number theory. Initially a study of systems of polynomial equations in several variables, the subject of algebraic geometry starts where equation solving leaves off, and it becomes even more important to understand the intrinsic properties of the totality of solutions of a system of equations, than to find a specific solution; this leads into some of the deepest areas in all of mathematics, both conceptually and in terms of technique.

In the 20th century, algebraic geometry has split into several subareas.

- The main stream of algebraic geometry is devoted to the study of the complex points of the algebraic varieties and more generally to the points with coordinates in an algebraically closed field.

- The study of the points of an algebraic variety with coordinates in the field of the rational numbers or in a number field became arithmetic geometry (or more classically Diophantine geometry), a subfield of algebraic number theory.

- The study of the real points of an algebraic variety is the subject of real algebraic geometry.

- A large part of singularity theory is devoted to the singularities of algebraic varieties.

- With the rise of the computers, a computational algebraic geometry area has emerged, which lies at the intersection of algebraic geometry and computer algebra. It consists essentially in developing algorithms and software for studying and finding the properties of explicitly given algebraic varieties.

Much of the development of the main stream of algebraic geometry in the 20th century occurred within an abstract algebraic framework, with increasing emphasis being placed on "intrinsic" properties of algebraic varieties not dependent on any particular way of embedding the variety in an ambient coordinate space; this parallels developments in topology, differential and complex geometry. One key achievement of this abstract algebraic geometry is Grothendieck's scheme

This Togliatti surface is an algebraic surface of degree five. The picture represents a portion of its real locus.

theory which allows one to use sheaf theory to study algebraic varieties in a way which is very similar to its use in the study of differential and analytic manifolds. This is obtained by extending the notion of point: In classical algebraic geometry, a point of an affine variety may be identified, through Hilbert's Nullstellensatz, with a maximal ideal of the coordinate ring, while the points of the corresponding affine scheme are all prime ideals of this ring. This means that a point of such a scheme may be either a usual point or a subvariety. This approach also enables a unification of the language and the tools of classical algebraic geometry, mainly concerned with complex points, and of algebraic number theory. Wiles's proof of the longstanding conjecture called Fermat's last theorem is an example of the power of this approach.

24.1 Basic notions

Further information: Algebraic variety

24.1.1   Zeros of simultaneous polynomials

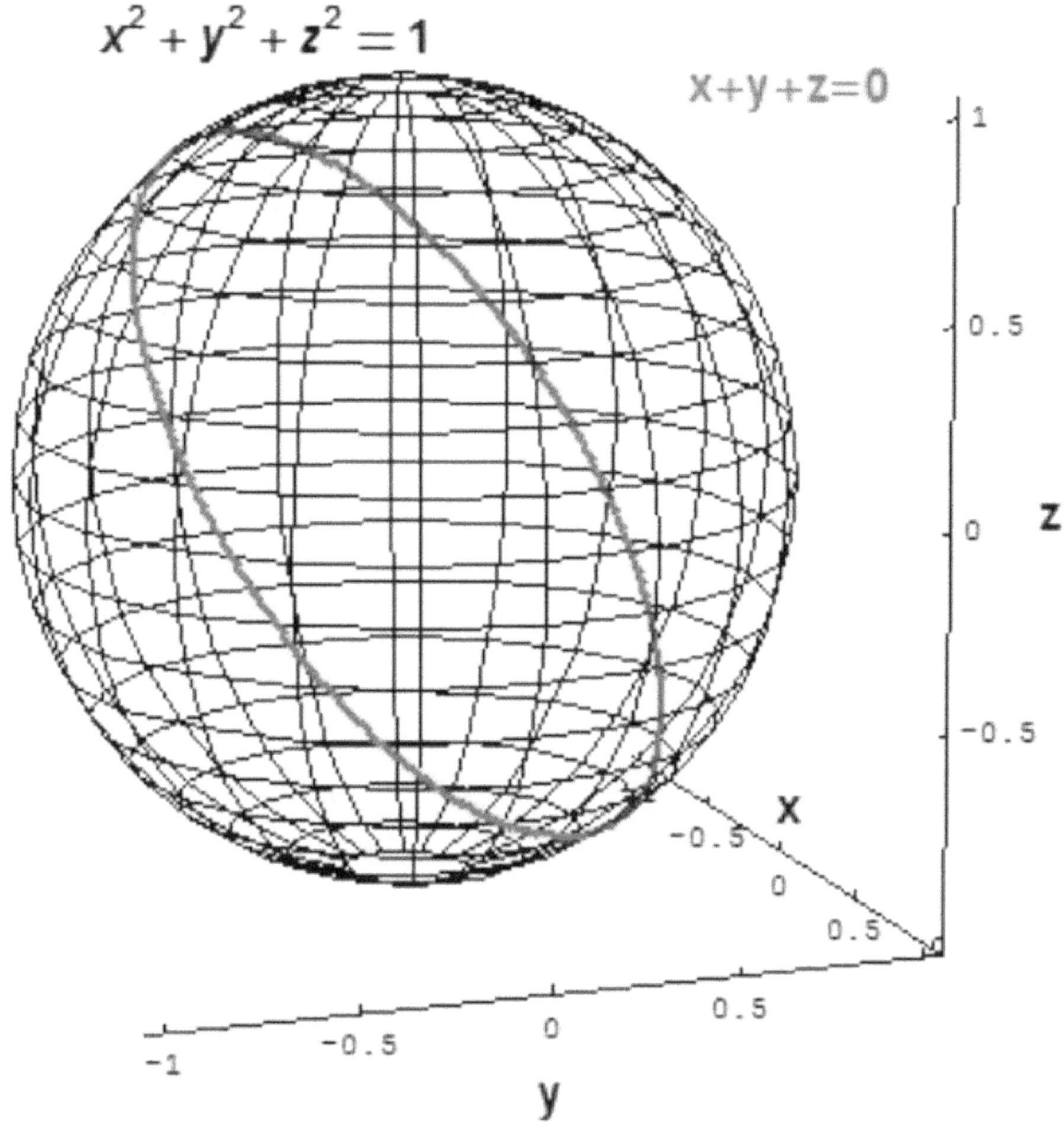

Sphere and slanted circle

In classical algebraic geometry, the main objects of interest are the vanishing sets of collections of polynomials, meaning the set of all points that simultaneously satisfy one or more polynomial equations. For instance, the two-dimensional sphere in three-dimensional Euclidean space $\mathbf{R}^3$ could be defined as the set of all points (x,y,z) with

$$x^2 + y^2 + z^2 - 1 = 0.$$

A "slanted" circle in $\mathbf{R}^3$ can be defined as the set of all points (x,y,z) which satisfy the two polynomial equations

$$x^2 + y^2 + z^2 - 1 = 0.$$

$x + y + z = 0.$

24.1.2   Affine varieties

Main article: Affine variety

First we start with a field k. In classical algebraic geometry, this field was always the complex numbers $\mathbf{C}$, but many of the same results are true if we assume only that k is algebraically closed. We consider the affine space of dimension n over k, denoted $\mathbf{A}^n(k)$ (or more simply $\mathbf{A}^n$, when k is clear from the context). When one fixes a coordinates system, one may identify $\mathbf{A}^n(k)$ with k^n. The purpose of not working with k^n is to emphasize that one "forgets" the vector space structure that k^n carries.

A function $f : \mathbf{A}^n \rightarrow \mathbf{A}^1$ is said to be *polynomial* (or *regular*) if it can be written as a polynomial, that is, if there is a polynomial p in $k[x_1,...,xn]$ such that $f(M) = p(t_1,...,tn)$ for every point M with coordinates $(t_1,...,tn)$ in $\mathbf{A}^n$. The property of a function to be polynomial (or regular) does not depend on the choice of a coordinate system in $\mathbf{A}^n$.

When a coordinate system is chosen, the regular functions on the affine n-space may be identified with the ring of polynomial functions in n variables over k. Therefore, the set of the regular functions on $\mathbf{A}^n$ is a ring, which is denoted $k[\mathbf{A}^n]$.

We say that a polynomial *vanishes* at a point if evaluating it at that point gives zero. Let S be a set of polynomials in $k[\mathbf{A}^n]$. The *vanishing set of S* (or *vanishing locus* or *zero set*) is the set $V(S)$ of all points in $\mathbf{A}^n$ where every polynomial in S vanishes. In other words,

$$V(S) = \{(t_1,\ldots,t_n)|\forall p \in S, p(t_1,\ldots,t_n) = 0\}.$$

A subset of $\mathbf{A}^n$ which is $V(S)$, for some S, is called an *algebraic set*. The V stands for *variety* (a specific type of algebraic set to be defined below).

Given a subset U of $\mathbf{A}^n$, can one recover the set of polynomials which generate it? If U is *any* subset of $\mathbf{A}^n$, define $I(U)$ to be the set of all polynomials whose vanishing set contains U. The I stands for ideal: if two polynomials f and g both vanish on U, then $f+g$ vanishes on U, and if h is any polynomial, then hf vanishes on U, so $I(U)$ is always an ideal of the polynomial ring $k[\mathbf{A}^n]$.

Two natural questions to ask are:

- Given a subset U of $\mathbf{A}^n$, when is $U = V(I(U))$?

- Given a set S of polynomials, when is $S = I(V(S))$?

The answer to the first question is provided by introducing the Zariski topology, a topology on $\mathbf{A}^n$ whose closed sets are the algebraic sets, and which directly reflects the algebraic structure of $k[\mathbf{A}^n]$. Then $U = V(I(U))$ if and only if U is an algebraic set or equivalently a Zariski-closed set. The answer to the second question is given by Hilbert's Nullstellensatz. In one of its forms, it says that $I(V(S))$ is the radical of the ideal generated by S. In more abstract language, there is a Galois connection, giving rise to two closure operators; they can be identified, and naturally play a basic role in the theory; the example is elaborated at Galois connection.

For various reasons we may not always want to work with the entire ideal corresponding to an algebraic set U. Hilbert's basis theorem implies that ideals in $k[\mathbf{A}^n]$ are always finitely generated.

An algebraic set is called *irreducible* if it cannot be written as the union of two smaller algebraic sets. Any algebraic set is a finite union of irreducible algebraic sets and this decomposition is unique. Thus its elements are called the *irreducible components* of the algebraic set. An irreducible algebraic set is also called a *variety*. It turns out that an algebraic set is a variety if and only if it may be defined as the vanishing set of a prime ideal of the polynomial ring.

Some authors do not make a clear distinction between algebraic sets and varieties and use *irreducible variety* to make the distinction when needed.

24.1.3   Regular functions

Main article: Regular function

Just as continuous functions are the natural maps on topological spaces and smooth functions are the natural maps on differentiable manifolds, there is a natural class of functions on an algebraic set, called *regular functions* or *polynomial functions*. A regular function on an algebraic set V contained in $\mathbf{A}^n$ is the restriction to V of a regular function on $\mathbf{A}^n$. For an algebraic set defined on the field of the complex numbers, the regular functions are smooth and even analytic.

It may seem unnaturally restrictive to require that a regular function always extend to the ambient space, but it is very similar to the situation in a normal topological space, where the Tietze extension theorem guarantees that a continuous function on a closed subset always extends to the ambient topological space.

Just as with the regular functions on affine space, the regular functions on V form a ring, which we denote by $k[V]$. This ring is called the *coordinate ring of V*.

Since regular functions on V come from regular functions on $\mathbf{A}^n$, there is a relationship between the coordinate rings. Specifically, if a regular function on V is the restriction of two functions f and g in $k[\mathbf{A}^n]$, then $f - g$ is a polynomial function which is null on V and thus belongs to $I(V)$. Thus $k[V]$ may be identified with $k[\mathbf{A}^n]/I(V)$.

24.1.4   Morphism of affine varieties

Using regular functions from an affine variety to $\mathbf{A}^1$, we can define regular maps from one affine variety to another. First we will define a regular map from a variety into affine space: Let V be a variety contained in $\mathbf{A}^n$. Choose m regular functions on V, and call them $f_1, ..., fm$. We define a *regular map* f from V to $\mathbf{A}^m$ by letting $f = (f_1, ..., fm)$. In other words, each fi determines one coordinate of the range of f.

If V' is a variety contained in $\mathbf{A}^m$, we say that f is a *regular map* from V to V' if the range of f is contained in V'.

The definition of the regular maps apply also to algebraic sets. The regular maps are also called *morphisms*, as they make the collection of all affine algebraic sets into a category, where the objects are the affine algebraic sets and the morphisms are the regular maps. The affine varieties is a subcategory of the category of the algebraic sets.

Given a regular map g from V to V' and a regular function f of $k[V']$, then $f \circ g \in k[V]$. The map $f \rightarrow f \circ g$ is a ring homomorphism from $k[V']$ to $k[V]$. Conversely, every ring homomorphism from $k[V']$ to $k[V]$ defines a regular map from V to V'. This defines an equivalence of categories between the category of algebraic sets and the opposite category of the finitely generated reduced k-algebras. This equivalence is one of the starting points of scheme theory.

24.1.5   Rational function and birational equivalence

Main article: Rational mapping

Contrarily to the preceding ones, this section concerns only varieties and not algebraic sets. On the other hand, the definitions extend naturally to projective varieties (next section), as an affine variety and its projective completion have the same field of functions.

If V is an affine variety, its coordinate ring is an integral domain and has thus a field of fractions which is denoted $k(V)$ and called the *field of the rational functions* on V or, shortly, the *function field* of V. Its elements are the restrictions to V of the rational functions over the affine space containing V. The domain of a rational function f is not V but the complement of the subvariety (a hypersurface) where the denominator of f vanishes.

Like for regular maps, one may define a *rational map* from a variety V to a variety V'. Like for the regular maps, the rational maps from V to V' may be identified to the field homomorphisms from $k(V')$ to $k(V)$.

Two affine varieties are *birationally equivalent* if there are two rational functions between them which are inverse one to the other in the regions where both are defined. Equivalently, they are birationally equivalent if their function fields are isomorphic.

An affine variety is a *rational variety* if it is birationally equivalent to an affine space. This means that the variety admits a rational parameterization. For example, the circle of equation $x^2 + y^2 - 1 = 0$ is a rational curve, as it has the parameterization

$$x = \frac{2t}{1 + t^2}$$

$$y = \frac{1 - t^2}{1 + t^2},$$

which may also be viewed as a rational map from the line to the circle.

The problem of resolution of singularities is to know if every algebraic variety is birationally equivalent to a variety whose projective completion is nonsingular (see also smooth completion). It has been positively solved in characteristic 0 by Heisuke Hironaka in 1964 and is yet unsolved in finite characteristic.

24.1.6 Projective variety

Main article: Algebraic geometry of projective spaces
 Just as the formulas for the roots of 2nd, 3rd and 4th degree polynomials suggest extending real numbers to the more

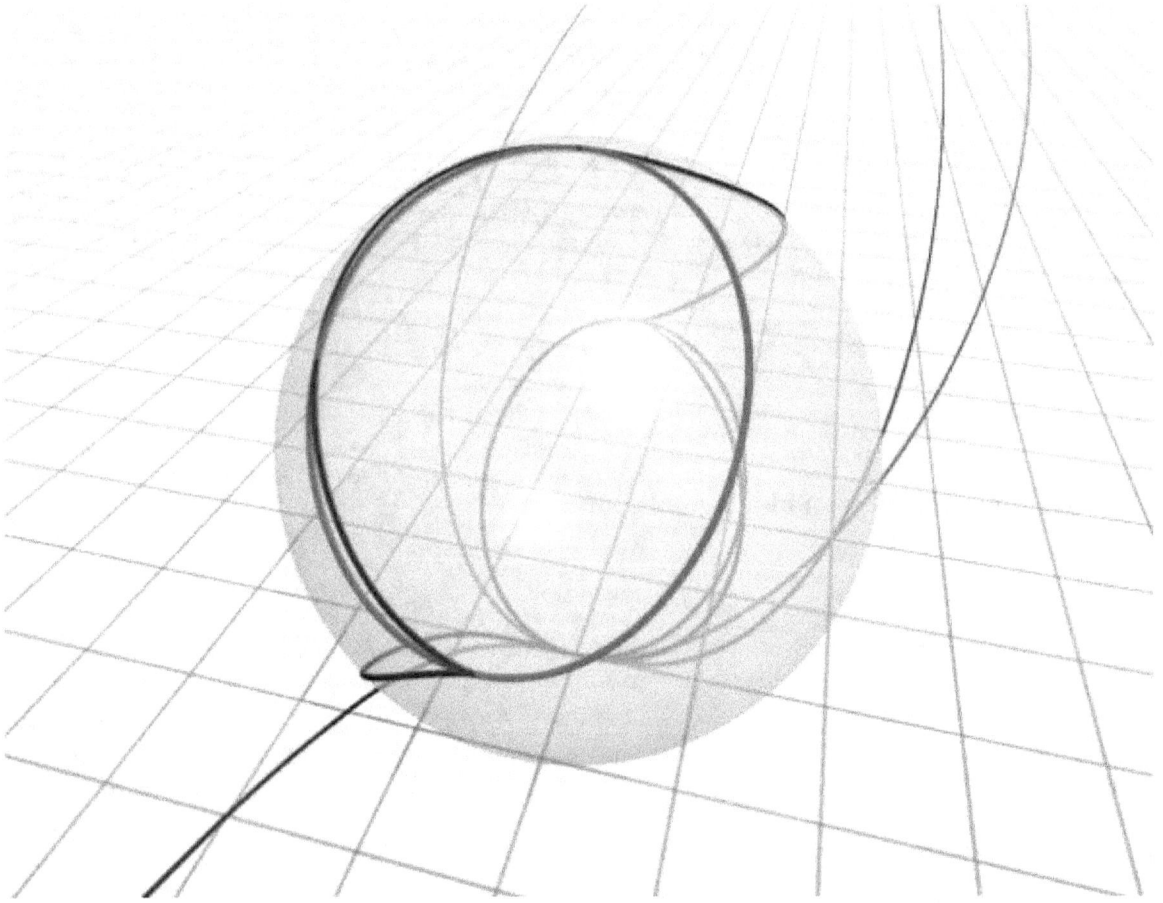

parabola ($y = x^2$, red) and cubic ($y = x^3$, blue) in projective space

algebraically complete setting of the complex numbers, many properties of algebraic varieties suggest extending affine space to a more geometrically complete projective space. Whereas the complex numbers are obtained by adding the number i, a root of the polynomial x^2 + 1, projective space is obtained by adding in appropriate points "at infinity", points where parallel lines may meet.

To see how this might come about, consider the variety $V(y - x^2)$. If we draw it, we get a parabola. As x goes to positive infinity, the slope of the line from the origin to the point (x, x^2) also goes to positive infinity. As x goes to negative infinity, the slope of the same line goes to negative infinity.

Compare this to the variety $V(y - x^3)$. This is a cubic curve. As x goes to positive infinity, the slope of the line from the origin to the point (x, x^3) goes to positive infinity just as before. But unlike before, as x goes to negative infinity, the slope of the same line goes to positive infinity as well; the exact opposite of the parabola. So the behavior "at infinity" of $V(y - x^3)$ is different from the behavior "at infinity" of $V(y - x^2)$.

The consideration of the *projective completion* of the two curves, which is their prolongation "at infinity" in the projective plane, allows to quantify this difference: the point at infinity of the parabola is a regular point, whose tangent is the line at infinity, while the point at infinity of the cubic curve is a cusp. Also, both curves are rational, as they are parameterized by x, and Riemann-Roch theorem implies that the cubic curve must have a singularity, which must be at infinity, as all its points in the affine space are regular.

Thus many of the properties of algebraic varieties, including birational equivalence and all the topological properties, depend on the behavior "at infinity" and so it is natural to study the varieties in projective space. Furthermore, the introduction of projective techniques made many theorems in algebraic geometry simpler and sharper: For example, Bézout's theorem on the number of intersection points between two varieties can be stated in its sharpest form only in projective space. For these reasons, projective space plays a fundamental role in algebraic geometry.

Nowadays, the *projective space* $\mathbf{P}^n$ of dimension n is usually defined as the set of the lines passing through a point, considered as the origin, in the affine space of dimension $n+1$, or equivalently to the set of the vector lines in a vector space of dimension $n+1$. When a coordinate system has been chosen in the space of dimension $n+1$, all the points of a line have the same set of coordinates, up to the multiplication by an element of k. This defines the homogeneous coordinates of a point of $\mathbf{P}^n$ as a sequence of $n+1$ elements of the base field k, defined up to the multiplication by a nonzero element of k (the same for the whole sequence).

Given a polynomial in $n+1$ variables, it vanishes at all the point of a line passing through the origin if and only if it is homogeneous. In this case, one says that the polynomial *vanishes* at the corresponding point of $\mathbf{P}^n$. This allows to define a *projective algebraic set* in $\mathbf{P}^n$ as the set $V(f_1, ..., fk)$ where a finite set of homogeneous polynomials $\{f_1, ..., fk\}$ vanishes. Like for affine algebraic sets, there is a bijection between the projective algebraic sets and the reduced homogeneous ideals which define them. The *projective varieties* are the projective algebraic sets whose defining ideal is prime. In other words, a projective variety is a projective algebraic set, whose homogeneous coordinate ring is an integral domain, the *projective coordinates ring* being defined as the quotient of the graded ring or the polynomials in $n+1$ variables by the homogeneous (reduced) ideal defining the variety. Every projective algebraic set may be uniquely decomposed into a finite union of projective varieties.

The only regular functions which may be defined properly on a projective variety are the constant functions. Thus this notion is not used in projective situations. On the other hand, the *field of the rational functions* or *function field* is a useful notion, which, similarly as in the affine case, is defined as the set of the quotients of two homogeneous elements of the same degree in the homogeneous coordinate ring.

24.2   Real algebraic geometry

Main article: Real algebraic geometry

The real algebraic geometry is the study of the real points of the algebraic geometry.

The fact that the field of the reals number is an ordered field may not be occulted in such a study. For example, the curve of equation $x^2 + y^2 - a = 0$ is a circle if $a > 0$, but does not have any real point if $a < 0$. It follows that real algebraic geometry is not only the study of the real algebraic varieties, but has been generalized to the study of the *semi-algebraic sets*, which are the solutions of systems of polynomial equations and polynomial inequalities. For example, a branch of the hyperbola of equation $xy - 1 = 0$ is not an algebraic variety, but is a semi-algebraic set defined by $xy - 1 = 0$ and

$x > 0$ or by $xy - 1 = 0$ and $x + y > 0$.

One of the challenging problems of real algebraic geometry is the unsolved Hilbert's sixteenth problem: Decide which respective positions are possible for the ovals of a nonsingular plane curve of degree 8.

24.3 Computational algebraic geometry

One may date the origin of computational algebraic geometry to meeting EUROSAM'79 (International Symposium on Symbolic and Algebraic Manipulation) held at Marseille, France in June 1979. At this meeting,

- Dennis S. Arnon showed that George E. Collins's Cylindrical algebraic decomposition (CAD) allows the computation of the topology of semi-algebraic sets.

- Bruno Buchberger presented the Gröbner bases and his algorithm to compute them.

- Daniel Lazard presented a new algorithm for solving systems of homogeneous polynomial equations with a computational complexity which is essentially polynomial in the expected number of solutions and thus simply exponential in the number of the unknowns. This algorithm is strongly related with Macaulay's multivariate resultant.

Since then, most results in this area are related to one or several of these items either by using or improving one of these algorithms, or by finding algorithms whose complexity is simply exponential in the number of the variables.

24.3.1 Gröbner basis

Main article: Gröbner basis

A Gröbner basis is a system of generators of a polynomial ideal whose computation allows the deduction of many properties of the affine algebraic variety defined by the ideal.

Given an ideal I defining an algebraic set V:

- V is empty (over an algebraically closed extension of the basis field), if and only if the Gröbner basis for any monomial ordering is reduced to $\{1\}$.

- By mean of the Hilbert series one may compute the dimension and the degree of V from any Gröbner basis of I for a monomial ordering refining the total degree.

- If the dimension of V is 0, one may compute the points (finite in number) of V from any Gröbner basis of I (see systems of polynomial equations).

- A Gröbner basis computation allows to remove from V all irreducible components which are contained in a given hyper surface.

- A Gröbner basis computation allows to compute the Zariski closure of the image of V by the projection on the k first coordinates, and the subset of the image where the projection is not proper.

- More generally Gröbner basis computations allows to compute the Zariski closure of the image and the critical points of a rational function of V into another affine variety.

Gröbner basis computations do not allow to compute directly the primary decomposition of I nor the prime ideals defining the irreducible components of V, but most algorithms for this involve Gröbner basis computation. The algorithms which are not based on Gröbner bases use regular chains but may need Gröbner bases in some exceptional situations.

Gröbner base are deemed to be difficult to compute. In fact they may contain, in the worst case, polynomials whose degree is doubly exponential in the number of variables and a number of polynomials which is also doubly exponential.

However, this is only a worst case complexity, and the complexity bound of Lazard's algorithm of 1979 may frequently apply. Faugère's F4 and F5 algorithms realize this complexity, as F5 algorithm may be viewed as an improvement of Lazard's 1979 algorithm. It follows that the best implementations allow to compute almost routinely with algebraic sets of degree more than 100. This means that, presently, the difficulty of computing a Gröbner basis is strongly related to the intrinsic difficulty of the problem.

24.3.2   Cylindrical Algebraic Decomposition (CAD)

CAD is an algorithm which had been introduced in 1973 by G. Collins to implement with an acceptable complexity the Tarski–Seidenberg theorem on quantifier elimination over the real numbers.

This theorem concerns the formulas of the first-order logic whose atomic formulas are polynomial equalities or inequalities between polynomials with real coefficients. These formulas are thus the formulas which may be constructed from the atomic formulas by the logical operators *and* ($\wedge$), *or* ($\vee$), *not* ($\neg$), *for all* ($\forall$) and *exists* ($\exists$). Tarski's theorem asserts that, from such a formula, one may compute an equivalent formula without quantifier ($\forall$, $\exists$).

The complexity of CAD is doubly exponential in the number of variables. This means that CAD allow, in theory, to solve every problem of real algebraic geometry which may be expressed by such a formula, that is almost every problem concerning explicitly given varieties and semi-algebraic sets.

While Gröbner basis computation has doubly exponential complexity only in rare cases, CAD has almost always this high complexity. This implies that, unless if most polynomials appearing in the input are linear, it may not solve problems with more than four variables.

Since 1973, most of the research on this subject is devoted either to improve CAD or to find alternate algorithms in special cases of general interest.

As an example of the state of art, there are efficient algorithms to find at least a point in every connected component of a semi-algebraic set, and thus to test if a semi-algebraic set is empty. On the other hand, CAD is yet, in practice, the best algorithm to count the number of connected components.

24.3.3   Asymptotic complexity vs. practical efficiency

The basic general algorithms of computational geometry have a double exponential worst case complexity. More precisely, if d is the maximal degree of the input polynomials and n the number of variables, their complexity is at most $d^{2^{cn}}$ for some constant c, and, for some inputs, the complexity is at least $d^{2^{c'n}}$ for another constant c'.

During the last 20 years of 20th century, various algorithms have been introduced to solve specific subproblems with a better complexity. Most of these algorithms have a complexity $d^{O(n^2)}$.

Among these algorithms which solve a sub problem of the problems solved by Gröbner bases, one may cite *testing if an affine variety is empty* and *solving nonhomogeneous polynomial systems which have a finite number of solutions*. Such algorithms are rarely implemented because, on most entries Faugère's F4 and F5 algorithms have a better practical efficiency and probably a similar or better complexity (*probably* because the evaluation of the complexity of Gröbner basis algorithms on a particular class of entries is a difficult task which has be done only in few special cases).

The main algorithms of real algebraic geometry which solve a problem solved by CAD are related to the topology of semi-algebraic sets. One may cite *counting the number of connected components, testing if two points are in the same components* or *computing a Whitney stratification of a real algebraic set*. They have a complexity of $d^{O(n^2)}$, but the constant involved by O notation is so high that using them to solve any nontrivial problem effectively solved by CAD, is impossible even if one could use all the existing computing power in the world. Therefore, these algorithms have never been implemented and this is an active research area to search for algorithms with have together a good asymptotic complexity and a good practical efficiency.

24.4 Abstract modern viewpoint

The modern approaches to algebraic geometry redefine and effectively extend the range of basic objects in various levels of generality to schemes, formal schemes, ind-schemes, algebraic spaces, algebraic stacks and so on. The need for this arises already from the useful ideas within theory of varieties, e.g. the formal functions of Zariski can be accommodated by introducing nilpotent elements in structure rings; considering spaces of loops and arcs, constructing quotients by group actions and developing formal grounds for natural intersection theory and deformation theory lead to some of the further extensions.

Most remarkably, in late 1950s, algebraic varieties were subsumed into Alexander Grothendieck's concept of a scheme. Their local objects are affine schemes or prime spectra which are locally ringed spaces which form a category which is antiequivalent to the category of commutative unital rings, extending the duality between the category of affine algebraic varieties over a field k, and the category of finitely generated reduced k-algebras. The gluing is along Zariski topology; one can glue within the category of locally ringed spaces, but also, using the Yoneda embedding, within the more abstract category of presheaves of sets over the category of affine schemes. The Zariski topology in the set theoretic sense is then replaced by a Grothendieck topology. Grothendieck introduced Grothendieck topologies having in mind more exotic but geometrically finer and more sensitive examples than the crude Zariski topology, namely the étale topology, and the two flat Grothendieck topologies: fppf and fpqc; nowadays some other examples became prominent including Nisnevich topology. Sheaves can be furthermore generalized to stacks in the sense of Grothendieck, usually with some additional representability conditions leading to Artin stacks and, even finer, Deligne-Mumford stacks, both often called algebraic stacks.

Sometimes other algebraic sites replace the category of affine schemes. For example, Nikolai Durov has introduced commutative algebraic monads as a generalization of local objects in a generalized algebraic geometry. Versions of a tropical geometry, of an absolute geometry over a field of one element and an algebraic analogue of Arakelov's geometry were realized in this setup.

Another formal generalization is possible to Universal algebraic geometry in which every variety of algebras has its own algebraic geometry. The term *variety of algebras* should not be confused with *algebraic variety*.

The language of schemes, stacks and generalizations has proved to be a valuable way of dealing with geometric concepts and became cornerstones of modern algebraic geometry.

Algebraic stacks can be further generalized and for many practical questions like deformation theory and intersection theory, this is often the most natural approach. One can extend the Grothendieck site of affine schemes to a higher categorical site of derived affine schemes, by replacing the commutative rings with an infinity category of differential graded commutative algebras, or of simplicial commutative rings or a similar category with an appropriate variant of a Grothendieck topology. One can also replace presheaves of sets by presheaves of simplicial sets (or of infinity groupoids). Then, in presence of an appropriate homotopic machinery one can develop a notion of derived stack as such a presheaf on the infinity category of derived affine schemes, which is satisfying certain infinite categorical version of a sheaf axiom (and to be algebraic, inductively a sequence of representability conditions). Quillen model categories, Segal categories and quasicategories are some of the most often used tools to formalize this yielding the *derived algebraic geometry*, introduced by the school of Carlos Simpson, including Andre Hirschowitz, Bertrand Toën, Gabrielle Vezzosi, Michel Vaquié and others; and developed further by Jacob Lurie, Bertrand Toën, and Gabrielle Vezzosi. Another (noncommutative) version of derived algebraic geometry, using A-infinity categories has been developed from early 1990-s by Maxim Kontsevich and followers.

24.5 History

24.5.1 Prehistory: before the 16th century

Some of the roots of algebraic geometry date back to the work of the Hellenistic Greeks from the 5th century BC. The Delian problem, for instance, was to construct a length x so that the cube of side x contained the same volume as the rectangular box $a^2 b$ for given sides a and b. Menaechmus (circa 350 BC) considered the problem geometrically by intersecting the pair of plane conics $ay = x^2$ and $xy = ab$.[11] The later work, in the 3rd century BC, of Archimedes and Apollonius studied more systematically problems on conic sections,[2] and also involved the use of coordinates.[1] The Arab mathematicians were able to solve by purely algebraic means certain cubic equations, and then to interpret the results geometrically. This was done, for instance, by Ibn al-Haytham in the 10th century AD.[3] Subsequently, Persian mathematician Omar Khayyám (born 1048 A.D.) discovered the general method of solving cubic equations by intersecting a parabola with a circle.[4] Each of these early developments in algebraic geometry dealt with questions of finding and describing the intersections of algebraic curves.

24.5.2   Renaissance

Such techniques of applying geometrical constructions to algebraic problems were also adopted by a number of Renaissance mathematicians such as Gerolamo Cardano and Niccolò Fontana "Tartaglia" on their studies of the cubic equation. The geometrical approach to construction problems, rather than the algebraic one, was favored by most 16th and 17th century mathematicians, notably Blaise Pascal who argued against the use of algebraic and analytical methods in geometry.[5] The French mathematicians Franciscus Vieta and later René Descartes and Pierre de Fermat revolutionized the conventional way of thinking about construction problems through the introduction of coordinate geometry. They were interested primarily in the properties of *algebraic curves*, such as those defined by Diophantine equations (in the case of Fermat), and the algebraic reformulation of the classical Greek works on conics and cubics (in the case of Descartes).

During the same period, Blaise Pascal and Gérard Desargues approached geometry from a different perspective, developing the synthetic notions of projective geometry. Pascal and Desargues also studied curves, but from the purely geometrical point of view: the analog of the Greek *ruler and compass construction*. Ultimately, the analytic geometry of Descartes and Fermat won out, for it supplied the 18th century mathematicians with concrete quantitative tools needed to study physical problems using the new calculus of Newton and Leibniz. However, by the end of the 18th century, most of the algebraic character of coordinate geometry was subsumed by the *calculus of infinitesimals* of Lagrange and Euler.

24.5.3   19th and early 20th century

It took the simultaneous 19th century developments of non-Euclidean geometry and Abelian integrals in order to bring the old algebraic ideas back into the geometrical fold. The first of these new developments was seized up by Edmond Laguerre and Arthur Cayley, who attempted to ascertain the generalized metric properties of projective space. Cayley introduced the idea of *homogeneous polynomial forms*, and more specifically quadratic forms, on projective space. Subsequently, Felix Klein studied projective geometry (along with other types of geometry) from the viewpoint that the geometry on a space is encoded in a certain class of transformations on the space. By the end of the 19th century, projective geometers were studying more general kinds of transformations on figures in projective space. Rather than the projective linear transformations which were normally regarded as giving the fundamental Kleinian geometry on projective space, they concerned themselves also with the higher degree birational transformations. This weaker notion of congruence would later lead members of the 20th century Italian school of algebraic geometry to classify algebraic surfaces up to birational isomorphism.

The second early 19th century development, that of Abelian integrals, would lead Bernhard Riemann to the development of Riemann surfaces.

In the same period began the algebraization of the algebraic geometry through commutative algebra. The prominent results in this direction are Hilbert's basis theorem and Hilbert's Nullstellensatz, which are the basis of the connexion between algebraic geometry and commutative algebra, and Macaulay's multivariate resultant, which is the basis of elimination theory. Probably because of the size of the computation which is implied by multivariate resultants, elimination theory was forgotten during the middle of the 20th century until it was renewed by singularity theory and computational algebraic geometry.[6]

24.5.4   20th century

B. L. van der Waerden, Oscar Zariski and André Weil developed a foundation for algebraic geometry based on contemporary commutative algebra, including valuation theory and the theory of ideals. One of the goals was to give a rigorous framework for proving the results of Italian school of algebraic geometry. In particular, this school used systematically the notion of generic point without any precise definition, which was first given by these authors during the 1930s.

In the 1950s and 1960s Jean-Pierre Serre and Alexander Grothendieck recast the foundations making use of sheaf theory. Later, from about 1960, and largely lead by Grothendieck, the idea of schemes was worked out, in conjunction with a very refined apparatus of homological techniques. After a decade of rapid development the field stabilized in the 1970s, and new applications were made, both to number theory and to more classical geometric questions on algebraic varieties, singularities and moduli.

An important class of varieties, not easily understood directly from their defining equations, are the abelian varieties, which are the projective varieties whose points form an abelian group. The prototypical examples are the elliptic curves, which have a rich theory. They were instrumental in the proof of Fermat's last theorem and are also used in elliptic curve cryptography.

In parallel with the abstract trend of the algebraic geometry, which is concerned with general statements about varieties, methods for effective computation with concretely-given varieties have also been developed, which lead to the new area of computational algebraic geometry. One of the founding methods of this area is the theory of Gröbner bases, introduced by Bruno Buchberger in 1965. Another founding method, more specially devoted to real algebraic geometry, is the cylindrical algebraic decomposition, introduced by George E. Collins in 1973.

24.6 Analytic geometry

An **analytic variety** is defined locally as the set of common solutions of several equations involving analytic functions. It is analogous to the included concept of real or complex algebraic variety. Any complex manifold is an analytic variety. Since analytic varieties may have singular points, not all analytic varieties are manifolds.

Modern analytic geometry is essentially equivalent to real and complex algebraic geometry, as has been shown by Jean-Pierre Serre in his paper *GAGA*, the name of which is French for *Algebraic geometry and analytic geometry*. Nevertheless, the two fields remain distinct, as the methods of proof are quite different and algebraic geometry includes also geometry in finite characteristic.

24.7 Applications

Algebraic geometry now finds applications in statistics,[7] control theory,[8][9] robotics,[10] error-correcting codes,[11] phylogenetics[12] and geometric modelling.[13] There are also connections to string theory,[14] game theory,[15] graph matchings,[16] solitons[17] and integer programming.[18]

24.8 See also

- Algebraic statistics

- Differential geometry

- Geometric algebra

- Glossary of classical algebraic geometry

- Intersection theory

- Important publications in algebraic geometry

- List of algebraic surfaces

- Noncommutative algebraic geometry

- Differential algebraic geometry

- Real algebraic geometry

24.9   Notes

[1] Dieudonné, Jean (1972). "The historical development of algebraic geometry". *The American Mathematical Monthly* **79** (8): 827–866. doi:10.2307/2317664. JSTOR 2317664.

[2] Kline, M. (1972) *Mathematical Thought from Ancient to Modern Times* (Volume 1). Oxford University Press. pp. 108, 90.

[3] Kline, M. (1972) *Mathematical Thought from Ancient to Modern Times* (Volume 1). Oxford University Press. p. 193.

[4] Kline, M. (1972) *Mathematical Thought from Ancient to Modern Times* (Volume 1). Oxford University Press. pp. 193–195.

[5] Kline, M. (1972) *Mathematical Thought from Ancient to Modern Times* (Volume 1). Oxford University Press. p. 279.

[6] A witness of this oblivion is the fact that Van der Waerden removed the chapter on elimination theory from the third edition (and all the subsequent ones) of his treatise *Moderne algebra* (in German).

[7] Drton, Mathias; Sturmfels, Bernd; Sullivant, Seth (2009). *Lectures on Algebraic Statistics*. Springer. ISBN 978-3-7643-8904-8.

[8] Falb, Peter (1990). *Methods of Algebraic Geometry in Control Theory Part II Multivariable Linear Systems and Projective Algebraic Geometry*. Springer. ISBN 978-0-8176-4113-9.

[9] Allen Tannenbaum (1982), Invariance and Systems Theory: Algebraic and Geometric Aspects, Lecture Notes in Mathematics, volume 845, Springer-Verlag, ISBN 9783540105657

[10] Selig, J.M. (2005). *Geometric Fundamentals of Robotics*. Springer. ISBN 978-0-387-20874-9.

[11] Tsfasman, Michael A.; Vlăduţ, Serge G.; Nogin, Dmitry (1990). *Algebraic Geometric Codes Basic Notions*. American Mathematical Soc. ISBN 978-0-8218-7520-9.

[12] Barry A. Cipra (2007), Algebraic Geometers See Ideal Approach to Biology, SIAM News, Volume 40, Number 6

[13] Jüttler, Bert; Piene, Ragni (2007). *Geometric Modeling and Algebraic Geometry*. Springer. ISBN 978-3-540-72185-7.

[14] Cox, David A.; Katz, Sheldon (1999). *Mirror Symmetry and Algebraic Geometry*. American Mathematical Soc. ISBN 978-0-8218-2127-5.

[15] Blume, L. E.; Zame, W. R. (1994). "The algebraic geometry of perfect and sequential equilibrium" (PDF). *Econometrica* **62** (4): 783–794. JSTOR 2951732.

[16] Kenyon, Richard; Okounkov, Andrei; Sheffield, Scott (2003). "Dimers and Amoebae". arXiv:math-ph/0311005 [math-ph].

[17] Fordy, Allan P. (1990). *Soliton Theory A Survey of Results*. Manchester University Press. ISBN 978-0-7190-1491-8.

[18] Cox, David A.; Sturmfels, Bernd. Manocha, Dinesh N., ed. *Applications of Computational Algebraic Geometry*. American Mathematical Soc. ISBN 978-0-8218-6758-7.

24.10   Further reading

Some classic textbooks that predate schemes

- van der Waerden, B. L. (1945). *Einfuehrung in die algebraische Geometrie*. Dover.

- Hodge, W. V. D.; Pedoe, Daniel (1994). *Methods of Algebraic Geometry Volume 1*. Cambridge University Press. ISBN 0-521-46900-7. Zbl 0796.14001.

- Hodge, W. V. D.; Pedoe, Daniel (1994). *Methods of Algebraic Geometry Volume 2*. Cambridge University Press. ISBN 0-521-46901-5. Zbl 0796.14002.

- Hodge, W. V. D.; Pedoe, Daniel (1994). *Methods of Algebraic Geometry Volume 3*. Cambridge University Press. ISBN 0-521-46775-6. Zbl 0796.14003.

Modern textbooks that do not use the language of schemes

- Garrity, Thomas et al. (2013). *Algebraic Geometry A Problem Solving Approach*. American Mathematical Society. ISBN 0-821-89396-3.

- Griffiths, Phillip; Harris, Joe (1994). *Principles of Algebraic Geometry*. Wiley-Interscience. ISBN 0-471-05059-8. Zbl 0836.14001.

- Harris, Joe (1995). *Algebraic Geometry A First Course*. Springer-Verlag. ISBN 0-387-97716-3. Zbl 0779.14001.

- Mumford, David (1995). *Algebraic Geometry I Complex Projective Varieties* (2nd ed.). Springer-Verlag. ISBN 3-540-58657-1. Zbl 0821.14001.

- Reid, Miles (1988). *Undergraduate Algebraic Geometry*. Cambridge University Press. ISBN 0-521-35662-8. Zbl 0701.14001.

- Shafarevich, Igor (1995). *Basic Algebraic Geometry 1 Varieties in Projective Space* (2nd ed.). Springer-Verlag. ISBN 0-387-54812-2. Zbl 0797.14001.

Textbooks in computational algebraic geometry

- Cox, David A.; Little, John; O'Shea, Donal (1997). *Ideals, Varieties, and Algorithms* (2nd ed.). Springer-Verlag. ISBN 0-387-94680-2. Zbl 0861.13012.

- Basu, Saugata; Pollack, Richard; Roy, Marie-Françoise (2006). *Algorithms in real algebraic geometry*. Springer-Verlag.

- González-Vega, Laureano; Recio, Tómas (1996). *Algorithms in algebraic geometry and applications*. Birkhaüser.

- Elkadi, Mohamed; Mourrain, Bernard; Piene, Ragni, eds. (2006). *Algebraic geometry and geometric modeling*. Springer-Verlag.

- Dickenstein, Alicia; Schreyer, Frank-Olaf; Sommese, Andrew J., eds. (2008). *Algorithms in Algebraic Geometry*. The IMA Volumes in Mathematics and its Applications **146**. Springer. ISBN 9780387751559. LCCN 2007938208.

- Cox, David A.; Little, John B.; O'Shea, Donal (1998). *Using algebraic geometry*. Springer-Verlag.

- Caviness, Bob F.; Johnson, Jeremy R. (1998). *Quantifier elimination and cylindrical algebraic decomposition*. Springer-Verlag.

Textbooks and references for schemes

- Eisenbud, David; Harris, Joe (1998). *The Geometry of Schemes*. Springer-Verlag. ISBN 0-387-98637-5. Zbl 0960.14002.

- Grothendieck, Alexander (1960). *Éléments de géométrie algébrique*. Publications Mathématiques de l'IHÉS. Zbl 0118.36206.

- Grothendieck, Alexander; Dieudonné, Jean Alexandre (1971). *Éléments de géométrie algébrique* **1** (2nd ed.). Springer-Verlag. ISBN 3-540-05113-9. Zbl 0203.23301.

- Hartshorne, Robin (1977). *Algebraic Geometry*. Springer-Verlag. ISBN 0-387-90244-9. Zbl 0367.14001.

- Mumford, David (1999). *The Red Book of Varieties and Schemes Includes the Michigan Lectures on Curves and Their Jacobians* (2nd ed.). Springer-Verlag. ISBN 3-540-63293-X. Zbl 0945.14001.

- Shafarevich, Igor (1995). *Basic Algebraic Geometry II Schemes and complex manifolds* (2nd ed.). Springer-Verlag. ISBN 3-540-57554-5. Zbl 0797.14002.

24.11   External links

- *Foundations of Algebraic Geometry* by Ravi Vakil, 764 pp.

- *Algebraic geometry* entry on PlanetMath

- English translation of the van der Waerden textbook

- The History of Algebraic Geometry (1.425 Gigabyte MOV file), a 1972 talk by Jean Dieudonné at the Department of Mathematics of the University of Wisconsin-Milwaukee

- The Stacks Project, an open source textbook and reference work on algebraic stacks and algebraic geometry

Chapter 25

Trigonometry

"Trig" redirects here. For other uses, see Trig (disambiguation).

Trigonometry (from Greek *trigōnon*, "triangle" and *metron*, "measure"[1]) is a branch of mathematics that studies

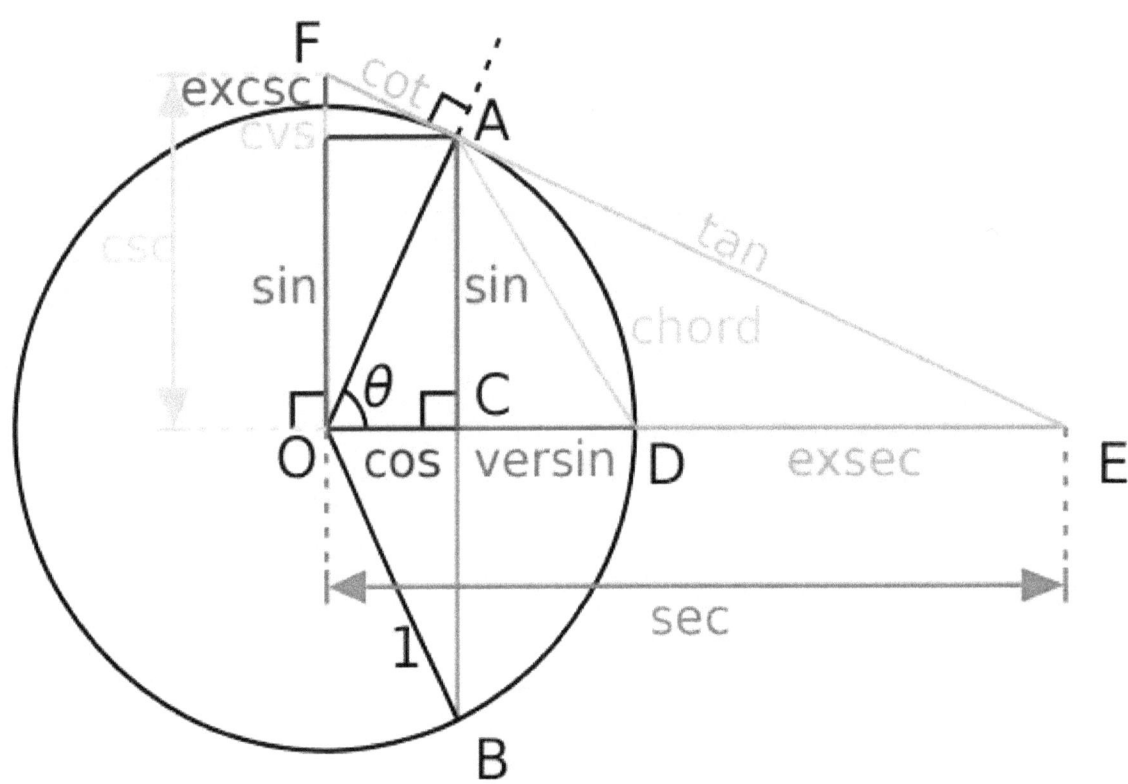

All of the trigonometric functions of an angle θ can be constructed geometrically in terms of a unit circle centered at O.

relationships involving lengths and angles of triangles. The field emerged in the Hellenistic world during the 3rd century BC from applications of geometry to astronomical studies.[2]

The 3rd-century astronomers first noted that the lengths of the sides of a right-angle triangle and the angles between those sides have fixed relationships: that is, if at least the length of one side and the value of one angle is known, then all other angles and lengths can be determined algorithmically. These calculations soon came to be defined as the trigonometric functions and today are pervasive in both pure and applied mathematics: fundamental methods of analysis such as the

Fourier transform, for example, or the wave equation, use trigonometric functions to understand cyclical phenomena across many applications in fields as diverse as physics, mechanical and electrical engineering, music and acoustics, astronomy, ecology, and biology. Trigonometry is also the foundation of surveying.

Trigonometry is most simply associated with planar right-angle triangles (each of which is a two-dimensional triangle with one angle equal to 90 degrees). The applicability to non-right-angle triangles exists, but, since any non-right-angle triangle (on a flat plane) can be bisected to create two right-angle triangles, most problems can be reduced to calculations on right-angle triangles. Thus the majority of applications relate to right-angle triangles. One exception to this is spherical trigonometry, the study of triangles on spheres, surfaces of constant positive curvature, in elliptic geometry (a fundamental part of astronomy and navigation). Trigonometry on surfaces of negative curvature is part of hyperbolic geometry.

Trigonometry basics are often taught in schools, either as a separate course or as a part of a precalculus course.

25.1   History

Main article: History of trigonometry

Sumerian astronomers studied angle measure, using a division of circles into 360 degrees.[4] They, and later the Babylonians, studied the ratios of the sides of similar triangles and discovered some properties of these ratios but did not turn that into a systematic method for finding sides and angles of triangles. The ancient Nubians used a similar method.[5]

In the 3rd century BCE, Hellenistic Greek mathematicians (such as Euclid (from Alexandria, Egypt) and Archimedes (from Syracuse, Sicily)) studied the properties of chords and inscribed angles in circles, and they proved theorems that are equivalent to modern trigonometric formulae, although they presented them geometrically rather than algebraically. In 140 BC Hipparchus (from Iznik, Turkey) gave the first tables of chords, analogous to modern tables of sine values, and used them to solve problems in trigonometry and spherical trigonometry.[6] In the 2nd century AD the Greco-Egyptian astronomer Ptolemy (from Alexandria, Egypt) printed detailed trigonometric tables (Ptolemy's table of chords) in Book 1, chapter 11 of his Almagest.[7] Ptolemy used chord length to define his trigonometric functions, a minor difference from the sine convention we use today.[8] (The value we call $\sin(\theta)$ can be found by looking up the chord length for twice the angle of interest (2θ) in Ptolemy's table, and then dividing that value by two.) Centuries passed before more detailed tables were produced, and Ptolemy's treatise remained in use for performing trigonometric calculations in astronomy throughout the next 1200 years in the medieval Byzantine, Islamic, and, later, Western European worlds.

The modern sine convention is first attested in the *Surya Siddhanta*, and its properties were further documented by the 5th century (CE) Indian mathematician and astronomer Aryabhata.[9] These Greek and Indian works were translated and expanded by medieval Islamic mathematicians. By the 10th century, Islamic mathematicians were using all six trigonometric functions, had tabulated their values, and were applying them to problems in spherical geometry. At about the same time, Chinese mathematicians developed trigonometry independently, although it was not a major field of study for them. Knowledge of trigonometric functions and methods reached Western Europe via Latin translations of Ptolemy's Greek Almagest as well as the works of Persian and Arabic astronomers such as Al Battani and Nasir al-Din al-Tusi.[10] One of the earliest works on trigonometry by a northern European mathematician is *De Triangulis* by the 15th century German mathematician Regiomontanus, who was encouraged to write, and provided with a copy of the Almagest, by the Byzantine Greek scholar cardinal Basilios Bessarion with whom he lived for several years.[11] At the same time another translation of the Almagest from Greek into Latin was completed by the Cretan George of Trebizond.[12] Trigonometry was still so little known in 16th-century northern Europe that Nicolaus Copernicus devoted two chapters of *De revolutionibus orbium coelestium* to explain its basic concepts.

Driven by the demands of navigation and the growing need for accurate maps of large geographic areas, trigonometry grew into a major branch of mathematics.[13] Bartholomaeus Pitiscus was the first to use the word, publishing his *Trigonometria* in 1595.[14] Gemma Frisius described for the first time the method of triangulation still used today in surveying. It was Leonhard Euler who fully incorporated complex numbers into trigonometry. The works of the Scottish mathematicians James Gregory in the 17th century and Colin Maclaurin in the 18th century were influential in the development of trigonometric series.[15] Also in the 18th century, Brook Taylor defined the general Taylor series.[16]

Hipparchus, credited with compiling the first trigonometric table, is known as "the father of trigonometry".[3]

25.2 Overview

Main article: Trigonometric function

If one angle of a triangle is 90 degrees and one of the other angles is known, the third is thereby fixed, because the three angles of any triangle add up to 180 degrees. The two acute angles therefore add up to 90 degrees: they are complementary

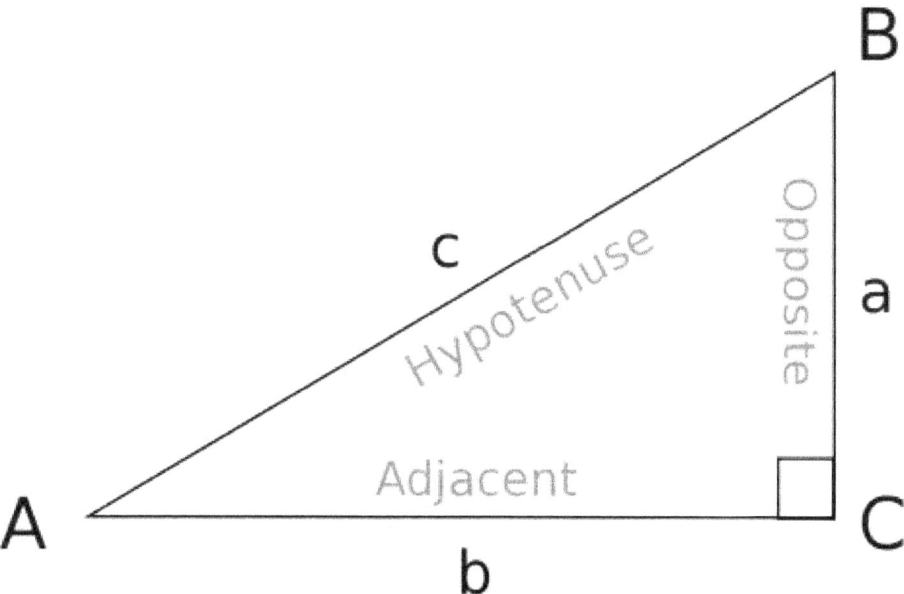

In this right triangle: sin A = a/c; cos A = b/c; tan A = a/b.

angles. The shape of a triangle is completely determined, except for similarity, by the angles. Once the angles are known, the ratios of the sides are determined, regardless of the overall size of the triangle. If the length of one of the sides is known, the other two are determined. These ratios are given by the following trigonometric functions of the known angle *A*, where *a*, *b* and *c* refer to the lengths of the sides in the accompanying figure:

- **Sine** function (sin), defined as the ratio of the side opposite the angle to the hypotenuse.

$$\sin A = \frac{\text{opposite}}{\text{hypotenuse}} = \frac{a}{c}.$$

- **Cosine** function (cos), defined as the ratio of the adjacent leg to the hypotenuse.

$$\cos A = \frac{\text{adjacent}}{\text{hypotenuse}} = \frac{b}{c}.$$

- **Tangent** function (tan), defined as the ratio of the opposite leg to the adjacent leg.

$$\tan A = \frac{\text{opposite}}{\text{adjacent}} = \frac{a}{b} = \frac{a}{c} * \frac{c}{b} = \frac{a}{c} / \frac{b}{c} = \frac{\sin A}{\cos A}.$$

The **hypotenuse** is the side opposite to the 90 degree angle in a right triangle; it is the longest side of the triangle and one of the two sides adjacent to angle A. The **adjacent leg** is the other side that is adjacent to angle A. The **opposite side** is the side that is opposite to angle A. The terms **perpendicular** and **base** are sometimes used for the opposite and adjacent sides respectively. Many people find it easy to remember what sides of the right triangle are equal to sine, cosine, or tangent, by memorizing the word SOH-CAH-TOA (see below under Mnemonics).

The reciprocals of these functions are named the **cosecant** (csc or cosec), **secant** (sec), and **cotangent** (cot), respectively:

$$\csc A = \frac{1}{\sin A} = \frac{\text{hypotenuse}}{\text{opposite}} = \frac{c}{a},$$

$$\sec A = \frac{1}{\cos A} = \frac{\text{hypotenuse}}{\text{adjacent}} = \frac{c}{b},$$

$$\cot A = \frac{1}{\tan A} = \frac{\text{adjacent}}{\text{opposite}} = \frac{\cos A}{\sin A} = \frac{b}{a}.$$

The inverse functions are called the **arcsine**, **arccosine**, and **arctangent**, respectively. There are arithmetic relations between these functions, which are known as trigonometric identities. The cosine, cotangent, and cosecant are so named because they are respectively the sine, tangent, and secant of the complementary angle abbreviated to "co-".

With these functions one can answer virtually all questions about arbitrary triangles by using the law of sines and the law of cosines. These laws can be used to compute the remaining angles and sides of any triangle as soon as two sides and their included angle or two angles and a side or three sides are known. These laws are useful in all branches of geometry, since every polygon may be described as a finite combination of triangles.

25.2.1   Extending the definitions

The above definitions only apply to angles between 0 and 90 degrees (0 and $\pi/2$ radians). Using the unit circle, one can extend them to all positive and negative arguments (see trigonometric function). The trigonometric functions are periodic, with a period of 360 degrees or 2π radians. That means their values repeat at those intervals. The tangent and cotangent functions also have a shorter period, of 180 degrees or π radians.

The trigonometric functions can be defined in other ways besides the geometrical definitions above, using tools from calculus and infinite series. With these definitions the trigonometric functions can be defined for complex numbers. The complex exponential function is particularly useful.

$$e^{x+iy} = e^x(\cos y + i \sin y).$$

See Euler's and De Moivre's formulas.

- Graphing process of $y = \sin(x)$ using a unit circle.
- Graphing process of $y = \csc(x)$, the reciprocal of sine, using a unit circle.
- Graphing process of $y = \tan(x)$ using a unit circle.

25.2.2   Mnemonics

Main article: Mnemonics in trigonometry

A common use of mnemonics is to remember facts and relationships in trigonometry. For example, the *sine, cosine,* and *tangent* ratios in a right triangle can be remembered by representing them and their corresponding sides as strings of letters. For instance, a mnemonic is SOH-CAH-TOA:[17]

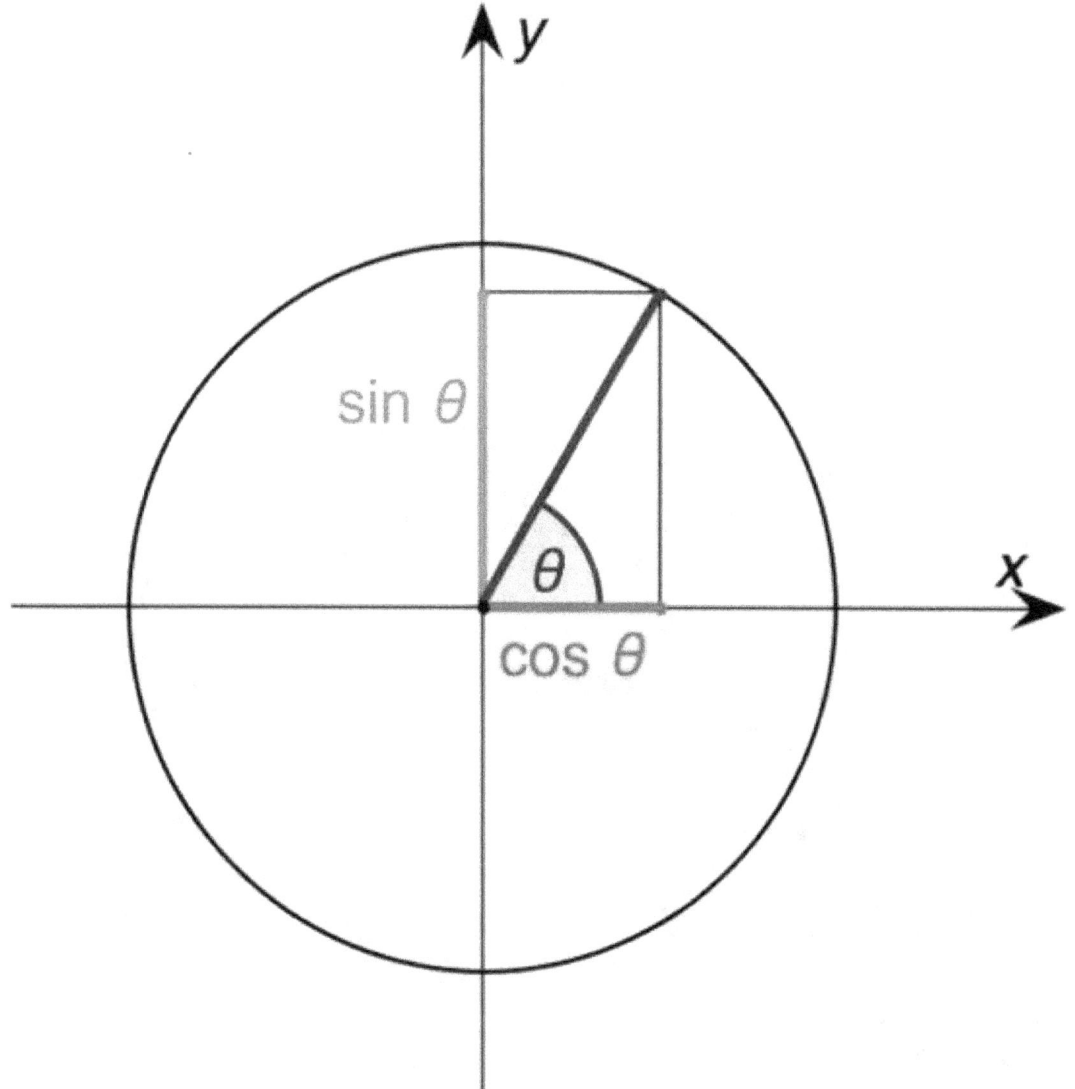

Fig. 1a – Sine and cosine of an angle θ defined using the unit circle.

Sine = Opposite ÷ Hypotenuse

Cosine = Adjacent ÷ Hypotenuse

Tangent = Opposite ÷ Adjacent

One way to remember the letters is to sound them out phonetically (i.e., *SOH-CAH-TOA*, which is pronounced 'so-kə-toe-uh' /soʊkə'toʊə/). Another method is to expand the letters into a sentence, such as "**S**ome **O**ld **H**ippie **C**aught **A**nother **H**ippie **T**rippin' **O**n **A**cid".[18]

25.2.3   Calculating trigonometric functions

Main article: Trigonometric tables

Trigonometric functions were among the earliest uses for mathematical tables. Such tables were incorporated into mathematics textbooks and students were taught to look up values and how to interpolate between the values listed to get higher accuracy. Slide rules had special scales for trigonometric functions.

Today scientific calculators have buttons for calculating the main trigonometric functions (sin, cos, tan, and sometimes cis and their inverses). Most allow a choice of angle measurement methods: degrees, radians, and sometimes gradians. Most computer programming languages provide function libraries that include the trigonometric functions. The floating point unit hardware incorporated into the microprocessor chips used in most personal computers has built-in instructions for calculating trigonometric functions.[19]

25.3 Applications of trigonometry

Sextants are used to measure the angle of the sun or stars with respect to the horizon. Using trigonometry and a marine chronometer, the position of the ship can be determined from such measurements.

Main article: Uses of trigonometry

There is an enormous number of uses of trigonometry and trigonometric functions. For instance, the technique of triangulation is used in astronomy to measure the distance to nearby stars, in geography to measure distances between landmarks, and in satellite navigation systems. The sine and cosine functions are fundamental to the theory of periodic functions such as those that describe sound and light waves.

Fields that use trigonometry or trigonometric functions include astronomy (especially for locating apparent positions of celestial objects, in which spherical trigonometry is essential) and hence navigation (on the oceans, in aircraft, and in space), music theory, audio synthesis, acoustics, optics, electronics, biology, medical imaging (CAT scans and ultrasound), pharmacy, chemistry, number theory (and hence cryptology), seismology, meteorology, oceanography, many physical sciences, land surveying and geodesy, architecture, image compression, phonetics, economics, electrical engineering, mechanical engineering, civil engineering, computer graphics, cartography, crystallography and game development.

25.4   Pythagorean identities

Identities are those equations that hold true for any value.

$$\sin^2 A + \cos^2 A = 1$$

(The following two can be derived from the first.)

$$\sec^2 A - \tan^2 A = 1$$
$$\csc^2 A - \cot^2 A = 1$$

25.5   Angle transformation formulae

$$\sin(A \pm B) = \sin A \, \cos B \pm \cos A \, \sin B$$
$$\cos(A \pm B) = \cos A \, \cos B \mp \sin A \, \sin B$$
$$\tan(A \pm B) = \frac{\tan A \pm \tan B}{1 \mp \tan A \, \tan B}$$
$$\cot(A \pm B) = \frac{\cot A \, \cot B \mp 1}{\cot B \pm \cot A}$$

25.6   Common formulae

Certain equations involving trigonometric functions are true for all angles and are known as *trigonometric identities*. Some identities equate an expression to a different expression involving the same angles. These are listed in List of trigonometric identities. Triangle identities that relate the sides and angles of a given triangle are listed below.

In the following identities, A, B and C are the angles of a triangle and a, b and c are the lengths of sides of the triangle opposite the respective angles (as shown in the diagram).

25.6.1   Law of sines

The **law of sines** (also known as the "sine rule") for an arbitrary triangle states:

$$\frac{a}{\sin A} = \frac{b}{\sin B} = \frac{c}{\sin C} = 2R = \frac{abc}{2\Delta}.$$

where Δ is the area of the triangle and R is the radius of the circumscribed circle of the triangle:

$$R = \frac{abc}{\sqrt{(a+b+c)(a-b+c)(a+b-c)(b+c-a)}}.$$

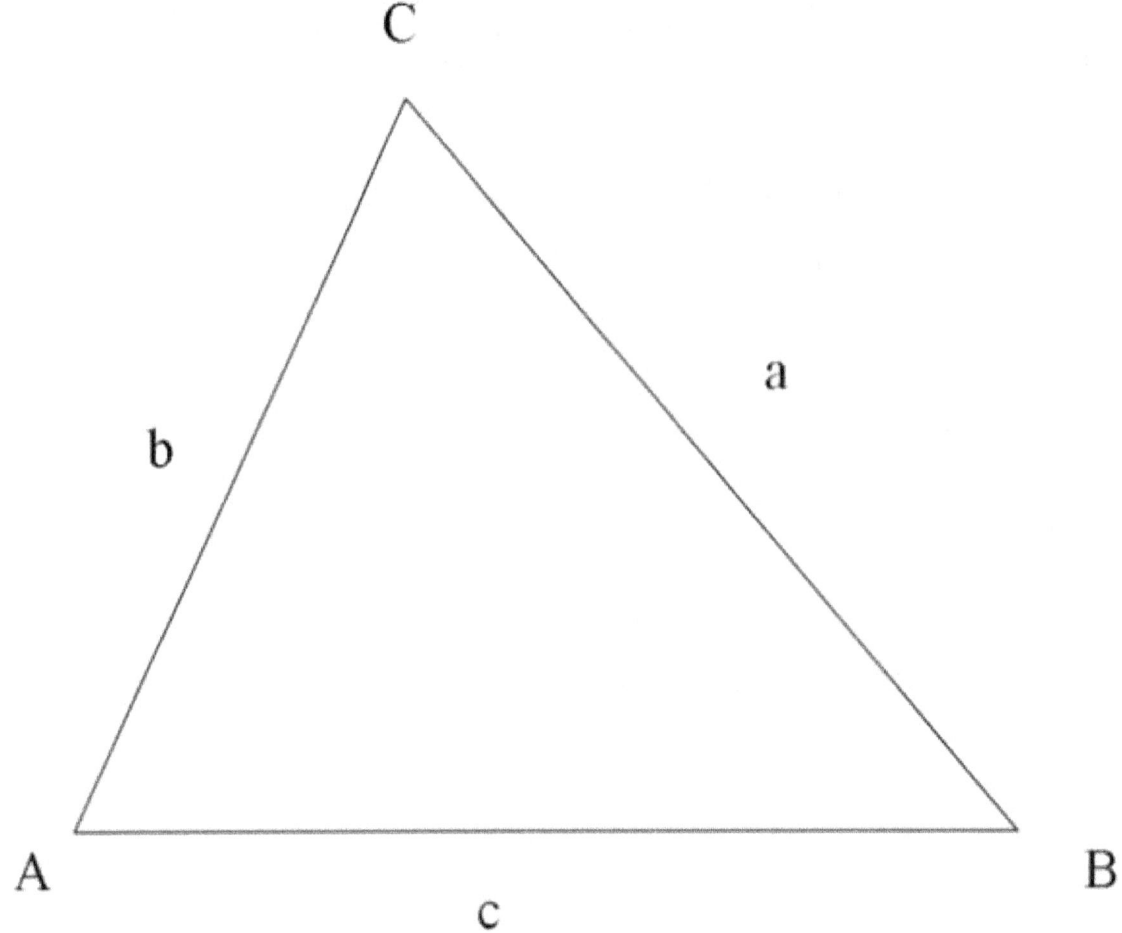

Triangle with sides a,b,c and respectively opposite angles A,B,C

Another law involving sines can be used to calculate the area of a triangle. Given two sides a and b and the angle between the sides C, the area of the triangle is given by half the product of the lengths of two sides and the sine of the angle between the two sides:

$$\text{Area} = \Delta = \frac{1}{2}ab\sin C.$$

25.6.2 Law of cosines

The **law of cosines** (known as the cosine formula, or the "cos rule") is an extension of the Pythagorean theorem to arbitrary triangles:

$$c^2 = a^2 + b^2 - 2ab\cos C,$$

or equivalently:

$$\cos C = \frac{a^2 + b^2 - c^2}{2ab}.$$

The law of cosines may be used to prove Heron's formula, which is another method that may be used to calculate the area of a triangle. This formula states that if a triangle has sides of lengths a, b, and c, and if the semiperimeter is

$$s = \frac{1}{2}(a + b + c),$$

then the area of the triangle is:

$$\text{Area} = \Delta = \sqrt{s(s - a)(s - b)(s - c)} = \frac{abc}{4R}$$

where R is the radius of the circumcircle of the triangle.

25.6.3   Law of tangents

The **law of tangents**:

$$\frac{a - b}{a + b} = \frac{\tan\left[\frac{1}{2}(A - B)\right]}{\tan\left[\frac{1}{2}(A + B)\right]}$$

25.6.4   Euler's formula

Euler's formula, which states that $e^{ix} = \cos x + i \sin x$, produces the following analytical identities for sine, cosine, and tangent in terms of e and the imaginary unit i:

$$\sin x = \frac{e^{ix} - e^{-ix}}{2i}, \qquad \cos x = \frac{e^{ix} + e^{-ix}}{2}, \qquad \tan x = \frac{i(e^{-ix} - e^{ix})}{e^{ix} + e^{-ix}}.$$

25.7   See also

- Aryabhata's sine table

- Generalized trigonometry

- Lénárt sphere

- List of triangle topics

- List of trigonometric identities

- Rational trigonometry

- Skinny triangle

- Small-angle approximation

- Trigonometric functions

- Trigonometry in Galois fields

- Unit circle

- Uses of trigonometry

25.8 References

[1] "trigonometry". Online Etymology Dictionary.

[2] R. Nagel (ed.), *Encyclopedia of Science*, 2nd Ed., The Gale Group (2002)

[3] Boyer (1991). "Greek Trigonometry and Mensuration". *A History of Mathematics*. p. 162.

[4] Aaboe, Asger. Episodes from the Early History of Astronomy. New York: Springer, 2001. ISBN 0-387-95136-9

[5] Otto Neugebauer (1975). *A history of ancient mathematical astronomy*. *1*. Springer-Verlag. pp. 744–. ISBN 978-3-540-06995-9.

[6] Thurston, pp. 235–236.

[7] Toomer, G. J. (1998). *Ptolemy's Almagest*. Princeton University Press. ISBN 0-691-00260-6

[8] Thurston, pp. 239–243.

[9] Boyer p. 215

[10] Boyer pp. 237, 274

[11] http://www-history.mcs.st-and.ac.uk/Biographies/Regiomontanus.html

[12] N.G. Wilson, *From Byzantium to Italy. Greek Studies in the Italian Renaissance*, London, 1992. ISBN 0-7156-2418-0

[13] Grattan-Guinness, Ivor (1997). *The Rainbow of Mathematics: A History of the Mathematical Sciences*. W.W. Norton. ISBN 0-393-32030-8.

[14] Robert E. Krebs (2004). *Groundbreaking Scientific Experiments, Inventions, and Discoveries of the Middle Ages and the Renaissance*. Greenwood Publishing Group. pp. 153–. ISBN 978-0-313-32433-8.

[15] William Bragg Ewald (2008). *From Kant to Hilbert: a source book in the foundations of mathematics*. Oxford University Press US. p. 93. ISBN 0-19-850535-3

[16] Kelly Dempski (2002). *Focus on Curves and Surfaces*. p. 29. ISBN 1-59200-007-X

[17] Weisstein, Eric W., "SOHCAHTOA", *MathWorld*.

[18] A sentence more appropriate for high schools is "'*Some Old Horse Came A*"Hopping Through Our Alley". Foster, Jonathan K. (2008). *Memory: A Very Short Introduction*. Oxford. p. 128. ISBN 0-19-280675-0.

[19] *Intel® 64 and IA-32 Architectures Software Developer's Manual Combined Volumes: 1, 2A, 2B, 2C, 3A, 3B and 3C* (PDF). Intel. 2013.

25.9 Bibliography

- Boyer, Carl B. (1991). *A History of Mathematics* (Second ed.). John Wiley & Sons, Inc. ISBN 0-471-54397-7.

- Hazewinkel, Michiel, ed. (2001), "Trigonometric functions", *Encyclopedia of Mathematics*, Springer, ISBN 978-1-55608-010-4

- Christopher M. Linton (2004). From Eudoxus to Einstein: A History of Mathematical Astronomy . Cambridge University Press.

- Weisstein, Eric W., "Trigonometric Addition Formulas", *MathWorld*.

25.10   External links

- Khan Academy: Trigonometry, free online micro lectures

- Trigonometry by Alfred Monroe Kenyon and Louis Ingold, The Macmillan Company, 1914. In images, full text presented.

- Benjamin Banneker's Trigonometry Puzzle at Convergence

- Dave's Short Course in Trigonometry by David Joyce of Clark University

- Trigonometry, by Michael Corral, Covers elementary trigonometry, Distributed under GNU Free Documentation License

Chapter 26

Differential geometry

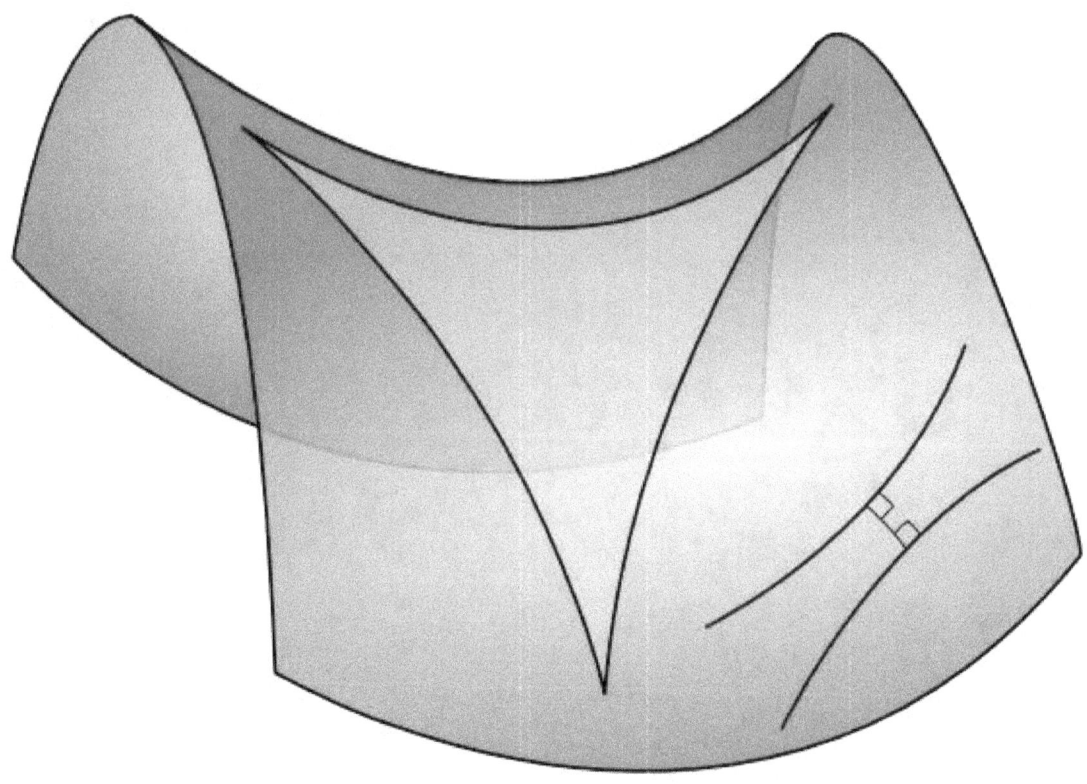

A triangle immersed in a saddle-shape plane (a hyperbolic paraboloid), as well as two diverging ultraparallel lines.

Differential geometry is a mathematical discipline that uses the techniques of differential calculus, integral calculus, linear algebra and multilinear algebra to study problems in geometry. The theory of plane and space curves and surfaces in the three-dimensional Euclidean space formed the basis for development of differential geometry during the 18th century and the 19th century.

Since the late 19th century, differential geometry has grown into a field concerned more generally with the geometric structures on differentiable manifolds. Differential geometry is closely related to differential topology and the geometric aspects of the theory of differential equations. The differential geometry of surfaces captures many of the key ideas and techniques characteristic of this field.

26.1   History of Development

Differential geometry arose and developed[1] as a result of and in connection to Mathematical Analysis of Curves and Surfaces. Mathematical Analysis of curves and surfaces had been developed to answer some of the nagging and unanswered questions, like the reasons for relationships between complex shapes and curves, series and analytic functions that appeared in Calculus. They indicated towards greater, hidden relationships and symmetries in nature that were still not unravelled at that time, and which the standard methods of analysis could not address.

When curves, surfaces enclosed by curves, and points on curves were found to be quantitatively, and generally, related by mathematical forms the formal study of the nature of curves and surfaces became a field of study in its own right, with Monge's paper in 1795, and especially, with Gauss's publication of his article, titled Disquisitiones Generales Circa Superficies Curvas, in Commentationes Societatis Regiae Scientiarum Gottingesis Recentiores[2] in 1827.

Initially applied to the Euclidean space, further explorations led to non-Euclidean space, and metric and topological spaces.

26.2   Branches of differential geometry

26.2.1   Riemannian geometry

Main article: Riemannian geometry

Riemannian geometry studies Riemannian manifolds, smooth manifolds with a *Riemannian metric*. This is a concept of distance expressed by means of a smooth positive definite symmetric bilinear form defined on the tangent space at each point. Riemannian geometry generalizes Euclidean geometry to spaces that are not necessarily flat, although they still resemble the Euclidean space at each point infinitesimally, i.e. in the first order of approximation. Various concepts based on length, such as the arc length of curves, area of plane regions, and volume of solids all possess natural analogues in Riemannian geometry. The notion of a directional derivative of a function from multivariable calculus is extended in Riemannian geometry to the notion of a covariant derivative of a tensor. Many concepts and techniques of analysis and differential equations have been generalized to the setting of Riemannian manifolds.

A distance-preserving diffeomorphism between Riemannian manifolds is called an isometry. This notion can also be defined *locally*, i.e. for small neighborhoods of points. Any two regular curves are locally isometric. However, the Theorema Egregium of Carl Friedrich Gauss showed that already for surfaces, the existence of a local isometry imposes strong compatibility conditions on their metrics: the Gaussian curvatures at the corresponding points must be the same. In higher dimensions, the Riemann curvature tensor is an important pointwise invariant associated to a Riemannian manifold that measures how close it is to being flat. An important class of Riemannian manifolds is the Riemannian symmetric spaces, whose curvature is not necessarily constant. These are the closest analogues to the "ordinary" plane and space considered in Euclidean and non-Euclidean geometry.

26.2.2   Pseudo-Riemannian geometry

Pseudo-Riemannian geometry generalizes Riemannian geometry to the case in which the metric tensor need not be positive-definite. A special case of this is a Lorentzian manifold, which is the mathematical basis of Einstein's general relativity theory of gravity.

26.2.3   Finsler geometry

Finsler geometry has the *Finsler manifold* as the main object of study. This is a differential manifold with a Finsler metric, i.e. a Banach norm defined on each tangent space. A Finsler metric is a much more general structure than a Riemannian metric. A Finsler structure on a manifold M is a function $F : \mathrm{T}M \to [0,\infty)$ such that:

1. $F(x, my) = |m|F(x,y)$ for all x, y in TM,

2. F is infinitely differentiable in $TM - \{0\}$.

3. The vertical Hessian of F^2 is positive definite.

26.2.4 Symplectic geometry

Main article: Symplectic geometry

Symplectic geometry is the study of symplectic manifolds. An **almost symplectic manifold** is a differentiable manifold equipped with a smoothly varying non-degenerate skew-symmetric bilinear form on each tangent space, i.e., a nonde-generate 2-form ω, called the *symplectic form*. A symplectic manifold is an almost symplectic manifold for which the symplectic form ω is closed: $d\omega = 0$.

A diffeomorphism between two symplectic manifolds which preserves the symplectic form is called a symplectomorphism. Non-degenerate skew-symmetric bilinear forms can only exist on even-dimensional vector spaces, so symplectic manifolds necessarily have even dimension. In dimension 2, a symplectic manifold is just a surface endowed with an area form and a symplectomorphism is an area-preserving diffeomorphism. The phase space of a mechanical system is a symplectic manifold and they made an implicit appearance already in the work of Joseph Louis Lagrange on analytical mechanics and later in Carl Gustav Jacobi's and William Rowan Hamilton's formulations of classical mechanics.

By contrast with Riemannian geometry, where the curvature provides a local invariant of Riemannian manifolds, Darboux's theorem states that all symplectic manifolds are locally isomorphic. The only invariants of a symplectic manifold are global in nature and topological aspects play a prominent role in symplectic geometry. The first result in symplectic topology is probably the Poincaré-Birkhoff theorem, conjectured by Henri Poincaré and then proved by G.D. Birkhoff in 1912. It claims that if an area preserving map of an annulus twists each boundary component in opposite directions, then the map has at least two fixed points.[3]

26.2.5 Contact geometry

Main article: Contact geometry

Contact geometry deals with certain manifolds of odd dimension. It is close to symplectic geometry and like the latter, it originated in questions of classical mechanics. A *contact structure* on a $(2n + 1)$ - dimensional manifold M is given by a smooth hyperplane field H in the tangent bundle that is as far as possible from being associated with the level sets of a differentiable function on M (the technical term is "completely nonintegrable tangent hyperplane distribution"). Near each point p, a hyperplane distribution is determined by a nowhere vanishing 1-form α, which is unique up to multiplication by a nowhere vanishing function:

$$H_p = \ker \alpha_p \subset T_p M.$$

A local 1-form on M is a *contact form* if the restriction of its exterior derivative to H is a non-degenerate two-form and thus induces a symplectic structure on Hp at each point. If the distribution H can be defined by a global one-form α then this form is contact if and only if the top-dimensional form

$$\alpha \wedge (d\alpha)^n$$

is a volume form on **M**, i.e. does not vanish anywhere. A contact analogue of the Darboux theorem holds: all contact structures on an odd-dimensional manifold are locally isomorphic and can be brought to a certain local normal form by a suitable choice of the coordinate system.

26.2.6   Complex and Kähler geometry

Complex differential geometry is the study of complex manifolds. An almost complex manifold is a *real* manifold M, endowed with a tensor of type $(1, 1)$, i.e. a vector bundle endomorphism (called an *almost complex structure*)

$$J : TM \to TM \text{, such that } J^2 = -1.$$

It follows from this definition that an almost complex manifold is even-dimensional.

An almost complex manifold is called *complex* if $N_J = 0$, where N_J is a tensor of type $(2, 1)$ related to J, called the Nijenhuis tensor (or sometimes the *torsion*). An almost complex manifold is complex if and only if it admits a holomorphic coordinate atlas. An *almost Hermitian structure* is given by an almost complex structure J, along with a Riemannian metric g, satisfying the compatibility condition

$$g(JX, JY) = g(X, Y)$$

An almost Hermitian structure defines naturally a differential two-form

$$\omega_{J,g}(X, Y) := g(JX, Y)$$

The following two conditions are equivalent:

1. $N_J = 0$ and $d\omega = 0$

2. $\nabla J = 0$

where ∇ is the Levi-Civita connection of g. In this case, (J, g) is called a *Kähler structure*, and a *Kähler manifold* is a manifold endowed with a Kähler structure. In particular, a Kähler manifold is both a complex and a symplectic manifold. A large class of Kähler manifolds (the class of Hodge manifolds) is given by all the smooth complex projective varieties.

26.2.7   CR geometry

CR geometry is the study of the intrinsic geometry of boundaries of domains in complex manifolds.

26.2.8   Differential topology

Differential topology is the study of (global) geometric invariants without a metric or symplectic form. It starts from the natural operations such as Lie derivative of natural vector bundles and de Rham differential of forms. Beside Lie algebroids, also Courant algebroids start playing a more important role.

26.2.9   Lie groups

A Lie group is a group in the category of smooth manifolds. Beside the algebraic properties this enjoys also differential geometric properties. The most obvious construction is that of a Lie algebra which is the tangent space at the unit endowed with the Lie bracket between left-invariant vector fields. Beside the structure theory there is also the wide field of representation theory.

26.3 Bundles and connections

The apparatus of vector bundles, principal bundles, and connections on bundles plays an extraordinarily important role in modern differential geometry. A smooth manifold always carries a natural vector bundle, the tangent bundle. Loosely speaking, this structure by itself is sufficient only for developing analysis on the manifold, while doing geometry requires, in addition, some way to relate the tangent spaces at different points, i.e. a notion of parallel transport. An important example is provided by affine connections. For a surface in $\mathbf{R}^3$, tangent planes at different points can be identified using a natural path-wise parallelism induced by the ambient Euclidean space, which has a well-known standard definition of metric and parallelism. In Riemannian geometry, the Levi-Civita connection serves a similar purpose. (The Levi-Civita connection defines path-wise parallelism in terms of a given arbitrary Riemannian metric on a manifold.) More generally, differential geometers consider spaces with a vector bundle and an arbitrary affine connection which is not defined in terms of a metric. In physics, the manifold may be the space-time continuum and the bundles and connections are related to various physical fields.

26.4 Intrinsic versus extrinsic

From the beginning and through the middle of the 18th century, differential geometry was studied from the *extrinsic* point of view: curves and surfaces were considered as lying in a Euclidean space of higher dimension (for example a surface in an ambient space of three dimensions). The simplest results are those in the differential geometry of curves and differential geometry of surfaces. Starting with the work of Riemann, the *intrinsic* point of view was developed, in which one cannot speak of moving "outside" the geometric object because it is considered to be given in a free-standing way. The fundamental result here is Gauss's theorema egregium, to the effect that Gaussian curvature is an intrinsic invariant.

The intrinsic point of view is more flexible. For example, it is useful in relativity where space-time cannot naturally be taken as extrinsic (what would be "outside" of it?). However, there is a price to pay in technical complexity: the intrinsic definitions of curvature and connections become much less visually intuitive.

These two points of view can be reconciled, i.e. the extrinsic geometry can be considered as a structure additional to the intrinsic one. (See the Nash embedding theorem.) In the formalism of geometric calculus both extrinsic and intrinsic geometry of a manifold can be characterized by a single bivector-valued one-form called the shape operator.[4]

26.5 Applications

Below are some examples of how differential geometry is applied to other fields of science and mathematics.

- In physics, four uses will be mentioned:

  - Differential geometry is the language in which Einstein's general theory of relativity is expressed. According to the theory, the universe is a smooth manifold equipped with a pseudo-Riemannian metric, which describes the curvature of space-time. Understanding this curvature is essential for the positioning of satellites into orbit around the earth. Differential geometry is also indispensable in the study of gravitational lensing and black holes.

  - Differential forms are used in the study of electromagnetism.

  - Differential geometry has applications to both Lagrangian mechanics and Hamiltonian mechanics. Symplectic manifolds in particular can be used to study Hamiltonian systems.

  - Riemannian geometry and contact geometry have been used to construct the formalism of geometrothermodynamics which has found applications in classical equilibrium thermodynamics.

- In economics, differential geometry has applications to the field of econometrics.[5]

- Geometric modeling (including computer graphics) and computer-aided geometric design draw on ideas from differential geometry.

- In engineering, differential geometry can be applied to solve problems in digital signal processing.[6]

- In control theory, differential geometry can be used to analyze nonlinear controllers, particularly geometric control[7]

- In probability, statistics, and information theory, one can interpret various structures as Riemannian manifolds, which yields the field of information geometry, particularly via the Fisher information metric.

- In structural geology, differential geometry is used to analyze and describe geologic structures.

- In computer vision, differential geometry is used to analyze shapes.[8]

- In image processing, differential geometry is used to process and analyse data on non-flat surfaces.[9]

- Grigori Perelman's proof of the Poincaré conjecture using the techniques of Ricci flows demonstrated the power of the differential-geometric approach to questions in topology and it highlighted the important role played by its analytic methods.

- In wireless communications, Grassmannian manifolds are used for beamforming techniques in multiple antenna systems.[10]

26.6 See also

- Abstract differential geometry

- Affine differential geometry

- Analysis on fractals

- Basic introduction to the mathematics of curved spacetime

- Discrete differential geometry

- Gauss

- Glossary of differential geometry and topology

- Integral geometry

- List of differential geometry topics

- Important publications in differential geometry

- Important publications in differential topology

- Noncommutative geometry

- Projective differential geometry

- Synthetic differential geometry

26.7 References

[1] http://www.encyclopediaofmath.org/index.php/Differential_geometry be referred to

[2] Disquisitiones Generales Circa Superficies Curvas (literal translation from Latin: General Investigations of Curved Surfaces), Commentationes Societatis Regiae Scientiarum Gottingesis Recentiores (literally, Recent Perspectives, Gottingen's Royal Society of Science). Volume VI, pp. 99–146. A translation of the work, by A.M.Hiltebeitel and J.C.Morehead, titled, "General Investigations of Curved Surfaces" was published 1965 by Raven Press, New York. A digitised version of the same is available at http://quod.lib.umich.edu/u/umhistmath/abr1255.0001.001 for free download, for non-commercial, personal use. In case of further information, the library could be contacted. Also, the Wikipedia article on Gauss's works in the year 1827 at https://en.wikipedia.org/wiki/Carl_Friedrich_Gauss#Writings could be looked at.

[3] It is easy to show that the area preserving condition (or the twisting condition) cannot be removed. Note that if one tries to extend such a theorem to higher dimensions, one would probably guess that a volume preserving map of a certain type must have fixed points. This is false in dimensions greater than 3.

[4] David Hestenes "The Shape of Differential Geometry in Geometric Calculus" http://geocalc.clas.asu.edu/pdf/Shape%20in%20GC-2012.pdf there is also a pdf available of a scientific talk on the subject http://staff.science.uva.nl/~{}leo/agacse2010/talks_world/Hestenes.pdf

[5] Paul Marriott and Mark Salmon (editors), "Applications of Differential Geometry to Econometrics", Cambridge University Press; 1 edition (September 18, 2000).

[6] Jonathan H. Manton, "On the role of differential geometry in signal processing" .

[7] Francesco Bullo and Andrew Lewis, "Geometric Control of Simple Mechanical Systems." Springer-Verlag, 2001.

[8] Mario Micheli, "The Differential Geometry of Landmark Shape Manifolds: Metrics, Geodesics, and Curvature", http://www.math.ucla.edu/~{}micheli/PUBLICATIONS/micheli_phd.pdf

[9] Anand A. Joshi, "Geometric methods for image processing and signal analysis",

[10] David J. Love and Robert W. Heath, Jr. "Grassmannian Beamforming for Multiple-Input Multiple-Output Wireless Systems," IEEE Transactions on Information Theory, Vol. 49, No. 10, October 2003

26.8 Further reading

- Wolfgang Kühnel (2002). *Differential Geometry: Curves - Surfaces - Manifolds* (2nd ed.). ISBN 0-8218-3988-8.

- Theodore Frankel (2004). *The geometry of physics: an introduction* (2nd ed.). ISBN 0-521-53927-7.

- Spivak, Michael (1999). *A Comprehensive Introduction to Differential Geometry (5 Volumes)* (3rd ed.).

- do Carmo, Manfredo (1976). *Differential Geometry of Curves and Surfaces*. ISBN 0-13-212589-7. Classical geometric approach to differential geometry without tensor analysis.

- Kreyszig, Erwin (1991). *Differential Geometry*. ISBN 0-486-66721-9. Good classical geometric approach to differential geometry with tensor machinery.

- do Carmo, Manfredo Perdigao (1994). *Riemannian Geometry*.

- McCleary, John (1994). *Geometry from a Differentiable Viewpoint*.

- Bloch, Ethan D. (1996). *A First Course in Geometric Topology and Differential Geometry*.

- Gray, Alfred (1998). *Modern Differential Geometry of Curves and Surfaces with Mathematica* (2nd ed.).

- Burke, William L. (1985). *Applied Differential Geometry*.

- ter Haar Romeny, Bart M. (2003). *Front-End Vision and Multi-Scale Image Analysis*. ISBN 1-4020-1507-0.

26.9 External links

- Hazewinkel, Michiel, ed. (2001), "Differential geometry", *Encyclopedia of Mathematics*, Springer, ISBN 978-1-55608-010-4

- B. Conrad. Differential Geometry handouts, Stanford University

- Michael Murray's online differential geometry course, 1996

- A Modern Course on Curves and Surface, Richard S Palais, 2003

- Richard Palais's 3DXM Surfaces Gallery

- Balázs Csikós's Notes on Differential Geometry

- N. J. Hicks, Notes on Differential Geometry, Van Nostrand.

- MIT OpenCourseWare: Differential Geometry, Fall 2008

Chapter 27

Topology

Not to be confused with topography.

This article is about the branch of mathematics. For other uses, see Topology (disambiguation).

In mathematics, **topology** (from the Greek τόπος, *place*, and λόγος, *study*), is the study of topological spaces. It

Möbius strips, which have only one surface and one edge, are a kind of object studied in topology.

is an area of mathematics concerned with the properties of space that are preserved under continuous deformations, such as stretching and bending, but not tearing or gluing. Important topological properties include connectedness and compactness.

Topology developed as a field of study out of geometry and set theory, through analysis of such concepts as space, dimension, and transformation. Such ideas go back to Gottfried Leibniz, who in the 17th century envisioned the *geometria situs* (Greek-Latin for "geometry of place") and *analysis situs* (Greek-Latin for "picking apart of place"). Leonhard Euler's Seven Bridges of Königsberg Problem and Polyhedron Formula are arguably the field's first theorems. The term *topology* was introduced by Johann Benedict Listing in the 19th century, although it was not until the first decades of the 20th century that the idea of a topological space was developed. By the middle of the 20th century, topology had become a major branch of mathematics.

Topology has many subfields:

- **General topology** establishes the foundational aspects of topology and investigates properties of topological spaces and investigates concepts inherent to topological spaces. It includes point-set topology, which is the foundational topology used in all other branches (including topics like compactness and connectedness).

- **Algebraic topology** tries to measure degrees of connectivity using algebraic constructs such as homology and homotopy groups.

- **Differential topology** is the field dealing with differentiable functions on differentiable manifolds. It is closely related to differential geometry and together they make up the geometric theory of differentiable manifolds.

- **Geometric topology** primarily studies manifolds and their embeddings (placements) in other manifolds. A particularly active area is **low dimensional topology**, which studies manifolds of four or fewer dimensions. This includes **knot theory**, the study of mathematical knots.

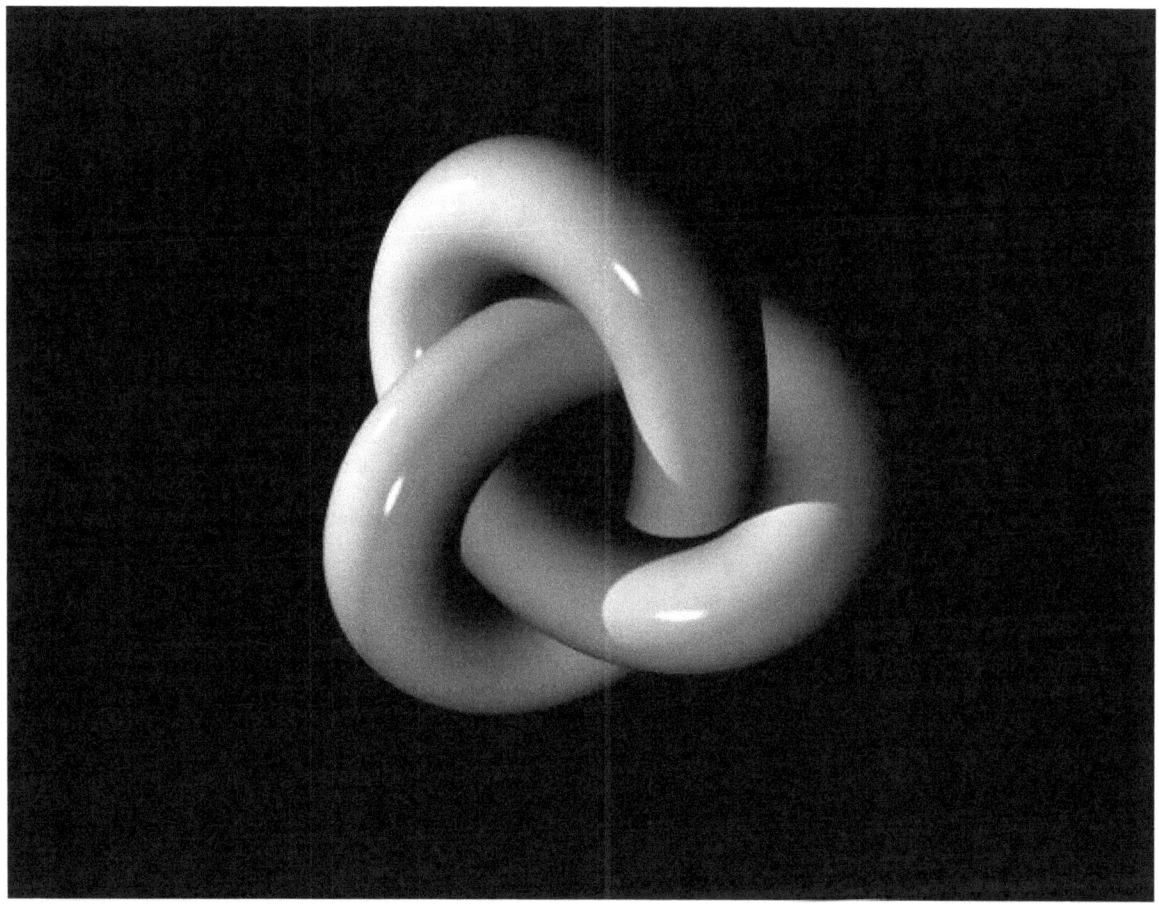

A three-dimensional depiction of a thickened trefoil knot, the simplest non-trivial knot

See also: topology glossary for definitions of some of the terms used in topology, and topological space for a more technical treatment of the subject.

The Seven Bridges of Königsberg was a problem solved by Euler.

27.1   History

Topology began with the investigation of certain questions in geometry. Leonhard Euler's 1736 paper on the Seven Bridges of Königsberg[1] is regarded as one of the first academic treatises in modern topology.

The term "Topologie" was introduced in German in 1847 by Johann Benedict Listing in *Vorstudien zur Topologie*,[2] who had used the word for ten years in correspondence before its first appearance in print. The English form topology was first used in 1883 in Listing's obituary in the journal *Nature*[3] to distinguish "...qualitative geometry from the ordinary geometry in which quantitative relations chiefly are treated." The term **topologist** in the sense of a specialist in topology was used in 1905 in the magazine *Spectator*. However, none of these uses corresponds exactly to the modern definition of topology.

Modern topology depends strongly on the ideas of set theory, developed by Georg Cantor in the later part of the 19th century. In addition to establishing the basic ideas of set theory, Cantor considered point sets in Euclidean space as part of his study of Fourier series.

Henri Poincaré published *Analysis Situs* in 1895,[4] introducing the concepts of homotopy and homology, which are now considered part of algebraic topology.

Unifying the work on function spaces of Georg Cantor, Vito Volterra, Cesare Arzelà, Jacques Hadamard, Giulio Ascoli and others, Maurice Fréchet introduced the metric space in 1906.[5] A metric space is now considered a special case of a general topological space. In 1914, Felix Hausdorff coined the term "topological space" and gave the definition for what is now called a Hausdorff space.[6] Currently, a topological space is a slight generalization of Hausdorff spaces, given in

1922 by Kazimierz Kuratowski.

For further developments, see point-set topology and algebraic topology.

27.2 Introduction

Topology can be formally defined as "the study of qualitative properties of certain objects (called topological spaces) that are invariant under a certain kind of transformation (called a continuous map), especially those properties that are invariant under a certain kind of transformation (called homeomorphism)."

Topology is also used to refer to a structure imposed upon a set X, a structure that essentially 'characterizes' the set X as a topological space by taking proper care of properties such as convergence, connectedness and continuity, upon transformation.

Topological spaces show up naturally in almost every branch of mathematics. This has made topology one of the great unifying ideas of mathematics.

The motivating insight behind topology is that some geometric problems depend not on the exact shape of the objects involved, but rather on the way they are put together. For example, the square and the circle have many properties in common: they are both one dimensional objects (from a topological point of view) and both separate the plane into two parts, the part inside and the part outside.

In one of the first papers in topology, Leonhard Euler demonstrated that it was impossible to find a route through the town of Königsberg (now Kaliningrad) that would cross each of its seven bridges exactly once. This result did not depend on the lengths of the bridges, nor on their distance from one another, but only on connectivity properties: which bridges connect to which islands or riverbanks. This problem in introductory mathematics called *Seven Bridges of Königsberg* led to the branch of mathematics known as graph theory.

Similarly, the hairy ball theorem of algebraic topology says that "one cannot comb the hair flat on a hairy ball without creating a cowlick." This fact is immediately convincing to most people, even though they might not recognize the more formal statement of the theorem, that there is no nonvanishing continuous tangent vector field on the sphere. As with the *Bridges of Königsberg*, the result does not depend on the shape of the sphere; it applies to any kind of smooth blob, as long as it has no holes.

To deal with these problems that do not rely on the exact shape of the objects, one must be clear about just what properties these problems *do* rely on. From this need arises the notion of homeomorphism. The impossibility of crossing each bridge just once applies to any arrangement of bridges homeomorphic to those in Königsberg, and the hairy ball theorem applies to any space homeomorphic to a sphere.

Intuitively, two spaces are homeomorphic if one can be deformed into the other without cutting or gluing. A traditional joke is that a topologist cannot distinguish a coffee mug from a doughnut, since a sufficiently pliable doughnut could be reshaped to a coffee cup by creating a dimple and progressively enlarging it, while shrinking the hole into a handle.

Homeomorphism can be considered the most basic *topological equivalence*. Another is homotopy equivalence. This is harder to describe without getting technical, but the essential notion is that two objects are homotopy equivalent if they both result from "squishing" some larger object.

An introductory exercise is to classify the uppercase letters of the English alphabet according to homeomorphism and homotopy equivalence. The result depends partially on the font used. The figures use the sans-serif Myriad font. Homotopy equivalence is a rougher relationship than homeomorphism; a homotopy equivalence class can contain several homeomorphism classes. The simple case of homotopy equivalence described above can be used here to show two letters are homotopy equivalent. For example, O fits inside P and the tail of the P can be squished to the "hole" part.

Homeomorphism classes are:

- no holes,
- no holes three tails,
- no holes four tails,

A continuous deformation (a type of homeomorphism) of a mug into a doughnut (torus) and back

- one hole no tail,

- one hole one tail,

- one hole two tails,

- two holes no tail, and

- a bar with four tails (the "bar" on the K is almost too short to see).

Homotopy classes are larger, because the tails can be squished down to a point. They are:

- one hole,

- two holes, and

- no holes.

To classify the letters correctly, we must show that two letters in the same class are equivalent and two letters in different classes are not equivalent. In the case of homeomorphism, this can be done by selecting points and showing their removal disconnects the letters differently. For example, X and Y are not homeomorphic because removing the center point of the X leaves four pieces; whatever point in Y corresponds to this point, its removal can leave at most three pieces. The case of homotopy equivalence is harder and requires a more elaborate argument showing an algebraic invariant, such as the fundamental group, is different on the supposedly differing classes.

Letter topology has practical relevance in stencil typography. For instance, Braggadocio font stencils are made of one connected piece of material.

27.3   Concepts

27.3.1   Topologies on Sets

Main article: Topological space

The term **topology** also refers to a specific mathematical idea central to the area of mathematics called topology. Informally, a topology tells how elements of a set relate spatially to each other. The same set can have different topologies. For instance, the real line, the complex plane, and the Cantor set can be thought of as the same set with different topologies.

Formally, let X be a set and let τ be a family of subsets of X. Then τ is called a *topology on X* if:

1. Both the empty set and X are elements of τ

2. Any union of elements of τ is an element of τ

3. Any intersection of finitely many elements of τ is an element of τ

If τ is a topology on X, then the pair (X, τ) is called a *topological space*. The notation $X\tau$ may be used to denote a set X endowed with the particular topology τ.

The members of τ are called *open sets* in X. A subset of X is said to be closed if its complement is in τ (i.e., its complement is open). A subset of X may be open, closed, both (clopen set), or neither. The empty set and X itself are always both closed and open. An open set containing a point x is called a 'neighborhood' of x.

A set with a topology is called a topological space.

27.3.2   Continuous functions and homeomorphisms

Main articles: Continuous function and homeomorphism

A function or map from one topological space to another is called *continuous* if the inverse image of any open set is open. If the function maps the real numbers to the real numbers (both spaces with the Standard Topology), then this definition of continuous is equivalent to the definition of continuous in calculus. If a continuous function is one-to-one and onto, and if the inverse of the function is also continuous, then the function is called a homeomorphism and the domain of the function is said to be homeomorphic to the range. Another way of saying this is that the function has a natural extension to the topology. If two spaces are homeomorphic, they have identical topological properties, and are considered topologically the same. The cube and the sphere are homeomorphic, as are the coffee cup and the doughnut. But the circle is not homeomorphic to the doughnut.

27.3.3 Manifolds

Main article: Manifold

While topological spaces can be extremely varied and exotic, many areas of topology focus on the more familiar class of spaces known as manifolds. A **manifold** is a topological space that resembles Euclidean space near each point. More precisely, each point of an *n*-dimensional manifold has a neighbourhood that is homeomorphic to the Euclidean space of dimension *n*. Lines and circles, but not figure eights, are one-dimensional manifolds. Two-dimensional manifolds are also called surfaces. Examples include the plane, the sphere, and the torus, which can all be realized in three dimensions, but also the Klein bottle and real projective plane, which cannot.

27.4 Topics

27.4.1 General topology

Main article: General topology

General topology is the branch of topology dealing with the basic set-theoretic definitions and constructions used in topology.[7][8] It is the foundation of most other branches of topology, including differential topology, geometric topology, and algebraic topology. Another name for general topology is **point-set topology**.

The fundamental concepts in point-set topology are *continuity*, *compactness*, and *connectedness*. Intuitively, continuous functions take nearby points to nearby points. Compact sets are those that can be covered by finitely many sets of arbitrarily small size. Connected sets are sets that cannot be divided into two pieces that are far apart. The words *nearby*, *arbitrarily small*, and *far apart* can all be made precise by using open sets. If we change the definition of *open set*, we change what continuous functions, compact sets, and connected sets are. Each choice of definition for *open set* is called a *topology*. A set with a topology is called a *topological space*.

Metric spaces are an important class of topological spaces where distances can be assigned a number called a *metric*. Having a metric simplifies many proofs, and many of the most common topological spaces are metric spaces.

27.4.2 Algebraic topology

Main article: Algebraic topology

Algebraic topology is a branch of mathematics that uses tools from abstract algebra to study topological spaces.[9] The basic goal is to find algebraic invariants that classify topological spaces up to homeomorphism, though usually most classify up to homotopy equivalence.

The most important of these invariants are homotopy groups, homology, and cohomology.

Although algebraic topology primarily uses algebra to study topological problems, using topology to solve algebraic problems is sometimes also possible. Algebraic topology, for example, allows for a convenient proof that any subgroup of a free group is again a free group.

27.4.3 Differential topology

Main article: Differential topology

Differential topology is the field dealing with differentiable functions on differentiable manifolds.[10] It is closely related to differential geometry and together they make up the geometric theory of differentiable manifolds.

More specifically, differential topology considers the properties and structures that require only a smooth structure on a manifold to be defined. Smooth manifolds are 'softer' than manifolds with extra geometric structures, which can act as obstructions to certain types of equivalences and deformations that exist in differential topology. For instance, volume and Riemannian curvature are invariants that can distinguish different geometric structures on the same smooth manifold—that is, one can smoothly "flatten out" certain manifolds, but it might require distorting the space and affecting the curvature or volume.

27.4.4 Geometric topology

Main article: Geometric topology

Geometric topology is a branch of topology that primarily focuses on low-dimensional manifolds (i.e. dimensions 2,3 and 4) and their interaction with geometry, but it also includes some higher-dimensional topology.[11] [12] Some examples of topics in geometric topology are orientability, handle decompositions, local flatness, and the planar and higher-dimensional Schönflies theorem.

In high-dimensional topology, characteristic classes are a basic invariant, and surgery theory is a key theory.

Low-dimensional topology is strongly geometric, as reflected in the uniformization theorem in 2 dimensions – every surface admits a constant curvature metric; geometrically, it has one of 3 possible geometries: positive curvature/spherical, zero curvature/flat, negative curvature/hyperbolic – and the geometrization conjecture (now theorem) in 3 dimensions – every 3-manifold can be cut into pieces, each of which has one of eight possible geometries.

2-dimensional topology can be studied as complex geometry in one variable (Riemann surfaces are complex curves) – by the uniformization theorem every conformal class of metrics is equivalent to a unique complex one, and 4-dimensional topology can be studied from the point of view of complex geometry in two variables (complex surfaces), though not every 4-manifold admits a complex structure.

27.4.5 Generalizations

Occasionally, one needs to use the tools of topology but a "set of points" is not available. In pointless topology one considers instead the lattice of open sets as the basic notion of the theory,[13] while Grothendieck topologies are structures defined on arbitrary categories that allow the definition of sheaves on those categories, and with that the definition of general cohomology theories.[14]

27.5   Applications

27.5.1   Biology

Knot theory, a branch of topology, is used in biology to study the effects of certain enzymes on DNA. These enzymes cut, twist, and reconnect the DNA, causing knotting with observable effects such as slower electrophoresis.[15] Topology is also used in evolutionary biology to represent the relationship between phenotype and genotype.[16] Phenotypic forms that appear quite different can be separated by only a few mutations depending on how genetic changes map to phenotypic changes during development.

27.5.2   Computer science

Topological data analysis uses techniques from algebraic topology to determine the large scale structure of a set (for instance, determining if a cloud of points is spherical or toroidal). The main method used by topological data analysis is:

1. Replace a set of data points with a family of simplicial complexes, indexed by a proximity parameter.

2. Analyse these topological complexes via algebraic topology — specifically, via the theory of persistent homology.[17]

3. Encode the persistent homology of a data set in the form of a parameterized version of a Betti number, which is called a **barcode**.[17]

27.5.3 Physics

In physics, topology is used in several areas such as quantum field theory and cosmology.

A **topological quantum field theory** (or **topological field theory** or **TQFT**) is a quantum field theory that computes topological invariants.

Although TQFTs were invented by physicists, they are also of mathematical interest, being related to, among other things, knot theory and the theory of four-manifolds in algebraic topology, and to the theory of moduli spaces in algebraic geometry. Donaldson, Jones, Witten, and Kontsevich have all won Fields Medals for work related to topological field theory.

In cosmology, topology can be used to describe the overall shape of the universe.[18] This area is known as spacetime topology.

27.5.4 Robotics

The various possible positions of a robot can be described by a manifold called configuration space.[19] In the area of motion planning, one finds paths between two points in configuration space. These paths represent a motion of the robot's joints and other parts into the desired location and pose.

27.6 See also

- Equivariant topology
- General topology
- List of algebraic topology topics
- List of examples in general topology
- List of general topology topics
- List of geometric topology topics
- List of topology topics
- Publications in topology
- Topology glossary

27.7 References

[1] Euler, Leonhard, Solutio problematis ad geometriam situs pertinentis

[2] Listing, Johann Benedict, "Vorstudien zur Topologie", Vandenhoeck und Ruprecht, Göttingen, p. 67, 1848

[3] Tait, Peter Guthrie, "Johann Benedict Listing (obituary)", Nature *27*, 1 February 1883, pp. 316–317

[4] Poincaré, Henri, "Analysis situs", Journal de l'École Polytechnique ser 2, 1 (1895) pp. 1–123

[5] Fréchet, Maurice. "Sur quelques points du calcul fonctionnel", PhD dissertation, 1906

[6] Hausdorff, Felix. "Grundzüge der Mengenlehre", Leipzig: Veit. In (Hausdorff Werke, II (2002), 91–576)

[7] Munkres, James R. Topology. Vol. 2. Upper Saddle River: Prentice Hall, 2000.

[8] Adams, Colin Conrad, and Robert David Franzosa. Introduction to topology: pure and applied. Pearson Prentice Hall, 2008.

[9] Allen Hatcher, *Algebraic topology*. (2002) Cambridge University Press, xii+544 pp. ISBN 0-521-79160-X and ISBN 0-521-79540-0.

[10] Lee, John M. (2006). *Introduction to Smooth Manifolds*. Springer-Verlag. ISBN 978-0-387-95448-6.

[11] Budney, Ryan (2011). "What is geometric topology?". *mathoverflow.net*. Retrieved 29 December 2013.

[12] R.B. Sher and R.J. Daverman (2002), *Handbook of Geometric Topology*, North-Holland. ISBN 0-444-82432-4

[13] Johnstone, Peter T., 1983, "The point of pointless topology," *Bulletin of the American Mathematical Society 8(1)*: 41-53.

[14] Artin, Michael (1962). *Grothendieck topologies*. Cambridge, MA: Harvard University, Dept. of Mathematics. Zbl 0208.48701.

[15] Adams, Colin (2004). *The Knot Book: An Elementary Introduction to the Mathematical Theory of Knots*. American Mathematical Society. ISBN 0-8218-3678-1

[16] Barble M R Stadler et al. "The Topology of the Possible: Formal Spaces Underlying Patterns of Evolutionary Change". *Journal of Theoretical Biology* **213**: 241–274. doi:10.1006/jtbi.2001.2423.

[17] Gunnar Carlsson (April 2009). "Topology and data" (PDF). *BULLETIN (New Series) OF THE AMERICAN MATHEMATICAL SOCIETY* **46** (2): 255–308. doi:10.1090/S0273-0979-09-01249-X.

[18] *The Shape of Space: How to Visualize Surfaces and Three-dimensional Manifolds* 2nd ed (Marcel Dekker, 1985, ISBN 0-8247-7437-X)

[19] John J. Craig, **Introduction to Robotics: Mechanics and Control**, 3rd Ed. Prentice-Hall, 2004

27.8   Further reading

- Ryszard Engelking, *General Topology*, Heldermann Verlag, Sigma Series in Pure Mathematics, December 1989, ISBN 3-88538-006-4.

- Bourbaki; *Elements of Mathematics: General Topology*, Addison–Wesley (1966).

- Breitenberger, E. (2006). "Johann Benedict Listing". In James, I. M. *History of Topology*. North Holland. ISBN 978-0-444-82375-5.

- Kelley, John L. (1975). *General Topology*. Springer-Verlag. ISBN 0-387-90125-6.

- Brown, Ronald (2006). *Topology and Groupoids*. Booksurge. ISBN 1-4196-2722-8. (Provides a well motivated, geometric account of general topology, and shows the use of groupoids in discussing van Kampen's theorem, covering spaces, and orbit spaces.)

- Wacław Sierpiński, *General Topology*, Dover Publications, 2000, ISBN 0-486-41148-6

- Pickover, Clifford A. (2006). *The Möbius Strip: Dr. August Möbius's Marvelous Band in Mathematics, Games, Literature, Art, Technology, and Cosmology*. Thunder's Mouth Press. ISBN 1-56025-826-8. (Provides a popular introduction to topology and geometry)

- Gemignani, Michael C. (1990) [1967], *Elementary Topology* (2nd ed.), Dover Publications Inc., ISBN 0-486-66522-4

27.9 External links

- Hazewinkel, Michiel, ed. (2001), "Topology, general", *Encyclopedia of Mathematics*, Springer, ISBN 978-1-55608-010-4

- Elementary Topology: A First Course Viro, Ivanov, Netsvetaev, Kharlamov.

- Topology at DMOZ

- The Topological Zoo at The Geometry Center.

- Topology Atlas

- Topology Course Lecture Notes Aisling McCluskey and Brian McMaster, Topology Atlas.

- Topology Glossary

- Moscow 1935: Topology moving towards America, a historical essay by Hassler Whitney.

Chapter 28

Fractal

For other uses, see Fractal (disambiguation).

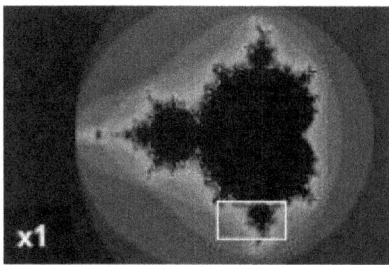

Mandelbrot set: Self-similarity illustrated by image enlargements. This panel, no magnification.

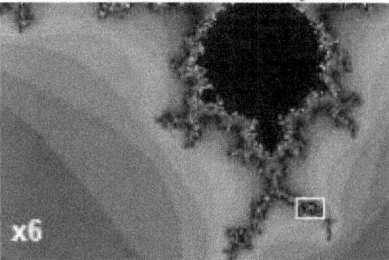

The same fractal as above, magnified 6-fold. Same patterns reappear, making the exact scale being examined difficult to determine.

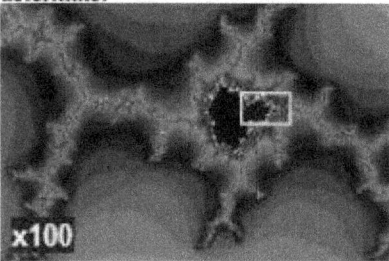

The same fractal as above, magnified a 100-fold.

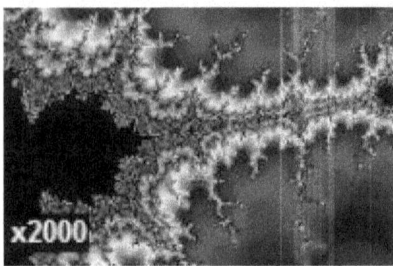

The same fractal as above, magnified a 2000-fold, where the Mandelbrot set fine detail resembles the detail at low magnification.

A **fractal** is a natural phenomenon or a mathematical set that exhibits a repeating pattern that displays at every scale. If the replication is exactly the same at every scale, it is called a self-similar pattern. An example of this is the Menger Sponge.[1] Fractals can also be nearly the same at different levels. This latter pattern is illustrated in the magnifications of the Mandelbrot set.[2][3][4][5] Fractals also include the idea of a detailed pattern that repeats itself.[2]:166; 18[3][6]

Fractals are different from other geometric figures because of the way in which they scale. Doubling the edge lengths of a polygon multiplies its area by four, which is two (the ratio of the new to the old side length) raised to the power of two (the dimension of the space the polygon resides in). Likewise, if the radius of a sphere is doubled, its volume scales by eight, which is two (the ratio of the new to the old radius) to the power of three (the dimension that the sphere resides in). But if a fractal's one-dimensional lengths are all doubled, the spatial content of the fractal scales by a power that is not necessarily an integer.[2] This power is called the fractal dimension of the fractal, and it usually exceeds the fractal's topological dimension.[7]

As mathematical equations, fractals are usually nowhere differentiable.[2][5][8] An infinite fractal curve can be conceived of as winding through space differently from an ordinary line, still being a 1-dimensional line yet having a fractal dimension indicating it also resembles a surface.[2]:15[7]:48

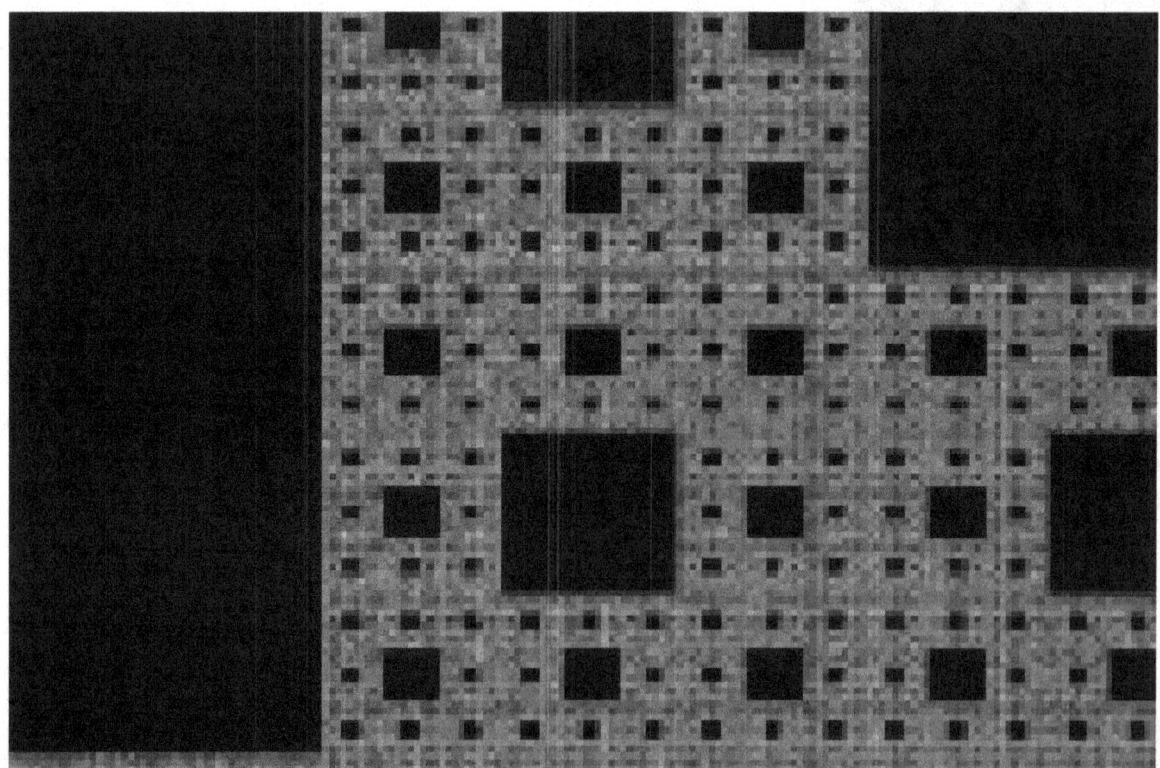

Animation of a Sierpinski carpet, a famous two-dimensional fractal.

The mathematical roots of the idea of fractals have been traced throughout the years as a formal path of published works, starting in the 17th century with notions of recursion, then moving through increasingly rigorous mathematical treatment of the concept to the study of continuous but not differentiable functions in the 19th century, and on to the coining of the word *fractal* in the 20th century with a subsequent burgeoning of interest in fractals and computer-based modelling in the 21st century.[9][10] The term "fractal" was first used by mathematician Benoît Mandelbrot in 1975. Mandelbrot based it on the Latin *frāctus* meaning "broken" or "fractured", and used it to extend the concept of theoretical fractional dimensions to geometric patterns in nature.[2]:405[6]

There is some disagreement amongst authorities about how the concept of a fractal should be formally defined. Mandelbrot himself summarized it as "beautiful, damn hard, increasingly useful. That's fractals."[11] The general consensus is that theoretical fractals are infinitely self-similar, iterated, and detailed mathematical constructs having fractal dimensions, of which many examples have been formulated and studied in great depth.[2][3][4] Fractals are not limited to geometric patterns, but can also describe processes in time.[1][5][12] Fractal patterns with various degrees of self-similarity have been rendered or studied in images, structures and sounds[13] and found in nature,[14][15][16][17][18] technology,[19][20][21][22] art,[23][24][25] and law.[26]

28.1   Introduction

The word "fractal" often has different connotations for laypeople than for mathematicians, where the layperson is more likely to be familiar with fractal art than a mathematical conception. The mathematical concept is difficult to define formally even for mathematicians, but key features can be understood with little mathematical background.

The feature of "self-similarity", for instance, is easily understood by analogy to zooming in with a lens or other device that zooms in on digital images to uncover finer, previously invisible, new structure. If this is done on fractals, however, no new detail appears; nothing changes and the same pattern repeats over and over, or for some fractals, nearly the same pattern reappears over and over. Self-similarity itself is not necessarily counter-intuitive (e.g., people have pondered self-similarity informally such as in the infinite regress in parallel mirrors or the homunculus, the little man inside the head of the little man inside the head...). The difference for fractals is that the pattern reproduced must be detailed.[2]:166; 18[3][6]

This idea of being detailed relates to another feature that can be understood without mathematical background: Having a fractional or fractal dimension greater than its topological dimension, for instance, refers to how a fractal scales compared to how geometric shapes are usually perceived. A regular line, for instance, is conventionally understood to be 1-dimensional; if such a curve is divided into pieces each 1/3 the length of the original, there are always 3 equal pieces. In contrast, consider the curve in Figure 2. It is also 1-dimensional for the same reason as the ordinary line, but it has, in addition, a fractal dimension greater than 1 because of how its detail can be measured. The fractal curve divided into parts 1/3 the length of the original line becomes 4 pieces rearranged to repeat the original detail, and this unusual relationship is the basis of its fractal dimension.

This also leads to understanding a third feature, that fractals as mathematical equations are "nowhere differentiable". In a concrete sense, this means fractals cannot be measured in traditional ways.[2][5][8] To elaborate, in trying to find the length of a wavy non-fractal curve, one could find straight segments of some measuring tool small enough to lay end to end over the waves, where the pieces could get small enough to be considered to conform to the curve in the normal manner of measuring with a tape measure. But in measuring a wavy fractal curve such as the one in Figure 2, one would never find a small enough straight segment to conform to the curve, because the wavy pattern would always re-appear, albeit at a smaller size, essentially pulling a little more of the tape measure into the total length measured each time one attempted to fit it tighter and tighter to the curve. This is perhaps counter-intuitive, but it is how fractals behave.[2]

28.2   History

The history of fractals traces a path from chiefly theoretical studies to modern applications in computer graphics, with several notable people contributing canonical fractal forms along the way.[9][10] According to Pickover, the mathematics behind fractals began to take shape in the 17th century when the mathematician and philosopher Gottfried Leibniz pondered recursive self-similarity (although he made the mistake of thinking that only the straight line was self-similar in this

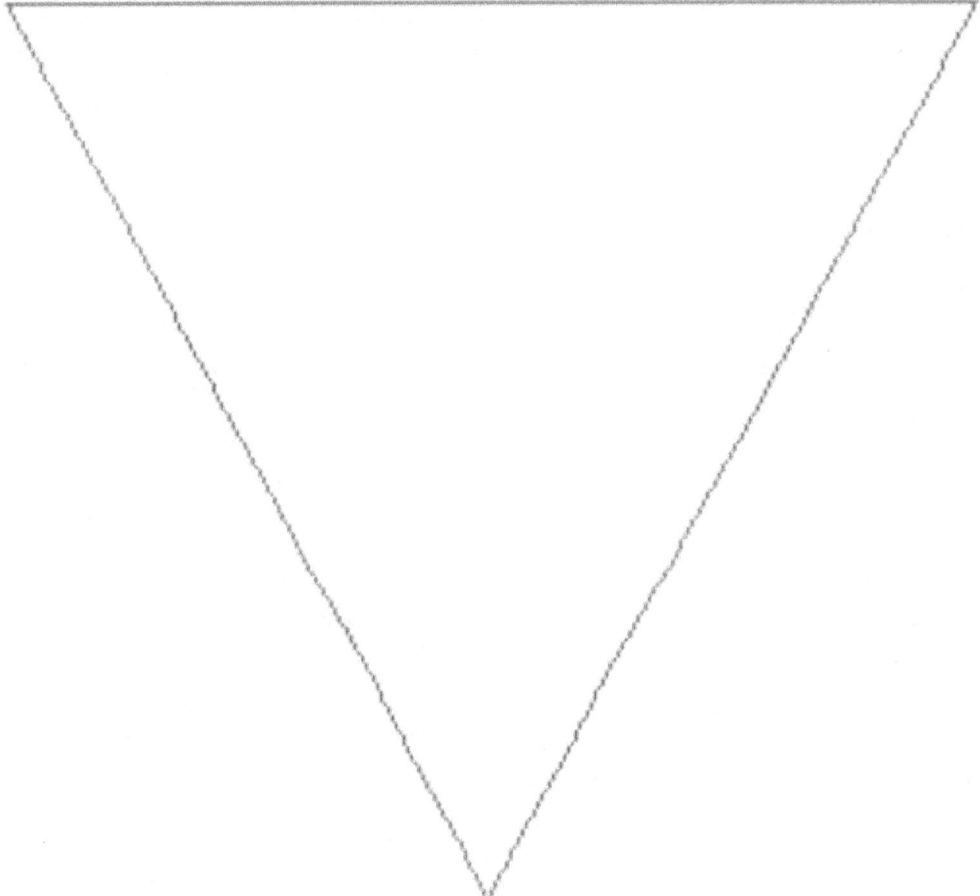

Figure 2a. Koch snowflake, a fractal that begins with an equilateral triangle and then replaces the middle third of every line segment with a pair of line segments that form an equilateral "bump"

sense).[27] In his writings, Leibniz used the term "fractional exponents", but lamented that "Geometry" did not yet know of them.[2]:405 Indeed, according to various historical accounts, after that point few mathematicians tackled the issues and the work of those who did remained obscured largely because of resistance to such unfamiliar emerging concepts, which were sometimes referred to as mathematical "monsters".[8][9][10] Thus, it was not until two centuries had passed that in 1872 Karl Weierstrass presented the first definition of a function with a graph that would today be considered fractal, having the non-intuitive property of being everywhere continuous but nowhere differentiable.[9]:7[10] Not long after that, in 1883, Georg Cantor, who attended lectures by Weierstrass,[10] published examples of subsets of the real line known as Cantor sets, which had unusual properties and are now recognized as fractals.[9]:11–24 Also in the last part of that century, Felix Klein and Henri Poincaré introduced a category of fractal that has come to be called "self-inverse" fractals.[2]:166

One of the next milestones came in 1904, when Helge von Koch, extending ideas of Poincaré and dissatisfied with

Figure 2b. Koch snowflake, a zoom out of the Koch Snowflake

Weierstrass's abstract and analytic definition, gave a more geometric definition including hand drawn images of a similar function, which is now called the Koch curve (see Figure 2).[9]:25[10] Another milestone came a decade later in 1915, when Wacław Sierpiński constructed his famous triangle then, one year later, his carpet. By 1918, two French mathematicians, Pierre Fatou and Gaston Julia, though working independently, arrived essentially simultaneously at results describing what are now seen as fractal behaviour associated with mapping complex numbers and iterative functions and leading to further ideas about attractors and repellors (i.e., points that attract or repel other points), which have become very important in the study of fractals (see Figure 3 and Figure 4).[5][9][10] Very shortly after that work was submitted, by March 1918, Felix Hausdorff expanded the definition of "dimension", significantly for the evolution of the definition of fractals, to allow for sets to have noninteger dimensions.[10] The idea of self-similar curves was taken further by Paul Lévy, who, in his 1938 paper *Plane or Space Curves and Surfaces Consisting of Parts Similar to the Whole* described a new fractal curve, the Lévy C curve.[notes 1]

Different researchers have postulated that without the aid of modern computer graphics, early investigators were limited to what they could depict in manual drawings, so lacked the means to visualize the beauty and appreciate some of the implications of many of the patterns they had discovered (the Julia set, for instance, could only be visualized through a few iterations as very simple drawings hardly resembling the image in Figure 3).[2]:179[8][10] That changed, however, in the 1960s, when Benoît Mandelbrot started writing about self-similarity in papers such as *How Long Is the Coast of Britain? Statistical Self-Similarity and Fractional Dimension*,[28] which built on earlier work by Lewis Fry Richardson. In 1975[6] Mandelbrot solidified hundreds of years of thought and mathematical development in coining the word "fractal" and illustrated his mathematical definition with striking computer-constructed visualizations. These images, such as of his canonical Mandelbrot set pictured in Figure 1, captured the popular imagination; many of them were based on recursion, leading to the popular meaning of the term "fractal".[29] Currently, fractal studies are essentially exclusively computer-based.[8][9][27]

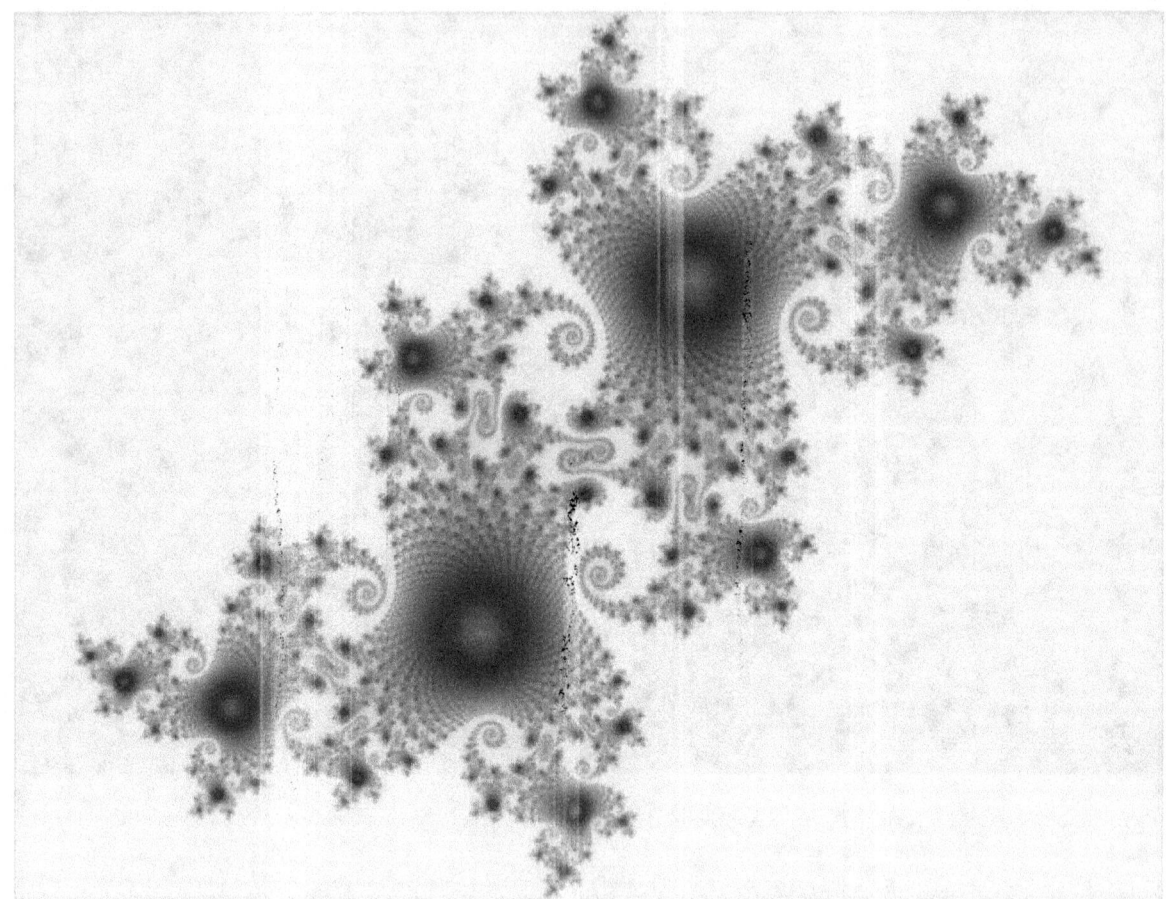

Figure 3. A Julia set, a fractal related to the Mandelbrot set

28.3 Characteristics

One often cited description that Mandelbrot published to describe geometric fractals is "a rough or fragmented geometric shape that can be split into parts, each of which is (at least approximately) a reduced-size copy of the whole";[2] this is generally helpful but limited. Authors disagree on the exact definition of *fractal*, but most usually elaborate on the basic ideas of self-similarity and an unusual relationship with the space a fractal is embedded in.[1][2][3][5][30] One point agreed on is that fractal patterns are characterized by fractal dimensions, but whereas these numbers quantify complexity (i.e., changing detail with changing scale), they neither uniquely describe nor specify details of how to construct particular fractal patterns.[31] In 1975 when Mandelbrot coined the word "fractal", he did so to denote an object whose Hausdorff–Besicovitch dimension is greater than its topological dimension.[6] It has been noted that this dimensional requirement is not met by fractal space-filling curves such as the Hilbert curve.[notes 2]

According to Falconer, rather than being strictly defined, fractals should, in addition to being nowhere differentiable and able to have a fractal dimension, be generally characterized by a gestalt of the following features:[3]

- Self-similarity, which may be manifested as:
  - Exact self-similarity: *identical at all scales; e.g. Koch snowflake*
  - Quasi self-similarity: *approximates the same pattern at different scales; may contain small copies of the entire fractal in distorted and degenerate forms; e.g., the Mandelbrot set's satellites are approximations of the entire set, but not exact copies, as shown in Figure 1*
  - Statistical self-similarity: *repeats a pattern stochastically so numerical or statistical measures are preserved across scales; e.g., randomly generated fractals; the well-known example of the*

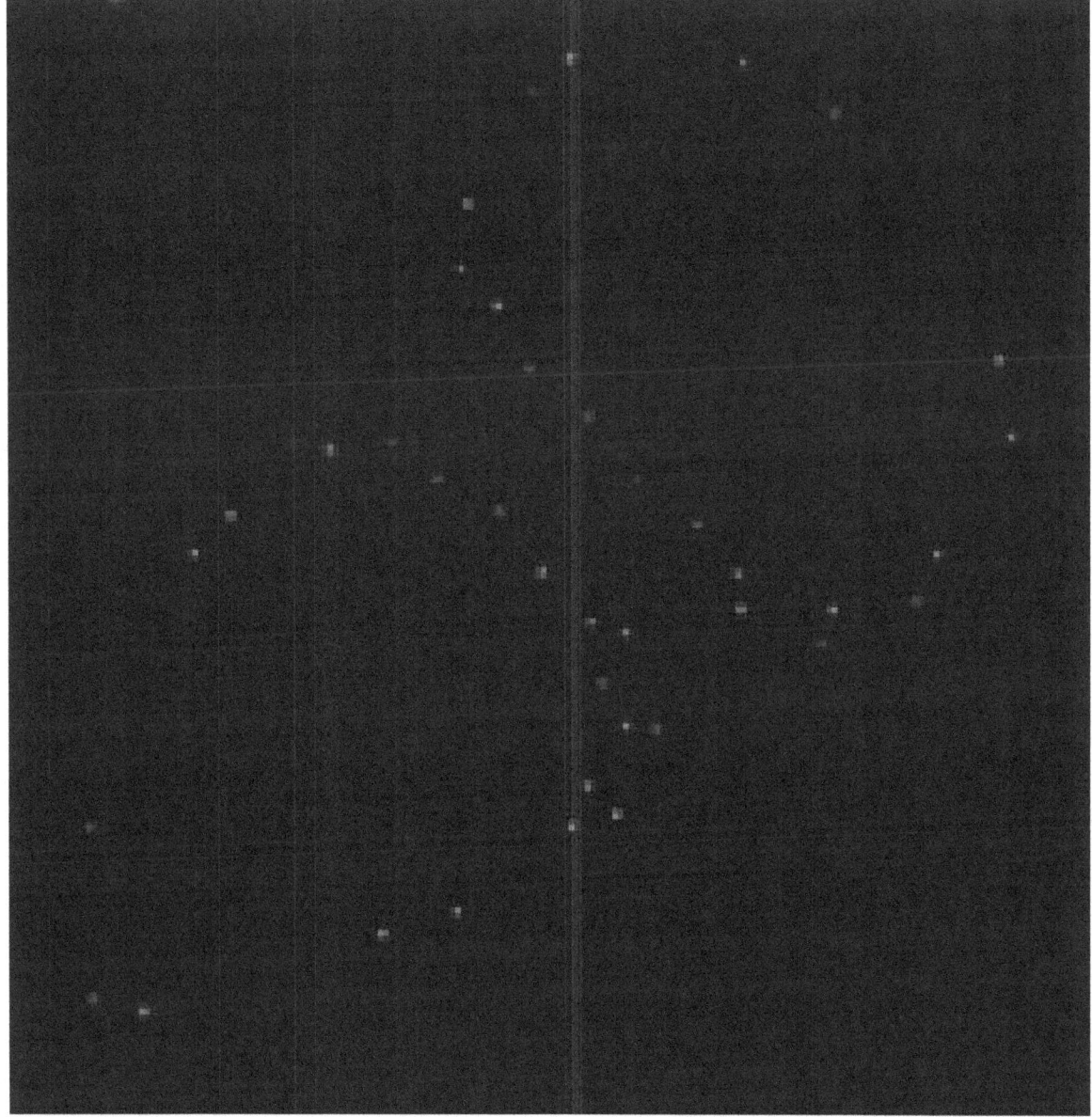

Figure 4. A strange attractor that exhibits multifractal scaling

coastline of Britain, for which one would not expect to find a segment scaled and repeated as neatly as the repeated unit that defines, for example, the Koch snowflake[5]

- Qualitative self-similarity: *as in a time series*[12]
- Multifractal scaling: *characterized by more than one fractal dimension or scaling rule*

- Fine or detailed structure at arbitrarily small scales. A consequence of this structure is fractals may have emergent properties[32] (related to the next criterion in this list).

- Irregularity locally and globally that is not easily described in traditional Euclidean geometric language. For images of fractal patterns, this has been expressed by phrases such as "smoothly piling up surfaces" and "swirls upon swirls".[7]

- Simple and "perhaps recursive" definitions *see Common techniques for generating fractals*

As a group, these criteria form guidelines for excluding certain cases, such as those that may be self-similar without having

Uniform Mass Center Triangle Fractal

other typically fractal features. A straight line, for instance, is self-similar but not fractal because it lacks detail, is easily described in Euclidean language, has the same Hausdorff dimension as topological dimension, and is fully defined without a need for recursion.[2][5]

28.4 Brownian motion

A path generated by a one dimensional Wiener process is a fractal curve of dimension 1.5, and Brownian motion is a finite version of this.[33]

28.5   Common techniques for generating fractals

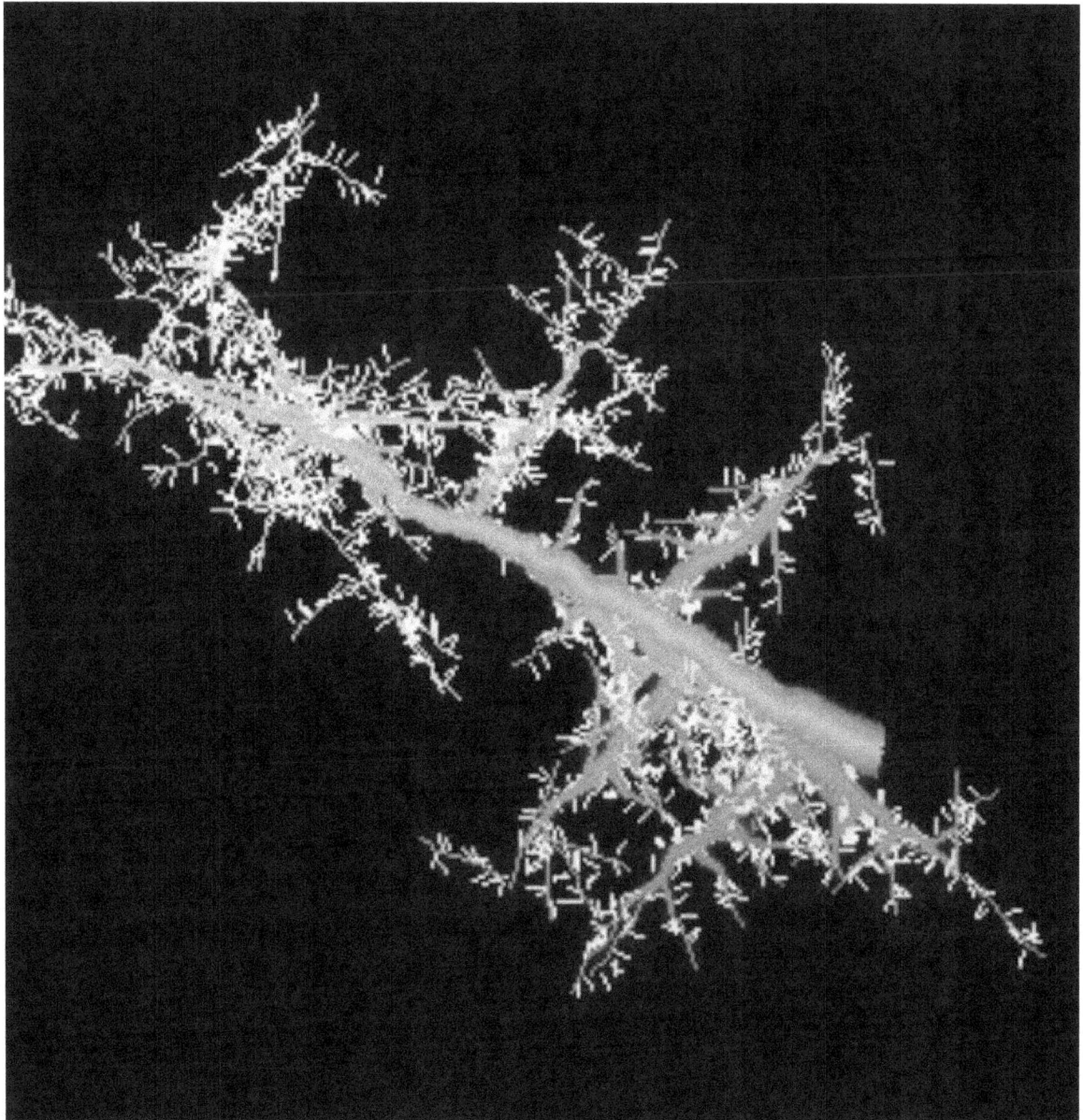

Figure 5. Self-similar branching pattern modeled in silico using L-systems principles[18]

Images of fractals can be created by fractal generating programs.

- *Iterated function systems* – use fixed geometric replacement rules; may be stochastic or deterministic;[34] e.g., Koch snowflake, Cantor set, Haferman carpet,[35] Sierpinski carpet, Sierpinski gasket, Peano curve, Harter-Heighway dragon curve, T-Square, Menger sponge

- *Strange attractors* – use iterations of a map or solutions of a system of initial-value differential equations that exhibit chaos (e.g., see multifractal image)

- *L-systems* – use string rewriting; may resemble branching patterns, such as in plants, biological cells (e.g., neurons and immune system cells[18]), blood vessels, pulmonary structure,[36] etc. (e.g., see Figure 5) or turtle graphics patterns such as space-filling curves and tilings

- *Escape-time fractals* – use a formula or recurrence relation at each point in a space (such as the complex plane); usually quasi-self-similar; also known as "orbit" fractals; e.g., the Mandelbrot set, Julia set, Burning Ship fractal, Nova fractal and Lyapunov fractal. The 2d vector fields that are generated by one or two iterations of escape-time formulae also give rise to a fractal form when points (or pixel data) are passed through this field repeatedly.

- *Random fractals* – use stochastic rules; e.g., Lévy flight, percolation clusters, self avoiding walks, fractal landscapes, trajectories of Brownian motion and the Brownian tree (i.e., dendritic fractals generated by modeling diffusion-limited aggregation or reaction-limited aggregation clusters).[5]

A fractal generated by a finite subdivision rule for an alternating link

- *Finite subdivision rules* use a recursive topological algorithm for refining tilings[37] and they are similar to the process of cell division.[38] The iterative processes used in creating the Cantor set and the Sierpinski carpet are examples of finite subdivision rules, as is barycentric subdivision.

28.6   Simulated fractals

A fractal flame

Fractal patterns have been modeled extensively, albeit within a range of scales rather than infinitely, owing to the practical limits of physical time and space. Models may simulate theoretical fractals or natural phenomena with fractal features. The outputs of the modelling process may be highly artistic renderings, outputs for investigation, or benchmarks for fractal analysis. Some specific applications of fractals to technology are listed elsewhere. Images and other outputs of modelling are normally referred to as being "fractals" even if they do not have strictly fractal characteristics, such as when it is possible to zoom into a region of the fractal image that does not exhibit any fractal properties. Also, these may include calculation or display artifacts which are not characteristics of true fractals.

Modeled fractals may be sounds,[113] digital images, electrochemical patterns, circadian rhythms,[139] etc. Fractal patterns have been reconstructed in physical 3-dimensional space[21]:10 and virtually, often called "in silico" modeling.[136] Models of fractals are generally created using fractal-generating software that implements techniques such as those outlined above.[5][12][21] As one illustration, trees, ferns, cells of the nervous system,[18] blood and lung vasculature,[36] and other branching patterns in nature can be modeled on a computer by using recursive algorithms and L-systems techniques.[18] The recursive nature of some patterns is obvious in certain examples—a branch from a tree or a frond from a fern is a miniature replica of the whole: not identical, but similar in nature. Similarly, random fractals have been used to describe/create many highly irregular real-world objects. A limitation of modeling fractals is that resemblance of a fractal model to a natural phenomenon does not prove that the phenomenon being modeled is formed by a process similar to the modeling algorithms.

28.7 Natural phenomena with fractal features

Further information: Patterns in nature

Approximate fractals found in nature display self-similarity over extended, but finite, scale ranges. The connection between fractals and leaves, for instance, is currently being used to determine how much carbon is contained in trees.[40] Phenomena known to have fractal features include:

28.8 In creative works

Further information: Fractal art

The paintings of American artist Jackson Pollock have a definite fractal dimension. While Pollock's paintings appear to be composed of chaotic dripping and splattering, computer analysis demonstrates a degree of self-similarity at different scales (levels of detail) in his work.[25]

Decalcomania, a technique used by artists such as Max Ernst, can produce fractal-like patterns.[56] It involves pressing paint between two surfaces and pulling them apart.

Cyberneticist Ron Eglash has suggested that fractal geometry and mathematics are prevalent in African art, games, divination, trade, and architecture. Circular houses appear in circles of circles, rectangular houses in rectangles of rectangles, and so on. Such scaling patterns can also be found in African textiles, sculpture, and even cornrow hairstyles.[24][57]

In a 1996 interview with Michael Silverblatt, David Foster Wallace admitted that the structure of the first draft of *Infinite Jest* he gave to his editor Michael Pietsch was inspired by fractals, specifically the Sierpinski triangle (a.k.a. Sierpinski gasket) but that the edited novel is "more like a lopsided Sierpinsky Gasket".[23]

28.9 Applications in technology

Main article: Fractal analysis

28.10 See also

28.10.1 Fractal-generating programs

There are many fractal generating programs available, both free and commercial. Some of the fractal generating programs include:

- Apophysis – open source software for Microsoft Windows based systems

- Electric Sheep – open source distributed computing software

- Fractint – freeware with available source code

- Sterling – Freeware software for Microsoft Windows based systems

- SpangFract – For Mac OS

- Ultra Fractal – A proprietary fractal generator for Microsoft Windows and Mac OS X based systems

- XaoS – A cross platform open source fractal zooming program

- Chaotica – a commercial software for Microsoft Windows, Linux and Mac OS

- Terragen – a fractal terrain generator.

- XENODREAM 3-D – A proprietary fractal generator for Microsoft Windows

Most of the above programs make two-dimensional fractals, with a few creating three-dimensional fractal objects, such as quaternions, mandelbulbs and mandelboxes.

28.11   Notes

[1] The original paper, Lévy, Paul (1938). "Les Courbes planes ou gauches et les surfaces composées de parties semblables au tout". *Journal de l'École Polytechnique*: 227–247, 249–291., is translated in Edgar, pages 181–239.

[2] The Hilbert curve map is not a homeomorphism, so it does not preserve topological dimension. The topological dimension and Hausdorff dimension of the image of the Hilbert map in $\mathbf{R}^2$ are both 2. Note, however, that the topological dimension of the *graph* of the Hilbert map (a set in $\mathbf{R}^3$) is 1.

28.12   References

[1] Gouyet, Jean-François (1996). *Physics and fractal structures*. Paris/New York: Masson Springer. ISBN 978-0-387-94153-0.

[2] Mandelbrot, Benoît B. (1983). *The fractal geometry of nature*. Macmillan. ISBN 978-0-7167-1186-5. Retrieved 1 February 2012.

[3] Falconer, Kenneth (2003). *Fractal Geometry: Mathematical Foundations and Applications*. John Wiley & Sons, Ltd. xxv. ISBN 0-470-84862-6.

[4] Briggs, John (1992). *Fractals:The Patterns of Chaos*. London, UK: Thames and Hudson. p. 148. ISBN 0-500-27693-5.

[5] Vicsek, Tamás (1992). *Fractal growth phenomena*. Singapore/New Jersey: World Scientific. pp. 31; 139–146. ISBN 978-981-02-0668-0.

[6] Albers, Donald J.; Alexanderson, Gerald L. (2008). "Benoît Mandelbrot: In his own words". *Mathematical people : profiles and interviews*. Wellesley, MA: AK Peters. p. 214. ISBN 978-1-56881-340-0.

[7] Mandelbrot, Benoît B. (2004). *Fractals and Professor Chaos*. Berlin: Springer. p. 38. ISBN 978-0-387-20158-0. A fractal set is one for which the fractal (Hausdorff-Besicovitch) dimension strictly exceeds the topological dimension

[8] Gordon, Nigel (2000). *Introducing fractal geometry*. Duxford, UK: Icon. p. 71. ISBN 978-1-84046-123-7.

[9] Edgar, Gerald (2004). *Classics on Fractals*. Boulder, CO: Westview Press. ISBN 978-0-8133-4153-8.

[10] Trochet, Holly (2009). "A History of Fractal Geometry". *MacTutor History of Mathematics*. Archived from the original on 4 February 2012.

[11] Mandelbrot, Benoit. "24/7 Lecture on Fractals". *2006 Ig Nobel Awards*. Improbable Research.

[12] Peters, Edgar (1996). *Chaos and order in the capital markets : a new view of cycles, prices, and market volatility*. New York: Wiley. ISBN 0-471-13938-6.

[13] Brothers, Harlan J. (2007). "Structural Scaling in Bach's Cello Suite No. 3". *Fractals* **15**: 89–95. doi:10.1142/S0218348X0700337X.

[14] Tan, Can Ozan; Cohen, Michael A.; Eckberg, Dwain L.; Taylor, J. Andrew (2009). "Fractal properties of human heart period variability: Physiological and methodological implications". *The Journal of Physiology* **587** (15): 3929. doi:10.1113/jphysiol.2009.

[15] Buldyrev, Sergey V.; Goldberger, Ary L.; Havlin, Shlomo; Peng, Chung-Kang; Stanley, H. Eugene (1995). Bunde, Armin; Havlin, Shlomo, eds. "Fractals in Science". Springer. lchapter= ignored (help)

[16]Liu, Jing Z.; Zhang, Lu D.; Yue, Guang H. (2003)."Fractal Dimension in Human Cerebellum Measured by Magnetic Resonae Imaging". *Biophysical Journal* **85** (6): 4041–4046. doi:10.1016/S0006-3495(03)74817-6. PMC 1303704. PMID 14645092.

[17] Karperien, Audrey L.; Jelinek, Herbert F.; Buchan, Alastair M. (2008). "Box-Counting Analysis of Microglia Form in Schizophrenia, Alzheimer's Disease and Affective Disorder". *Fractals* **16** (2): 103. doi:10.1142/S0218348X08003880.

[18] Jelinek, Herbert F.; Karperien, Audrey; Cornforth, David; Cesar, Roberto; Leandro, Jorge de Jesus Gomes (2002). "MicroModan L-systems approach to neural modelling". In Sarker, Ruhul. *Workshop proceedings: the Sixth Australia-Japan Joint Workshop on Intelligent and Evolutionary Systems, University House, ANU,*. University of New South Wales. ISBN 9780731705054. OCLC 224846454.http://researchoutput.csu.edu.au/R/-?func=dbin-jump-full&object_id=6595&local_base=GEN01-C
1.Retrieved 3 February 2012. Event location: Canberra, Australia

[19] Hu, Shougeng; Cheng, Qiuming; Wang, Le; Xie, Shuyun (2012). "Multifractal characterization of urban residential land price in space and time". *Applied Geography* **34**: 161. doi:10.1016/j.apgeog.2011.10.016.

[20] Karperien, Audrey; Jelinek, Herbert F.; Leandro, Jorge de Jesus Gomes; Soares, João V. B.; Cesar Jr, Roberto M.; Luckie, Alan (2008). "Automated detection of proliferative retinopathy in clinical practice". *Clinical ophthalmology (Auckland, N.Z.)* **2** (1): 109–122. doi:10.2147/OPTH.S1579. PMC 2698675. PMID 19668394.

[21] Losa, Gabriele A.; Nonnenmacher, Theo F. (2005). *Fractals in biology and medicine*. Springer. ISBN 978-3-7643-7172-2. Retrieved 1 February 2012.

[22] Vannucchi, Paola; Leoni, Lorenzo (2007). "Structural characterization of the Costa Rica décollement: Evidence for seismically-induced fluid pulsing". *Earth and Planetary Science Letters* **262** (3–4): 413. Bibcode:2007E&PSL.262..413V. doi:10.07.056.

[23] Wallace, David Foster. "Bookworm on KCRW". Kcrw.com. Retrieved 2010-10-17.

[24] Eglash, Ron (1999). "African Fractals: Modern Computing and Indigenous Design". New Brunswick: Rutgers University Press. Retrieved 2010-10-17.

[25] Taylor, Richard; Micolich, Adam P.; Jonas, David. "Fractal Expressionism: Can Science Be Used To Further Our Understanding Of Art?". Phys.unsw.edu.au. Retrieved 2010-10-17.

[26] Stumpff, Andrew (2013). "The Law is a Fractal: The Attempt to Anticipate Everything" **44**. Loyola University Chicago Law Journal. p. 649.

[27] Pickover, Clifford A. (2009). *The Math Book: From Pythagoras to the 57th Dimension, 250 Milestones in the History of Mathematics*. Sterling Publishing Company, Inc. p. 310. ISBN 978-1-4027-5796-9. Retrieved 2011-02-05.

[28] Batty, Michael (1985-04-04). "Fractals – Geometry Between Dimensions". *New Scientist* (Holborn Publishing Group) **105** (1450): 31.

[29] Russ, John C. (1994). *Fractal surfaces* **1**. Springer. p. 1. ISBN 978-0-306-44702-0. Retrieved 2011-02-05.

[30] Edgar, Gerald (2008). *Measure, topology, and fractal geometry*. New York, NY: Springer-Verlag. p. 1. ISBN 978-0-387-74748-4.

[31] Karperien, Audrey (2004). http://www.webcitation.org/65DyLbmF1 *Defining microglial morphology: Form, Function, and Fractal Dimension*. Charles Sturt University. Retrieved 2012-02-05.

[32] Spencer, John; Thomas, Michael S. C.; McClelland, James L. (2009). *Toward a unified theory of development : connectionism and dynamic systems theory re-considered*. Oxford/New York: Oxford University Press. ISBN 978-0-19-530059-8.

[33] Falconer, Kenneth (2013). *Fractals, A Very Short Introduction*. Oxford University Press.

[34] Frame, Angus (3 August 1998). "Iterated Function Systems". In Pickover, Clifford A. *Chaos and fractals: a computer graphical journey : ten year compilation of advanced research*. Elsevier. pp. 349–351. ISBN 978-0-444-50002-1. Retrieved 4 February 2012.

[35] "Haferman Carpet". WolframAlpha. Retrieved 18 October 2012.

[36] Hahn, Horst K.; Georg, Manfred; Peitgen, Heinz-Otto (2005). "Fractal aspects of three-dimensional vascular constructive optimization". In Losa, Gabriele A.; Nonnenmacher, Theo F. *Fractals in biology and medicine*. Springer. pp. 55–66. ISBN 978-3-7643-7172-2.

[37] J. W. Cannon, W. J. Floyd, W. R. Parry. *Finite subdivision rules.* Conformal Geometry and Dynamics, vol. 5 (2001), pp. 153–196.

[38] J. W. Cannon, W. Floyd and W. Parry. *Crystal growth, biological cell growth and geometry.* Pattern Formation in Biology, Vision and Dynamics, pp. 65–82. World Scientific, 2000. ISBN 981-02-3792-8,ISBN 978-981-02-3792-9.

[39] Fathallah-Shaykh, Hassan M. (2011). "Fractal Dimension of the Drosophila Circadian Clock". *Fractals* **19** (4): 423–430. doi:10.1142/S0218348X11005476.

[40] "Hunting the Hidden Dimension." *Nova.* PBS. WPMB-Maryland. 28 October 2008.

[41] Sweet, D.; Ott, E.; Yorke, J. A. (1999), "Complex topology in Chaotic scattering: A Laboratory Observation", *Nature* **399** (6734): 315, doi:10.1038/20573

[42] Sornette, Didier (2004). *Critical phenomena in natural sciences: chaos, fractals, selforganization, and disorder : concepts and tools.* Springer. pp. 128–140. ISBN 978-3-540-40754-6.

[43] Meyer, Yves; Roques, Sylvie (1993). *Progress in wavelet analysis and applications: proceedings of the International Conference "Wavelets and Applications," Toulouse, France – June 1992.* Atlantica Séguier Frontières. p. 25. ISBN 978-2-86332-130-0. Retrieved 2011-02-05.

[44] Pincus, David (September 2009). "The Chaotic Life: Fractal Brains Fractal Thoughts". *psychologytoday.com.*

[45] Carbone, Alessandra; Gromov, Mikhael; Prusinkiewicz, Przemyslaw (2000). *Pattern formation in biology, vision and dynamics.* World Scientific. p. 78. ISBN 978-981-02-3792-9.

[46] Addison, Paul S. (1997). *Fractals and chaos: an illustrated course.* CRC Press. pp. 44–46. ISBN 978-0-7503-0400-9. Retrieved 2011-02-05.

[47] Ozhovan M.I., Dmitriev I.E., Batyukhnova O.G. Fractal structure of pores of clay soil. Atomic Energy, 74, 241–243 (1993)

[48] Takayasu, H. (1990). *Fractals in the physical sciences.* Manchester: Manchester University Press. p. 36. ISBN 9780719034343.

[49] Jun, Li; Ostoja-Starzewski, Martin. "Saturn's Rings are Fractal" (PDF). Retrieved 28 June 2014.

[50] Enright, Matthew B.; Leitner, David M. (27 January 2005). "Mass fractal dimension and the compactness of proteins". *Physical Review E* **71** (1). doi:10.1103/PhysRevE.71.011912.

[51] Reuveni, Shlomi; Granek, Rony; Klafter, Joseph (19 May 2008). "Proteins: Coexistence of Stability and Flexibility". *Physical Review Letters* **100** (20). doi:10.1103/PhysRevLett.100.208101.

[52] de Leeuw, Marina; Reuveni, Shlomi; Klafter, Joseph; Granek, Rony (9 October 2009). "Coexistence of Flexibility and Stability of Proteins: An Equation of State". *PLoS ONE* **4** (10): e7296. doi:10.1371/journal.pone.0007296.

[53] Reuveni, S.; Granek, R.; Klafter, J. (16 July 2010). "Anomalies in the vibrational dynamics of proteins are a consequence of fractal-like structure". *Proceedings of the National Academy of Sciences* **107** (31): 13696–13700. doi:10.1073/pnas.1002018107.

[54] Reuveni, Shlomi; Klafter, Joseph; Granek, Rony (7 February 2012). "Dynamic Structure Factor of Vibrating Fractals". *Physical Review Letters* **108** (6). doi:10.1103/PhysRevLett.108.068101.

[55] Reuveni, Shlomi; Klafter, Joseph; Granek, Rony (10 January 2012). "Dynamic structure factor of vibrating fractals: Proteins as a case study". *Physical Review E* **85** (1). doi:10.1103/PhysRevE.85.011906.

[56] Frame, Michael; and Mandelbrot, Benoît B.; *A Panorama of Fractals and Their Uses*

[57] Nelson, Bryn; *Sophisticated Mathematics Behind African Village Designs Fractal patterns use repetition on large, small scale,* San Francisco Chronicle, Wednesday, February 23, 2009

[58] Hohlfeld, Robert G.; Cohen, Nathan (1999). "Self-similarity and the geometric requirements for frequency independence in Antennae". *Fractals* **7** (1): 79–84. doi:10.1142/S0218348X99000098.

[59] Reiner, Richard; Waltereit, Patrick; Benkhelifa, Fouad; Müller, Stefan; Walcher, Herbert; Wagner, Sandrine; Quay, Rüdiger; Schlechtweg, Michael; Ambacher, Oliver; Ambacher, O. (2012). "Fractal structures for low-resistance large area AlGaN/GaN power transistors" (PDF). *Proceedings of ISPSD*: 341. doi:10.1109/ISPSD.2012.6229091. ISBN 978-1-4577-1596-9.

[60] Chen, Yanguang (2011). "Modeling Fractal Structure of City-Size Distributions Using Correlation Functions". *PLoS ONE* **6** (9): e24791. doi:10.1371/journal.pone.0024791. PMC 3176775. PMID 21949753.

[61] "Applications". Retrieved 2007-10-21.

[62] Smith, Robert F.; Mohr, David N.; Torres, Vicente E.; Offord, Kenneth P.; Melton III, L. Joseph (1989). "Renal insufficiency in community patients with mild asymptomatic microhematuria". *Mayo Clinic proceedings. Mayo Clinic* **64** (4): 409–414. doi:10.1016/s0025-6196(12)65730-9. PMID 2716356.

[63] Landini, Gabriel (2011). "Fractals in microscopy". *Journal of Microscopy* **241** (1): 1–8. doi:10.1111/j.1365-2818.2010.03454.x. PMID 21118245.

[64] Cheng, Qiuming (1997). "Multifractal Modeling and Lacunarity Analysis". *Mathematical Geology* **29**(7): 919–932. doi:1781.

[65] Chen, Yanguang (2011). "Modeling Fractal Structure of City-Size Distributions Using Correlation Functions". *PLoS ONE* **6** (9): e24791. doi:10.1371/journal.pone.0024791. PMC 3176775. PMID 21949753.

[66] Burkle-Elizondo, Gerardo; Valdéz-Cepeda, Ricardo David (2006). "Fractal analysis of Mesoamerican pyramids". *Nonlinear dynamics, psychology, and life sciences* **10** (1): 105–122. PMID 16393505.

[67] Brown, Clifford T.; Witschey, Walter R. T.; Liebovitch, Larry S. (2005). "The Broken Past: Fractals in Archaeology". *Journal of Archaeological Method and Theory* **12**: 37. doi:10.1007/s10816-005-2396-6.

[68] Saeedi, Panteha; Sorensen, Soren A. "An Algorithmic Approach to Generate After-disaster Test Fields for Search and Rescue Agents" (PDF). *Proceedings of the World Congress on Engineering 2009*: 93–98. ISBN 978-988-17-0125-1.

[69] Bunde, A.; Havlin, S. (2009). "Fractal Geometry, A Brief Introduction to". *Encyclopedia of Complexity and Systems Science*. p. 3700. doi:10.1007/978-0-387-30440-3_218. ISBN 978-0-387-75888-6.

28.13 Further reading

- Barnsley, Michael F.; and Rising, Hawley; *Fractals Everywhere*. Boston: Academic Press Professional, 1993. ISBN 0-12-079061-0

- Duarte, German A.; *Fractal Narrative. About the Relationship Between Geometries and Technology and Its Impact on Narrative Spaces*. Bielefeld: Transcript, 2014. ISBN 978-3-8376-2829-6

- Falconer, Kenneth; *Techniques in Fractal Geometry*. John Wiley and Sons, 1997. ISBN 0-471-92287-0

- Jürgens, Hartmut; Peitgen, Heins-Otto; and Saupe, Dietmar; *Chaos and Fractals: New Frontiers of Science*. New York: Springer-Verlag, 1992. ISBN 0-387-97903-4

- Mandelbrot, Benoit B.; *The Fractal Geometry of Nature*. New York: W. H. Freeman and Co., 1982. ISBN 0-7167-1186-9

- Peitgen, Heinz-Otto; and Saupe, Dietmar; eds.; *The Science of Fractal Images*. New York: Springer-Verlag, 1988. ISBN 0-387-96608-0

- Pickover, Clifford A.; ed.; *Chaos and Fractals: A Computer Graphical Journey – A 10 Year Compilation of Advanced Research*. Elsevier, 1998. ISBN 0-444-50002-2

- Jones, Jesse; *Fractals for the Macintosh*, Waite Group Press, Corte Madera, CA, 1993. ISBN 1-878739-46-8.

- Lauwerier, Hans; *Fractals: Endlessly Repeated Geometrical Figures*, Translated by Sophia Gill-Hoffstadt, Princeton University Press, Princeton NJ, 1991. ISBN 0-691-08551-X, cloth. ISBN 0-691-02445-6 paperback. "This book has been written for a wide audience..." Includes sample BASIC programs in an appendix.

- Sprott, Julien Clinton (2003). *Chaos and Time-Series Analysis*. Oxford University Press. ISBN 978-0-19-850839-7.

- Wahl, Bernt; Van Roy, Peter; Larsen, Michael; and Kampman, Eric; *Exploring Fractals on the Macintosh*, Addison Wesley, 1995. ISBN 0-201-62630-6

- Lesmoir-Gordon, Nigel; "The Colours of Infinity: The Beauty, The Power and the Sense of Fractals." ISBN 1-904555-05-5 (The book comes with a related DVD of the Arthur C. Clarke documentary introduction to the fractal concept and the Mandelbrot set).

- Liu, Huajie; *Fractal Art*, Changsha: Hunan Science and Technology Press, 1997, ISBN 9787535722348.

- Gouyet, Jean-François; *Physics and Fractal Structures* (Foreword by B. Mandelbrot); Masson, 1996. ISBN 2-225-85130-1, and New York: Springer-Verlag, 1996. ISBN 978-0-387-94153-0. Out-of-print. Available in PDF version at."Physics and Fractal Structures" (in French). Jfgouyet.fr. Retrieved 2010-10-17.

- Bunde, Armin; Havlin, Shlomo (1996). *Fractals and Disordered Systems*. Springer.

- Bunde, Armin; Havlin, Shlomo (1995). *Fractals in Science*. Springer.

- ben-Avraham, Daniel; Havlin, Shlomo (2000). *Diffusion and Reactions in Fractals and Disordered Systems*. Cambridge University Press.

- Falconer, Kenneth (2013). *Fractals, A Very Short Introduction*. Oxford University Press.

28.14   External links

- Fractals at DMOZ

- Scaling and Fractals presented by Shlomo Havlin, Bar-Ilan University

- Hunting the Hidden Dimension, *PBS NOVA*, first aired August 24, 2011

- Benoit Mandelbrot: Fractals and the Art of Roughness, TED (conference), February 2010

- Technical Library on Fractals for controlling fluid

- Equations of self-similar fractal measure based on the fractional-order calculus(2007)

Chapter 29

Calculus

This article is about the branch of mathematics. For other uses, see Calculus (disambiguation).

Calculus is the mathematical study of change, in the same way that geometry is the study of shape and algebra is the study of operations and their application to solving equations. It has two major branches, differential calculus (concerning rates of change and slopes of curves),[1] and integral calculus (concerning accumulation of quantities and the areas under and between curves);[2] these two branches are related to each other by the fundamental theorem of calculus. Both branches make use of the fundamental notions of convergence of infinite sequences and infinite series to a well-defined limit. Generally, modern calculus is considered to have been developed in the 17th century by Isaac Newton and Gottfried Leibniz. Today, calculus has widespread uses in science, engineering and economics[3] and can solve many problems that algebra alone cannot.

Calculus is a part of modern mathematics education. A course in calculus is a gateway to other, more advanced courses in mathematics devoted to the study of functions and limits, broadly called mathematical analysis. Calculus has historically been called "the calculus of infinitesimals", or "infinitesimal calculus". The word "calculus" comes from Latin (*calculus*) and refers to a small stone used for counting. More generally, *calculus* (plural *calculi*) refers to any method or system of calculation guided by the symbolic manipulation of expressions. Some examples of other well-known calculi are propositional calculus, calculus of variations, lambda calculus, and process calculus.

29.1 History

Main article: History of calculus

Modern calculus was developed in 17th century Europe by Isaac Newton and Gottfried Wilhelm Leibniz, but elements of it have appeared in ancient India, Greece, China, medieval Europe, and the Middle East.

29.1.1 Ancient

The ancient period introduced some of the ideas that led to integral calculus, but does not seem to have developed these ideas in a rigorous and systematic way. Calculations of volume and area, one goal of integral calculus, can be found in the Egyptian Moscow papyrus (c. 1820 BC), but the formulas are simple instructions, with no indication as to method, and some of them lack major components.[4] From the age of Greek mathematics, Eudoxus (c. 408–355 BC) used the method of exhaustion, which foreshadows the concept of the limit, to calculate areas and volumes, while Archimedes (c. 287–212 BC) developed this idea further, inventing heuristics which resemble the methods of integral calculus.[5] The method of exhaustion was later reinvented in China by Liu Hui in the 3rd century AD in order to find the area of a circle.[6] In the 5th century AD, Zu Chongzhi established a method that would later be called Cavalieri's principle to find the volume of a sphere.[7]

29.1.2   Medieval

Alexander the Great's invasion of northern India brought Greek Trigonometry, using the chord, to India where the sine, cosine, and tangent were conceived. Indian mathematicians gave a semi-rigorous method of differentiation of some trigonometric functions. In the Middle East, Alhazen derived a formula for the sum of fourth powers. He used the results to carry out what would now be called an integration, where the formulas for the sums of integral squares and fourth powers allowed him to calculate the volume of a paraboloid.[8] In the 14th century, Indian mathematician Madhava of Sangamagrama and the Kerala school of astronomy and mathematics stated components of calculus such as the Taylor series and infinite series approximations.[9] However, they were not able to "combine many differing ideas under the two unifying themes of the derivative and the integral, show the connection between the two, and turn calculus into the great problem-solving tool we have today".[8]

29.1.3   Modern

In Europe, the foundational work was a treatise due to Bonaventura Cavalieri, who argued that volumes and areas should be computed as the sums of the volumes and areas of infinitesimally thin cross-sections. The ideas were similar to Archimedes' in The Method, but this treatise is believed to have been lost in the 13th century, and was only rediscovered in the early 20th century, and so would have been unknown to Cavalieri. Cavalieri's work was not well respected since his methods could lead to erroneous results, and the infinitesimal quantities he introduced were disreputable at first.

The formal study of calculus brought together Cavalieri's infinitesimals with the calculus of finite differences developed in Europe at around the same time. Pierre de Fermat, claiming that he borrowed from Diophantus, introduced the concept of adequality, which represented equality up to an infinitesimal error term.[11] The combination was achieved by John Wallis, Isaac Barrow, and James Gregory, the latter two proving the second fundamental theorem of calculus around 1670.

The product rule and chain rule, the notion of higher derivatives, Taylor series, and analytical functions were introduced by Isaac Newton in an idiosyncratic notation which he used to solve problems of mathematical physics.[12] In his works, Newton rephrased his ideas to suit the mathematical idiom of the time, replacing calculations with infinitesimals by equivalent geometrical arguments which were considered beyond reproach. He used the methods of calculus to solve the problem of planetary motion, the shape of the surface of a rotating fluid, the oblateness of the earth, the motion of a weight sliding on a cycloid, and many other problems discussed in his *Principia Mathematica* (1687). In other work, he developed series expansions for functions, including fractional and irrational powers, and it was clear that he understood the principles of the Taylor series. He did not publish all these discoveries, and at this time infinitesimal methods were still considered disreputable.

These ideas were arranged into a true calculus of infinitesimals by Gottfried Wilhelm Leibniz, who was originally accused of plagiarism by Newton.[13] He is now regarded as an independent inventor of and contributor to calculus. His contribution was to provide a clear set of rules for working with infinitesimal quantities, allowing the computation of second and higher derivatives, and providing the product rule and chain rule, in their differential and integral forms. Unlike Newton, Leibniz paid a lot of attention to the formalism, often spending days determining appropriate symbols for concepts.

Leibniz and Newton are usually both credited with the invention of calculus. Newton was the first to apply calculus to general physics and Leibniz developed much of the notation used in calculus today. The basic insights that both Newton and Leibniz provided were the laws of differentiation and integration, second and higher derivatives, and the notion of an approximating polynomial series. By Newton's time, the fundamental theorem of calculus was known.

When Newton and Leibniz first published their results, there was great controversy over which mathematician (and therefore which country) deserved credit. Newton derived his results first (later to be published in his *Method of Fluxions*), but Leibniz published his *Nova Methodus pro Maximis et Minimis* first. Newton claimed Leibniz stole ideas from his unpublished notes, which Newton had shared with a few members of the Royal Society. This controversy divided English-speaking mathematicians from continental European mathematicians for many years, to the detriment of English mathematics. A careful examination of the papers of Leibniz and Newton shows that they arrived at their results independently, with Leibniz starting first with integration and Newton with differentiation. Today, both Newton and Leibniz are given credit for developing calculus independently. It is Leibniz, however, who gave the new discipline its name. Newton called his calculus "the science of fluxions".

Since the time of Leibniz and Newton, many mathematicians have contributed to the continuing development of calculus. One of the first and most complete works on finite and infinitesimal analysis was written in 1748 by Maria Gaetana Agnesi.[14]

29.1.4 Foundations

In calculus, *foundations* refers to the rigorous development of a subject from precise axioms and definitions. In early calculus the use of infinitesimal quantities was thought unrigorous, and was fiercely criticized by a number of authors, most notably Michel Rolle and Bishop Berkeley. Berkeley famously described infinitesimals as the ghosts of departed quantities in his book *The Analyst* in 1734. Working out a rigorous foundation for calculus occupied mathematicians for much of the century following Newton and Leibniz, and is still to some extent an active area of research today.

Several mathematicians, including Maclaurin, tried to prove the soundness of using infinitesimals, but it would not be until 150 years later when, due to the work of Cauchy and Weierstrass, a way was finally found to avoid mere "notions" of infinitely small quantities.[15] The foundations of differential and integral calculus had been laid. In Cauchy's Cours d'Analyse, we find a broad range of foundational approaches, including a definition of continuity in terms of infinitesimals, and a (somewhat imprecise) prototype of an (ε, δ)-definition of limit in the definition of differentiation. In his work Weierstrass formalized the concept of limit and eliminated infinitesimals. Following the work of Weierstrass, it eventually became common to base calculus on limits instead of infinitesimal quantities, though the subject is still occasionally called "infinitesimal calculus". Bernhard Riemann used these ideas to give a precise definition of the integral. It was also during this period that the ideas of calculus were generalized to Euclidean space and the complex plane.

In modern mathematics, the foundations of calculus are included in the field of real analysis, which contains full definitions and proofs of the theorems of calculus. The reach of calculus has also been greatly extended. Henri Lebesgue invented measure theory and used it to define integrals of all but the most pathological functions. Laurent Schwartz introduced distributions, which can be used to take the derivative of any function whatsoever.

Limits are not the only rigorous approach to the foundation of calculus. Another way is to use Abraham Robinson's non-standard analysis. Robinson's approach, developed in the 1960s, uses technical machinery from mathematical logic to augment the real number system with infinitesimal and infinite numbers, as in the original Newton-Leibniz conception. The resulting numbers are called hyperreal numbers, and they can be used to give a Leibniz-like development of the usual rules of calculus.

29.1.5 Significance

While many of the ideas of calculus had been developed earlier in Greece, China, India, Iraq, Persia, and Japan, the use of calculus began in Europe, during the 17th century, when Isaac Newton and Gottfried Wilhelm Leibniz built on the work of earlier mathematicians to introduce its basic principles. The development of calculus was built on earlier concepts of instantaneous motion and area underneath curves.

Applications of differential calculus include computations involving velocity and acceleration, the slope of a curve, and optimization. Applications of integral calculus include computations involving area, volume, arc length, center of mass, work, and pressure. More advanced applications include power series and Fourier series.

Calculus is also used to gain a more precise understanding of the nature of space, time, and motion. For centuries, mathematicians and philosophers wrestled with paradoxes involving division by zero or sums of infinitely many numbers. These questions arise in the study of motion and area. The ancient Greek philosopher Zeno of Elea gave several famous examples of such paradoxes. Calculus provides tools, especially the limit and the infinite series, which resolve the paradoxes.

29.2 Principles

29.2.1 Limits and infinitesimals

Main articles: Limit of a function and Infinitesimal

Calculus is usually developed by working with very small quantities. Historically, the first method of doing so was by infinitesimals. These are objects which can be treated like numbers but which are, in some sense, "infinitely small". An infinitesimal number dx could be greater than 0, but less than any number in the sequence 1, 1/2, 1/3, ... and less than any positive real number. Any integer multiple of an infinitesimal is still infinitely small, i.e., infinitesimals do not satisfy the Archimedean property. From this point of view, calculus is a collection of techniques for manipulating infinitesimals. This approach fell out of favor in the 19th century because it was difficult to make the notion of an infinitesimal precise. However, the concept was revived in the 20th century with the introduction of non-standard analysis and smooth infinitesimal analysis, which provided solid foundations for the manipulation of infinitesimals.

In the 19th century, infinitesimals were replaced by the epsilon, delta approach to limits. Limits describe the value of a function at a certain input in terms of its values at a nearby input. They capture small-scale behavior in the context of the real number system. In this treatment, calculus is a collection of techniques for manipulating certain limits. Infinitesimals get replaced by very small numbers, and the infinitely small behavior of the function is found by taking the limiting behavior for smaller and smaller numbers. Limits were the first way to provide rigorous foundations for calculus, and for this reason they are the standard approach.

29.2.2 Differential calculus

Main article: Differential calculus

Differential calculus is the study of the definition, properties, and applications of the derivative of a function. The process of finding the derivative is called *differentiation*. Given a function and a point in the domain, the derivative at that point is a way of encoding the small-scale behavior of the function near that point. By finding the derivative of a function at every point in its domain, it is possible to produce a new function, called the *derivative function* or just the *derivative* of the original function. In mathematical jargon, the derivative is a linear operator which inputs a function and outputs a second function. This is more abstract than many of the processes studied in elementary algebra, where functions usually input a number and output another number. For example, if the doubling function is given the input three, then it outputs six, and if the squaring function is given the input three, then it outputs nine. The derivative, however, can take the squaring function as an input. This means that the derivative takes all the information of the squaring function—such as that two is sent to four, three is sent to nine, four is sent to sixteen, and so on—and uses this information to produce another function. (The function it produces turns out to be the doubling function.)

The most common symbol for a derivative is an apostrophe-like mark called prime. Thus, the derivative of the function of f is f', pronounced "f prime." For instance, if $f(x) = x^2$ is the squaring function, then $f'(x) = 2x$ is its derivative, the doubling function.

If the input of the function represents time, then the derivative represents change with respect to time. For example, if f is a function that takes a time as input and gives the position of a ball at that time as output, then the derivative of f is how the position is changing in time, that is, it is the velocity of the ball.

If a function is linear (that is, if the graph of the function is a straight line), then the function can be written as $y = mx + b$, where x is the independent variable, y is the dependent variable, b is the y-intercept, and:

$$m = \frac{\text{rise}}{\text{run}} = \frac{\text{in change} y}{\text{in change} x} = \frac{\Delta y}{\Delta x}.$$

This gives an exact value for the slope of a straight line. If the graph of the function is not a straight line, however, then the change in y divided by the change in x varies. Derivatives give an exact meaning to the notion of change in output with respect to change in input. To be concrete, let f be a function, and fix a point a in the domain of f. $(a, f(a))$ is a point on the graph of the function. If h is a number close to zero, then $a + h$ is a number close to a. Therefore, $(a + h, f(a + h))$ is close to $(a, f(a))$. The slope between these two points is

$$m = \frac{f(a+h) - f(a)}{(a+h) - a} = \frac{f(a+h) - f(a)}{h}.$$

This expression is called a *difference quotient*. A line through two points on a curve is called a *secant line*, so m is the slope of the secant line between $(a, f(a))$ and $(a + h, f(a + h))$. The secant line is only an approximation to the behavior of the function at the point a because it does not account for what happens between a and $a + h$. It is not possible to discover the behavior at a by setting h to zero because this would require dividing by zero, which is undefined. The derivative is defined by taking the limit as h tends to zero, meaning that it considers the behavior of f for all small values of h and extracts a consistent value for the case when h equals zero:

$$\lim_{h \to 0} \frac{f(a+h) - f(a)}{h}.$$

Geometrically, the derivative is the slope of the tangent line to the graph of f at a. The tangent line is a limit of secant lines just as the derivative is a limit of difference quotients. For this reason, the derivative is sometimes called the slope of the function f.

Here is a particular example, the derivative of the squaring function at the input 3. Let $f(x) = x^2$ be the squaring function.

$$
\begin{aligned}
f'(3) &= \lim_{h \to 0} \frac{(3+h)^2 - 3^2}{h} \\
&= \lim_{h \to 0} \frac{9 + 6h + h^2 - 9}{h} \\
&= \lim_{h \to 0} \frac{6h + h^2}{h} \\
&= \lim_{h \to 0} (6 + h) \\
&= 6.
\end{aligned}
$$

The slope of the tangent line to the squaring function at the point $(3, 9)$ is 6, that is to say, it is going up six times as fast as it is going to the right. The limit process just described can be performed for any point in the domain of the squaring function. This defines the *derivative function* of the squaring function, or just the *derivative* of the squaring function for short. A similar computation to the one above shows that the derivative of the squaring function is the doubling function.

29.2.3 Leibniz notation

Main article: Leibniz's notation

A common notation, introduced by Leibniz, for the derivative in the example above is

$$
\begin{aligned}
y &= x^2 \\
\frac{dy}{dx} &= 2x.
\end{aligned}
$$

In an approach based on limits, the symbol dy/dx is to be interpreted not as the quotient of two numbers but as a shorthand for the limit computed above. Leibniz, however, did intend it to represent the quotient of two infinitesimally small numbers, dy being the infinitesimally small change in y caused by an infinitesimally small change dx applied to x. We can also think of d/dx as a differentiation operator, which takes a function as an input and gives another function, the derivative, as the output. For example:

$$\frac{d}{dx}(x^2) = 2x.$$

In this usage, the dx in the denominator is read as "with respect to x". Even when calculus is developed using limits rather than infinitesimals, it is common to manipulate symbols like dx and dy as if they were real numbers; although it is possible to avoid such manipulations, they are sometimes notationally convenient in expressing operations such as the total derivative.

29.2.4   Integral calculus

Main article: Integral

Integral calculus is the study of the definitions, properties, and applications of two related concepts, the *indefinite integral* and the *definite integral*. The process of finding the value of an integral is called *integration*. In technical language, integral calculus studies two related linear operators.

The *indefinite integral* is the *antiderivative*, the inverse operation to the derivative. F is an indefinite integral of f when f is a derivative of F. (This use of lower- and upper-case letters for a function and its indefinite integral is common in calculus.)

The *definite integral* inputs a function and outputs a number, which gives the algebraic sum of areas between the graph of the input and the x-axis. The technical definition of the definite integral is the limit of a sum of areas of rectangles, called a Riemann sum.

A motivating example is the distances traveled in a given time.

$$\text{Distance} = \text{Speed} \cdot \text{Time}$$

If the speed is constant, only multiplication is needed, but if the speed changes, a more powerful method of finding the distance is necessary. One such method is to approximate the distance traveled by breaking up the time into many short intervals of time, then multiplying the time elapsed in each interval by one of the speeds in that interval, and then taking the sum (a Riemann sum) of the approximate distance traveled in each interval. The basic idea is that if only a short time elapses, then the speed will stay more or less the same. However, a Riemann sum only gives an approximation of the distance traveled. We must take the limit of all such Riemann sums to find the exact distance traveled.

When velocity is constant, the total distance traveled over the given time interval can be computed by multiplying velocity and time. For example, travelling a steady 50 mph for 3 hours results in a total distance of 150 miles. In the diagram on the left, when constant velocity and time are graphed, these two values form a rectangle with height equal to the velocity and width equal to the time elapsed. Therefore, the product of velocity and time also calculates the rectangular area under the (constant) velocity curve. This connection between the area under a curve and distance traveled can be extended to *any* irregularly shaped region exhibiting a fluctuating velocity over a given time period. If $f(x)$ in the diagram on the right represents speed as it varies over time, the distance traveled (between the times represented by a and b) is the area of the shaded region s.

To approximate that area, an intuitive method would be to divide up the distance between a and b into a number of equal segments, the length of each segment represented by the symbol Δx. For each small segment, we can choose one value of the function $f(x)$. Call that value h. Then the area of the rectangle with base Δx and height h gives the distance (time Δx multiplied by speed h) traveled in that segment. Associated with each segment is the average value of the function above it, $f(x) = h$. The sum of all such rectangles gives an approximation of the area between the axis and the curve, which is an approximation of the total distance traveled. A smaller value for Δx will give more rectangles and in most cases a better approximation, but for an exact answer we need to take a limit as Δx approaches zero.

The symbol of integration is $\int$, an elongated S (the S stands for "sum"). The definite integral is written as:

$$\int_a^b f(x)\,dx.$$

and is read "the integral from a to b of f-of-x with respect to x." The Leibniz notation dx is intended to suggest dividing the area under the curve into an infinite number of rectangles, so that their width Δx becomes the infinitesimally small dx. In a formulation of the calculus based on limits, the notation

$$\int_a^b \cdots\, dx$$

is to be understood as an operator that takes a function as an input and gives a number, the area, as an output. The terminating differential, dx, is not a number, and is not being multiplied by $f(x)$, although, serving as a reminder of the Δx limit definition, it can be treated as such in symbolic manipulations of the integral. Formally, the differential indicates the variable over which the function is integrated and serves as a closing bracket for the integration operator.

The indefinite integral, or antiderivative, is written:

$$\int f(x)\,dx.$$

Functions differing by only a constant have the same derivative, and it can be shown that the antiderivative of a given function is actually a family of functions differing only by a constant. Since the derivative of the function $y = x^2 + C$, where C is any constant, is $y' = 2x$, the antiderivative of the latter given by:

$$\int 2x\,dx = x^2 + C.$$

The unspecified constant C present in the indefinite integral or antiderivative is known as the constant of integration.

29.2.5 Fundamental theorem

Main article: Fundamental theorem of calculus

The fundamental theorem of calculus states that differentiation and integration are inverse operations. More precisely, it relates the values of antiderivatives to definite integrals. Because it is usually easier to compute an antiderivative than to apply the definition of a definite integral, the fundamental theorem of calculus provides a practical way of computing definite integrals. It can also be interpreted as a precise statement of the fact that differentiation is the inverse of integration.

The fundamental theorem of calculus states: If a function f is continuous on the interval $[a, b]$ and if F is a function whose derivative is f on the interval (a, b), then

$$\int_a^b f(x)\,dx = F(b) - F(a).$$

Furthermore, for every x in the interval (a, b),

$$\frac{d}{dx}\int_a^x f(t)\,dt = f(x).$$

This realization, made by both Newton and Leibniz, who based their results on earlier work by Isaac Barrow, was key to the proliferation of analytic results after their work became known. The fundamental theorem provides an algebraic method of computing many definite integrals—without performing limit processes—by finding formulas for antiderivatives. It is also a prototype solution of a differential equation. Differential equations relate an unknown function to its derivatives, and are ubiquitous in the sciences.

29.3   Applications

Calculus is used in every branch of the physical sciences, actuarial science, computer science, statistics, engineering, economics, business, medicine, demography, and in other fields wherever a problem can be mathematically modeled and an optimal solution is desired. It allows one to go from (non-constant) rates of change to the total change or vice versa, and many times in studying a problem we know one and are trying to find the other.

Physics makes particular use of calculus; all concepts in classical mechanics and electromagnetism are related through calculus. The mass of an object of known density, the moment of inertia of objects, as well as the total energy of an object within a conservative field can be found by the use of calculus. An example of the use of calculus in mechanics is Newton's second law of motion: historically stated it expressly uses the term "rate of change" which refers to the derivative saying *The* **rate of change** *of momentum of a body is equal to the resultant force acting on the body and is in the same direction.* Commonly expressed today as Force = Mass × acceleration, it involves differential calculus because acceleration is the time derivative of velocity or second time derivative of trajectory or spatial position. Starting from knowing how an object is accelerating, we use calculus to derive its path.

Maxwell's theory of electromagnetism and Einstein's theory of general relativity are also expressed in the language of differential calculus. Chemistry also uses calculus in determining reaction rates and radioactive decay. In biology, population dynamics starts with reproduction and death rates to model population changes.

Calculus can be used in conjunction with other mathematical disciplines. For example, it can be used with linear algebra to find the "best fit" linear approximation for a set of points in a domain. Or it can be used in probability theory to determine the probability of a continuous random variable from an assumed density function. In analytic geometry, the study of graphs of functions, calculus is used to find high points and low points (maxima and minima), slope, concavity and inflection points.

Green's Theorem, which gives the relationship between a line integral around a simple closed curve C and a double integral over the plane region D bounded by C, is applied in an instrument known as a planimeter, which is used to calculate the area of a flat surface on a drawing. For example, it can be used to calculate the amount of area taken up by an irregularly shaped flower bed or swimming pool when designing the layout of a piece of property.

Discrete Green's Theorem, which gives the relationship between a double integral of a function around a simple closed rectangular curve C and a linear combination of the antiderivative's values at corner points along the edge of the curve, allows fast calculation of sums of values in rectangular domains. For example, it can be used to efficiently calculate sums of rectangular domains in images, in order to rapidly extract features and detect object; another algorithm that could be used is the summed area table.

In the realm of medicine, calculus can be used to find the optimal branching angle of a blood vessel so as to maximize flow. From the decay laws for a particular drug's elimination from the body, it is used to derive dosing laws. In nuclear medicine, it is used to build models of radiation transport in targeted tumor therapies.

In economics, calculus allows for the determination of maximal profit by providing a way to easily calculate both marginal cost and marginal revenue.

Calculus is also used to find approximate solutions to equations; in practice it is the standard way to solve differential equations and do root finding in most applications. Examples are methods such as Newton's method, fixed point iteration, and linear approximation. For instance, spacecraft use a variation of the Euler method to approximate curved courses within zero gravity environments.

29.4 Varieties

Over the years, many reformulations of calculus have been investigated for different purposes.

29.4.1 Non-standard calculus

Main article: Non-standard calculus

Imprecise calculations with infinitesimals were widely replaced with the rigorous (ε, δ)-definition of limit starting in the 1870s. Meanwhile, calculations with infinitesimals persisted and often led to correct results. This led Abraham Robinson to investigate if were possible to develop a number system with infinitesimal quantities over which the theorems of calculus were still valid. In 1960, building upon the work of Edwin Hewitt and Jerzy Łoś, he succeeded in developing non-standard analysis. The theory of non-standard analysis is rich enough to be applied in many branches of mathematics. As such, books and articles dedicated solely to the traditional theorems of calculus often go by the title non-standard calculus.

29.4.2 Smooth infinitesimal analysis

Main article: Smooth infinitesimal analysis

This is another reformulation of the calculus in terms of infinitesimals. Based on the ideas of F. W. Lawvere and employing the methods of category theory, it views all functions as being continuous and incapable of being expressed in terms of discrete entities. One aspect of this formulation is that the law of excluded middle does not hold in this formulation.

29.4.3 Constructive analysis

Main article: Constructive analysis

Constructive mathematics is a branch of mathematics that insists that proofs of the existence of a number, function, or other mathematical object should give a construction of the object. As such constructive mathematics also rejects the law of excluded middle. Reformulations of calculus in a constructive framework are generally part of the subject of constructive analysis.

29.5 See also

Main article: Outline of calculus

29.5.1 Lists

- List of calculus topics
- List of derivatives and integrals in alternative calculi
- List of differentiation identities
- Publications in calculus
- Table of integrals

29.5.2 Other related topics

- Calculus of finite differences
- Calculus with polynomials
- Complex analysis
- Differential equation
- Differential geometry
- *Elementary Calculus: An Infinitesimal Approach*
- Fourier series
- Integral equation
- Mathematical analysis
- Multivariable calculus
- Non-classical analysis
- Non-standard analysis
- Non-standard calculus
- Precalculus (mathematical education)
- Product integral
- Stochastic calculus
- Taylor series

29.6   References

29.6.1 Notes

[1] http://www.merriam-webster.com/dictionary/differential%20calculus

[2] http://www.merriam-webster.com/dictionary/integral+calculus?show=0&t=1421520369

[3] Fisher, Irving (1897). *A brief introduction to the infinitesimal calculus*. New York: The Macmillan Company.

[4] Morris Kline, *Mathematical thought from ancient to modern times*, Vol. I

[5] Archimedes, *Method*, in *The Works of Archimedes* ISBN 978-0-521-66160-7

[6] Dun, Liu; Fan, Dainian; Cohen, Robert Sonné (1966). "A comparison of Archimdes' and Liu Hui's studies of circles". Chinese studies in the history and philosophy of science and technology **130**. Springer. p. 279. ISBN 0-7923-3463-9.,Chapter , p. 279

[7] Zill, Dennis G.; Wright, Scott; Wright, Warren S. (2009). *Calculus: Early Transcendentals* (3 ed.). Jones & Bartlett Learning. p. xxvii. ISBN 0-7637-5995-3., Extract of page 27

[8] Katz, V. J. 1995. "Ideas of Calculus in Islam and India." *Mathematics Magazine* (Mathematical Association of America). 68(3):163-174.

[9] Indian mathematics

[10] von Neumann, J., "The Mathematician", in Heywood, R. B., ed., *The Works of the Mind*, University of Chicago Press, 1947, pp. 180–196. Reprinted in Bródy, F., Vámos, T., eds., *The Neumann Compedium*, World Scientific Publishing Co. Pte. Ltd., 1995, ISBN 981-02-2201-7, pp. 618–626.

[11] André Weil: Number theory. An approach through history. From Hammurapi to Legendre. Birkhauser Boston, Inc., Boston, MA, 1984, ISBN 0-8176-4565-9, p. 28.

[12] Donald Allen: Calculus, http://www.math.tamu.edu/~{}dallen/history/calc1/calc1.html

[13] Leibniz, Gottfried Wilhelm. The Early Mathematical Manuscripts of Leibniz. Cosimo, Inc., 2008. Page 228. Copy

[14] Unlu, Elif (April 1995). "Maria Gaetana Agnesi". Agnes Scott College.

[15] Russell, Bertrand (1946). *History of Western Philosophy*. London: George Allen & Unwin Ltd. p. 857. The great mathematicians of the seventeenth century were optimistic and anxious for quick results; consequently they left the foundations of analytical geometry and the infinitesimal calculus insecure. Leibniz believed in actual infinitesimals, but although this belief suited his metaphysics it had no sound basis in mathematics. Weierstrass, soon after the middle of the nineteenth century, showed how to establish the calculus without infinitesimals, and thus at last made it logically secure. Next came Georg Cantor, who developed the theory of continuity and infinite number. "Continuity" had been, until he defined it, a vague word, convenient for philosophers like Hegel, who wished to introduce metaphysical muddles into mathematics. Cantor gave a precise significance to the word, and showed that continuity, as he defined it, was the concept needed by mathematicians and physicists. By this means a great deal of mysticism, such as that of Bergson, was rendered antiquated.

29.6.2   Books

- Larson, Ron, Bruce H. Edwards (2010). *Calculus*, 9th ed., Brooks Cole Cengage Learning. ISBN 978-0-547-16702-2

- McQuarrie, Donald A. (2003). *Mathematical Methods for Scientists and Engineers*, University Science Books. ISBN 978-1-891389-24-5

- Salas, Saturnino L.; Hille, Einar; Etgen, Garret J. (2007). *Calculus: One and Several Variables* (10th ed.). Wiley. ISBN 978-0-471-69804-3.

- Stewart, James (2012). *Calculus: Early Transcendentals*, 7th ed., Brooks Cole Cengage Learning. ISBN 978-0-538-49790-9

- Thomas, George B., Maurice D. Weir, Joel Hass, Frank R. Giordano (2008), *Calculus*, 11th ed., Addison-Wesley. ISBN 0-321-48987-X

29.7   Other resources

29.7.1   Further reading

- Boyer, Carl Benjamin (1949). *The History of the Calculus and its Conceptual Development*. Hafner. Dover edition 1959, ISBN 0-486-60509-4

- Courant, Richard ISBN 978-3-540-65058-4 *Introduction to calculus and analysis 1*.

- Edmund Landau. ISBN 0-8218-2830-4 *Differential and Integral Calculus*, American Mathematical Society.

- Robert A. Adams. (1999). ISBN 978-0-201-39607-2 *Calculus: A complete course*.

- Albers, Donald J.; Richard D. Anderson and Don O. Loftsgaarden, ed. (1986) *Undergraduate Programs in the Mathematics and Computer Sciences: The 1985-1986 Survey*, Mathematical Association of America No. 7.

- John Lane Bell: *A Primer of Infinitesimal Analysis*, Cambridge University Press, 1998. ISBN 978-0-521-62401-5. Uses synthetic differential geometry and nilpotent infinitesimals.

- Florian Cajori, "The History of Notations of the Calculus." *Annals of Mathematics*, 2nd Ser., Vol. 25, No. 1 (Sep., 1923), pp. 1–46.

- Leonid P. Lebedev and Michael J. Cloud: "Approximating Perfection: a Mathematician's Journey into the World of Mechanics, Ch. 1: The Tools of Calculus", Princeton Univ. Press, 2004.

- Cliff Pickover. (2003). ISBN 978-0-471-26987-8 *Calculus and Pizza: A Math Cookbook for the Hungry Mind*.

- Michael Spivak. (September 1994). ISBN 978-0-914098-89-8 *Calculus*. Publish or Perish publishing.

- Tom M. Apostol. (1967). ISBN 978-0-471-00005-1 *Calculus, Volume 1, One-Variable Calculus with an Introduction to Linear Algebra*. Wiley.

- Tom M. Apostol. (1969). ISBN 978-0-471-00007-5 *Calculus, Volume 2, Multi-Variable Calculus and Linear Algebra with Applications*. Wiley.

- Silvanus P. Thompson and Martin Gardner. (1998). ISBN 978-0-312-18548-0 *Calculus Made Easy*.

- Mathematical Association of America. (1988). *Calculus for a New Century; A Pump, Not a Filter*. The Association. Stony Brook, NY. ED 300 252.

- Thomas/Finney. (1996). ISBN 978-0-201-53174-9 *Calculus and Analytic geometry 9th*. Addison Wesley.

- Weisstein, Eric W. "Second Fundamental Theorem of Calculus." From MathWorld—A Wolfram Web Resource.

- Howard Anton,Irl Bivens,Stephen Davis:"Calculus",John Willey and Sons Pte. Ltd.,2002.ISBN 978-81-265-1259-1

29.7.2 Online books

- Boelkins, M. (2012). "*Active Calculus: a free, open text*". Retrieved 1 Feb 2013 from http://gvsu.edu/s/km

- Crowell, B. (2003). "*Calculus*" Light and Matter, Fullerton. Retrieved 6 May 2007 from http://www.lightandmatter.com/calc/calc.pdf

- Garrett, P. (2006). "*Notes on first year calculus*" University of Minnesota. Retrieved 6 May 2007 from http://www.math.umn.edu/~{}garrett/calculus/first_year/notes.pdf

- Faraz, H. (2006). "*Understanding Calculus*" Retrieved 6 May 2007 from Understanding Calculus. URL http://www.understandingcalculus.com/ (HTML only)

- Keisler, H. J. (2000). "*Elementary Calculus: An Approach Using Infinitesimals*" Retrieved 29 August 2010 from http://www.math.wisc.edu/~{}keisler/calc.html

- Mauch, S. (2004). "*Sean's Applied Math Book*" California Institute of Technology. Retrieved 6 May 2007 from http://www.cacr.caltech.edu/~{}sean/applied_math.pdf

- Sloughter, Dan (2000). "*Difference Equations to Differential Equations: An introduction to calculus*". Retrieved 17 March 2009 from http://synechism.org/drupal/de2de/

- Stroyan, K.D. (2004). "*A brief introduction to infinitesimal calculus*" University of Iowa. Retrieved 6 May 2007 from http://www.math.uiowa.edu/~{}stroyan/InfsmlCalculus/InfsmlCalc.htm (HTML only)

- Strang, G. (1991). "*Calculus*" Massachusetts Institute of Technology. Retrieved 6 May 2007 from http://ocw.mit.edu/ans7870/resources/Strang/strangtext.htm

- Smith, William V. (2001). "*The Calculus*" Retrieved 4 July 2008 (HTML only).

29.8 External links

- Hazewinkel, Michiel, ed. (2001), "Calculus", *Encyclopedia of Mathematics*, Springer, ISBN 978-1-55608-010-4

- Weisstein, Eric W., "Calculus", *MathWorld*.

- Topics on Calculus at PlanetMath.org.‹The template *PlanetMath* is being considered for deletion.›

- Calculus Made Easy (1914) by Silvanus P. Thompson Full text in PDF

-

- Calculus on *In Our Time* at the BBC. (listen now)

- Calculus.org: The Calculus page at University of California, Davis – contains resources and links to other sites

- COW: Calculus on the Web at Temple University – contains resources ranging from pre-calculus and associated algebra

- Earliest Known Uses of Some of the Words of Mathematics: Calculus & Analysis

- Online Integrator (WebMathematica) from Wolfram Research

- The Role of Calculus in College Mathematics from ERICDigests.org

- OpenCourseWare Calculus from the Massachusetts Institute of Technology

- Infinitesimal Calculus – an article on its historical development, in *Encyclopedia of Mathematics*, ed. Michiel Hazewinkel.

- Calculus for Beginners and Artists by Daniel Kleitman, MIT

- Calculus Problems and Solutions by D. A. Kouba

- Donald Allen's notes on calculus

- Calculus training materials at imomath.com

- (English) (Arabic) The Excursion of Calculus, 1772

Isaac Newton developed the use of calculus in his laws of motion and gravitation.

Gottfried Wilhelm Leibniz was the first to publish his results on the development of calculus.

Maria Gaetana Agnesi

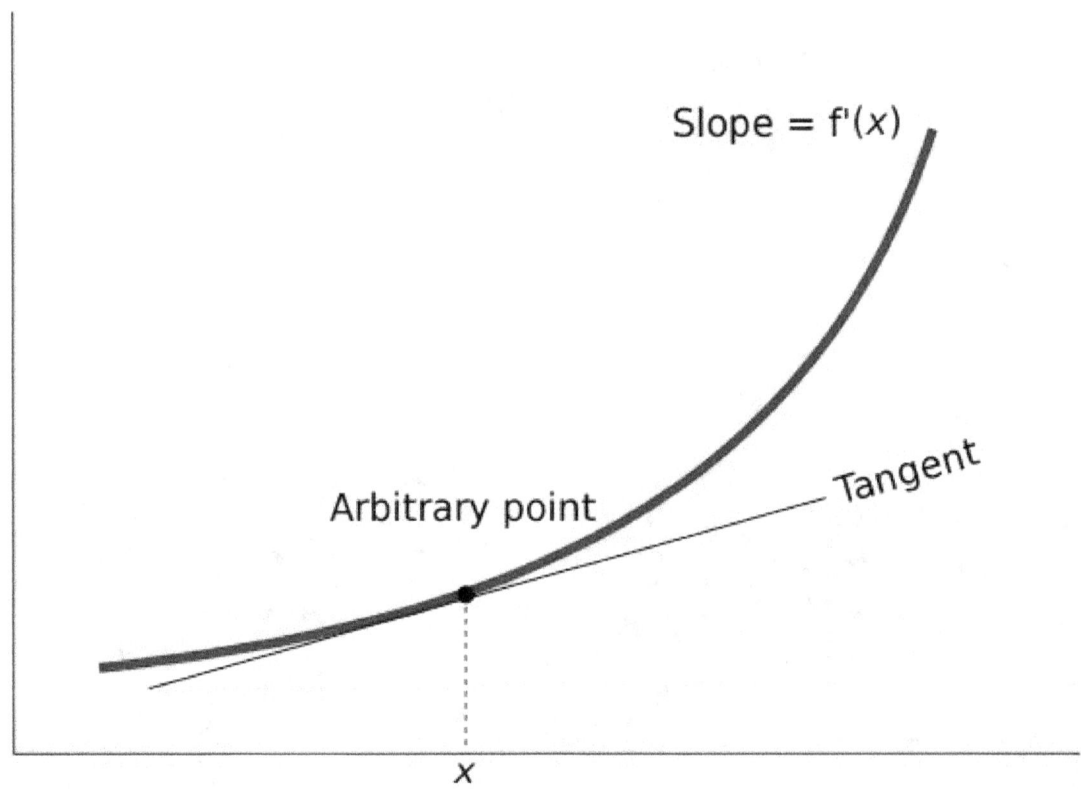

Tangent line at (x, f(x)). The derivative f'(x) of a curve at a point is the slope (rise over run) of the line tangent to that curve at that point.

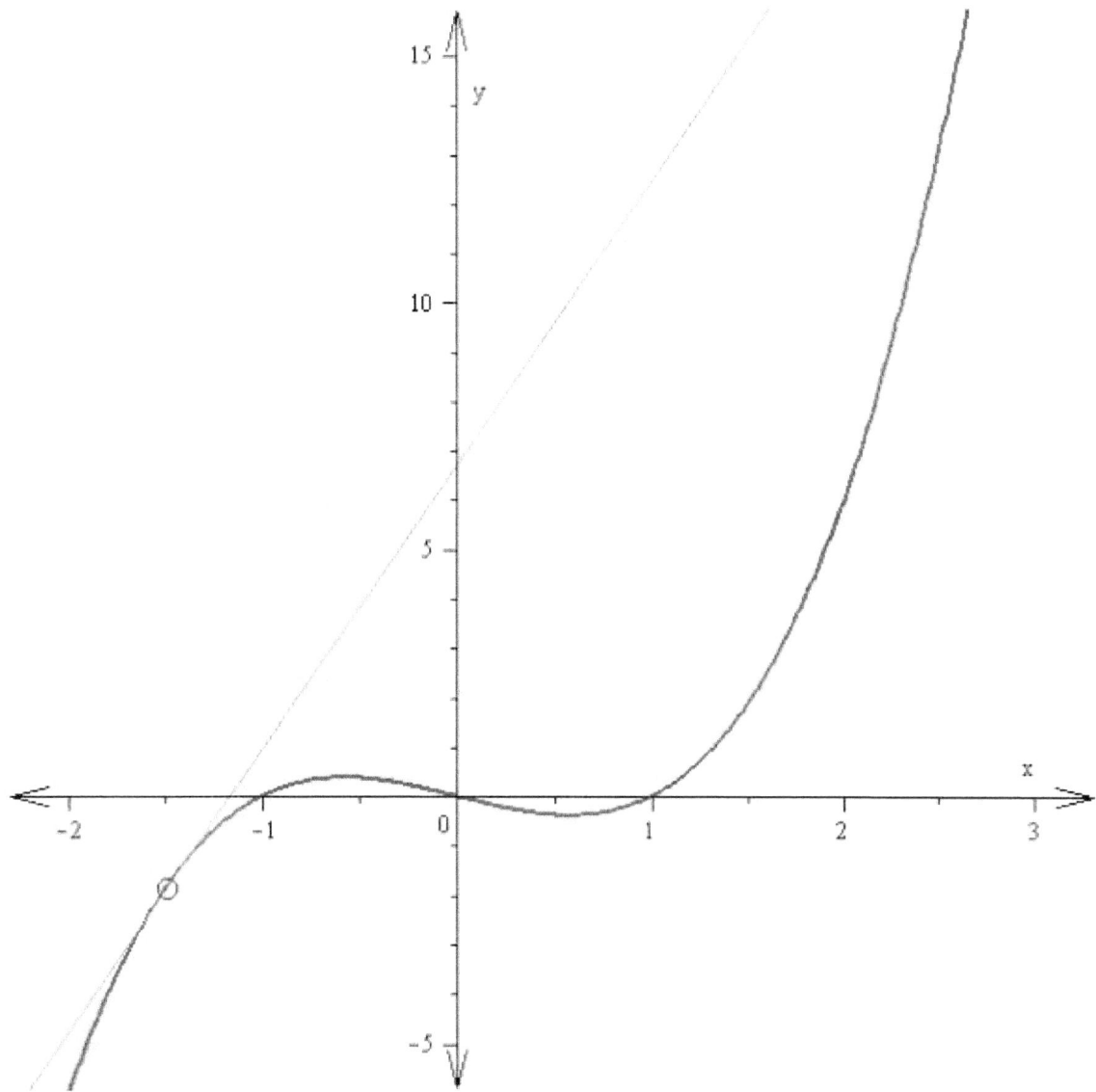

The derivative f'(x) of a curve at a point is the slope of the line tangent to that curve at that point. This slope is determined by considering the limiting value of the slopes of secant lines. Here the function involved (drawn in red) is $f(x) = x^3 - x$. The tangent line (in green) which passes through the point $(-3/2, -15/8)$ has a slope of $23/4$. Note that the vertical and horizontal scales in this image are different.

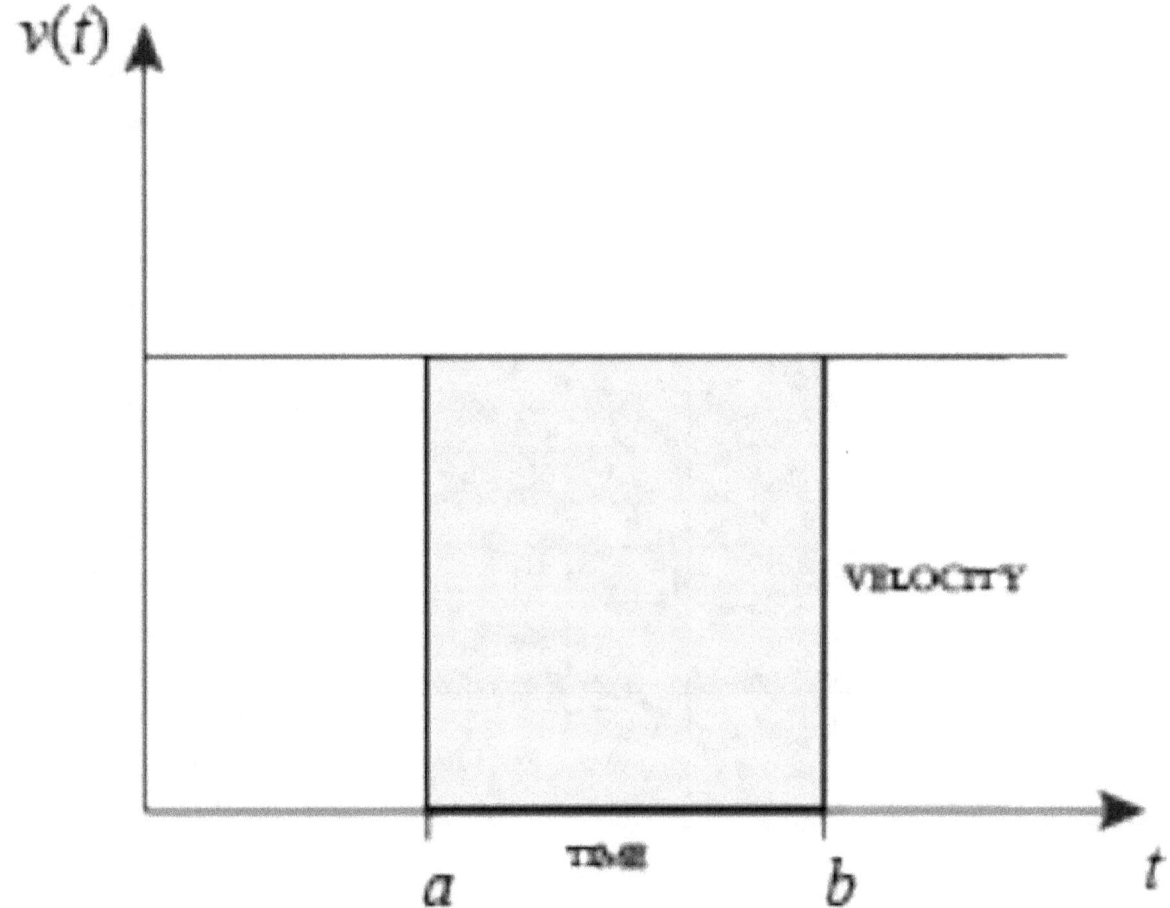

Constant Velocity

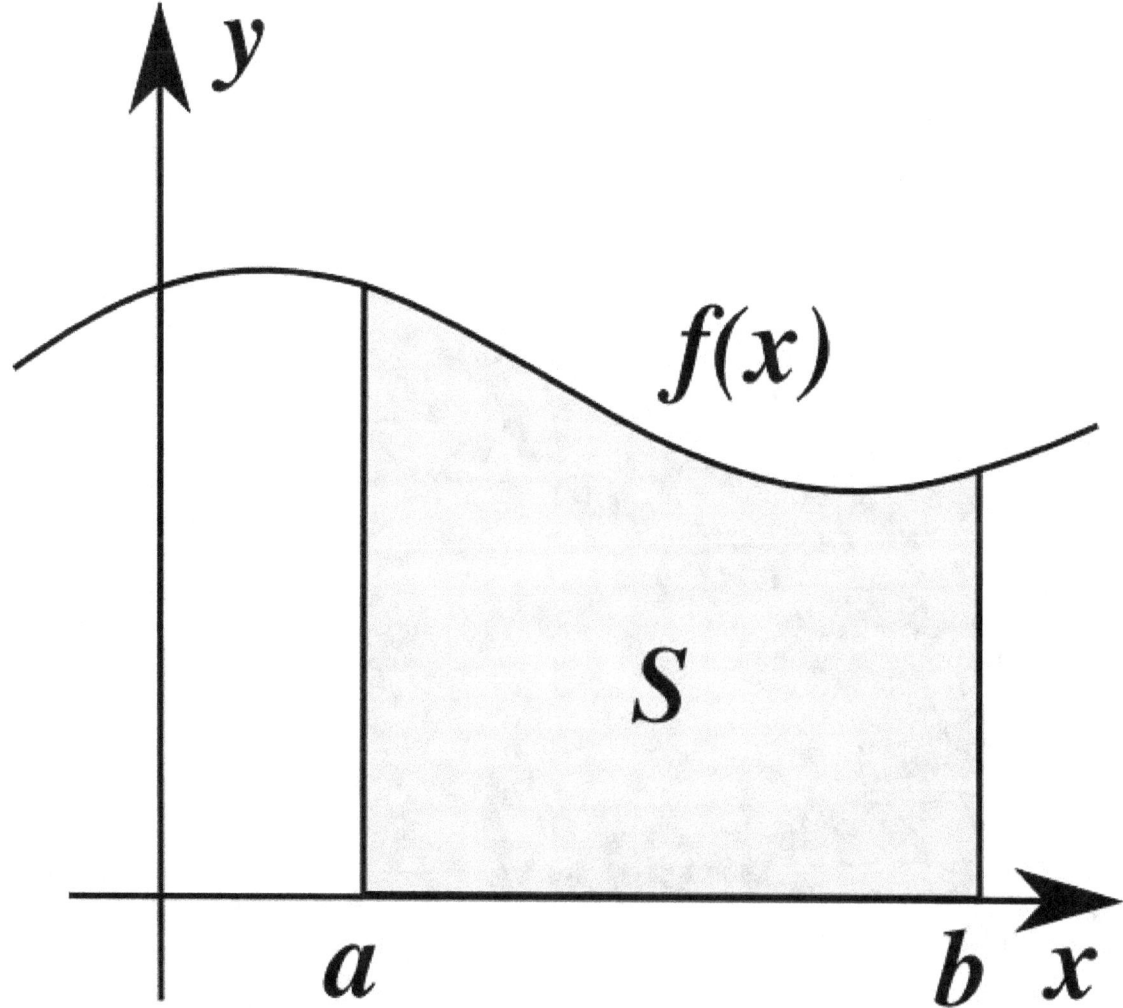

Integration can be thought of as measuring the area under a curve, defined by f(x), between two points (here a and b).

The logarithmic spiral of the Nautilus shell is a classical image used to depict the growth and change related to calculus

Chapter 30

Vector calculus

Not to be confused with geometric calculus or matrix calculus.

Vector calculus (or **vector analysis**) is a branch of mathematics concerned with differentiation and integration of vector fields, primarily in 3-dimensional Euclidean space $\mathbb{R}^3$. The term "vector calculus" is sometimes used as a synonym for the broader subject of multivariable calculus, which includes vector calculus as well as partial differentiation and multiple integration. Vector calculus plays an important role in differential geometry and in the study of partial differential equations. It is used extensively in physics and engineering, especially in the description of electromagnetic fields, gravitational fields and fluid flow.

Vector calculus was developed from quaternion analysis by J. Willard Gibbs and Oliver Heaviside near the end of the 19th century, and most of the notation and terminology was established by Gibbs and Edwin Bidwell Wilson in their 1901 book, *Vector Analysis*. In the conventional form using cross products, vector calculus does not generalize to higher dimensions, while the alternative approach of geometric algebra, which uses exterior products does generalize, as discussed below.

30.1 Basic objects

30.1.1 Scalar fields

Main article: Scalar field

A scalar field associates a scalar value to every point in a space. The scalar may either be a mathematical number or a physical quantity. Examples of scalar fields in applications include the temperature distribution throughout space, the pressure distribution in a fluid, and spin-zero quantum fields, such as the Higgs field. These fields are the subject of scalar field theory.

30.1.2 Vector fields

Main article: Vector field

A vector field is an assignment of a vector to each point in a subset of space.[1] A vector field in the plane, for instance, can be visualized as a collection of arrows with a given magnitude and direction each attached to a point in the plane. Vector fields are often used to model, for example, the speed and direction of a moving fluid throughout space, or the strength and direction of some force, such as the magnetic or gravitational force, as it changes from point to point.

30.1.3 Vectors and pseudovectors

In more advanced treatments, one further distinguishes pseudovector fields and pseudoscalar fields, which are identical to vector fields and scalar fields except that they change sign under an orientation-reversing map: for example, the curl of a vector field is a pseudovector field, and if one reflects a vector field, the curl points in the opposite direction. This distinction is clarified and elaborated in geometric algebra, as described below.

30.2 Vector operations

30.2.1 Algebraic operations

Main article: Euclidean vector § Basic properties

The basic algebraic (non-differential) operations in vector calculus are referred to as **vector algebra**, being defined for a vector space and then globally applied to a vector field, and consist of:

scalar multiplication multiplication of a scalar field and a vector field, yielding a vector field: $a\mathbf{v}$;

vector addition addition of two vector fields, yielding a vector field: $\mathbf{v}_1 + \mathbf{v}_2$;

dot product multiplication of two vector fields, yielding a scalar field: $\mathbf{v}_1 \cdot \mathbf{v}_2$;

cross product multiplication of two vector fields, yielding a vector field: $\mathbf{v}_1 \times \mathbf{v}_2$;

There are also two triple products:

scalar triple product the dot product of a vector and a cross product of two vectors: $\mathbf{v}_1 \cdot (\mathbf{v}_2 \times \mathbf{v}_3)$;

vector triple product the cross product of a vector and a cross product of two vectors: $\mathbf{v}_1 \times (\mathbf{v}_2 \times \mathbf{v}_3)$ or $(\mathbf{v}_3 \times \mathbf{v}_2) \times \mathbf{v}_1$
  ;

although these are less often used as basic operations, as they can be expressed in terms of the dot and cross products.

30.2.2 Differential operations

Main articles: Gradient, Curl (mathematics), Divergence and Laplacian

Vector calculus studies various differential operators defined on scalar or vector fields, which are typically expressed in terms of the del operator (∇), also known as "nabla". The five most important differential operations in vector calculus are:

where the curl and divergence differ because the former uses a cross product and the latter a dot product. f denotes a scalar field and F denotes a vector field. A quantity called the Jacobian is useful for studying functions when both the domain and range of the function are multivariable, such as a change of variables during integration.

30.3 Theorems

Likewise, there are several important theorems related to these operators which generalize the fundamental theorem of calculus to higher dimensions:

30.4 Applications

30.4.1 Linear approximations

Main article: Linear approximation

Linear approximations are used to replace complicated functions with linear functions that are almost the same. Given a differentiable function $f(x, y)$ with real values, one can approximate $f(x, y)$ for (x, y) close to (a, b) by the formula

$$f(x, y) \approx f(a, b) + \frac{\partial f}{\partial x}(a, b)(x - a) + \frac{\partial f}{\partial y}(a, b)(y - b).$$

The right-hand side is the equation of the plane tangent to the graph of $z = f(x, y)$ at (a, b).

30.4.2 Optimization

Main article: Mathematical optimization

For a continuously differentiable function of several real variables, a point P (that is a set of values for the input variables, which is viewed as a point in $\mathbf{R}^n$) is **critical** if all of the partial derivatives of the function are zero at P, or, equivalently, if its gradient is zero. The critical values are the values of the function at the critical points.

If the function is smooth, or, at least twice continuously differentiable, a critical point may be either a local maximum, a local minimum or a saddle point. The different cases may be distinguished by considering the eigenvalues of the Hessian matrix of second derivatives.

By Fermat's theorem, all local maxima and minima of a differentiable function occur at critical points. Therefore, to find the local maxima and minima, it suffices, theoretically, to compute the zeros of the gradient and the eigenvalues of the Hessian matrix at these zeros.

30.4.3 Physics and engineering

Vector calculus is particularly useful in studying:

- Center of mass
- Field theory
- Kinematics

30.5 Generalizations

30.5.1 Different 3-manifolds

Vector calculus is initially defined for Euclidean 3-space, $\mathbb{R}^3$, which has additional structure beyond simply being a 3-dimensional real vector space, namely: a norm (giving a notion of length) defined via an inner product (the dot product), which in turn gives a notion of angle, and an orientation, which gives a notion of left-handed and right-handed. These structures give rise to a volume form, and also the cross product, which is used pervasively in vector calculus.

The gradient and divergence require only the inner product, while the curl and the cross product also requires the handedness of the coordinate system to be taken into account (see cross product and handedness for more detail).

Vector calculus can be defined on other 3-dimensional real vector spaces if they have an inner product (or more generally a symmetric nondegenerate form) and an orientation; note that this is less data than an isomorphism to Euclidean space, as it does not require a set of coordinates (a frame of reference), which reflects the fact that vector calculus is invariant under rotations (the special orthogonal group SO(3)).

More generally, vector calculus can be defined on any 3-dimensional oriented Riemannian manifold, or more generally pseudo-Riemannian manifold. This structure simply means that the tangent space at each point has an inner product (more generally, a symmetric nondegenerate form) and an orientation, or more globally that there is a symmetric nondegenerate metric tensor and an orientation, and works because vector calculus is defined in terms of tangent vectors at each point.

30.5.2 Other dimensions

Most of the analytic results are easily understood, in a more general form, using the machinery of differential geometry, of which vector calculus forms a subset. Grad and div generalize immediately to other dimensions, as do the gradient theorem, divergence theorem, and Laplacian (yielding harmonic analysis), while curl and cross product do not generalize as directly.

From a general point of view, the various fields in (3-dimensional) vector calculus are uniformly seen as being k-vector fields: scalar fields are 0-vector fields, vector fields are 1-vector fields, pseudovector fields are 2-vector fields, and pseudoscalar fields are 3-vector fields. In higher dimensions there are additional types of fields (scalar/vector/pseudovector/pseudo scalar corresponding to $0/1/n-1/n$ dimensions, which is exhaustive in dimension 3), so one cannot only work with (pseudo)scalars and (pseudo)vectors.

In any dimension, assuming a nondegenerate form, grad of a scalar function is a vector field, and div of a vector field is a scalar function, but only in dimension 3 and 7 (and, trivially, dimension 0) is the curl of a vector field a vector field, and only in 3 or 7 dimensions can a cross product be defined (generalizations in other dimensionalities either require $n-1$ vectors to yield 1 vector, or are alternative Lie algebras, which are more general antisymmetric bilinear products). The generalization of grad and div, and how curl may be generalized is elaborated at Curl: Generalizations; in brief, the curl of a vector field is a bivector field, which may be interpreted as the special orthogonal Lie algebra of infinitesimal rotations; however, this cannot be identified with a vector field because the dimensions differ - there are 3 dimensions of rotations in 3 dimensions, but 6 dimensions of rotations in 4 dimensions (and more generally $\binom{n}{2} = \frac{1}{2}n(n-1)$ dimensions of rotations in n dimensions).

There are two important alternative generalizations of vector calculus. The first, geometric algebra, uses k-vector fields instead of vector fields (in 3 or fewer dimensions, every k-vector field can be identified with a scalar function or vector field, but this is not true in higher dimensions). This replaces the cross product, which is specific to 3 dimensions, taking in two vector fields and giving as output a vector field, with the exterior product, which exists in all dimensions and takes in two vector fields, giving as output a bivector (2-vector) field. This product yields Clifford algebras as the algebraic structure on vector spaces (with an orientation and nondegenerate form). Geometric algebra is mostly used in generalizations of physics and other applied fields to higher dimensions.

The second generalization uses differential forms (k-covector fields) instead of vector fields or k-vector fields, and is widely used in mathematics, particularly in differential geometry, geometric topology, and harmonic analysis, in particular yielding Hodge theory on oriented pseudo-Riemannian manifolds. From this point of view, grad, curl, and div correspond to the exterior derivative of 0-forms, 1-forms, and 2-forms, respectively, and the key theorems of vector calculus are all special cases of the general form of Stokes' theorem.

From the point of view of both of these generalizations, vector calculus implicitly identifies mathematically distinct objects, which makes the presentation simpler but the underlying mathematical structure and generalizations less clear. From the point of view of geometric algebra, vector calculus implicitly identifies k-vector fields with vector fields or scalar functions: 0-vectors and 3-vectors with scalars, 1-vectors and 2-vectors with vectors. From the point of view of differential forms, vector calculus implicitly identifies k-forms with scalar fields or vector fields: 0-forms and 3-forms with scalar fields, 1-forms and 2-forms with vector fields. Thus for example the curl naturally takes as input a vector field, but naturally has as output a 2-vector field or 2-form (hence pseudovector field), which is then interpreted as a vector field, rather than directly taking a vector field to a vector field; this is reflected in the curl of a vector field in higher dimensions not having as output a vector field.

30.6 See also

- Real-valued function
- Function of a real variable
- Real multivariable function
- Vector calculus identities
- Del in cylindrical and spherical coordinates
- Directional derivative
- Irrotational vector field
- Solenoidal vector field
- Laplacian vector field
- Helmholtz decomposition
- Orthogonal coordinates
- Skew coordinates
- Curvilinear coordinates
- Tensor

30.7 Notes

- There is also the **perp dot product**,[2] which is essentially the dot product of two vectors, one vector rotated by $\pi/2$ rads, equivalently the magnitude of the cross product:

$$\mathbf{v}_1 \perp \cdot \mathbf{v}_2 = |\mathbf{v}_1 \times \mathbf{v}_2| = |\mathbf{v}_1||\mathbf{v}_2|\sin\theta$$

where θ is the included angle between v_1 and v_2. It is rarely used, since the dot and cross product both incorporate it.

30.8 References

[1] Galbis, Antonio & Maestre, Manuel (2012). *Vector Analysis Versus Vector Calculus*. Springer. p. 12. ISBN 978-1-4614-2199-3.

[2] Weisstein, Eric W. "Perp Dot Product." From MathWorld--A Wolfram Web Resource.

- Sandro Caparrini (2002) "The discovery of the vector representation of moments and angular velocity", Archive for History of Exact Sciences 56:151–81.

- Michael J. Crowe (1967). *A History of Vector Analysis : The Evolution of the Idea of a Vectorial System*. Dover Publications; Reprint edition. ISBN 0-486-67910-1.

- J.E. Marsden (1976). *Vector Calculus*. W. H. Freeman & Company. ISBN 0-7167-0462-5.

- H. M. Schey (2005). *Div Grad Curl and all that: An informal text on vector calculus*. W. W. Norton & Company. ISBN 0-393-92516-1.

- Barry Spain (1965) Vector Analysis, 2nd edition, link from Internet Archive.

- Chen-To Tai (1995). *A historical study of vector analysis*. Technical Report RL 915, Radiation Laboratory, University of Michigan.

30.9 External links

- Hazewinkel, Michiel, ed. (2001), "Vector analysis", *Encyclopedia of Mathematics*, Springer, ISBN 978-1-55608-010-4

- Hazewinkel, Michiel, ed. (2001), "Vector algebra", *Encyclopedia of Mathematics*, Springer, ISBN 978-1-55608-010-4

- Vector Calculus Video Lectures from University of New South Wales on Academic Earth

- A survey of the improper use of ∇ in vector analysis (1994) Tai, Chen

- Expanding vector analysis to an oblique coordinate system

- Vector Analysis: A Text-book for the Use of Students of Mathematics and Physics, (based upon the lectures of Willard Gibbs) by Edwin Bidwell Wilson, published 1902.

- Earliest Known Uses of Some of the Words of Mathematics: Vector Analysis

Chapter 31

Differential equation

Not to be confused with Difference equation.

A **differential equation** is a mathematical equation that relates some function with its derivatives. In applications, the

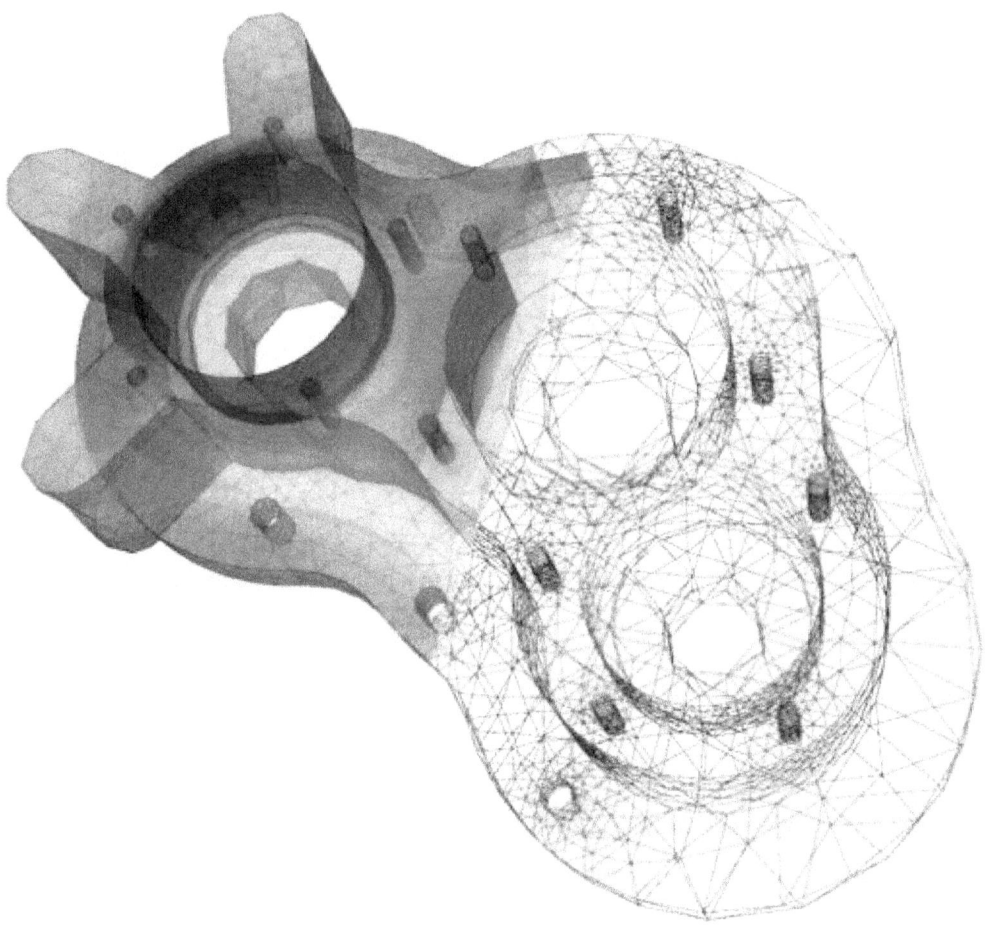

Visualization of heat transfer in a pump casing, created by solving the heat equation. Heat is being generated internally in the casing and being cooled at the boundary, providing a steady state temperature distribution.

functions usually represent physical quantities, the derivatives represent their rates of change, and the equation defines a relationship between the two. Because such relations are extremely common, differential equations play a prominent role in many disciplines including engineering, physics, economics, and biology.

In pure mathematics, differential equations are studied from several different perspectives, mostly concerned with their solutions—the set of functions that satisfy the equation. Only the simplest differential equations are solvable by explicit formulas; however, some properties of solutions of a given differential equation may be determined without finding their exact form.

If a self-contained formula for the solution is not available, the solution may be numerically approximated using computers. The theory of dynamical systems puts emphasis on qualitative analysis of systems described by differential equations, while many numerical methods have been developed to determine solutions with a given degree of accuracy.

31.1 History

Differential equations first came into existence with the invention of calculus by Newton and Leibniz. In Chapter 2 of his 1671 work "Methodus fluxionum et Serierum Infinitarum",[1] Isaac Newton listed three kinds of differential equations:

$$\frac{dy}{dx} = f(x)$$

$$\frac{dy}{dx} = f(x, y)$$

$$x_1 \frac{\partial y}{\partial x_1} + x_2 \frac{\partial y}{\partial x_2} = y$$

He solves these examples and others using infinite series and discusses the non-uniqueness of solutions.

Jacob Bernoulli proposed the Bernoulli differential equation in 1695.[2] This is an ordinary differential equation of the form

$$y' + P(x)y = Q(x)y^n$$

for which the following year Leibniz obtained solutions by simplifying it.[3]

Historically, the problem of a vibrating string such as that of a musical instrument was studied by Jean le Rond d'Alembert, Leonhard Euler, Daniel Bernoulli, and Joseph-Louis Lagrange.[4][5][6][7] In 1746, d'Alembert discovered the one-dimensional wave equation, and within ten years Euler discovered the three-dimensional wave equation.[8]

The Euler–Lagrange equation was developed in the 1750s by Euler and Lagrange in connection with their studies of the tautochrone problem. This is the problem of determining a curve on which a weighted particle will fall to a fixed point in a fixed amount of time, independent of the starting point.

Lagrange solved this problem in 1755 and sent the solution to Euler. Both further developed Lagrange's method and applied it to mechanics, which led to the formulation of Lagrangian mechanics.

Fourier published his work on heat flow in *Théorie analytique de la chaleur* (The Analytic Theory of Heat),[9] in which he based his reasoning on Newton's law of cooling, namely, that the flow of heat between two adjacent molecules is proportional to the extremely small difference of their temperatures. Contained in this book was Fourier's proposal of his heat equation for conductive diffusion of heat. This partial differential equation is now taught to every student of mathematical physics.

31.2 Example

For example, in classical mechanics, the motion of a body is described by its position and velocity as the time value varies. Newton's laws allow (given the position, velocity, acceleration and various forces acting on the body) one to express these variables dynamically as a differential equation for the unknown position of the body as a function of time.

In some cases, this differential equation (called an equation of motion) may be solved explicitly.

An example of modelling a real world problem using differential equations is the determination of the velocity of a ball falling through the air, considering only gravity and air resistance. The ball's acceleration towards the ground is the acceleration due to gravity minus the acceleration due to air resistance.

Gravity is considered constant, and air resistance may be modeled as proportional to the ball's velocity. This means that the ball's acceleration, which is a derivative of its velocity, depends on the velocity (and the velocity depends on time). Finding the velocity as a function of time involves solving a differential equation and verifying its validity.

31.3   Main topics

31.3.1   Ordinary differential equations

Main article: Ordinary differential equation

An ordinary differential equation (**ODE**) is an equation containing a function of one independent variable and its derivatives. The term "*ordinary*" is used in contrast with the term partial differential equation which may be with respect to *more than* one independent variable.

Linear differential equations, which have solutions that can be added and multiplied by coefficients, are well-defined and understood, and exact closed-form solutions are obtained. By contrast, ODEs that lack additive solutions are nonlinear, and solving them is far more intricate, as one can rarely represent them by elementary functions in closed form: Instead, exact and analytic solutions of ODEs are in series or integral form. Graphical and numerical methods, applied by hand or by computer, may approximate solutions of ODEs and perhaps yield useful information, often sufficing in the absence of exact, analytic solutions.

31.3.2   Partial differential equations

Main article: Partial differential equation

A partial differential equation (**PDE**) is a differential equation that contains unknown multivariable functions and their partial derivatives. (This is in contrast to ordinary differential equations, which deal with functions of a single variable and their derivatives.) PDEs are used to formulate problems involving functions of several variables, and are either solved by hand, or used to create a relevant computer model.

PDEs can be used to describe a wide variety of phenomena such as sound, heat, electrostatics, electrodynamics, fluid flow, elasticity, or quantum mechanics. These seemingly distinct physical phenomena can be formalised similarly in terms of PDEs. Just as ordinary differential equations often model one-dimensional dynamical systems, partial differential equations often model multidimensional systems. PDEs find their generalisation in stochastic partial differential equations.

31.4   Linear and non-linear

Both ordinary and partial differential equations are broadly classified as **linear** and **nonlinear**.

- A differential equation is **linear** if the unknown function and its derivatives appear to the power 1 (products of the unknown function and its derivatives are not allowed) and **nonlinear** otherwise. The characteristic property of linear equations is that their solutions form an affine subspace of an appropriate function space, which results in much more developed theory of linear differential equations. **Homogeneous** linear differential equations are a further subclass for which the space of solutions is a linear subspace i.e. the sum of any set of solutions or multiples of solutions is also a solution. The coefficients of the unknown function and its derivatives in a linear differential equation are allowed to be (known) functions of the independent variable or variables; if these coefficients are constants then one speaks of a **constant coefficient linear differential equation**.

- There are very few methods of solving nonlinear differential equations exactly; those that are known typically depend on the equation having particular symmetries. Nonlinear differential equations can exhibit very complicated behavior over extended time intervals, characteristic of chaos. Even the fundamental questions of existence, uniqueness, and extendability of solutions for nonlinear differential equations, and well-posedness of initial and boundary value problems for nonlinear PDEs are hard problems and their resolution in special cases is considered to be a significant advance in the mathematical theory (cf. Navier–Stokes existence and smoothness). However, if the differential equation is a correctly formulated representation of a meaningful physical process, then one expects it to have a solution.[10]

Linear differential equations frequently appear as approximations to nonlinear equations. These approximations are only valid under restricted conditions. For example, the harmonic oscillator equation is an approximation to the nonlinear pendulum equation that is valid for small amplitude oscillations (see below).

31.4.1 Examples

In the first group of examples, let u be an unknown function of x, and c and ω are known constants.

- Inhomogeneous first-order linear constant coefficient ordinary differential equation:

$$\frac{du}{dx} = cu + x^2.$$

- Homogeneous second-order linear ordinary differential equation:

$$\frac{d^2u}{dx^2} - x\frac{du}{dx} + u = 0.$$

- Homogeneous second-order linear constant coefficient ordinary differential equation describing the harmonic oscillator:

$$\frac{d^2u}{dx^2} + \omega^2 u = 0.$$

- Inhomogeneous first-order nonlinear ordinary differential equation:

$$\frac{du}{dx} = u^2 + 4.$$

- Second-order nonlinear (due to sine function) ordinary differential equation describing the motion of a pendulum of length L:

$$L\frac{d^2u}{dx^2} + g\sin u = 0.$$

In the next group of examples, the unknown function u depends on two variables x and t or x and y.

- Homogeneous first-order linear partial differential equation:

$$\frac{\partial u}{\partial t} + t\frac{\partial u}{\partial x} = 0.$$

- Homogeneous second-order linear constant coefficient partial differential equation of elliptic type, the Laplace equation:

$$\frac{\partial^2 u}{\partial x^2} + \frac{\partial^2 u}{\partial y^2} = 0.$$

- Third-order nonlinear partial differential equation, the Korteweg–de Vries equation:

$$\frac{\partial u}{\partial t} = 6u\frac{\partial u}{\partial x} - \frac{\partial^3 u}{\partial x^3}.$$

31.5   Existence of solutions

Solving differential equations is not like solving algebraic equations. Not only are their solutions oftentimes unclear, but whether solutions are unique or exist at all are also notable subjects of interest.

For first order initial value problems, it is easy to tell whether a unique solution exists. Given any point (a, b) in the xy-plane, define some rectangular region Z, such that $Z = [l, m] \times [n, p]$ and (a, b) is in Z. If we are given a differential equation $\frac{dx}{dt} = g(x, t)$ and an initial condition $x(t_0) = x_0$, then there is a unique solution to this initial value problem if $g(x, t)$ and $\frac{\partial g}{\partial x}$ are both continuous on Z. This unique solution exists on some interval with its center at a.

However, this only helps us with first order initial value problems. Suppose we had a linear initial value problem of the nth order:

$$f_n(x)\frac{d^n y}{dx^n} + \cdots + f_1(x)\frac{dy}{dx} + f_0(x)y = h(x)$$

such that

$$y(x_0) = y_0, y'(x_0) = y'_0, y''(x_0) = y''_0, \cdots$$

For any nonzero $f_n(x)$, if $\{f_0, f_1, \cdots\}$ and g are continuous on some interval containing x_0, y is unique and exists.[11]

31.6   Related concepts

- A delay differential equation (DDE) is an equation for a function of a single variable, usually called **time**, in which the derivative of the function at a certain time is given in terms of the values of the function at earlier times.

- A stochastic differential equation (SDE) is an equation in which the unknown quantity is a stochastic process and the equation involves some known stochastic processes, for example, the Wiener process in the case of diffusion equations.

- A differential algebraic equation (DAE) is a differential equation comprising differential and algebraic terms, given in implicit form.

31.7 Connection to difference equations

See also: Time scale calculus

The theory of differential equations is closely related to the theory of difference equations, in which the coordinates assume only discrete values, and the relationship involves values of the unknown function or functions and values at nearby coordinates. Many methods to compute numerical solutions of differential equations or study the properties of differential equations involve approximation of the solution of a differential equation by the solution of a corresponding difference equation.

31.8 Applications

The study of differential equations is a wide field in pure and applied mathematics, physics, and engineering. All of these disciplines are concerned with the properties of differential equations of various types. Pure mathematics focuses on the existence and uniqueness of solutions, while applied mathematics emphasizes the rigorous justification of the methods for approximating solutions. Differential equations play an important role in modelling virtually every physical, technical, or biological process, from celestial motion, to bridge design, to interactions between neurons. Differential equations such as those used to solve real-life problems may not necessarily be directly solvable, i.e. do not have closed form solutions. Instead, solutions can be approximated using numerical methods.

Many fundamental laws of physics and chemistry can be formulated as differential equations. In biology and economics, differential equations are used to model the behavior of complex systems. The mathematical theory of differential equations first developed together with the sciences where the equations had originated and where the results found application. However, diverse problems, sometimes originating in quite distinct scientific fields, may give rise to identical differential equations. Whenever this happens, mathematical theory behind the equations can be viewed as a unifying principle behind diverse phenomena. As an example, consider propagation of light and sound in the atmosphere, and of waves on the surface of a pond. All of them may be described by the same second-order partial differential equation, the wave equation, which allows us to think of light and sound as forms of waves, much like familiar waves in the water. Conduction of heat, the theory of which was developed by Joseph Fourier, is governed by another second-order partial differential equation, the heat equation. It turns out that many diffusion processes, while seemingly different, are described by the same equation; the Black–Scholes equation in finance is, for instance, related to the heat equation.

31.8.1 Physics

- Euler–Lagrange equation in classical mechanics

- Hamilton's equations in classical mechanics

- Radioactive decay in nuclear physics

- Newton's law of cooling in thermodynamics

- The wave equation

- The heat equation in thermodynamics

- Laplace's equation, which defines harmonic functions

- Poisson's equation

- The geodesic equation

- The Navier–Stokes equations in fluid dynamics

- The Diffusion equation in stochastic processes

- The Convection–diffusion equation in fluid dynamics

- The Cauchy–Riemann equations in complex analysis

- The Poisson–Boltzmann equation in molecular dynamics

- The shallow water equations

- Universal differential equation

- The Lorenz equations whose solutions exhibit chaotic flow.

Classical mechanics

So long as the force acting on a particle is known, Newton's second law is sufficient to describe the motion of a particle. Once independent relations for each force acting on a particle are available, they can be substituted into Newton's second law to obtain an ordinary differential equation, which is called the *equation of motion*.

Electrodynamics

Maxwell's equations are a set of partial differential equations that, together with the Lorentz force law, form the foundation of classical electrodynamics, classical optics, and electric circuits. These fields in turn underlie modern electrical and communications technologies. Maxwell's equations describe how electric and magnetic fields are generated and altered by each other and by charges and currents. They are named after the Scottish physicist and mathematician James Clerk Maxwell, who published an early form of those equations between 1861 and 1862.

General relativity

The Einstein field equations (EFE; also known as "Einstein's equations") are a set of ten partial differential equations in Albert Einstein's general theory of relativity which describe the fundamental interaction of gravitation as a result of spacetime being curved by matter and energy.[12] First published by Einstein in 1915[13] as a tensor equation, the EFE equate local spacetime curvature (expressed by the Einstein tensor) with the local energy and momentum within that spacetime (expressed by the stress–energy tensor).[14]

Quantum mechanics

In quantum mechanics, the analogue of Newton's law is Schrödinger's equation (a partial differential equation) for a quantum system (usually atoms, molecules, and subatomic particles whether free, bound, or localized). It is not a simple algebraic equation, but in general a linear partial differential equation, describing the time-evolution of the system's wave function (also called a "state function").[15]:1-2

31.8.2   Biology

- Verhulst equation – biological population growth

- von Bertalanffy model – biological individual growth

- Replicator dynamics – found in theoretical biology

- Hodgkin–Huxley model – neural action potentials

Predator-prey equations

The Lotka–Volterra equations, also known as the predator–prey equations, are a pair of first-order, non-linear, differential equations frequently used to describe the dynamics of biological systems in which two species interact, one as a predator and the other as prey.

31.8.3 Chemistry

The *rate law* or rate equation for a chemical reaction is a differential equation that links the reaction rate with concentrations or pressures of reactants and constant parameters (normally rate coefficients and partial reaction orders).[16] To determine the rate equation for a particular system one combines the reaction rate with a mass balance for the system.[17]

31.8.4 Economics

- The key equation of the Solow–Swan model is $\frac{\partial k(t)}{\partial t} = s[k(t)]^\alpha - \delta k(t)$

- The Black–Scholes PDE

- Malthusian growth model

- The Vidale–Wolfe advertising model

31.9 See also

- Complex differential equation

- Exact differential equation

- Initial condition

- Integral equations

- Numerical methods

- Picard–Lindelöf theorem on existence and uniqueness of solutions

- Recurrence relation, also known as 'Difference Equation'

31.10 References

[1] Newton, Isaac. (c.1671). Methodus Fluxionum et Serierum Infinitarum (The Method of Fluxions and Infinite Series), published in 1736 [Opuscula, 1744, Vol. I. p. 66].

[2] Bernoulli, Jacob (1695). "Explicationes, Annotationes & Additiones ad ea, quae in Actis sup. de Curva Elastica, Isochrona Paracentrica, & Velaria, hinc inde memorata, & paratim controversa legundur; ubi de Linea mediarum directionum, alliisque novis", *Acta Eruditorum*

[3] Hairer, Ernst; Nørsett, Syvert Paul; Wanner, Gerhard (1993), *Solving ordinary differential equations I: Nonstiff problems*, Berlin, New York: Springer-Verlag, ISBN 978-3-540-56670-0

[4] Cannon, John T.; Dostrovsky, Sigalia (1981). "The evolution of dynamics, vibration theory from 1687 to 1742". Studies in the History of Mathematics and Physical Sciences **6**. New York: Springer-Verlag. pp. ix + 184 pp. ISBN 0-3879-0626-6. GRAY, JW (July 1983). "BOOK REVIEWS". *BULLETIN (New Series) OF THE AMERICAN MATHEMATICAL SOCIETY* **9** (1). (retrieved 13 Nov 2012).

[5] Wheeler, Gerard F.; Crummett, William P. (1987). "The Vibrating String Controversy". *Am.   J. Phys.* **55** (1): 33–37. doi:10.1119/1.15311.

[6] For a special collection of the 9 groundbreaking papers by the three authors, see First Appearance of the wave equation: D'Alembert, Leonhard Euler, Daniel Bernoulli. - the controversy about vibrating strings (retrieved 13 Nov 2012). Herman HJ Lynge and Son.

[7] For de Lagrange's contributions to the acoustic wave equation, can consult Acoustics: An Introduction to Its Physical Principles and Applications Allan D. Pierce, Acoustical Soc of America, 1989; page 18.(retrieved 9 Dec 2012)

[8] Speiser, David. *Discovering the Principles of Mechanics 1600-1800*, p. 191 (Basel: Birkhäuser, 2008).

[9] Fourier, Joseph (1822). *Théorie analytique de la chaleur* (in French). Paris: Firmin Didot Père et Fils. OCLC 2688081.

[10] Boyce, William E.; DiPrima, Richard C. (1967). *Elementary Differential Equations and Boundary Value Problems* (4th ed.). John Wiley & Sons. p. 3.

[11] Zill, Dennis G. *A First Course in Differential Equations* (5th ed.). Brooks/Cole. ISBN 0-534-37388-7.

[12] Einstein, Albert (1916). "The Foundation of the General Theory of Relativity" (PDF). *Annalen der Physik* **354** (7): 769. Bibcode:1916AnP...354..769E. doi:10.1002/andp.19163540702.

[13] Einstein, Albert (November 25, 1915). "Die Feldgleichungen der Gravitation". *Sitzungsberichte der Preussischen Akademie der Wissenschaften zu Berlin*: 844–847. Retrieved 2006-09-12.

[14] Misner, Charles W.; Thorne, Kip S.; Wheeler, John Archibald (1973). *Gravitation*. San Francisco: W. H. Freeman. ISBN 978-0-7167-0344-0 Chapter 34, p. 916.

[15] Griffiths, David J. (2004). *Introduction to Quantum Mechanics (2nd ed.)*. Prentice Hall. ISBN 0-13-111892-7

[16] IUPAC Gold Book definition of rate law. See also: According to IUPAC Compendium of Chemical Terminology.

[17] Kenneth A. Connors *Chemical Kinetics, the study of reaction rates in solution*, 1991, VCH Publishers.

31.11   Further reading

- P. Abbott and H. Neill, *Teach Yourself Calculus*, 2003 pages 266-277

- P. Blanchard, R. L. Devaney, G. R. Hall, *Differential Equations*, Thompson, 2006

- E. A. Coddington and N. Levinson, *Theory of Ordinary Differential Equations*, McGraw-Hill, 1955

- E. L. Ince, *Ordinary Differential Equations*, Dover Publications, 1956

- W. Johnson, *A Treatise on Ordinary and Partial Differential Equations*, John Wiley and Sons, 1913, in University of Michigan Historical Math Collection

- A. D. Polyanin and V. F. Zaitsev, *Handbook of Exact Solutions for Ordinary Differential Equations (2nd edition)*, Chapman & Hall/CRC Press, Boca Raton, 2003. ISBN 1-58488-297-2.

- R. I. Porter, *Further Elementary Analysis*, 1978, chapter XIX Differential Equations

- Teschl, Gerald (2012). *Ordinary Differential Equations and Dynamical Systems*. Providence: American Mathematical Society. ISBN 978-0-8218-8328-0.

- D. Zwillinger, *Handbook of Differential Equations (3rd edition)*, Academic Press, Boston, 1997.

31.12 External links

- Lectures on Differential Equations MIT Open CourseWare Videos

- Online Notes / Differential Equations Paul Dawkins, Lamar University

- Differential Equations, S.O.S. Mathematics

- Introduction to modeling via differential equations Introduction to modeling by means of differential equations, with critical remarks.

- Mathematical Assistant on Web Symbolic ODE tool, using Maxima

- Exact Solutions of Ordinary Differential Equations

- Collection of ODE and DAE models of physical systems MATLAB models

- Notes on Diffy Qs: Differential Equations for Engineers An introductory textbook on differential equations by Jiri Lebl of UIUC

- Khan Academy Video playlist on differential equations Topics covered in a first year course in differential equations.

- MathDiscuss Video playlist on differential equations

Chapter 32

Dynamical system

This article is about the general aspects of dynamical systems. For technical details, see Dynamical system (definition). For the study, see Dynamical systems theory.

"Dynamical" redirects here. For other uses, see Dynamical (disambiguation).

In mathematics, a **dynamical system** is a set of relationships among two or more measurable quantities, in which a fixed rule describes how the quantities evolve over time in response to their own values. Examples include the mathematical models that describe the swinging of a clock pendulum, the flow of water in a pipe, and the number of fish each springtime in a lake.

At any given time a dynamical system has a state given by a set of real numbers (a vector) that can be represented by a point in an appropriate state space (a geometrical manifold). The *evolution rule* of the dynamical system is a function that describes what future states follow from the current state. Often the function is deterministic; in other words, for a given time interval only one future state follows from the current state;[1][2] however, some systems are stochastic, in that random events also affect the evolution of the state variables.

32.1   Overview

The concept of a dynamical system has its origins in Newtonian mechanics. There, as in other natural sciences and engineering disciplines, the evolution rule of dynamical systems is an implicit relation that gives the state of the system for only a short time into the future. (The relation is either a differential equation, difference equation or other time scale.) To determine the state for all future times requires iterating the relation many times—each advancing time a small step. The iteration procedure is referred to as *solving the system* or *integrating the system*. If the system can be solved, given an initial point it is possible to determine all its future positions, a collection of points known as a *trajectory* or *orbit*.

Before the advent of computers, finding an orbit required sophisticated mathematical techniques and could be accomplished only for a small class of dynamical systems. Numerical methods implemented on electronic computing machines have simplified the task of determining the orbits of a dynamical system.

For simple dynamical systems, knowing the trajectory is often sufficient, but most dynamical systems are too complicated to be understood in terms of individual trajectories. The difficulties arise because:

- The systems studied may only be known approximately—the parameters of the system may not be known precisely or terms may be missing from the equations. The approximations used bring into question the validity or relevance of numerical solutions. To address these questions several notions of stability have been introduced in the study of dynamical systems, such as Lyapunov stability or structural stability. The stability of the dynamical system implies that there is a class of models or initial conditions for which the trajectories would be equivalent. The operation for comparing orbits to establish their equivalence changes with the different notions of stability.

- The type of trajectory may be more important than one particular trajectory. Some trajectories may be periodic, whereas others may wander through many different states of the system. Applications often require enumerating

The Lorenz attractor arises in the study of the Lorenz Oscillator, a dynamical system.

these classes or maintaining the system within one class. Classifying all possible trajectories has led to the qualitative study of dynamical systems, that is, properties that do not change under coordinate changes. Linear dynamical systems and systems that have two numbers describing a state are examples of dynamical systems where the possible classes of orbits are understood.

- The behavior of trajectories as a function of a parameter may be what is needed for an application. As a parameter is varied, the dynamical systems may have bifurcation points where the qualitative behavior of the dynamical system changes. For example, it may go from having only periodic motions to apparently erratic behavior, as in the transition to turbulence of a fluid.

- The trajectories of the system may appear erratic, as if random. In these cases it may be necessary to compute averages using one very long trajectory or many different trajectories. The averages are well defined for ergodic systems and a more detailed understanding has been worked out for hyperbolic systems. Understanding the probabilistic aspects of dynamical systems has helped establish the foundations of statistical mechanics and of chaos.

32.2   History

Many people regard Henri Poincaré as the founder of dynamical systems.[3] Poincaré published two now classical monographs, "New Methods of Celestial Mechanics" (1892–1899) and "Lectures on Celestial Mechanics" (1905–1910). In them, he successfully applied the results of their research to the problem of the motion of three bodies and studied in detail the behavior of solutions (frequency, stability, asymptotic, and so on). These papers included the Poincaré recurrence theorem, which states that certain systems will, after a sufficiently long but finite time, return to a state very close to the initial state.

Aleksandr Lyapunov developed many important approximation methods. His methods, which he developed in 1899, make it possible to define the stability of sets of ordinary differential equations. He created the modern theory of the stability of a dynamic system.

In 1913, George David Birkhoff proved Poincaré's "Last Geometric Theorem," a special case of the three-body problem, a result that made him world famous. In 1927, he published his *Dynamical Systems*Birkhoff's most durable result has been his 1931 discovery of what is now called the ergodic theorem. Combining insights from physics on the ergodic hypothesis with measure theory, this theorem solved, at least in principle, a fundamental problem of statistical mechanics. The ergodic theorem has also had repercussions for dynamics.

Stephen Smale made significant advances as well. His first contribution is the Smale horseshoe that jumpstarted significant research in dynamical systems. He also outlined a research program carried out by many others.

Oleksandr Mykolaiovych Sharkovsky developed Sharkovsky's Theorem on the periods of discrete dynamical systems in 1964. One of the implications of the theorem is that if a discrete dynamical system on the real line has a periodic point of period 3, then it must have periodic points of every other period.

32.3   Basic definitions

Main article: Dynamical system (definition)

A dynamical system is a manifold M called the phase (or state) space endowed with a family of smooth evolution functions Φ^t that for any element of $t \in T$, the time, map a point of the phase space back into the phase space. The notion of smoothness changes with applications and the type of manifold. There are several choices for the set T. When T is taken to be the reals, the dynamical system is called a *flow*; and if T is restricted to the non-negative reals, then the dynamical system is a *semi-flow*. When T is taken to be the integers, it is a *cascade* or a *map*; and the restriction to the non-negative integers is a *semi-cascade*.

32.3.1   Examples

The evolution function Φ^t is often the solution of a *differential equation of motion*

$$\dot{x} = v(x).$$

The equation gives the time derivative, represented by the dot, of a trajectory $x(t)$ on the phase space starting at some point x_0. The *vector field* $v(x)$ is a smooth function that at every point of the phase space M provides the velocity vector of the dynamical system at that point. (These vectors are not vectors in the phase space M, but in the tangent space T_xM of the point x.) Given a smooth Φ^t, an autonomous vector field can be derived from it.

There is no need for higher order derivatives in the equation, nor for time dependence in $v(x)$ because these can be eliminated by considering systems of higher dimensions. Other types of differential equations can be used to define the evolution rule:

$$G(x, \dot{x}) = 0$$

is an example of an equation that arises from the modeling of mechanical systems with complicated constraints.

The differential equations determining the evolution function Φ^t are often ordinary differential equations: in this case the phase space M is a finite dimensional manifold. Many of the concepts in dynamical systems can be extended to infinite-dimensional manifolds—those that are locally Banach spaces—in which case the differential equations are partial differential equations. In the late 20th century the dynamical system perspective to partial differential equations started gaining popularity.

32.3.2 Further examples

- Logistic map

- Complex quadratic polynomial

- Dyadic transformation

- Tent map

- Double pendulum

- Arnold's cat map

- Horseshoe map

- Baker's map is an example of a chaotic piecewise linear map

- Billiards and outer billiards

- Hénon map

- Lorenz system

- Circle map

- Rössler map

- Kaplan-Yorke map

- List of chaotic maps

- Swinging Atwood's machine

- Quadratic map simulation system

- Bouncing ball dynamics

32.4 Linear dynamical systems

Main article: Linear dynamical system

Linear dynamical systems can be solved in terms of simple functions and the behavior of all orbits classified. In a linear system the phase space is the N-dimensional Euclidean space, so any point in phase space can be represented by a vector with N numbers. The analysis of linear systems is possible because they satisfy a superposition principle: if $u(t)$ and $w(t)$ satisfy the differential equation for the vector field (but not necessarily the initial condition), then so will $u(t) + w(t)$.

32.4.1   Flows

For a flow, the vector field $\Phi(x)$ is an affine function of the position in the phase space, that is,

$$\dot{x} = \phi(x) = Ax + b.$$

with A a matrix, b a vector of numbers and x the position vector. The solution to this system can be found by using the superposition principle (linearity). The case $b \neq 0$ with $A = 0$ is just a straight line in the direction of b:

$$\Phi^t(x_1) = x_1 + bt.$$

When b is zero and $A \neq 0$ the origin is an equilibrium (or singular) point of the flow, that is, if $x_0 = 0$, then the orbit remains there. For other initial conditions, the equation of motion is given by the exponential of a matrix: for an initial point x_0.

$$\Phi^t(x_0) = e^{tA} x_0.$$

When $b = 0$, the eigenvalues of A determine the structure of the phase space. From the eigenvalues and the eigenvectors of A it is possible to determine if an initial point will converge or diverge to the equilibrium point at the origin.

The distance between two different initial conditions in the case $A \neq 0$ will change exponentially in most cases, either converging exponentially fast towards a point, or diverging exponentially fast. Linear systems display sensitive dependence on initial conditions in the case of divergence. For nonlinear systems this is one of the (necessary but not sufficient) conditions for chaotic behavior.

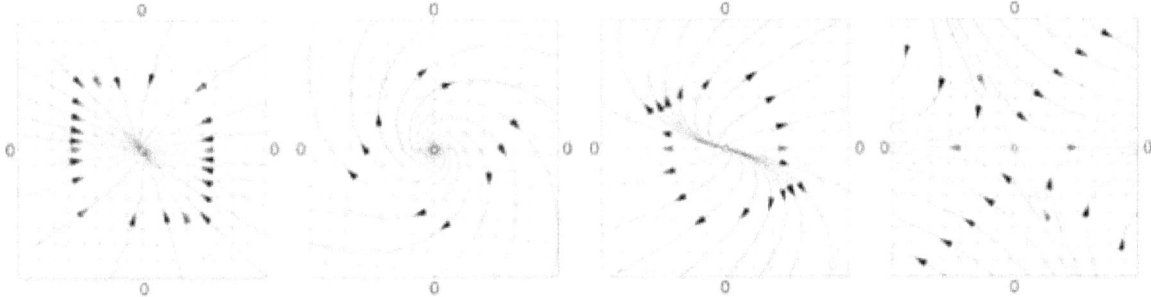

Linear vector fields and a few trajectories.

32.4.2   Maps

A discrete-time, affine dynamical system has the form of a matrix difference equation:

$$x_{n+1} = Ax_n + b.$$

with A a matrix and b a vector. As in the continuous case, the change of coordinates $x \rightarrow x + (1 - A)^{-1}b$ removes the term b from the equation. In the new coordinate system, the origin is a fixed point of the map and the solutions are of the linear system $A^n x_0$. The solutions for the map are no longer curves, but points that hop in the phase space. The orbits are organized in curves, or fibers, which are collections of points that map into themselves under the action of the map.

As in the continuous case, the eigenvalues and eigenvectors of A determine the structure of phase space. For example, if u_1 is an eigenvector of A, with a real eigenvalue smaller than one, then the straight lines given by the points along αu_1, with $\alpha \in \mathbf{R}$, is an invariant curve of the map. Points in this straight line run into the fixed point.

There are also many other discrete dynamical systems.

32.5 Local dynamics

The qualitative properties of dynamical systems do not change under a smooth change of coordinates (this is sometimes taken as a definition of qualitative): a *singular point* of the vector field (a point where $v(x) = 0$) will remain a singular point under smooth transformations; a *periodic orbit* is a loop in phase space and smooth deformations of the phase space cannot alter it being a loop. It is in the neighborhood of singular points and periodic orbits that the structure of a phase space of a dynamical system can be well understood. In the qualitative study of dynamical systems, the approach is to show that there is a change of coordinates (usually unspecified, but computable) that makes the dynamical system as simple as possible.

32.5.1 Rectification

A flow in most small patches of the phase space can be made very simple. If y is a point where the vector field $v(y) \neq 0$, then there is a change of coordinates for a region around y where the vector field becomes a series of parallel vectors of the same magnitude. This is known as the rectification theorem.

The rectification theorem says that away from singular points the dynamics of a point in a small patch is a straight line. The patch can sometimes be enlarged by stitching several patches together, and when this works out in the whole phase space M the dynamical system is *integrable*. In most cases the patch cannot be extended to the entire phase space. There may be singular points in the vector field (where $v(x) = 0$); or the patches may become smaller and smaller as some point is approached. The more subtle reason is a global constraint, where the trajectory starts out in a patch, and after visiting a series of other patches comes back to the original one. If the next time the orbit loops around phase space in a different way, then it is impossible to rectify the vector field in the whole series of patches.

32.5.2 Near periodic orbits

In general, in the neighborhood of a periodic orbit the rectification theorem cannot be used. Poincaré developed an approach that transforms the analysis near a periodic orbit to the analysis of a map. Pick a point x_0 in the orbit γ and consider the points in phase space in that neighborhood that are perpendicular to $v(x_0)$. These points are a Poincaré section $S(\gamma, x_0)$, of the orbit. The flow now defines a map, the Poincaré map $F : S \rightarrow S$, for points starting in S and returning to S. Not all these points will take the same amount of time to come back, but the times will be close to the time it takes x_0.

The intersection of the periodic orbit with the Poincaré section is a fixed point of the Poincaré map F. By a translation, the point can be assumed to be at $x = 0$. The Taylor series of the map is $F(x) = J \cdot x + O(x^2)$, so a change of coordinates h can only be expected to simplify F to its linear part

$$h^{-1} \circ F \circ h(x) = J \cdot x.$$

This is known as the conjugation equation. Finding conditions for this equation to hold has been one of the major tasks of research in dynamical systems. Poincaré first approached it assuming all functions to be analytic and in the process discovered the non-resonant condition. If $\lambda_1, ..., \lambda\nu$ are the eigenvalues of J they will be resonant if one eigenvalue is an integer linear combination of two or more of the others. As terms of the form $\lambda i - \sum$ (multiples of other eigenvalues) occurs in the denominator of the terms for the function h, the non-resonant condition is also known as the small divisor problem.

32.5.3 Conjugation results

The results on the existence of a solution to the conjugation equation depend on the eigenvalues of J and the degree of smoothness required from h. As J does not need to have any special symmetries, its eigenvalues will typically be complex numbers. When the eigenvalues of J are not in the unit circle, the dynamics near the fixed point x_0 of F is called *hyperbolic* and when the eigenvalues are on the unit circle and complex, the dynamics is called *elliptic*.

In the hyperbolic case, the Hartman–Grobman theorem gives the conditions for the existence of a continuous function that maps the neighborhood of the fixed point of the map to the linear map $J \cdot x$. The hyperbolic case is also *structurally stable*. Small changes in the vector field will only produce small changes in the Poincaré map and these small changes will reflect in small changes in the position of the eigenvalues of J in the complex plane, implying that the map is still hyperbolic.

The Kolmogorov–Arnold–Moser (KAM) theorem gives the behavior near an elliptic point.

32.6   Bifurcation theory

Main article: Bifurcation theory

When the evolution map Φ^t (or the vector field it is derived from) depends on a parameter μ, the structure of the phase space will also depend on this parameter. Small changes may produce no qualitative changes in the phase space until a special value μ_0 is reached. At this point the phase space changes qualitatively and the dynamical system is said to have gone through a bifurcation.

Bifurcation theory considers a structure in phase space (typically a fixed point, a periodic orbit, or an invariant torus) and studies its behavior as a function of the parameter μ. At the bifurcation point the structure may change its stability, split into new structures, or merge with other structures. By using Taylor series approximations of the maps and an understanding of the differences that may be eliminated by a change of coordinates, it is possible to catalog the bifurcations of dynamical systems.

The bifurcations of a hyperbolic fixed point x_0 of a system family $F\mu$ can be characterized by the eigenvalues of the first derivative of the system $DF\mu(x_0)$ computed at the bifurcation point. For a map, the bifurcation will occur when there are eigenvalues of $DF\mu$ on the unit circle. For a flow, it will occur when there are eigenvalues on the imaginary axis. For more information, see the main article on Bifurcation theory.

Some bifurcations can lead to very complicated structures in phase space. For example, the Ruelle–Takens scenario describes how a periodic orbit bifurcates into a torus and the torus into a strange attractor. In another example, Feigenbaum period-doubling describes how a stable periodic orbit goes through a series of period-doubling bifurcations.

32.7   Ergodic systems

Main article: Ergodic theory

In many dynamical systems, it is possible to choose the coordinates of the system so that the volume (really a ν-dimensional volume) in phase space is invariant. This happens for mechanical systems derived from Newton's laws as long as the coordinates are the position and the momentum and the volume is measured in units of (position) × (momentum). The flow takes points of a subset A into the points $\Phi^t(A)$ and invariance of the phase space means that

$$\text{vol}(A) = \text{vol}(\Phi^t(A)).$$

In the Hamiltonian formalism, given a coordinate it is possible to derive the appropriate (generalized) momentum such that the associated volume is preserved by the flow. The volume is said to be computed by the Liouville measure.

In a Hamiltonian system, not all possible configurations of position and momentum can be reached from an initial condition. Because of energy conservation, only the states with the same energy as the initial condition are accessible. The states with the same energy form an energy shell Ω, a sub-manifold of the phase space. The volume of the energy shell, computed using the Liouville measure, is preserved under evolution.

For systems where the volume is preserved by the flow, Poincaré discovered the recurrence theorem: Assume the phase space has a finite Liouville volume and let F be a phase space volume-preserving map and A a subset of the phase space.

Then almost every point of A returns to A infinitely often. The Poincaré recurrence theorem was used by Zermelo to object to Boltzmann's derivation of the increase in entropy in a dynamical system of colliding atoms.

One of the questions raised by Boltzmann's work was the possible equality between time averages and space averages, what he called the ergodic hypothesis. The hypothesis states that the length of time a typical trajectory spends in a region A is $\text{vol}(A)/\text{vol}(\Omega)$.

The ergodic hypothesis turned out not to be the essential property needed for the development of statistical mechanics and a series of other ergodic-like properties were introduced to capture the relevant aspects of physical systems. Koopman approached the study of ergodic systems by the use of functional analysis. An observable a is a function that to each point of the phase space associates a number (say instantaneous pressure, or average height). The value of an observable can be computed at another time by using the evolution function φ^t. This introduces an operator U^t, the transfer operator,

$$(U^t a)(x) = a(\Phi^{-t}(x)).$$

By studying the spectral properties of the linear operator U it becomes possible to classify the ergodic properties of Φ^t. In using the Koopman approach of considering the action of the flow on an observable function, the finite-dimensional nonlinear problem involving Φ^t gets mapped into an infinite-dimensional linear problem involving U.

The Liouville measure restricted to the energy surface Ω is the basis for the averages computed in equilibrium statistical mechanics. An average in time along a trajectory is equivalent to an average in space computed with the Boltzmann factor $\exp(-\beta H)$. This idea has been generalized by Sinai, Bowen, and Ruelle (SRB) to a larger class of dynamical systems that includes dissipative systems. SRB measures replace the Boltzmann factor and they are defined on attractors of chaotic systems.

32.7.1 Nonlinear dynamical systems and chaos

Main article: Chaos theory

Simple nonlinear dynamical systems and even piecewise linear systems can exhibit a completely unpredictable behavior, which might seem to be random, despite the fact that they are fundamentally deterministic. This seemingly unpredictable behavior has been called *chaos*. Hyperbolic systems are precisely defined dynamical systems that exhibit the properties ascribed to chaotic systems. In hyperbolic systems the tangent space perpendicular to a trajectory can be well separated into two parts: one with the points that converge towards the orbit (the *stable manifold*) and another of the points that diverge from the orbit (the *unstable manifold*).

This branch of mathematics deals with the long-term qualitative behavior of dynamical systems. Here, the focus is not on finding precise solutions to the equations defining the dynamical system (which is often hopeless), but rather to answer questions like "Will the system settle down to a steady state in the long term, and if so, what are the possible attractors?" or "Does the long-term behavior of the system depend on its initial condition?"

Note that the chaotic behavior of complex systems is not the issue. Meteorology has been known for years to involve complex—even chaotic—behavior. Chaos theory has been so surprising because chaos can be found within almost trivial systems. The logistic map is only a second-degree polynomial; the horseshoe map is piecewise linear.

32.7.2 Geometrical definition

A dynamical system is the tuple $\langle \mathcal{M}, f, \mathcal{T} \rangle$, with $\mathcal{M}$ a manifold (locally a Banach space or Euclidean space), $\mathcal{T}$ the domain for time (non-negative reals, the integers, ...) and f an evolution rule $t \to f^t$ (with $t \in \mathcal{T}$) such that f^t is a diffeomorphism of the manifold to itself. So, f is a mapping of the time-domain $\mathcal{T}$ into the space of diffeomorphisms of the manifold to itself. In other terms, $f(t)$ is a diffeomorphism, for every time t in the domain $\mathcal{T}$.

32.7.3 Measure theoretical definition

Main article: Measure-preserving dynamical system

A dynamical system may be defined formally, as a measure-preserving transformation of a sigma-algebra, the quadruplet (X, Σ, μ, τ). Here, X is a set, and Σ is a sigma-algebra on X, so that the pair (X, Σ) is a measurable space. μ is a finite measure on the sigma-algebra, so that the triplet (X, Σ, μ) is a probability space. A map τ: $X \to X$ is said to be Σ-measurable if and only if, for every $\sigma \in \Sigma$, one has $\tau^{-1}\sigma \in \Sigma$. A map τ is said to **preserve the measure** if and only if, for every $\sigma \in \Sigma$, one has $\mu(\tau^{-1}\sigma) = \mu(\sigma)$. Combining the above, a map τ is said to be a **measure-preserving transformation of** X, if it is a map from X to itself, it is Σ-measurable, and is measure-preserving. The quadruple (X, Σ, μ, τ), for such a τ, is then defined to be a **dynamical system**.

The map τ embodies the time evolution of the dynamical system. Thus, for discrete dynamical systems the iterates $\tau^n = \tau \circ \tau \circ \cdots \circ \tau$ for integer n are studied. For continuous dynamical systems, the map τ is understood to be a finite time evolution map and the construction is more complicated.

32.8 Examples of dynamical systems

- Arnold's cat map

- Baker's map is an example of a chaotic piecewise linear map

- Circle map

- Double pendulum

- Billiards and Outer billiards

- Hénon map

- Horseshoe map

- Irrational rotation

- List of chaotic maps

- Logistic map

- Lorenz system

- Rossler map

32.9 Multidimensional generalization

Dynamical systems are defined over a single independent variable, usually thought of as time. A more general class of systems are defined over multiple independent variables and are therefore called multidimensional systems. Such systems are useful for modeling, for example, image processing.

32.10 See also

- Behavioral modeling

- Cognitive modeling

- Dynamical systems theory

- Feedback passivation

- Infinite compositions of analytic functions

- List of dynamical system topics

- Oscillation

- People in systems and control

- Sharkovskii's theorem

- System dynamics

- Systems theory

32.11   References

[1] Strogatz, S. H. (2001). Nonlinear dynamics and chaos: with applications to physics, biology and chemistry. Perseus publishing.

[2] Katok, A., & Hasselblatt, B. (1995). Introduction to the modern theory of dynamical systems. Cambridge, Cambridge.

[3] Holmes, Philip. "Poincaré, celestial mechanics, dynamical-systems theory and "chaos"." *Physics Reports* 193.3 (1990): 137-163.

32.12   Further reading

Works providing a broad coverage:

- Ralph Abraham and Jerrold E. Marsden (1978). *Foundations of mechanics*. Benjamin–Cummings. ISBN 0-8053-0102-X. (available as a reprint: ISBN 0-201-40840-6)

- *Encyclopaedia of Mathematical Sciences* (ISSN 0938-0396) has a sub-series on dynamical systems with reviews of current research.

- Christian Bonatti, Lorenzo J. Díaz, Marcelo Viana (2005). *Dynamics Beyond Uniform Hyperbolicity: A Global Geometric and Probabilistic Perspective*. Springer. ISBN 3-540-22066-6.

- Stephen Smale (1967). "Differentiable dynamical systems". *Bulletin of the American Mathematical Society* **73** (6): 747–817. doi:10.1090/S0002-9904-1967-11798-1.

Introductory texts with a unique perspective:

- V. I. Arnold (1982). *Mathematical methods of classical mechanics*. Springer-Verlag. ISBN 0-387-96890-3.

- Jacob Palis and Wellington de Melo (1982). *Geometric theory of dynamical systems: an introduction*. Springer-Verlag. ISBN 0-387-90668-1.

- David Ruelle (1989). *Elements of Differentiable Dynamics and Bifurcation Theory*. Academic Press. ISBN 0-12-601710-7.

- Tim Bedford, Michael Keane and Caroline Series, *eds.* (1991). *Ergodic theory, symbolic dynamics and hyperbolic spaces*. Oxford University Press. ISBN 0-19-853390-X.

- Ralph H. Abraham and Christopher D. Shaw (1992). *Dynamics—the geometry of behavior, 2nd edition*. Addison-Wesley. ISBN 0-201-56716-4.

Textbooks

- Kathleen T. Alligood, Tim D. Sauer and James A. Yorke (2000). *Chaos. An introduction to dynamical systems*. Springer Verlag. ISBN 0-387-94677-2.

- Oded Galor (2011). Discrete Dynamical Systems. Springer. ISBN 978-3-642-07185-0.

- Anatole Katok and Boris Hasselblatt (1996). *Introduction to the modern theory of dynamical systems*. Cambridge. ISBN 0-521-57557-5.

- Guenter Ludyk (1985). *Stability of Time-variant Discrete-Time Systems*. Springer. ISBN 3-528-08911-3.

- Stephen Lynch (2010). *Dynamical Systems with Applications using Maple 2nd Ed*. Springer. ISBN 0-8176-4389-3.

- Stephen Lynch (2007). *Dynamical Systems with Applications using Mathematica*. Springer. ISBN 0-8176-4482-2.

- Stephen Lynch (2004). *Dynamical Systems with Applications using MATLAB*. Springer. ISBN 0-8176-4321-4.

- James Meiss (2007). *Differential Dynamical Systems*. SIAM. ISBN 0-89871-635-7.

- Morris W. Hirsch, Stephen Smale and Robert Devaney (2003). *Differential Equations, dynamical systems, and an introduction to chaos*. Academic Press. ISBN 0-12-349703-5.

- Julien Clinton Sprott (2003). Chaos and time-series analysis. Oxford University Press. ISBN 0-19-850839-5.

- Steven H. Strogatz (1994). *Nonlinear dynamics and chaos: with applications to physics, biology chemistry and engineering*. Addison Wesley. ISBN 0-201-54344-3.

- Teschl, Gerald (2012). *Ordinary Differential Equations and Dynamical Systems*. Providence: American Mathematical Society. ISBN 978-0-8218-8328-0.

- Stephen Wiggins (2003). *Introduction to Applied Dynamical Systems and Chaos*. Springer. ISBN 0-387-00177-8.

Popularizations:

- Florin Diacu and Philip Holmes (1996). *Celestial Encounters*. Princeton. ISBN 0-691-02743-9.

- James Gleick (1988). *Chaos: Making a New Science*. Penguin. ISBN 0-14-009250-1.

- Ivar Ekeland (1990). *Mathematics and the Unexpected (Paperback)*. University Of Chicago Press. ISBN 0-226-19990-8.

- Ian Stewart (1997). *Does God Play Dice? The New Mathematics of Chaos*. Penguin. ISBN 0-14-025602-4.

32.13   External links

- Interactive applet for the Standard and Henon Maps by A. Luhn

- A collection of dynamic and non-linear system models and demo applets (in Monash University's Virtual Lab)

- Arxiv preprint server has daily submissions of (non-refereed) manuscripts in dynamical systems.

- DSWeb provides up-to-date information on dynamical systems and its applications.

- Encyclopedia of dynamical systems A part of Scholarpedia — peer reviewed and written by invited experts.

- Nonlinear Dynamics. Models of bifurcation and chaos by Elmer G. Wiens

- Oliver Knill has a series of examples of dynamical systems with explanations and interactive controls.

- Sci.Nonlinear FAQ 2.0 (Sept 2003) provides definitions, explanations and resources related to nonlinear science

Online books or lecture notes:

- Geometrical theory of dynamical systems. Nils Berglund's lecture notes for a course at ETH at the advanced undergraduate level.

- Dynamical systems. George D. Birkhoff's 1927 book already takes a modern approach to dynamical systems.

- Chaos: classical and quantum. An introduction to dynamical systems from the periodic orbit point of view.

- Modeling Dynamic Systems. An introduction to the development of mathematical models of dynamic systems.

- Learning Dynamical Systems. Tutorial on learning dynamical systems.

- Ordinary Differential Equations and Dynamical Systems. Lecture notes by Gerald Teschl

Research groups:

- Dynamical Systems Group Groningen, IWI, University of Groningen.

- Chaos @ UMD. Concentrates on the applications of dynamical systems.

- Dynamical Systems, SUNY Stony Brook. Lists of conferences, researchers, and some open problems.

- Center for Dynamics and Geometry, Penn State.

- Control and Dynamical Systems, Caltech.

- Laboratory of Nonlinear Systems, Ecole Polytechnique Fédérale de Lausanne (EPFL).

- Center for Dynamical Systems, University of Bremen

- Systems Analysis, Modelling and Prediction Group, University of Oxford

- Non-Linear Dynamics Group, Instituto Superior Técnico, Technical University of Lisbon

- Dynamical Systems, IMPA, Instituto Nacional de Matemática Pura e Applicada.

- Nonlinear Dynamics Workgroup, Institute of Computer Science, Czech Academy of Sciences.

Simulation software based on Dynamical Systems approach:

- FyDiK

- iDMC, simulation and dynamical analysis of nonlinear models

Chapter 33

Chaos theory

For other uses, see Chaos Theory (disambiguation).
 Chaos theory is the field of study in mathematics that studies the behavior of dynamical systems that are highly sensitive to initial conditions—a response popularly referred to as the butterfly effect.[1] Small differences in initial conditions (such as those due to rounding errors in numerical computation) yield widely diverging outcomes for such dynamical systems, rendering long-term prediction impossible in general.[2] This happens even though these systems are deterministic, meaning that their future behavior is fully determined by their initial conditions, with no random elements involved.[3] In other words, the deterministic nature of these systems does not make them predictable.[4][5] This behavior is known as **deterministic chaos**, or simply **chaos**. The theory was summarized by Edward Lorenz as:[6]

> Chaos: When the present determines the future, but the approximate present does not approximately determine the future.

Chaotic behavior exists in many natural systems, such as weather and climate.[7][8] This behavior can be studied through analysis of a chaotic mathematical model, or through analytical techniques such as recurrence plots and Poincaré maps. Chaos theory has applications in several disciplines, including meteorology, sociology, physics, engineering, economics, biology, and philosophy.

33.1   Introduction

Chaos theory concerns deterministic systems whose behavior can in principle be predicted. Chaotic systems are predictable for a while and then 'appear' to become random. The amount of time for which the behavior of a chaotic system can be effectively predicted depends on three things: How much uncertainty we are willing to tolerate in the forecast, how accurately we are able to measure its current state, and a time scale depending on the dynamics of the system, called the Lyapunov time. Some examples of Lyapunov times are: chaotic electrical circuits, about 1 millisecond; weather systems, a few days (unproven); the solar system, 50 million years. In chaotic systems, the uncertainty in a forecast increases exponentially with elapsed time. Hence, doubling the forecast time more than squares the proportional uncertainty in the forecast. This means, in practice, a meaningful prediction cannot be made over an interval of more than two or three times the Lyapunov time. When meaningful predictions cannot be made, the system appears to be random.[9]

33.2   Chaotic dynamics

In common usage, "chaos" means "a state of disorder".[10] However, in chaos theory, the term is defined more precisely. Although no universally accepted mathematical definition of chaos exists, a commonly used definition says that, for a dynamical system to be classified as chaotic, it must have these properties:[11]

A plot of the Lorenz attractor for values r = 28, σ = 10, b = 8/3

1. it must be sensitive to initial conditions

2. it must be topologically mixing

3. it must have dense periodic orbits

33.2.1   Sensitivity to initial conditions

Main article: Butterfly effect

Sensitivity to initial conditions means that each point in a chaotic system is arbitrarily closely approximated by other points with significantly different future paths, or trajectories. Thus, an arbitrarily small change, or perturbation, of the current trajectory may lead to significantly different future behavior.

A double rod pendulum animation showing chaotic behavior. Starting the pendulum from a slightly different initial condition would result in a completely different trajectory. The double rod pendulum is one of the simplest dynamical systems that has chaotic solutions.

In some cases, the last two properties in the above have been shown to actually imply sensitivity to initial conditions,[12][13] and if attention is restricted to intervals, the second property implies the other two[14] (an alternative, and in general weaker, definition of chaos uses only the first two properties in the above list).[15] The most practically significant property, sensitivity to initial conditions, is redundant in the definition, since it is implied by two (or for intervals, one) purely topological properties, which are therefore of greater interest to mathematicians.

Sensitivity to initial conditions is popularly known as the "butterfly effect", so-called because of the title of a paper given by Edward Lorenz in 1972 to the American Association for the Advancement of Science in Washington, D.C., entitled *Predictability: Does the Flap of a Butterfly's Wings in Brazil set off a Tornado in Texas?*.[16] The flapping wing represents a small change in the initial condition of the system, which causes a chain of events leading to large-scale phenomena. Had the butterfly not flapped its wings, the trajectory of the system might have been vastly different.

A consequence of sensitivity to initial conditions is that if we start with only a finite amount of information about the system (as is usually the case in practice), then beyond a certain time the system will no longer be predictable. This is most familiar in the case of weather, which is generally predictable only about a week ahead.[17] Of course, this does not mean that we cannot say anything about events far in the future; some restrictions on the system are present. With weather, we know that the temperature will never reach 100 °C or fall to −130 °C on earth, but we are not able to say exactly what day we will have the hottest temperature of the year.

In more mathematical terms, the Lyapunov exponent measures the sensitivity to initial conditions. Given two starting trajectories in the phase space that are infinitesimally close, with initial separation $\delta \mathbf{Z}_0$ end up diverging at a rate given by

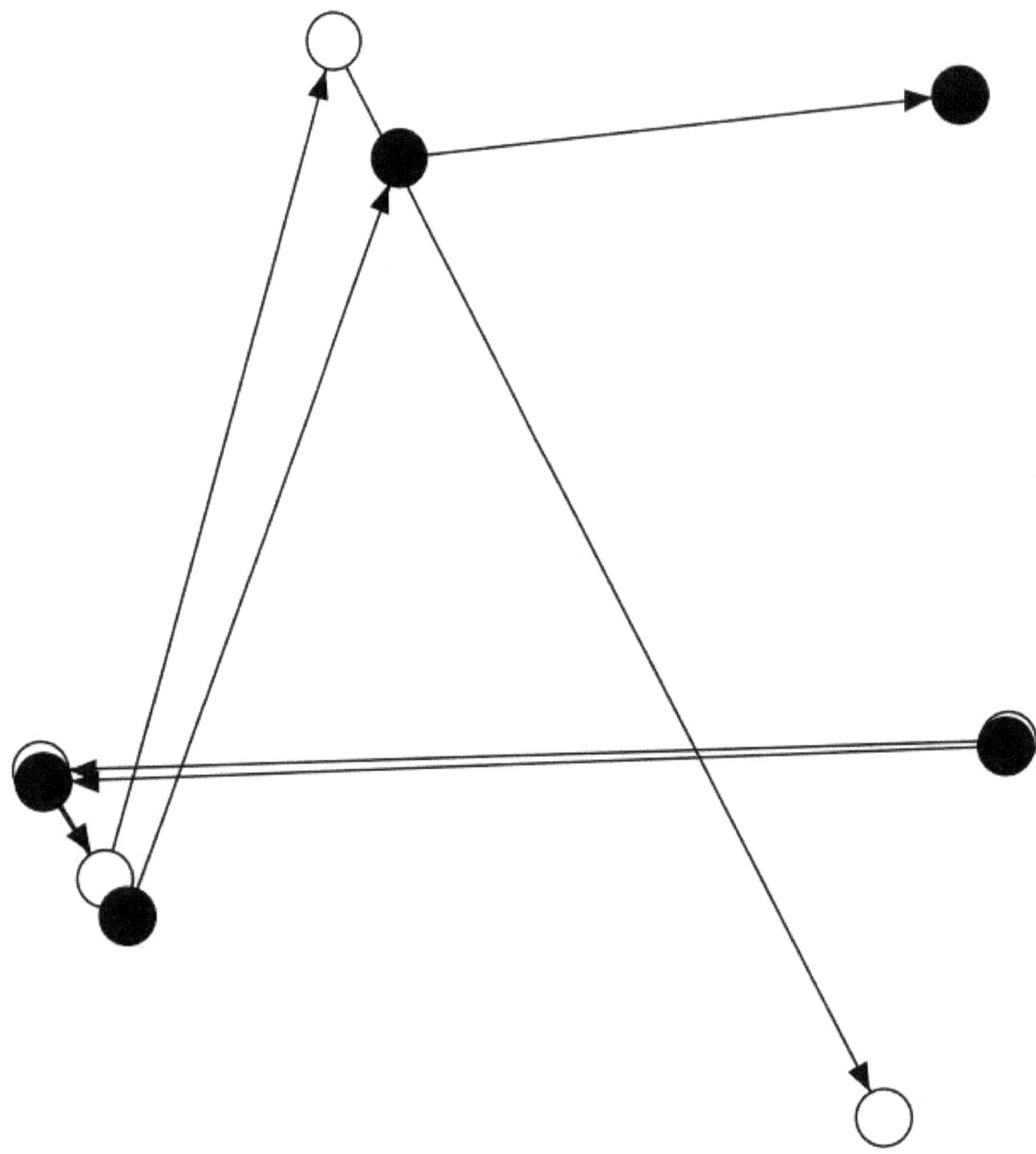

The map defined by x → 4 x (1 − x) and y → x + y mod 1 displays sensitivity to initial conditions. Here, two series of x and y values diverge markedly over time from a tiny initial difference. Note, however, that the y coordinate is effectively only defined modulo one, so the square region is actually depicting a cylinder, and the two points are closer than they look

$$|\delta \mathbf{Z}(t)| \approx e^{\lambda t} |\delta \mathbf{Z}_0|$$

where t is the time and λ is the Lyapunov exponent. The rate of separation depends on the orientation of the initial separation vector, so a whole spectrum of Lyapunov exponents exist. The number of Lyapunov exponents is equal to the number of dimensions of the phase space, though it is common to just refer to the largest one. For example, the maximal Lyapunov exponent (MLE) is most often used because it determines the overall predictability of the system. A positive MLE is usually taken as an indication that the system is chaotic.

Also, other properties relate to sensitivity of initial conditions, such as measure-theoretical mixing (as discussed in ergodic theory) and properties of a K-system.[5]

33.2.2   Topological mixing

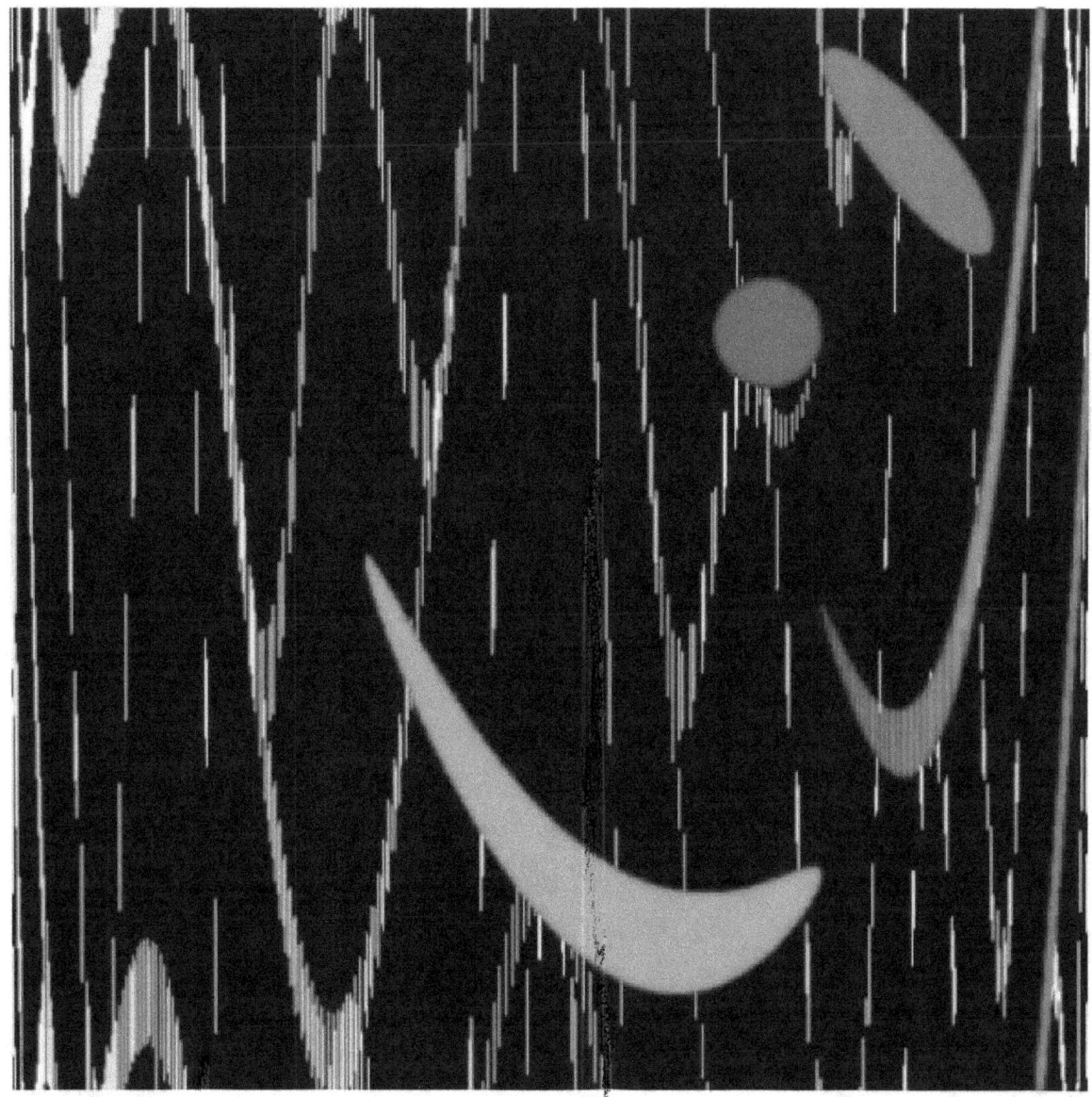

The map defined by x → 4 x (1 − x) and y → x + y mod 1 also displays topological mixing. Here, the blue region is transformed by the dynamics first to the purple region, then to the pink and red regions, and eventually to a cloud of points scattered across the space.

Topological mixing (or **topological transitivity**) means that the system will evolve over time so that any given region or open set of its phase space will eventually overlap with any other given region. This mathematical concept of "mixing" corresponds to the standard intuition, and the mixing of colored dyes or fluids is an example of a chaotic system.

Topological mixing is often omitted from popular accounts of chaos, which equate chaos with only sensitivity to initial conditions. However, sensitive dependence on initial conditions alone does not give chaos. For example, consider the simple dynamical system produced by repeatedly doubling an initial value. This system has sensitive dependence on initial conditions everywhere, since any pair of nearby points will eventually become widely separated. However, this example has no topological mixing, and therefore has no chaos. Indeed, it has extremely simple behavior: all points except 0 will tend to positive or negative infinity.

33.2.3 Density of periodic orbits

For a chaotic system to have a dense periodic orbit means that every point in the space is approached arbitrarily closely by periodic orbits.[18] The one-dimensional logistic map defined by $x \rightarrow 4\,x\,(1-x)$ is one of the simplest systems with density of periodic orbits. For example, $\frac{5-\sqrt{5}}{8} \rightarrow \frac{5+\sqrt{5}}{8} \rightarrow \frac{5-\sqrt{5}}{8}$ (or approximately $0.3454915 \rightarrow 0.9045085 \rightarrow 0.3454915$) is an (unstable) orbit of period 2, and similar orbits exist for periods 4, 8, 16, etc. (indeed, for all the periods specified by Sharkovskii's theorem).[19]

Sharkovskii's theorem is the basis of the Li and Yorke[20] (1975) proof that any one-dimensional system that exhibits a regular cycle of period three will also display regular cycles of every other length, as well as completely chaotic orbits.

33.2.4 Strange attractors

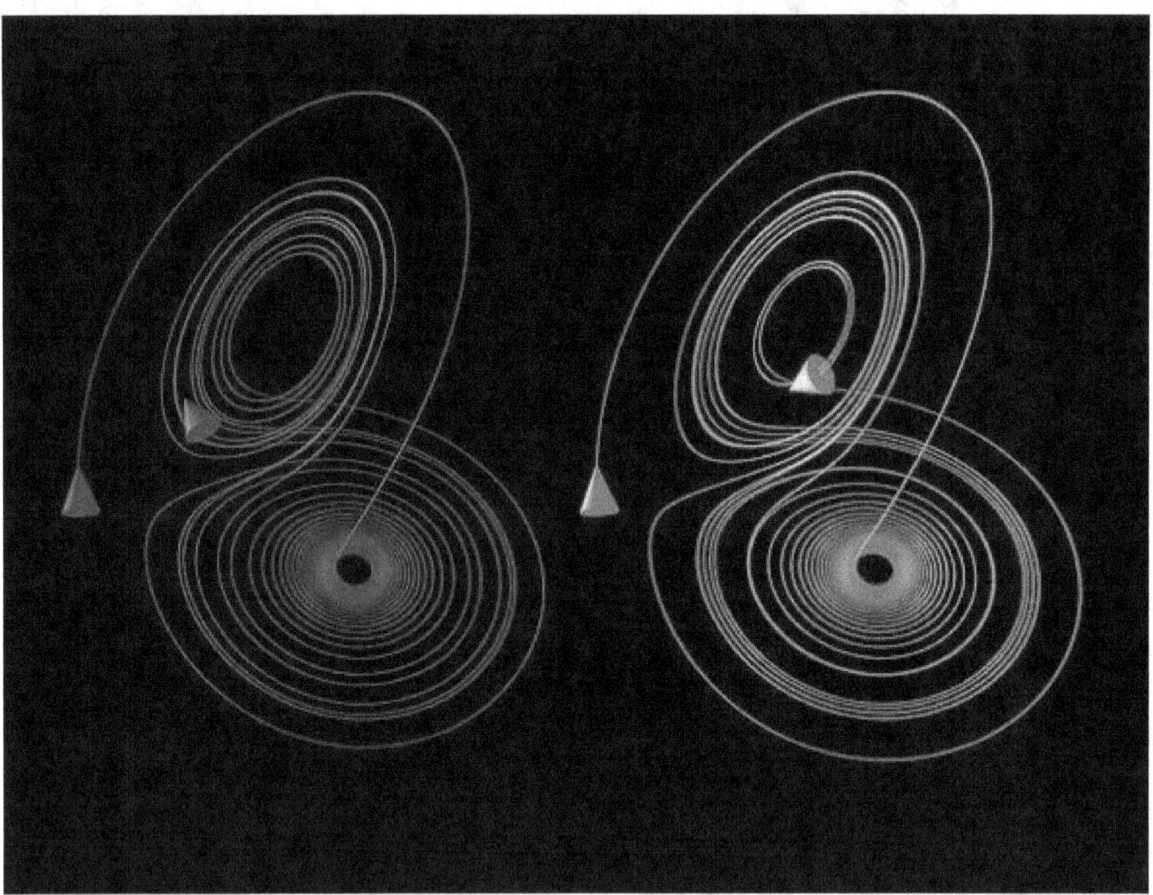

The Lorenz attractor displays chaotic behavior. These two plots demonstrate sensitive dependence on initial conditions within the region of phase space occupied by the attractor.

Some dynamical systems, like the one-dimensional logistic map defined by $x \rightarrow 4\,x\,(1-x)$, are chaotic everywhere, but in many cases chaotic behavior is found only in a subset of phase space. The cases of most interest arise when the chaotic behavior takes place on an attractor, since then a large set of initial conditions will lead to orbits that converge to this chaotic region.

An easy way to visualize a chaotic attractor is to start with a point in the basin of attraction of the attractor, and then simply plot its subsequent orbit. Because of the topological transitivity condition, this is likely to produce a picture of the entire final attractor, and indeed both orbits shown in the figure on the right give a picture of the general shape of the Lorenz attractor. This attractor results from a simple three-dimensional model of the Lorenz weather system. The Lorenz attractor is perhaps one of the best-known chaotic system diagrams, probably because it was not only one of the first, but it is also one of the most complex and as such gives rise to a very interesting pattern, that with a little imagination, looks like the wings of a butterfly.

Unlike fixed-point attractors and limit cycles, the attractors that arise from chaotic systems, known as strange attractors, have great detail and complexity. Strange attractors occur in both continuous dynamical systems (such as the Lorenz system) and in some discrete systems (such as the Hénon map). Other discrete dynamical systems have a repelling structure called a Julia set which forms at the boundary between basins of attraction of fixed points – Julia sets can be thought of as strange repellers. Both strange attractors and Julia sets typically have a fractal structure, and the fractal dimension can be calculated for them.

33.2.5   Minimum complexity of a chaotic system

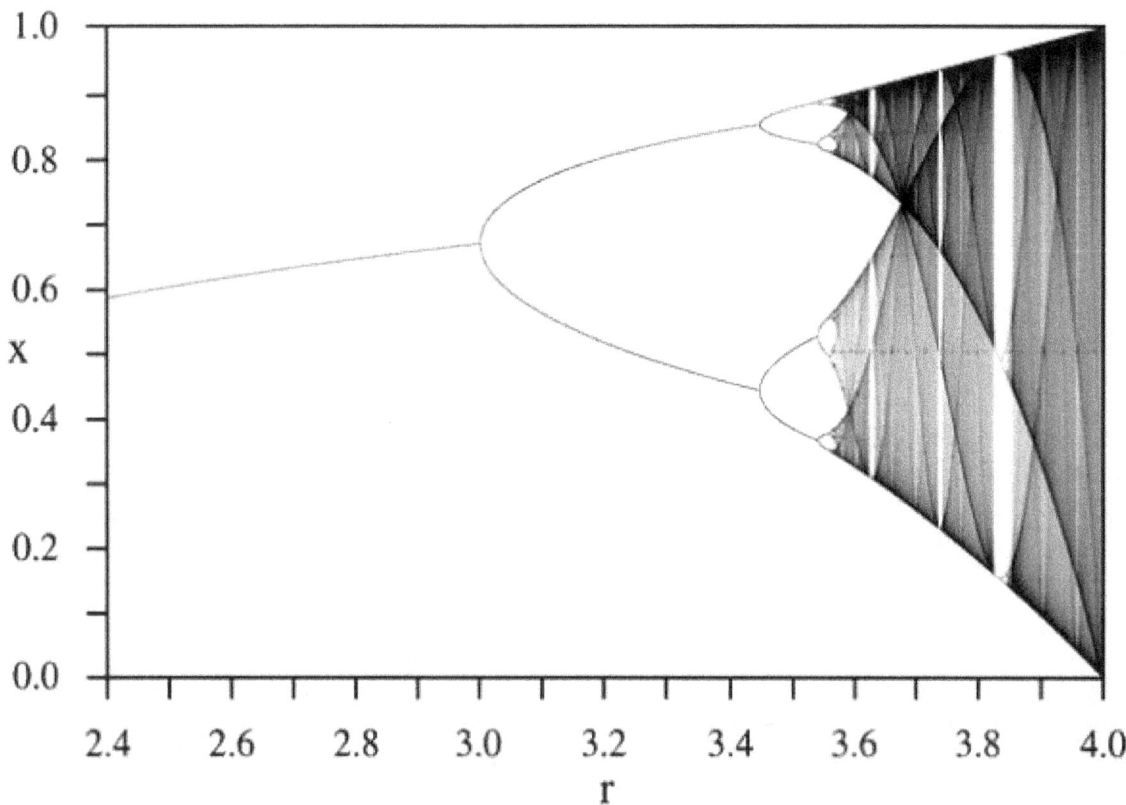

Bifurcation diagram of the logistic map x → r x (1 − x). *Each vertical slice shows the attractor for a specific value of* r. *The diagram displays period-doubling as* r *increases, eventually producing chaos.*

Discrete chaotic systems, such as the logistic map, can exhibit strange attractors whatever their dimensionality. In contrast, for continuous dynamical systems, the Poincaré–Bendixson theorem shows that a strange attractor can only arise in three or more dimensions. Finite-dimensional linear systems are never chaotic; for a dynamical system to display chaotic behavior, it has to be either nonlinear or infinite-dimensional.

The Poincaré–Bendixson theorem states that a two-dimensional differential equation has very regular behavior. The Lorenz attractor discussed above is generated by a system of three differential equations such as:

$$\frac{dx}{dt} = \sigma y - \sigma x,$$

$$\frac{dy}{dt} = \rho x - xz - y,$$

$$\frac{dz}{dt} = xy - \beta z.$$

where x, y, and z make up the system state, t is time, and σ, ρ, β are the system parameters. Five of the terms on the right hand side are linear, while two are quadratic; a total of seven terms. Another well-known chaotic attractor is generated by the Rossler equations which have only one nonlinear term out of seven. Sprott [21] found a three-dimensional system with just five terms, that had only one nonlinear term, which exhibits chaos for certain parameter values. Zhang and Heidel [22][23] showed that, at least for dissipative and conservative quadratic systems, three-dimensional quadratic systems with only three or four terms on the right-hand side cannot exhibit chaotic behavior. The reason is, simply put, that solutions to such systems are asymptotic to a two-dimensional surface and therefore solutions are well behaved.

While the Poincaré–Bendixson theorem shows that a continuous dynamical system on the Euclidean plane cannot be chaotic, two-dimensional continuous systems with non-Euclidean geometry can exhibit chaotic behavior.[24] Perhaps surprisingly, chaos may occur also in linear systems, provided they are infinite dimensional.[25] A theory of linear chaos is being developed in a branch of mathematical analysis known as functional analysis.

33.2.6 Jerk systems

In physics, jerk is the third derivative of position, and as such, in mathematics differential equations of the form

$$J\left(\dddot{x}, \ddot{x}, \dot{x}, x\right) = 0$$

are sometimes called *Jerk equations*. It has been shown, that a jerk equation, which is equivalent to a system of three first order, ordinary, non-linear differential equation is in a certain sense the minimal setting for solutions showing chaotic behaviour. This motivates mathematical interest in jerk systems. Systems involving a fourth or higher derivative are called accordingly hyperjerk systems.[26]

A jerk system's behavior is described by a jerk equation, and for certain jerk equations, simple electronic circuits may be designed which model the solutions to this equation. These circuits are known as jerk circuits.

One of the most interesting properties of jerk circuits is the possibility of chaotic behavior. In fact, certain well-known chaotic systems, such as the Lorenz attractor and the Rössler map, are conventionally described as a system of three first-order differential equations, but which may be combined into a single (although rather complicated) jerk equation. Nonlinear jerk systems are in a sense minimally complex systems to show chaotic behaviour, there is no chaotic system involving only two first-order, ordinary differential equations (the system resulting in an equation of second order only).

An example of a jerk equation with nonlinearity in the magnitude of x is:

$$\frac{d^3 x}{dt^3} + A\frac{d^2 x}{dt^2} + \frac{dx}{dt} - |x| + 1 = 0.$$

Here, A is an adjustable parameter. This equation has a chaotic solution for $A=3/5$ and can be implemented with the following jerk circuit; the required nonlinearity is brought about by the two diodes:

In the above circuit, all resistors are of equal value, except $R_A = R/A = 5R/3$, and all capacitors are of equal size. The dominant frequency will be $1/2\pi RC$. The output of op amp 0 will correspond to the x variable, the output of 1 will correspond to the first derivative of x and the output of 2 will correspond to the second derivative.

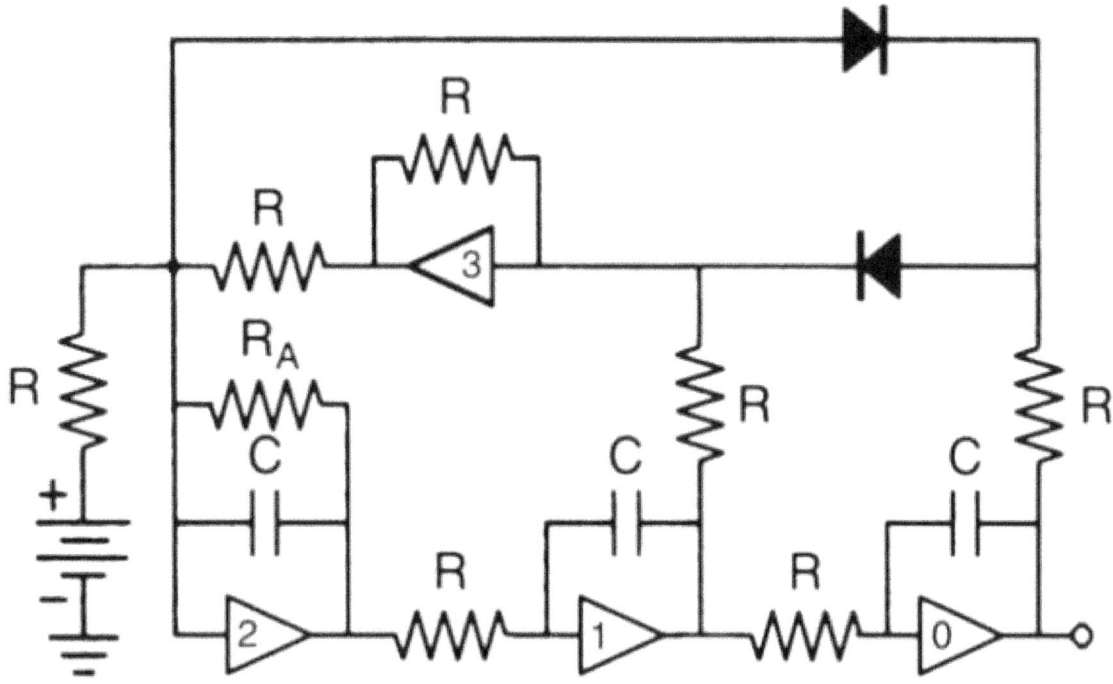

33.3 Spontaneous order

Under the right conditions, chaos will spontaneously evolve into a lockstep pattern. In the Kuramoto model, four conditions suffice to produce synchronization in a chaotic system. Examples include the coupled oscillation of Christiaan Huygens' pendulums, fireflies, neurons, the London Millenium Bridge resonance, and large arrays of Josephson junctions.[27]

33.4 History

An early proponent of chaos theory was Henri Poincaré. In the 1880s, while studying the three-body problem, he found that there can be orbits that are nonperiodic, and yet not forever increasing nor approaching a fixed point.[28][29] In 1898 Jacques Hadamard published an influential study of the chaotic motion of a free particle gliding frictionlessly on a surface of constant negative curvature, called "Hadamard's billiards".[30] Hadamard was able to show that all trajectories are unstable, in that all particle trajectories diverge exponentially from one another, with a positive Lyapunov exponent.

Chaos theory got its start in the field of ergodic theory. Later studies, also on the topic of nonlinear differential equations, were carried out by George David Birkhoff,[31] Andrey Nikolaevich Kolmogorov,[32][33][34] Mary Lucy Cartwright and John Edensor Littlewood,[35] and Stephen Smale.[36] Except for Smale, these studies were all directly inspired by physics: the three-body problem in the case of Birkhoff, turbulence and astronomical problems in the case of Kolmogorov, and radio engineering in the case of Cartwright and Littlewood. Although chaotic planetary motion had not been observed, experimentalists had encountered turbulence in fluid motion and nonperiodic oscillation in radio circuits without the benefit of a theory to explain what they were seeing.

Despite initial insights in the first half of the twentieth century, chaos theory became formalized as such only after mid-century, when it first became evident to some scientists that linear theory, the prevailing system theory at that time, simply could not explain the observed behavior of certain experiments like that of the logistic map. What had been attributed to measure imprecision and simple "noise" was considered by chaos theorists as a full component of the studied systems.

The main catalyst for the development of chaos theory was the electronic computer. Much of the mathematics of chaos

Barnsley fern created using the chaos game. Natural forms (ferns, clouds, mountains, etc.) may be recreated through an Iterated function system (IFS).

theory involves the repeated iteration of simple mathematical formulas, which would be impractical to do by hand. Electronic computers made these repeated calculations practical, while figures and images made it possible to visualize these systems. As a graduate student in Chihiro Hayashi's laboratory at Kyoto University, Yoshisuke Ueda was experimenting with analog computers and noticed, on Nov. 27, 1961, what he called "randomly transitional phenomena". Yet his advisor did not agree with his conclusions at the time, and did not allow him to report his findings until 1970.[37][38]

An early pioneer of the theory was Edward Lorenz whose interest in chaos came about accidentally through his work on weather prediction in 1961.[7] Lorenz was using a simple digital computer, a Royal McBee LGP-30, to run his weather simulation. He wanted to see a sequence of data again and to save time he started the simulation in the middle of its course. He was able to do this by entering a printout of the data corresponding to conditions in the middle of his simulation which he had calculated last time. To his surprise the weather that the machine began to predict was completely different from the weather calculated before. Lorenz tracked this down to the computer printout. The computer worked with 6-digit precision, but the printout rounded variables off to a 3-digit number, so a value like 0.506127 was printed as 0.506. This difference is tiny and the consensus at the time would have been that it should have had practically no effect. However, Lorenz had discovered that small changes in initial conditions produced large changes in the long-term outcome.[39]

Turbulence in the tip vortex from an airplane wing. Studies of the critical point beyond which a system creates turbulence were important for chaos theory, analyzed for example by the Soviet physicist Lev Landau, who developed the Landau-Hopf theory of turbulence. David Ruelle and Floris Takens later predicted, against Landau, that fluid turbulence could develop through a strange attractor, a main concept of chaos theory.

Lorenz's discovery, which gave its name to Lorenz attractors, showed that even detailed atmospheric modelling cannot, in general, make precise long-term weather predictions.

In 1963, Benoit Mandelbrot found recurring patterns at every scale in data on cotton prices.[40] Beforehand he had studied information theory and concluded noise was patterned like a Cantor set: on any scale the proportion of noise-containing periods to error-free periods was a constant – thus errors were inevitable and must be planned for by incorporating redundancy.[41] Mandelbrot described both the "Noah effect" (in which sudden discontinuous changes can occur) and the "Joseph effect" (in which persistence of a value can occur for a while, yet suddenly change afterwards).[42][43] This challenged the idea that changes in price were normally distributed. In 1967, he published "How long is the coast of Britain? Statistical self-similarity and fractional dimension", showing that a coastline's length varies with the scale of the measuring instrument, resembles itself at all scales, and is infinite in length for an infinitesimally small measuring device.[44] Arguing that a ball of twine appears to be a point when viewed from far away (0-dimensional), a ball when viewed from fairly near (3-dimensional), or a curved strand (1-dimensional), he argued that the dimensions of an object are relative to the observer and may be fractional. An object whose irregularity is constant over different scales ("self-similarity") is a fractal (examples include the Menger sponge, the Sierpiński gasket, and the Koch curve or "snowflake", which is infinitely long yet encloses a finite space and has a fractal dimension of circa 1.2619). In 1982 Mandelbrot published *The Fractal Geometry of Nature*, which became a classic of chaos theory. Biological systems such as the branching of the circulatory and bronchial systems proved to fit a fractal model.[45]

In December 1977, the New York Academy of Sciences organized the first symposium on Chaos, attended by David Ruelle, Robert May, James A. Yorke (coiner of the term "chaos" as used in mathematics), Robert Shaw, and the meteorologist Edward Lorenz. The following year, independently Pierre Coullet and Charles Tresser with the article "Iterations d'endomorphismes et groupe de renormalisation" and Mitchell Feigenbaum with the article "Quantitative Universality for a Class of Nonlinear Transformations" described logistic maps.[46][47] They notably discovered the universality in chaos, permitting the application of chaos theory to many different phenomena.

In 1979, Albert J. Libchaber, during a symposium organized in Aspen by Pierre Hohenberg, presented his experimental observation of the bifurcation cascade that leads to chaos and turbulence in Rayleigh–Bénard convection systems. He was awarded the Wolf Prize in Physics in 1986 along with Mitchell J. Feigenbaum for their inspiring achievements.[48]

In 1986, the New York Academy of Sciences co-organized with the National Institute of Mental Health and the Office of Naval Research the first important conference on chaos in biology and medicine. There, Bernardo Huberman presented a mathematical model of the eye tracking disorder among schizophrenics.[49] This led to a renewal of physiology in the 1980s through the application of chaos theory, for example, in the study of pathological cardiac cycles.

In 1987, Per Bak, Chao Tang and Kurt Wiesenfeld published a paper in *Physical Review Letters*[50] describing for the first time self-organized criticality (SOC), considered to be one of the mechanisms by which complexity arises in nature.

Alongside largely lab-based approaches such as the Bak–Tang–Wiesenfeld sandpile, many other investigations have focused on large-scale natural or social systems that are known (or suspected) to display scale-invariant behavior. Although these approaches were not always welcomed (at least initially) by specialists in the subjects examined, SOC has nevertheless become established as a strong candidate for explaining a number of natural phenomena, including earthquakes (which, long before SOC was discovered, were known as a source of scale-invariant behavior such as the Gutenberg–Richter law describing the statistical distribution of earthquake sizes, and the Omori law[51] describing the frequency of aftershocks), solar flares, fluctuations in economic systems such as financial markets (references to SOC are common in econophysics), landscape formation, forest fires, landslides, epidemics, and biological evolution (where SOC has been invoked, for example, as the dynamical mechanism behind the theory of "punctuated equilibria" put forward by Niles Eldredge and Stephen Jay Gould). Given the implications of a scale-free distribution of event sizes, some researchers have suggested that another phenomenon that should be considered an example of SOC is the occurrence of wars. These investigations of SOC have included both attempts at modelling (either developing new models or adapting existing ones to the specifics of a given natural system), and extensive data analysis to determine the existence and/or characteristics of natural scaling laws.

In the same year, James Gleick published *Chaos: Making a New Science*, which became a best-seller and introduced the general principles of chaos theory as well as its history to the broad public, though his history under-emphasized important Soviet contributions.[52] Initially the domain of a few, isolated individuals, chaos theory progressively emerged as a transdisciplinary and institutional discipline, mainly under the name of nonlinear systems analysis. Alluding to Thomas Kuhn's concept of a paradigm shift exposed in *The Structure of Scientific Revolutions* (1962), many "chaologists" (as some described themselves) claimed that this new theory was an example of such a shift, a thesis upheld by Gleick.

The availability of cheaper, more powerful computers broadens the applicability of chaos theory. Currently, chaos theory continues to be a very active area of research,[53] involving many different disciplines (mathematics, topology, physics, social systems, population modeling, biology, meteorology, astrophysics, information theory, computational neuroscience, etc.).

33.5 Distinguishing random from chaotic data

It can be difficult to tell from data whether a physical or other observed process is random or chaotic, because in practice no time series consists of a pure "signal". There will always be some form of corrupting noise, even if it is present as round-off or truncation error. Thus any real time series, even if mostly deterministic, will contain some (pseudo-)randomness.[54][55]

All methods for distinguishing deterministic and stochastic processes rely on the fact that a deterministic system always evolves in the same way from a given starting point.[54][56] Thus, given a time series to test for determinism, one can

1. pick a test state;

2. search the time series for a similar or nearby state; and

3. compare their respective time evolutions.

Define the error as the difference between the time evolution of the test state and the time evolution of the nearby state. A deterministic system will have an error that either remains small (stable, regular solution) or increases exponentially with time (chaos). A stochastic system will have a randomly distributed error.[57]

Essentially, all measures of determinism taken from time series rely upon finding the closest states to a given test state (e.g., correlation dimension, Lyapunov exponents, etc.). To define the state of a system, one typically relies on phase space embedding methods such as Poincaré plots.[58] Typically one chooses an embedding dimension and investigates the propagation of the error between two nearby states. If the error looks random, one increases the dimension. If the dimension can be increased to obtain a deterministically looking error, then analysis is done. Though it may sound simple, one complication is that as the dimension increases, the search for a nearby state requires a lot more computation time and a lot of data (the amount of data required increases exponentially with embedding dimension) to find a suitably close candidate. If the embedding dimension (number of measures per state) is chosen too small (less than the "true" value), deterministic data can appear to be random, but in theory there is no problem choosing the dimension too large – the method will work.

When a nonlinear deterministic system is attended by external fluctuations, its trajectories present serious and permanent distortions. Furthermore, the noise is amplified due to the inherent nonlinearity and reveals totally new dynamical properties. Statistical tests attempting to separate noise from the deterministic skeleton or inversely isolate the deterministic part risk failure. Things become worse when the deterministic component is a nonlinear feedback system.[59] In presence of interactions between nonlinear deterministic components and noise, the resulting nonlinear series can display dynamics that traditional tests for nonlinearity are sometimes not able to capture.[60]

The question of how to distinguish deterministic chaotic systems from stochastic systems has also been discussed in philosophy. It has been shown that they might be observationally equivalent.[61]

33.6 Applications

Chaos theory was born from observing weather patterns, but it has become applicable to a variety of other situations. Some areas benefiting from chaos theory today are geology, mathematics, microbiology, biology, computer science, economics,[63][64][65] engineering,[66] finance,[67][68] algorithmic trading,[69][70][71] meteorology, philosophy, physics, politics, population dynamics,[72] psychology, and robotics. A few categories are listed below with examples, but this is by no means a comprehensive list as new applications are appearing every day.

33.6.1 Computer science

Chaos theory is not new to computer science and has been used for many years in cryptography. One type of encryption, secret key or symmetric key, relies on diffusion and confusion, which is modeled well by chaos theory.[73] Another type of computing, DNA computing, when paired with chaos theory, offers a more efficient way to encrypt images and other information.[74] Robotics is another area that has recently benefited from chaos theory. Instead of robots acting in a trial-and-error type of refinement to interact with their environment, chaos theory has been used to build a predictive model.[75] Chaotic dynamics have been exhibited by passive walking biped robots.[76]

33.6.2 Biology

For over a hundred years, biologists have been keeping track of populations of different species with population models. Most models are continuous, but recently scientists have been able to implement chaotic models in certain populations.[77] For example, a study on models of Canadian lynx showed there was chaotic behavior in the population growth.[78] Chaos can also be found in ecological systems, such as hydrology. While a chaotic model for hydrology has its shortcomings, there is still much to be learned from looking at the data through the lens of chaos theory.[79] Another biological application

A conus textile shell, similar in appearance to Rule 30, a cellular automaton with chaotic behaviour.[62]

is found in cardiotocography. Fetal surveillance is a delicate balance of obtaining accurate information while being as noninvasive as possible. Better models of warning signs of fetal hypoxia can be obtained through chaotic modeling.[80]

33.6.3 Other areas

In chemistry, predicting gas solubility is essential to manufacturing polymers, but models using particle swarm optimization (PSO) tend to converge to the wrong points. An improved version of PSO has been created by introducing chaos, which keeps the simulations from getting stuck.[81] In celestial mechanics, especially when observing asteroids, applying chaos theory leads to better predictions about when these objects will come in range of Earth and other planets.[82] In quantum physics and electrical engineering, the study of large arrays of Josephson junctions benefitted greatly from chaos theory.[83] Closer to home, coal mines have always been dangerous places where frequent natural gas leaks cause many deaths. Until recently, there was no reliable way to predict when they would occur. But these gas leaks have chaotic tendencies that, when properly modeled, can be predicted fairly accurately.[84]

Chaos theory can be applied outside of the natural sciences. By adapting a model of career counseling to include a chaotic interpretation of the relationship between employees and the job market, better suggestions can be made to people struggling with career decisions.[85] Modern organizations are increasingly seen as open complex adaptive systems, with fundamental natural nonlinear structures, subject to internal and external forces which may be sources of chaos. The chaos metaphor—used in verbal theories—grounded on mathematical models and psychological aspects of human behavior provides helpful insights to describing the complexity of small work groups, that go beyond the metaphor itself.[86]

It is possible that economic models can also be improved through an application of chaos theory, but predicting the health of an economic system and what factors influence it most is an extremely complex task.[87] Economic and financial systems

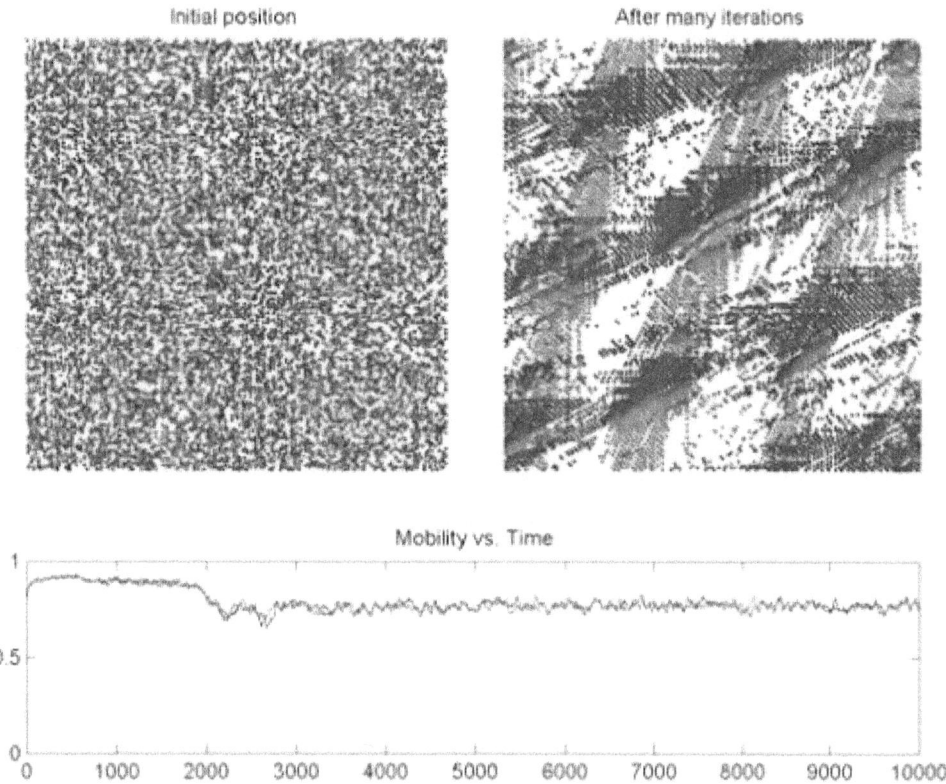

The red cars and blue cars take turns to move; the red ones only move upwards, and the blue ones move rightwards. Every time, all the cars of the same colour try to move one step if there is no car in front of it. Here, the model has self-organized in a somewhat geometric pattern where there are some traffic jams and some areas where cars can move at top speed.
Source: https://en.wikipedia.org/wiki/File:BML_N%3D200_P%3D32.png

are fundamentally different from those in the physical and natural sciences since the former are inherently stochastic in nature, as they result from the interactions of people, and thus pure deterministic models are unlikely to provide accurate representations of the data. The empirical literature that tests for chaos in economics and finance presents very mixed results, in part due to confusion between specific tests for chaos and more general tests for non-linear relationships.[88]

Traffic forecasting is another area that greatly benefits from applications of chaos theory. Better predictions of when traffic will occur would allow measures to be taken for it to be dispersed before the traffic starts, rather than after. Combining chaos theory principles with a few other methods has led to a more accurate short-term prediction model (see the plot of the BML traffic model at right).[89]

Chaos theory also finds applications in psychology. For example, in modeling group behavior in which heterogeneous members may behave as if sharing to different degrees what in Wilfred Bion's theory is a basic assumption, the group dynamics is the result of the individual dynamics of the members: each individual reproduces the group dynamics in a different scale, and the chaotic behavior of the group is reflected in each member.[90]

33.7 See also

33.8 References

[1] Boeing (2015). "Chaos Theory and the Logistic Map". Retrieved 2015-07-16.

[2] Kellert, Stephen H. (1993). *In the Wake of Chaos: Unpredictable Order in Dynamical Systems*. University of Chicago Press. p. 32. ISBN 0-226-42976-8.

[3] Kellert 1993, p. 56

[4] Kellert 1993, p. 62

[5] Werndl, Charlotte (2009). "What are the New Implications of Chaos for Unpredictability?". *The British Journal for the Philosophy of Science* **60** (1): 195–220. doi:10.1093/bjps/axn053.

[6] Danforth, Christopher M. (April 2013). "Chaos in an Atmosphere Hanging on a Wall". *Mathematics of Planet Earth 2013*. Retrieved 4 April 2013.

[7] Lorenz, Edward N. (1963). "Deterministic non-periodic flow".*Journal of the Atmospheric Sciences***20**(2): 130–141.BibcoL. doi:10.1175/1520-0469(1963)020<0130:DNF>2.0.CO;2.

[8] Ivancevic, Vladimir G.; Tijana T. Ivancevic (2008). *Complex nonlinearity: chaos, phase transitions, topology change, and path integrals*. Springer. ISBN 978-3-540-79356-4.

[9] *Sync: The Emerging Science of Spontaneous Order*, Steven Strogatz, Hyperion, New York, 2003, pages 189-190.

[10] Definition of chaos at Wiktionary;

[11] Hasselblatt, Boris; Anatole Katok (2003). *A First Course in Dynamics: With a Panorama of Recent Developments*. Cambridge University Press. ISBN 0-521-58750-6.

[12] Elaydi, Saber N. (1999). *Discrete Chaos*. Chapman & Hall/CRC. p. 117. ISBN 1-58488-002-3.

[13] Basener, William F. (2006). *Topology and its applications*. Wiley. p. 42. ISBN 0-471-68755-3.

[14] Vellekoop, Michel; Berglund, Raoul (April 1994). "On Intervals, Transitivity = Chaos". *The American Mathematical Monthly* **101** (4): 353–5. doi:10.2307/2975629. JSTOR 2975629.

[15] Medio, Alfredo; Lines, Marji (2001). *Nonlinear Dynamics: A Primer*. Cambridge University Press. p. 165. ISBN 0-521-55874-3.

[16] Wikiversity (28 July 2011). "1972/Lorenz". Wikipedia. Retrieved 8 April 2014.

[17] Watts, Robert G. (2007). *Global Warming and the Future of the Earth*. Morgan & Claypool. p. 17.

[18] Devaney 2003

[19] Alligood, Sauer & Yorke 1997

[20] Li, T.Y.; Yorke, J.A. (1975). "Period Three Implies Chaos" (PDF). *American Mathematical Monthly* **82** (10): 985–92. doi:10.2307/2318254.

[21] Sprott, J.C. (1997). "Simplest dissipative chaotic flow". *Physics Letters A* **228** (4–5): 271. Bibcode:1997PhLA..228..271S. doi:10.1016/S0375-9601(97)00088-1.

[22] Fu, Z.; Heidel, J. (1997). "Non-chaotic behaviour in three-dimensional quadratic systems". *Nonlinearity* **10** (5): 1289. Bibcode:1997Nonli..10.1289F. doi:10.1088/0951-7715/10/5/014.

[23] Heidel, J.; Fu, Z. (1999). "Nonchaotic behaviour in three-dimensional quadratic systems II. The conservative case". *Nonlinearity* **12** (3): 617. Bibcode:1999Nonli..12..617H. doi:10.1088/0951-7715/12/3/012.

[24] Rosario, Pedro (2006). *Underdetermination of Science: Part I*. Lulu.com. ISBN 1411693914.

[25] Bonet, J.; Martínez-Giménez, F.; Peris, A. (2001). "A Banach space which admits no chaotic operator". *Bulletin of the London Mathematical Society* **33** (2): 196–8. doi:10.1112/blms/33.2.196.

[26] K. E. Chlouverakis and J. C. Sprott, Chaos Solitons & Fractals 28, 739-746 (2005), Chaotic Hyperjerk Systems, http://sprott. physics.wisc.edu/pubs/paper297.htm

[27] Steven Strogatz, *Sync: The Emerging Science of Spontaneous Order*, Hyperion, 2003.

[28] Poincaré, Jules Henri (1890). "Sur le problème des trois corps et les équations de la dynamique. Divergence des séries de M. Lindstedt". *Acta Mathematica* **13**: 1–270. doi:10.1007/BF02392506.

[29] Diacu, Florin; Holmes, Philip (1996). *Celestial Encounters: The Origins of Chaos and Stability*. Princeton University Press.

[30] Hadamard, Jacques (1898). "Les surfaces à courbures opposées et leurs lignes géodesiques". *Journal de Mathématiques Pures et Appliquées* **4**: 27–73.

[31] George D. Birkhoff, *Dynamical Systems*, vol. 9 of the American Mathematical Society Colloquium Publications (Providence, Rhode Island: American Mathematical Society, 1927)

[32] Kolmogorov, Andrey Nikolaevich (1941). "Local structure of turbulence in an incompressible fluid for very large Reynolds numbers". *Doklady Akademii Nauk SSSR* **30** (4): 301–5. Bibcode:1941DoSSR..30..301K. Reprinted in: Kolmogorov, A. N. (1991). "The Local Structure of Turbulence in Incompressible Viscous Fluid for Very Large Reynolds Numbers". *Proceedings of the Royal Society A* **434** (1890): 9–13. Bibcode:1991RSPSA.434....9K. doi:10.1098/rspa.1991.0075.

[33] Kolmogorov, A. N. (1941). "On degeneration of isotropic turbulence in an incompressible viscous liquid". *Doklady Akademii Nauk SSSR* **31** (6): 538–540. Reprinted in: Kolmogorov, A. N. (1991). "Dissipation of Energy in the Locally Isotropic Turbulence". *Proceedings of the Royal Society A* **434** (1890): 15–17. Bibcode:1991RSPSA.434...15K. doi:10.1098/rspa.1991.0076.

[34] Kolmogorov, A. N. (1954). "Preservation of conditionally periodic movements with small change in the Hamiltonian function". *Doklady Akademii Nauk SSSR*. Lecture Notes in Physics **98**: 527–530. Bibcode:1979LNP....93...51K. doi:10.1007/BFb0021737. ISBN 3-540-09120-3. See also Kolmogorov–Arnold–Moser theorem

[35] Cartwright, Mary L.; Littlewood, John E. (1945). "On non-linear differential equations of the second order, I: The equation $y'' + k(1-y^2)y' + y = b\lambda k\cos(\lambda t + a)$, k large". *Journal of the London Mathematical Society* **20** (3): 180–9. doi:10.1112/jlms/s1-20.3.180. See also: Van der Pol oscillator

[36] Smale, Stephen (January 1960). "Morse inequalities for a dynamical system". *Bulletin of the American Mathematical Society* **66**: 43–49. doi:10.1090/S0002-9904-1960-10386-2.

[37] Abraham & Ueda 2001, See Chapters 3 and 4

[38] Sprott 2003, p. 89

[39] Gleick, James (1987). *Chaos: Making a New Science*. London: Cardinal. p. 17. ISBN 0-434-29554-X.

[40] Mandelbrot, Benoît (1963). "The variation of certain speculative prices". *Journal of Business* **36** (4): 394–419. doi:10.1086/294632.

[41] Berger J.M., Mandelbrot B. (1963). "A new model for error clustering in telephone circuits". *IBM Journal of Research and Development* **7**: 224–236. doi:10.1147/rd.73.0224.

[42] Mandelbrot, B. (1977). *The Fractal Geometry of Nature*. New York: Freeman. p. 248.

[43] See also: Mandelbrot, Benoît B.; Hudson, Richard L. (2004). *The (Mis)behavior of Markets: A Fractal View of Risk, Ruin, and Reward*. New York: Basic Books. p. 201.

[44] Mandelbrot, Benoît (5 May 1967). "How Long Is the Coast of Britain? Statistical Self-Similarity and Fractional Dimension". *Science* **156** (3775): 636–8. Bibcode:1967Sci...156..636M. doi:10.1126/science.156.3775.636. PMID 17837158.

[45] Buldyrev, S.V.; Goldberger, A.L.; Havlin, S.; Peng, C.K.; Stanley, H.E. (1994). "Fractals in Biology and Medicine: From DNA to the Heartbeat". In Bunde, Armin; Havlin, Shlomo. *Fractals in Science*. Springer. pp. 49–89. ISBN 3-540-56220-6.

[46] Feigenbaum, Mitchell (July 1978). "Quantitative universality for a class of nonlinear transformations". *Journal of Statistical Physics* **19** (1): 25–52. Bibcode:1978JSP....19...25F. doi:10.1007/BF01020332.

[47] Coullet, Pierre, and Charles Tresser. "Iterations d'endomorphismes et groupe de renormalisation." Le Journal de Physique Colloques 39.C5 (1978): C5-25

[48] "The Wolf Prize in Physics in 1986.".

[49] Huberman, B.A. (July 1987). "A Model for Dysfunctions in Smooth Pursuit Eye Movement". *Annals of the New York Academy of Sciences*. 504 Perspectives in Biological Dynamics and Theoretical Medicine: 260–273. Bibcode:1987NYASA.504..260H. doi:10.1111/j.1749-6632.1987.tb48737.x.

[50] Bak, Per; Tang, Chao; Wiesenfeld, Kurt; Tang; Wiesenfeld (27 July 1987). "Self-organized criticality: An explanation of the 1/f noise". *Physical Review Letters* **59** (4): 381–4. Bibcode:1987PhRvL..59..381B. doi:10.1103/PhysRevLett.59.381. However, the conclusions of this article have been subject to dispute. "?".. See especially: Laurson, Lasse; Alava, Mikko J.; Zapperi, Stefano (15 September 2005). "Letter: Power spectra of self-organized critical sand piles". *Journal of Statistical Mechanics: Theory and Experiment* **0511**. L001.

[51] Omori, F. (1894). "On the aftershocks of earthquakes". *Journal of the College of Science, Imperial University of Tokyo* **7**: 111–200.

[52] Gleick, James (August 26, 2008). *Chaos: Making a New Science*. Penguin Books. ISBN 0143113453.

[53] Motter A. E. and Campbell D. K., Chaos at fifty, Phys. Today 66(5), 27-33 (2013).

[54] Provenzale, A.; Smith; Vio; Murante et al. (1992). "Distinguishing between low-dimensional dynamics and randomness in measured time-series". *Physica D* **58**: 31–49. Bibcode:1992PhyD...58...31P. doi:10.1016/0167-2789(92)90100-2.

[55] Brock, W.A. (October 1986). "Distinguishing random and deterministic systems: Abridged version". *Journal of Economic Theory* **40**: 168–195. doi:10.1016/0022-0531(86)90014-1.

[56] Sugihara G., May R.; May (1990). "Nonlinear forecasting as a way of distinguishing chaos from measurement error in time series" (PDF). *Nature* **344** (6268): 734–741. Bibcode:1990Natur.344..734S. doi:10.1038/344734a0. PMID 2330029.

[57] Casdagli, Martin (1991). "Chaos and Deterministic *versus* Stochastic Non-linear Modelling". *Journal of the Royal Statistical Society, Series B* **54** (2): 303–328. JSTOR 2346130.

[58] Broomhead, D.S.; King, G.P.; King (June–July 1986). "Extracting qualitative dynamics from experimental data". *Physica D* **20** (2–3): 217–236. Bibcode:1986PhyD...20..217B. doi:10.1016/0167-2789(86)90031-X.

[59] Kyrtsou C (2008). "Re-examining the sources of heteroskedasticity: the paradigm of noisy chaotic models". *Physica A* **387** (27): 6785–9. Bibcode:2008PhyA..387.6785K. doi:10.1016/j.physa.2008.09.008.

[60] Kyrtsou, C. (2005). "Evidence for neglected linearity in noisy chaotic models". *International Journal of Bifurcation and Chaos* **15** (10): 3391–4. Bibcode:2005IJBC...15.3391K. doi:10.1142/S0218127405013964.

[61] Werndl, Charlotte (2009). "Are Deterministic Descriptions and Indeterministic Descriptions Observationally Equivalent?". *Studies in History and Philosophy of Modern Physics* **40** (3): 232–242. doi:10.1016/j.shpsb.2009.06.004.

[62] Stephen Coombes (February 2009). "The Geometry and Pigmentation of Seashells" (PDF). *www.maths.nottingham.ac.uk*. University of Nottingham. Retrieved 2013-04-10.

[63] Kyrtsou C., Labys W. (2006). "Evidence for chaotic dependence between US inflation and commodity prices". *Journal of Macroeconomics* **28** (1): 256–266. doi:10.1016/j.jmacro.2005.10.019.

[64] Kyrtsou C., Labys W.; Labys (2007). "Detecting positive feedback in multivariate time series: the case of metal prices and US inflation". *Physica A* **377** (1): 227–229. Bibcode:2007PhyA..377..227K. doi:10.1016/j.physa.2006.11.002.

[65] Kyrtsou, C.; Vorlow, C. (2005). "Complex dynamics in macroeconomics: A novel approach". In Diebolt, C.; Kyrtsou, C. *New Trends in Macroeconomics*. Springer Verlag.

[66] Applying Chaos Theory to Embedded Applications

[67] Hristu-Varsakelis, D.; Kyrtsou, C. (2008). "Evidence for nonlinear asymmetric causality in US inflation, metal and stock returns". *Discrete Dynamics in Nature and Society* **2008**: 1. doi:10.1155/2008/138547. 138547.

[68] Kyrtsou, C. and M. Terraza, (2003). "Is it possible to study chaotic and ARCH behaviour jointly? Application of a noisy Mackey-Glass equation with heteroskedastic errors to the Paris Stock Exchange returns series". *Computational Economics* **21** (3): 257–276. doi:10.1023/A:1023939610962.

[69] Williams, Bill Williams, Justine (2004). *Trading chaos : maximize profits with proven technical techniques* (2nd ed.). New York: Wiley. ISBN 9780471463085.

[70] Peters, Edgar E. (1994). *Fractal market analysis : applying chaos theory to investment and economics* (2. print. ed.). New York u.a.: Wiley. ISBN 978-0471585244.

[71] Peters, / Edgar E. (1996). *Chaos and order in the capital markets : a new view of cycles, prices, and market volatility* (2nd ed.). New York: John Wiley & Sons. ISBN 978-0471139386.

[72] Dilão, R.; Domingos, T. (2001). "Periodic and Quasi-Periodic Behavior in Resource Dependent Age Structured Population Models". *Bulletin of Mathematical Biology* **63** (2): 207–230. doi:10.1006/bulm.2000.0213. PMID 11276524.

[73] Wang, Xingyuan; Zhao, Jianfeng (2012). "An improved key agreement protocol based on chaos". *Commun. Nonlinear Sci. Numer. Simul.* **15** (12): 4052–4057. Bibcode:2010CNSNS..15.4052W. doi:10.1016/j.cnsns.2010.02.014.

[74] Babaei, Majid (2013). "A novel text and image encryption method based on chaos theory and DNA computing". *Natural Computing, an International Journal* **12** (1): 101–107. doi:10.1007/s11047-012-9334-9.

[75] Nehmzow, Ulrich; Keith Walker (Dec 2005). "Quantitative description of robot–environment interaction using chaos theory". *Robotics and Autonomous Systems* **53** (3–4): 177–193. doi:10.1016/j.robot.2005.09.009.

[76] Goswami, Ambarish; Thuilot, Benoit; Espiau, Bernard (1998). "A Study of the Passive Gait of a Compass-Like Biped Robot: Symmetry and Chaos". *The International Journal of Robotics Research* **17** (12): 1282–1301. doi:10.1177/027836499801701202.

[77] Eduardo, Liz; Ruiz-Herrera, Alfonso (2012). "Chaos in discrete structured population models". *SIAM Journal on Applied Dynamical Systems* **11** (4): 1200–1214. doi:10.1137/120868980.

[78] Lai, Dejian (1996). "Comparison study of AR models on the Canadian lynx data: a close look at BDS statistic". *Computational Statistics \& Data Analysis* **22** (4): 409–423. doi:10.1016/0167-9473(95)00056-9.

[79] Sivakumar, B (31 January 2000). "Chaos theory in hydrology: important issues and interpretations". *Journal of Hydrology* **227** (1–4): 1–20. Bibcode:2000JHyd..227....1S. doi:10.1016/S0022-1694(99)00186-9.

[80] Bozóki, Zsolt (February 1997). "Chaos theory and power spectrum analysis in computerized cardiotocography". *European Journal of Obstetrics & Gynecology and Reproductive Biology* **71** (2): 163–168. doi:10.1016/s0301-2115(96)02628-0.

[81] Li, Mengshan; Xingyuan Huanga; Hesheng Liua; Bingxiang Liub; Yan Wub; Aihua Xiongc; Tianwen Dong (25 October 2013). "Prediction of gas solubility in polymers by back propagation artificial neural network based on self-adaptive particle swarm optimization algorithm and chaos theory". *Fluid Phase Equilibria* **356**: 11–17. doi:10.1016/j.fluid.2013.07.017.

[82] Morbidelli, A. (2001). "Chaotic diffusion in celestial mechanics". *Regular & Chaotic Dynamics, International Scientific Journal* **6** (4): 339–353.

[83] Steven Strogatz, *Sync: The Emerging Science of Spontaneous Order*, Hyperion, 2003

[84] Dingqi, Li; Yuanping Chenga; Lei Wanga; Haifeng Wanga; Liang Wanga; Hongxing Zhou (May 2011). "Prediction method for risks of coal and gas outbursts based on spatial chaos theory using gas desorption index of drill cuttings". *Mining Science and Technology* **21** (3): 439–443.

[85] Pryor, Robert G. L.; Norman E. Aniundson; Jim E. H. Bright (June 2008). "Probabilities and Possibilities: The Strategic Counseling Implications of the Chaos Theory of Careers". *The Career Development Quarterly* **56**: 309–318. doi:10.1002/j.2161-0045.2008.tb00096.x.

[86] Dal Forno, Arianna; Merlone, Ugo (2013). "Chaotic Dynamics in Organization Theory". In Bischi, Gian Italo; Chiarella, Carl; Shusko, Irina. *Global Analysis of Dynamic Models in Economics and Finance*. Springer-Verlag. pp. 185–204. ISBN 978-3-642-29503-4.

[87] Juárez, Fernando (2011). "Applying the theory of chaos and a complex model of health to establish relations among financial indicators". *Procedia Computer Science* **3**: 982–986. doi:10.1016/j.procs.2010.12.161.

[88] Brooks, Chris (1998). "Chaos in foreign exchange markets: a sceptical view". *Computational Economics* **11**: 265–281. doi:10.1023/A:1008650024944. ISSN 1572-9974.

[89] Wang, Jin; Qixin Shi (February 2013). "Short-term traffic speed forecasting hybrid model based on Chaos–Wavelet Analysis-Support Vector Machine theory". *Transportation Research Part C: Emerging Technologies* **27**: 219–232. doi:10.1016/j.trc.2012.08.004.

[90] Dal Forno, Arianna; Merlone, Ugo (2013). "Nonlinear dynamics in work groups with Bion's basic assumptions". *Nonlinear Dynamics, Psychology, and Life Sciences* **17** (2): 295–315. ISSN 1090-0578.

33.9 Scientific literature

33.9.1 Articles

- Sharkovskii, A.N. (1964). "Co-existence of cycles of a continuous mapping of the line into itself". *Ukrainian Math. J.* **16**: 61–71.

- Li, T.Y.; Yorke, J.A. (1975). "Period Three Implies Chaos". *American Mathematical Monthly* **82** (10): 985–92. Bibcode:1975AmMM...82..985L. doi:10.2307/2318254.

- Crutchfield; Tucker; Morrison; J.D.; Packard; N.H.; Shaw; R.S (December 1986). "Chaos". *Scientific American* **255** (6): 38–49 (bibliography p.136). Bibcode:1986SciAm.255...38T. Online version (Note: the volume and page citation cited for the online text differ from that cited here. The citation here is from a photocopy, which is consistent with other citations found online, but which don't provide article views. The online content is identical to the hardcopy text. Citation variations will be related to country of publication).

- Kolyada, S.F. (2004). "Li-Yorke sensitivity and other concepts of chaos". *Ukrainian Math. J.* **56** (8): 1242–57. doi:10.1007/s11253-005-0055-4.

- Strelioff, C.; Hübler, A. (2006). "Medium-Term Prediction of Chaos" (PDF). *Phys. Rev. Lett.* **96** (4): 044101. Bibcode:2006PhRvL..96d4101S. doi:10.1103/PhysRevLett.96.044101. PMID 16486826. 044101.

- Hübler, A.; Foster, G.; Phelps, K. (2007). "Managing Chaos: Thinking out of the Box" (PDF). *Complexity* **12** (3): 10–13. doi:10.1002/cplx.20159.

33.9.2 Textbooks

- Alligood, K.T.; Sauer, T.; Yorke, J.A. (1997). *Chaos: an introduction to dynamical systems*. Springer-Verlag. ISBN 0-387-94677-2.

- Baker, G. L. (1996). *Chaos, Scattering and Statistical Mechanics*. Cambridge University Press. ISBN 0-521-39511-9.

- Badii, R.; Politi A. (1997). *Complexity: hierarchical structures and scaling in physics*. Cambridge University Press. ISBN 0-521-66385-7.

- Bunde; Havlin, Shlomo, eds. (1996). *Fractals and Disordered Systems*. Springer. ISBN 3642848702. and Bunde; Havlin, Shlomo, eds. (1994). *Fractals in Science*. Springer. ISBN 3-540-56220-6.

- Collet, Pierre, and Eckmann, Jean-Pierre (1980). *Iterated Maps on the Interval as Dynamical Systems*. Birkhauser. ISBN 0-8176-4926-3.

- Devaney, Robert L. (2003). *An Introduction to Chaotic Dynamical Systems* (2nd ed.). Westview Press. ISBN 0-8133-4085-3.

- Gollub, J. P.; Baker, G. L. (1996). *Chaotic dynamics*. Cambridge University Press. ISBN 0-521-47685-2.

- Guckenheimer, John; Holmes, Philip (1983). *Nonlinear Oscillations, Dynamical Systems, and Bifurcations of Vector Fields*. Springer-Verlag. ISBN 0-387-90819-6.

- Gulick, Denny (1992). *Encounters with Chaos*. McGraw-Hill. ISBN 0-07-025203-3.

- Gutzwiller, Martin (1990). *Chaos in Classical and Quantum Mechanics*. Springer-Verlag. ISBN 0-387-97173-4.

- Hoover, William Graham (2001) [1999]. *Time Reversibility, Computer Simulation, and Chaos*. World Scientific. ISBN 981-02-4073-2.

- Kautz, Richard (2011). *Chaos: The Science of Predictable Random Motion*. Oxford University Press. ISBN 978-0-19-959458-0.

- Kiel, L. Douglas; Elliott, Euel W. (1997). *Chaos Theory in the Social Sciences*. Perseus Publishing. ISBN 0-472-08472-0.

- Moon, Francis (1990). *Chaotic and Fractal Dynamics*. Springer-Verlag. ISBN 0-471-54571-6.

- Ott, Edward (2002). *Chaos in Dynamical Systems*. Cambridge University Press. ISBN 0-521-01084-5.

- Strogatz, Steven (2000). *Nonlinear Dynamics and Chaos*. Perseus Publishing. ISBN 0-7382-0453-6.

- Sprott, Julien Clinton (2003). *Chaos and Time-Series Analysis*. Oxford University Press. ISBN 0-19-850840-9.

- Tél, Tamás; Gruiz, Márton (2006). *Chaotic dynamics: An introduction based on classical mechanics*. Cambridge University Press. ISBN 0-521-83912-2.

- Teschl, Gerald (2012). *Ordinary Differential Equations and Dynamical Systems*. Providence: American Mathematical Society. ISBN 978-0-8218-8328-0.

- Thompson J M T, Stewart H B (2001). *Nonlinear Dynamics And Chaos*. John Wiley and Sons Ltd. ISBN 0-471-87645-3.

- Tufillaro; Reilly (1992). *An experimental approach to nonlinear dynamics and chaos*. Addison-Wesley. ISBN 0-201-55441-0.

- Wiggins, Stephen (2003). *Introduction to Applied Dynamical Systems and Chaos*. Springer. ISBN 0-387-00177-8.

- Zaslavsky, George M. (2005). *Hamiltonian Chaos and Fractional Dynamics*. Oxford University Press. ISBN 0-19-852604-0.

33.9.3   Semitechnical and popular works

- Christophe Letellier, *Chaos in Nature*, World Scientific Publishing Company, 2012, ISBN 978-981-4374-42-2.

- Abraham, Ralph H.; Ueda, Yoshisuke, eds. (2000). *The Chaos Avant-Garde: Memoirs of the Early Days of Chaos Theory*. World Scientific. ISBN 978-981-238-647-2.

- Barnsley, Michael F. (2000). *Fractals Everywhere*. Morgan Kaufmann. ISBN 978-0-12-079069-2.

- Bird, Richard J. (2003). *Chaos and Life: Complexit and Order in Evolution and Thought*. Columbia University Press. ISBN 978-0-231-12662-5.

- John Briggs and David Peat, *Turbulent Mirror: : An Illustrated Guide to Chaos Theory and the Science of Wholeness*, Harper Perennial 1990, 224 pp.

- John Briggs and David Peat, *Seven Life Lessons of Chaos: Spiritual Wisdom from the Science of Change*, Harper Perennial 2000, 224 pp.

- Cunningham, Lawrence A. (1994). "From Random Walks to Chaotic Crashes: The Linear Genealogy of the Efficient Capital Market Hypothesis". *George Washington Law Review* **62**: 546.

- Predrag Cvitanović, *Universality in Chaos*, Adam Hilger 1989, 648 pp.

- Leon Glass and Michael C. Mackey, *From Clocks to Chaos: The Rhythms of Life*, Princeton University Press 1988, 272 pp.

- James Gleick, *Chaos: Making a New Science*, New York: Penguin, 1988, 368 pp.

- John Gribbin. *Deep Simplicity*. Penguin Press Science. Penguin Books.

- L Douglas Kiel, Euel W Elliott (ed.), *Chaos Theory in the Social Sciences: Foundations and Applications*, University of Michigan Press, 1997, 360 pp.

- Arvind Kumar, *Chaos, Fractals and Self-Organisation; New Perspectives on Complexity in Nature* , National Book Trust, 2003.

- Hans Lauwerier, *Fractals*, Princeton University Press, 1991.

- Edward Lorenz, *The Essence of Chaos*, University of Washington Press, 1996.

- Alan Marshall (2002) The Unity of Nature: Wholeness and Disintegration in Ecology and Science, Imperial College Press: London

- Heinz-Otto Peitgen and Dietmar Saupe (Eds.), *The Science of Fractal Images*, Springer 1988, 312 pp.

- Clifford A. Pickover, *Computers, Pattern, Chaos, and Beauty: Graphics from an Unseen World* , St Martins Pr 1991.

- Ilya Prigogine and Isabelle Stengers, *Order Out of Chaos*, Bantam 1984.

- Heinz-Otto Peitgen and P. H. Richter, *The Beauty of Fractals : Images of Complex Dynamical Systems*, Springer 1986, 211 pp.

- David Ruelle, *Chance and Chaos*, Princeton University Press 1993.

- Ivars Peterson, *Newton's Clock: Chaos in the Solar System*, Freeman, 1993.

- Ian Roulstone and John Norbury (2013). *Invisible in the Storm: the role of mathematics in understanding weather*. Princeton University Press. ISBN 0691152721.

- David Ruelle, *Chaotic Evolution and Strange Attractors*, Cambridge University Press, 1989.

- Peter Smith, *Explaining Chaos*, Cambridge University Press, 1998.

- Ian Stewart, *Does God Play Dice?: The Mathematics of Chaos* , Blackwell Publishers, 1990.

- Steven Strogatz, *Sync: The emerging science of spontaneous order*, Hyperion, 2003.

- Yoshisuke Ueda, *The Road To Chaos*, Aerial Pr, 1993.

- M. Mitchell Waldrop, *Complexity : The Emerging Science at the Edge of Order and Chaos*, Simon & Schuster, 1992.

- Sawaya, Antonio (2010). *Financial time series analysis : Chaos and neurodynamics approach*.

33.10   External links

- Hazewinkel, Michiel, ed. (2001), "Chaos", *Encyclopedia of Mathematics*, Springer, ISBN 978-1-55608-010-4

- Nonlinear Dynamics Research Group with Animations in Flash

- The Chaos group at the University of Maryland

- The Chaos Hypertextbook. An introductory primer on chaos and fractals

- ChaosBook.org An advanced graduate textbook on chaos (no fractals)

- Society for Chaos Theory in Psychology & Life Sciences

- Nonlinear Dynamics Research Group at CSDC, Florence Italy

- Interactive live chaotic pendulum experiment, allows users to interact and sample data from a real working damped driven chaotic pendulum

- Nonlinear dynamics: how science comprehends chaos, talk presented by Sunny Auyang, 1998.

- Nonlinear Dynamics. Models of bifurcation and chaos by Elmer G. Wiens

- Gleick's *Chaos* (excerpt)

- Systems Analysis, Modelling and Prediction Group at the University of Oxford

- A page about the Mackey-Glass equation

- High Anxieties — The Mathematics of Chaos (2008) BBC documentary directed by David Malone

- The chaos theory of evolution - article published in Newscientist featuring similarities of evolution and non-linear systems including fractal nature of life and chaos.

- Jos Leys, Étienne Ghys et Aurélien Alvarez, *Chaos, A Mathematical Adventure*. Nine films about dynamical systems, the butterfly effect and chaos theory, intended for a wide audience.

Chapter 34

Mathematical analysis

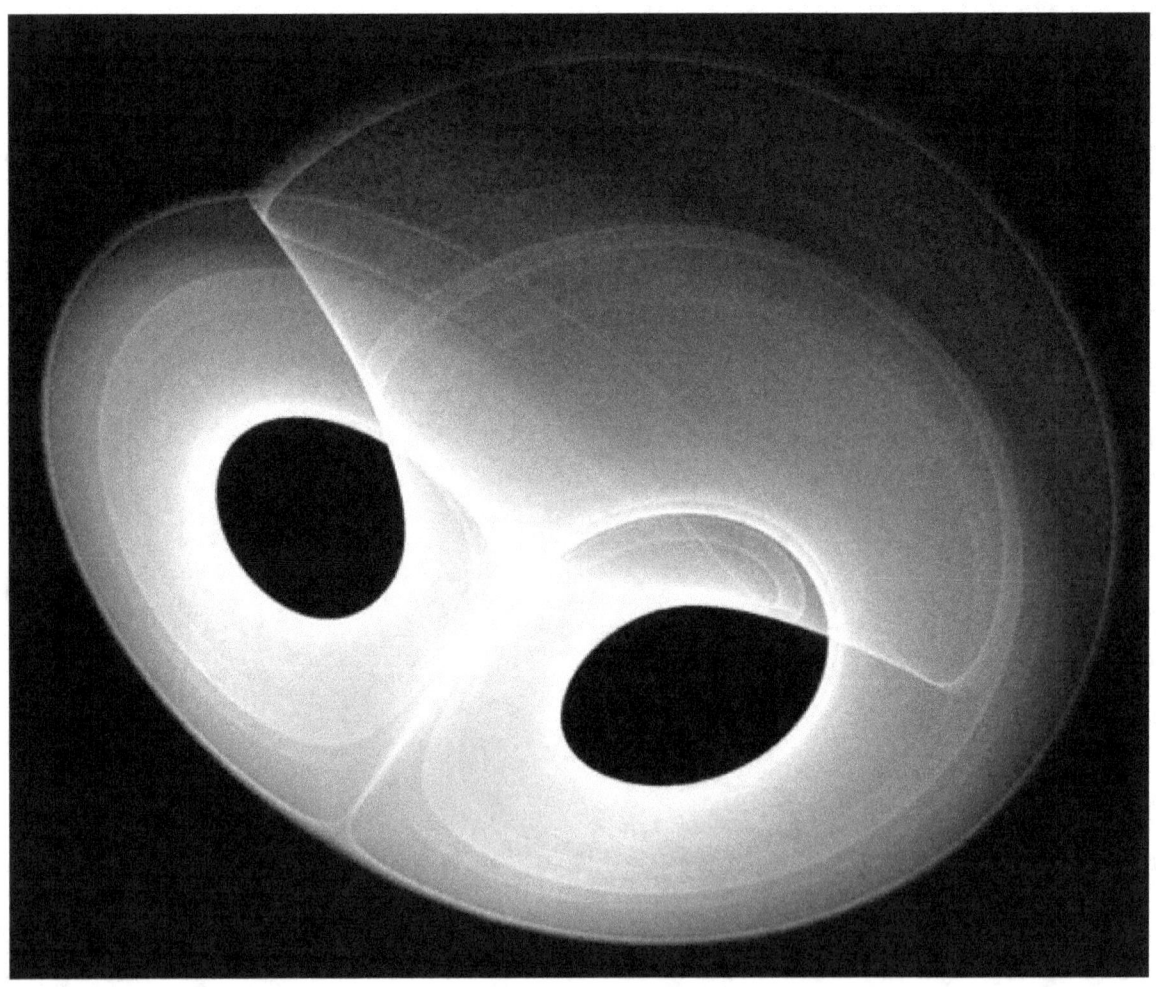

A strange attractor arising from a differential equation. Differential equations are an important area of mathematical analysis with many applications to science and engineering.

Mathematical analysis is a branch of mathematics that studies continuous change and includes the theories of differentiation, integration, measure, limits, infinite series, and analytic functions.[1][2]

These theories are usually studied in the context of real and complex numbers and functions. Analysis evolved from calculus, which involves the elementary concepts and techniques of analysis. Analysis may be distinguished from geometry; however, it can be applied to any space of mathematical objects that has a definition of nearness (a topological space) or specific distances between objects (a metric space).

34.1   History

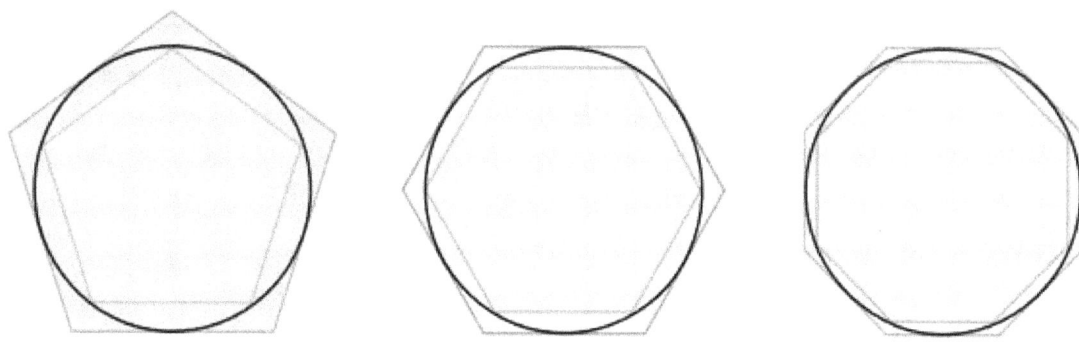

Archimedes used the method of exhaustion to compute the area inside a circle by finding the area of regular polygons with more and more sides. This was an early but informal example of a limit, one of the most basic concepts in mathematical analysis.

Mathematical analysis formally developed in the 17th century during the Scientific Revolution,[3] but many of its ideas can be traced back to earlier mathematicians. Early results in analysis were implicitly present in the early days of ancient Greek mathematics. For instance, an infinite geometric sum is implicit in Zeno's paradox of the dichotomy.[4] Later, Greek mathematicians such as Eudoxus and Archimedes made more explicit, but informal, use of the concepts of limits and convergence when they used the method of exhaustion to compute the area and volume of regions and solids.[5] The explicit use of infinitesimals appears in Archimedes' *The Method of Mechanical Theorems*, a work rediscovered in the 20th century.[6] In Asia, the Chinese mathematician Liu Hui used the method of exhaustion in the 3rd century AD to find the area of a circle.[7] Zu Chongzhi established a method that would later be called Cavalieri's principle to find the volume of a sphere in the 5th century.[8] The Indian mathematician Bhāskara II gave examples of the derivative and used what is now known as Rolle's theorem in the 12th century.[9]

In the 14th century, Madhava of Sangamagrama developed infinite series expansions, like the power series and the Taylor series, of functions such as sine, cosine, tangent and arctangent.[10] Alongside his development of the Taylor series of the trigonometric functions, he also estimated the magnitude of the error terms created by truncating these series and gave a rational approximation of an infinite series. His followers at the Kerala school of astronomy and mathematics further expanded his works, up to the 16th century.

The modern foundations of mathematical analysis were established in 17th century Europe.[3] Descartes and Fermat independently developed analytic geometry, and a few decades later Newton and Leibniz independently developed infinitesimal calculus, which grew, with the stimulus of applied work that continued through the 18th century, into analysis topics such as the calculus of variations, ordinary and partial differential equations, Fourier analysis, and generating functions. During this period, calculus techniques were applied to approximate discrete problems by continuous ones.

In the 18th century, Euler introduced the notion of mathematical function.[11] Real analysis began to emerge as an independent subject when Bernard Bolzano introduced the modern definition of continuity in 1816,[12] but Bolzano's work did not become widely known until the 1870s. In 1821, Cauchy began to put calculus on a firm logical foundation by rejecting the principle of the generality of algebra widely used in earlier work, particularly by Euler. Instead, Cauchy formulated calculus in terms of geometric ideas and infinitesimals. Thus, his definition of continuity required an infinitesimal change in x to correspond to an infinitesimal change in y. He also introduced the concept of the Cauchy sequence, and started the formal theory of complex analysis. Poisson, Liouville, Fourier and others studied partial differential equations and harmonic analysis. The contributions of these mathematicians and others, such as Weierstrass, developed the (ε, δ)-definition of limit approach, thus founding the modern field of mathematical analysis.

In the middle of the 19th century Riemann introduced his theory of integration. The last third of the century saw the arithmetization of analysis by Weierstrass, who thought that geometric reasoning was inherently misleading, and introduced the "epsilon-delta" definition of limit. Then, mathematicians started worrying that they were assuming the existence of a continuum of real numbers without proof. Dedekind then constructed the real numbers by Dedekind cuts, in which irrational numbers are formally defined, which serve to fill the "gaps" between rational numbers, thereby creating a complete set: the continuum of real numbers, which had already been developed by Simon Stevin in terms of decimal expansions. Around that time, the attempts to refine the theorems of Riemann integration led to the study of the "size" of the set of discontinuities of real functions.

Also, "monsters" (nowhere continuous functions, continuous but nowhere differentiable functions, space-filling curves) began to be investigated. In this context, Jordan developed his theory of measure, Cantor developed what is now called naive set theory, and Baire proved the Baire category theorem. In the early 20th century, calculus was formalized using an axiomatic set theory. Lebesgue solved the problem of measure, and Hilbert introduced Hilbert spaces to solve integral equations. The idea of normed vector space was in the air, and in the 1920s Banach created functional analysis.

34.2 Important concepts

34.2.1 Metric spaces

Main article: Metric space

In mathematics, a **metric space** is a set where a notion of distance (called a metric) between elements of the set is defined.

Much of analysis happens in some metric space; the most commonly used are the real line, the complex plane, Euclidean space, other vector spaces, and the integers. Examples of analysis without a metric include measure theory (which describes size rather than distance) and functional analysis (which studies topological vector spaces that need not have any sense of distance).

Formally, A **metric space** is an ordered pair (M, d) where M is a set and d is a metric on M, i.e., a function

$$d : M \times M \to \mathbb{R}$$

such that for any $x, y, z \in M$, the following holds:

1. $d(x, y) = 0$ if and only if $x = y$ (*identity of indiscernibles*),

2. $d(x, y) = d(y, x)$ (*symmetry*) and

3. $d(x, z) \leq d(x, y) + d(y, z)$ (*triangle inequality*).

By taking the third property and letting $z = x$, it can be shown that $d(x, y) \geq 0$ (*non-negative*).

34.2.2 Sequences and limits

Main article: Sequence

A **sequence** is an ordered list. Like a set, it contains members (also called *elements*, or *terms*). Unlike a set, order matters, and exactly the same elements can appear multiple times at different positions in the sequence. Most precisely, a sequence can be defined as a function whose domain is a countable totally ordered set, such as the natural numbers.

One of the most important properties of a sequence is *convergence*. Informally, a sequence converges if it has a *limit*. Continuing informally, a (singly-infinite) sequence has a limit if it approaches some point x, called the limit, as n becomes very large. That is, for an abstract sequence (a_n) (with n running from 1 to infinity understood) the distance between a_n and x approaches 0 as $n \to \infty$, denoted

$$\lim_{n \to \infty} a_n = x.$$

34.3   Main branches

34.3.1   Real analysis

Main article: Real analysis

Real analysis (traditionally, the **theory of functions of a real variable**) is a branch of mathematical analysis dealing with the real numbers and real-valued functions of a real variable.[113][114] In particular, it deals with the analytic properties of real functions and sequences, including convergence and limits of sequences of real numbers, the calculus of the real numbers, and continuity, smoothness and related properties of real-valued functions.

34.3.2   Complex analysis

Main article: Complex analysis

Complex analysis, traditionally known as the **theory of functions of a complex variable**, is the branch of mathematical analysis that investigates functions of complex numbers.[115] It is useful in many branches of mathematics, including algebraic geometry, number theory, applied mathematics; as well as in physics, including hydrodynamics, thermodynamics, mechanical engineering, electrical engineering, and particularly, quantum field theory.

Complex analysis is particularly concerned with the analytic functions of complex variables (or, more generally, meromorphic functions). Because the separate real and imaginary parts of any analytic function must satisfy Laplace's equation, complex analysis is widely applicable to two-dimensional problems in physics.

34.3.3   Functional analysis

Main article: Functional analysis

Functional analysis is a branch of mathematical analysis, the core of which is formed by the study of vector spaces endowed with some kind of limit-related structure (e.g. inner product, norm, topology, etc.) and the linear operators acting upon these spaces and respecting these structures in a suitable sense.[116][117] The historical roots of functional analysis lie in the study of spaces of functions and the formulation of properties of transformations of functions such as the Fourier transform as transformations defining continuous, unitary etc. operators between function spaces. This point of view turned out to be particularly useful for the study of differential and integral equations.

34.3.4   Differential equations

Main article: Differential equations

A **differential equation** is a mathematical equation for an unknown function of one or several variables that relates the values of the function itself and its derivatives of various orders.[18][19][20] Differential equations play a prominent role in engineering, physics, economics, biology, and other disciplines.

Differential equations arise in many areas of science and technology, specifically whenever a deterministic relation involving some continuously varying quantities (modeled by functions) and their rates of change in space and/or time (expressed as derivatives) is known or postulated. This is illustrated in classical mechanics, where the motion of a body is described by its position and velocity as the time value varies. Newton's laws allow one (given the position, velocity, acceleration and various forces acting on the body) to express these variables dynamically as a differential equation for the unknown position of the body as a function of time. In some cases, this differential equation (called an equation of motion) may be solved explicitly.

34.3.5 Measure theory

Main article: Measure (mathematics)

A **measure** on a set is a systematic way to assign a number to each suitable subset of that set, intuitively interpreted as its size.[21] In this sense, a measure is a generalization of the concepts of length, area, and volume. A particularly important example is the Lebesgue measure on a Euclidean space, which assigns the conventional length, area, and volume of Euclidean geometry to suitable subsets of the n -dimensional Euclidean space $\mathbb{R}^n$. For instance, the Lebesgue measure of the interval $[0, 1]$ in the real numbers is its length in the everyday sense of the word – specifically, 1.

Technically, a measure is a function that assigns a non-negative real number or $+\infty$ to (certain) subsets of a set X . It must assign 0 to the empty set and be (countably) additive: the measure of a 'large' subset that can be decomposed into a finite (or countable) number of 'smaller' disjoint subsets, is the sum of the measures of the "smaller" subsets. In general, if one wants to associate a *consistent* size to *each* subset of a given set while satisfying the other axioms of a measure, one only finds trivial examples like the counting measure. This problem was resolved by defining measure only on a sub-collection of all subsets; the so-called *measurable* subsets, which are required to form a σ -algebra. This means that countable unions, countable intersections and complements of measurable subsets are measurable. Non-measurable sets in a Euclidean space, on which the Lebesgue measure cannot be defined consistently, are necessarily complicated in the sense of being badly mixed up with their complement. Indeed, their existence is a non-trivial consequence of the axiom of choice.

34.3.6 Numerical analysis

Main article: Numerical analysis

Numerical analysis is the study of algorithms that use numerical approximation (as opposed to general symbolic manipulations) for the problems of mathematical analysis (as distinguished from discrete mathematics).[22]

Modern numerical analysis does not seek exact answers, because exact answers are often impossible to obtain in practice. Instead, much of numerical analysis is concerned with obtaining approximate solutions while maintaining reasonable bounds on errors.

Numerical analysis naturally finds applications in all fields of engineering and the physical sciences, but in the 21st century, the life sciences and even the arts have adopted elements of scientific computations. Ordinary differential equations appear in celestial mechanics (planets, stars and galaxies); numerical linear algebra is important for data analysis; stochastic differential equations and Markov chains are essential in simulating living cells for medicine and biology.

34.4 Other topics in mathematical analysis

- Calculus of variations deals with extremizing functionals, as opposed to ordinary calculus which deals with functions.

- Harmonic analysis deals with Fourier series and their abstractions.

- Geometric analysis involves the use of geometrical methods in the study of partial differential equations and the application of the theory of partial differential equations to geometry.

- Clifford analysis, the study of Clifford valued functions that are annihilated by Dirac or Dirac-like operators, termed in general as monogenic or Clifford analytic functions.

- *p*-adic analysis, the study of analysis within the context of *p*-adic numbers, which differs in some interesting and surprising ways from its real and complex counterparts.

- Non-standard analysis, which investigates the hyperreal numbers and their functions and gives a rigorous treatment of infinitesimals and infinitely large numbers.

- Computable analysis, the study of which parts of analysis can be carried out in a computable manner.

- Stochastic calculus – analytical notions developed for stochastic processes.

- Set-valued analysis – applies ideas from analysis and topology to set-valued functions.

- Convex analysis, the study of convex sets and functions.

- Tropical analysis (or idempotent analysis) – analysis in the context of the semiring of the max-plus algebra where the lack of an additive inverse is compensated somewhat by the idempotent rule $A + A = A$. When transferred to the tropical setting, many nonlinear problems become linear.[23]

34.5   Applications

Techniques from analysis are also found in other areas such as:

34.5.1   Physical sciences

The vast majority of classical mechanics, relativity, and quantum mechanics is based on applied analysis, and differential equations in particular. Examples of important differential equations include Newton's second law, the Schrödinger equation, and the Einstein field equations.

Functional analysis is also a major factor in quantum mechanics.

34.5.2   Signal processing

When processing signals, such as audio, radio waves, light waves, seismic waves, and even images, Fourier analysis can isolate individual components of a compound waveform, concentrating them for easier detection and/or removal. A large family of signal processing techniques consist of Fourier-transforming a signal, manipulating the Fourier-transformed data in a simple way, and reversing the transformation.[24]

34.5.3   Other areas of mathematics

Techniques from analysis are used in many areas of mathematics, including:

- Analytic number theory

- Analytic combinatorics

- Continuous probability

- Differential entropy in information theory

- Differential games

- Differential geometry, the application of calculus to specific mathematical spaces known as manifolds that possess a complicated internal structure but behave in a simple manner locally.

- Differential topology

- Mathematical finance

34.6 See also

- Constructive analysis

- History of calculus

- Non-classical analysis

- Paraconsistent mathematics

- Smooth infinitesimal analysis

- Timeline of calculus and mathematical analysis

34.7 Notes

[1] Edwin Hewitt and Karl Stromberg, "Real and Abstract Analysis", Springer-Verlag, 1965

[2] Stillwell, John Colin. "analysis | mathematics". Encyclopedia Britannica. Retrieved 2015-07-31.

[3] Jahnke, Hans Niels (2003). *A History of Analysis*. American Mathematical Society. p. 7. ISBN 978-0-8218-2623-2.

[4] Stillwell (2004). "Infinite Series". Mathematics and its History (2nd ed.). Springer Science + Business Media Inc. p. 170. ISBN 0-387-95336-1. Infinite series were present in Greek mathematics, [...] There is no question that Zeno's paradox of the dichotomy (Section 4.1), for example, concerns the decomposition of the number 1 into the infinite series $\frac{1}{2} + \frac{1}{2}^2 + \frac{1}{2}^3 + \frac{1}{2}^4 + ...$ and that Archimedes found the area of the parabolic segment (Section 4.4) essentially by summing the infinite series $1 + \frac{1}{4} + \frac{1}{4}^2 + \frac{1}{4}^3 + ... = \frac{4}{3}$. Both these examples are special cases of the result we express as summation of a geometric series

[5] (Smith, 1958)

[6] Pinto, J. Sousa (2004). *Infinitesimal Methods of Mathematical Analysis*. Horwood Publishing. p. 8. ISBN 978-1-898563-99-0.

[7] Dun, Liu; Fan, Dainian; Cohen, Robert Sonné (1966). "A comparison of Archimedes' and Liu Hui's studies of circles". Chinese studies in the history and philosophy of science and technology **130**. Springer. p. 279. ISBN 0-7923-3463-9., Chapter , p. 279

[8] Zill, Dennis G.; Wright, Scott; Wright, Warren S. (2009). *Calculus: Early Transcendentals* (3 ed.). Jones & Bartlett Learning. p. xxvii. ISBN 0-7637-5995-3., Extract of page 27

[9] Seal, Sir Brajendranath (1915), *The positive sciences of the ancient Hindus*, Longmans, Green and co.

[10] C. T. Rajagopal and M. S. Rangachari (June 1978). "On an untapped source of medieval Keralese Mathematics". *Archive for History of Exact Sciences* **18** (2): 89–102. doi:10.1007/BF00348142.

[11] Dunham, William (1999). *Euler: The Master of Us All*. The Mathematical Association of America. p. 17.

[12]   - Cooke, Roger (1997). "Beyond the Calculus". *The History of Mathematics: A Brief Course*. Wiley-Interscience. p. 379. ISBN 0-471-18082-3. Real analysis began its growth as an independent subject with the introduction of the modern definition of continuity in 1816 by the Czech mathematician Bernard Bolzano (1781–1848)

[13] Rudin, Walter. *Principles of Mathematical Analysis*. Walter Rudin Student Series in Advanced Mathematics (3rd ed.). McGraw–Hill. ISBN 978-0-07-054235-8.

[14] Abbott, Stephen (2001). *Understanding Analysis*. Undergraduate Texts in Mathematics. New York: Springer-Verlag. ISBN 0-387-95060-5.

[15] Ahlfors..*Complex Analysis* (McGraw-Hill)

[16] Rudin, W.: *Functional Analysis*, McGraw-Hill Science, 1991

[17] Conway, J. B.: *A Course in Functional Analysis*, 2nd edition, Springer-Verlag, 1994, ISBN 0-387-97245-5

[18] E. L. Ince, *Ordinary Differential Equations*, Dover Publications, 1958, ISBN 0-486-60349-0

[19] Witold Hurewicz, *Lectures on Ordinary Differential Equations*, Dover Publications, ISBN 0-486-49510-8

[20] Evans, L. C. (1998), *Partial Differential Equations*, Providence: American Mathematical Society, ISBN 0-8218-0772-2

[21] Terence Tao, 2011. *An Introduction to Measure Theory*. American Mathematical Society.

[22] Hildebrand, F. B. (1974). *Introduction to Numerical Analysis* (2nd ed.). McGraw-Hill. ISBN 0-07-028761-9.

[23] THE MASLOV DEQUANTIZATION, IDEMPOTENT AND TROPICAL MATHEMATICS: A BRIEF INTRODUCTION

[24] Theory and application of digital signal processing Rabiner, L. R.; Gold, B. Englewood Cliffs, N.J., Prentice-Hall, Inc., 1975.

34.8   References

- Aleksandrov, A. D., Kolmogorov, A. N., Lavrent'ev, M. A. (eds.). 1984. *Mathematics, its Content, Methods, and Meaning*. 2nd ed. Translated by S. H. Gould, K. A. Hirsch and T. Bartha; translation edited by S. H. Gould. MIT Press; published in cooperation with the American Mathematical Society.

- Apostol, Tom M. 1974. *Mathematical Analysis*. 2nd ed. Addison–Wesley. ISBN 978-0-201-00288-1.

- Binmore, K.G. 1980–1981. *The foundations of analysis: a straightforward introduction*. 2 volumes. Cambridge University Press.

- Johnsonbaugh, Richard, & W. E. Pfaffenberger. 1981. *Foundations of mathematical analysis*. New York: M. Dekker.

- Nikol'skii, S. M. 2002. "Mathematical analysis". In *Encyclopaedia of Mathematics*, Michiel Hazewinkel (editor). Springer-Verlag. ISBN 1-4020-0609-8.

- Rombaldi, Jean-Étienne. 2004. *Éléments d'analyse réelle : CAPES et agrégation interne de mathématiques*. EDP Sciences. ISBN 2-86883-681-X.

- Rudin, Walter. 1976. *Principles of Mathematical Analysis*. McGraw–Hill Publishing Co.; 3rd revised edition (September 1, 1976), ISBN 978-0-07-085613-4.

- Smith, David E. 1958. *History of Mathematics*. Dover Publications. ISBN 0-486-20430-8.

- Whittaker, E. T. and Watson, G. N.. 1927. *A Course of Modern Analysis*. 4th edition. Cambridge University Press. ISBN 0-521-58807-3.

- Real Analysis - Course Notes

34.9   External links

- Earliest Known Uses of Some of the Words of Mathematics: Calculus & Analysis

- Basic Analysis: Introduction to Real Analysis by Jiri Lebl (Creative Commons BY-NC-SA)

- Mathematical Analysis-Encyclopedia Britannica

- Calculus and Analysis

Chapter 35

Category theory

Category theory[1] formalizes mathematical structure and its concepts in terms of a collection of *objects* and of *arrows* (also called morphisms). A category has two basic properties: the ability to compose the arrows associatively and the existence of an identity arrow for each object. Category theory can be used to formalize concepts of other high-level abstractions such as sets, rings, and groups.

Several terms used in category theory, including the term "morphism", are used differently from their uses in the rest of mathematics. In category theory, a "morphism" obeys a set of conditions specific to category theory itself. Thus, care must be taken to understand the context in which statements are made.

35.1 An abstraction of other mathematical concepts

Many significant areas of mathematics can be formalised by category theory as categories. Category theory is an abstraction of mathematics itself that allows many intricate and subtle mathematical results in these fields to be stated, and proved, in a much simpler way than without the use of categories.[2]

The most accessible example of a category is the category of sets, where the objects are sets and the arrows are functions from one set to another. However, the objects of a category need not be sets, and the arrows need not be functions; any way of formalising a mathematical concept such that it meets the basic conditions on the behaviour of objects and arrows is a valid category, and all the results of category theory will apply to it.

The "arrows" of category theory are often said to represent a process connecting two objects, or in many cases a "structure-preserving" transformation connecting two objects. There are, however, many applications where much more abstract concepts are represented by objects and morphisms. The most important property of the arrows is that they can be "composed", in other words, arranged in a sequence to form a new arrow.

Categories now appear in most branches of mathematics, some areas of theoretical computer science where they can correspond to types, and mathematical physics where they can be used to describe vector spaces. Categories were first introduced by Samuel Eilenberg and Saunders Mac Lane in 1942–45, in connection with algebraic topology.

Category theory has several faces known not just to specialists, but to other mathematicians. A term dating from the 1940s, "general abstract nonsense", refers to its high level of abstraction, compared to more classical branches of mathematics. Homological algebra is category theory in its aspect of organising and suggesting manipulations in abstract algebra.

35.2 Utility

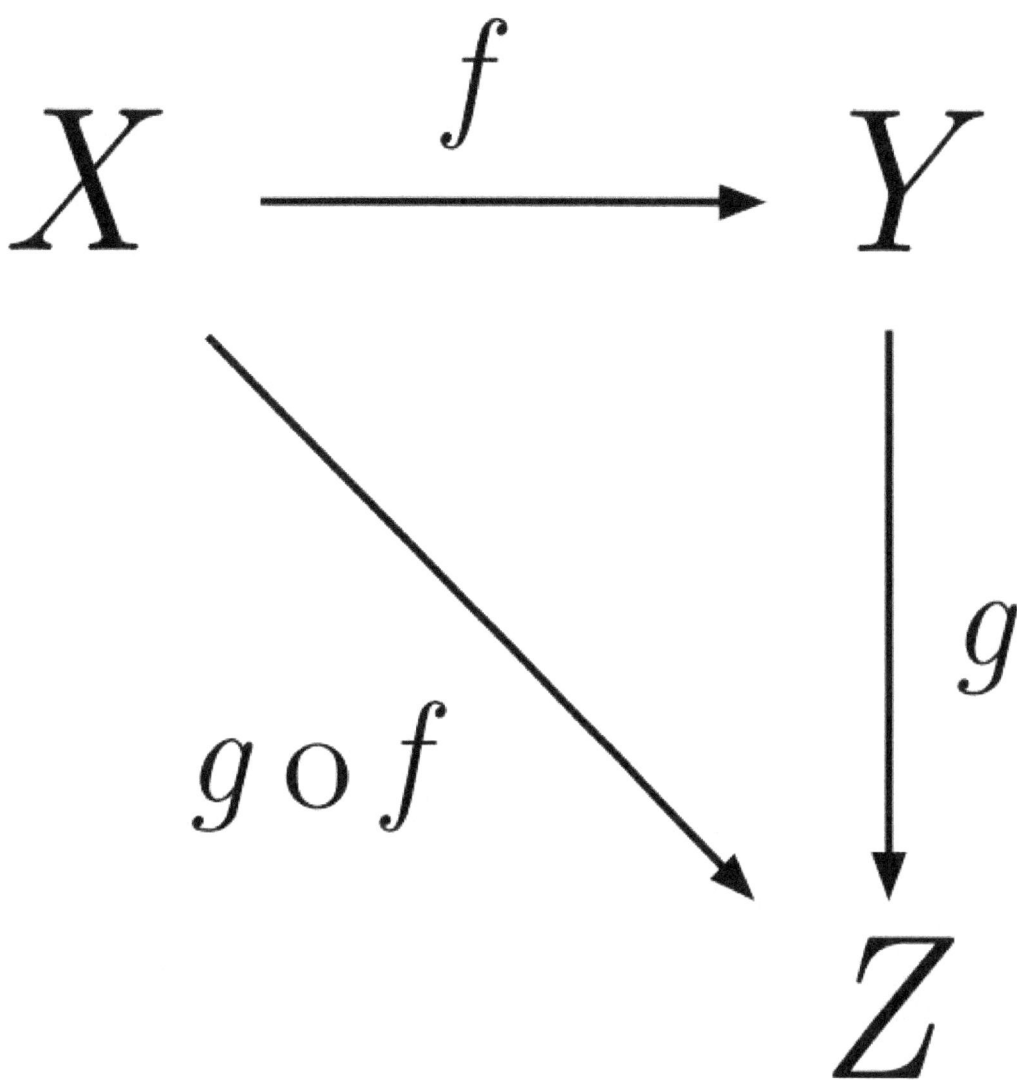

Schematic representation of a category with objects X, Y, Z and morphisms f, g, g ∘ f. (The category's three identity morphisms 1X, 1Y and 1Z, if explicitly represented, would appear as three arrows, next to the letters X, Y, and Z, respectively, each having as its "shaft" a circular arc measuring almost 360 degrees.)

35.2.1   Categories, objects, and morphisms

The study of categories is an attempt to *axiomatically* capture what is commonly found in various classes of related mathematical structures by relating them to the *structure-preserving functions* between them. A systematic study of category theory then allows us to prove general results about any of these types of mathematical structures from the axioms of a category.

Consider the following example. The class **Grp** of groups consists of all objects having a "group structure". One can proceed to prove theorems about groups by making logical deductions from the set of axioms. For example, it is immediately proven from the axioms that the identity element of a group is unique.

Instead of focusing merely on the individual objects (e.g., groups) possessing a given structure, category theory emphasizes the morphisms – the structure-preserving mappings – *between* these objects; by studying these morphisms, we are able to learn more about the structure of the objects. In the case of groups, the morphisms are the group homomorphisms. A group homomorphism between two groups "preserves the group structure" in a precise sense – it is a "process" taking one group to another, in a way that carries along information about the structure of the first group into the second group. The study of group homomorphisms then provides a tool for studying general properties of groups and consequences of the group axioms.

A similar type of investigation occurs in many mathematical theories, such as the study of continuous maps (morphisms) between topological spaces in topology (the associated category is called **Top**), and the study of smooth functions (morphisms) in manifold theory.

Not all categories arise as "structure preserving (set) functions", however; the standard example is the category of homotopies between pointed topological spaces.

If one axiomatizes relations instead of functions, one obtains the theory of allegories.

35.2.2 Functors

Main article: Functor
See also: Adjoint functors § Motivation

A category is *itself* a type of mathematical structure, so we can look for "processes" which preserve this structure in some sense; such a process is called a functor.

Diagram chasing is a visual method of arguing with abstract "arrows" joined in diagrams. Functors are represented by arrows between categories, subject to specific defining commutativity conditions. Functors can define (construct) categorical diagrams and sequences (viz. Mitchell, 1965). A functor associates to every object of one category an object of another category, and to every morphism in the first category a morphism in the second.

In fact, what we have done is define a category *of categories and functors* – the objects are categories, and the morphisms (between categories) are functors.

By studying categories and functors, we are not just studying a class of mathematical structures and the morphisms between them; we are studying the *relationships between various classes of mathematical structures*. This is a fundamental idea, which first surfaced in algebraic topology. Difficult *topological* questions can be translated into *algebraic* questions which are often easier to solve. Basic constructions, such as the fundamental group or the fundamental groupoid of a topological space, can be expressed as functors to the category of groupoids in this way, and the concept is pervasive in algebra and its applications.

35.2.3 Natural transformations

Main article: Natural transformation

Abstracting yet again, some diagrammatic and/or sequential constructions are often "naturally related" – a vague notion, at first sight. This leads to the clarifying concept of natural transformation, a way to "map" one functor to another. Many important constructions in mathematics can be studied in this context. "Naturality" is a principle, like general covariance in physics, that cuts deeper than is initially apparent. An arrow between two functors is a natural transformation when it is subject to certain naturality or commutativity conditions.

Functors and natural transformations ('naturality') are the key concepts in category theory.[3]

35.3   Categories, objects, and morphisms

Main articles: Category (mathematics) and Morphism

35.3.1   Categories

A *category* C consists of the following three mathematical entities:

- A class ob(C), whose elements are called *objects*;

- A class hom(C), whose elements are called morphisms or maps or *arrows*. Each morphism f has a *source object a* and *target object b*.
  The expression $f : a \to b$, would be verbally stated as "f is a morphism from a to b".
  The expression **hom(a, b)** — alternatively expressed as **homC(a, b)**, **mor(a, b)**, or **C(a, b)** — denotes the *hom-class* of all morphisms from a to b.

- A binary operation ∘, called *composition of morphisms*, such that for any three objects a, b, and c, we have hom(b, c) × hom(a, b) → hom(a, c). The composition of $f : a \to b$ and $g : b \to c$ is written as $g \circ f$ or gf,[4] governed by two axioms:

  - Associativity: If $f : a \to b$, $g : b \to c$ and $h : c \to d$ then $h \circ (g \circ f) = (h \circ g) \circ f$, and
  - Identity: For every object x, there exists a morphism $1x : x \to x$ called the *identity morphism for x*, such that for every morphism $f : a \to b$, we have $1b \circ f = f = f \circ 1a$.

    From the axioms, it can be proved that there is exactly one identity morphism for every object. Some authors deviate from the definition just given by identifying each object with its identity morphism.

35.3.2   Morphisms

Relations among morphisms (such as $fg = h$) are often depicted using commutative diagrams, with "points" (corners) representing objects and "arrows" representing morphisms.

Morphisms can have any of the following properties. A morphism $f : a \to b$ is a:

- monomorphism (or *monic*) if $f \circ g_1 = f \circ g_2$ implies $g_1 = g_2$ for all morphisms $g_1, g_2 : x \to a$.

- epimorphism (or *epic*) if $g_1 \circ f = g_2 \circ f$ implies $g_1 = g_2$ for all morphisms $g_1, g_2 : b \to x$.

- *bimorphism* if f is both epic and monic.

- isomorphism if there exists a morphism $g : b \to a$ such that $f \circ g = 1b$ and $g \circ f = 1a$.[5]

- endomorphism if $a = b$. end(a) denotes the class of endomorphisms of a.

- automorphism if f is both an endomorphism and an isomorphism. aut(a) denotes the class of automorphisms of a.

- retraction if a right inverse of f exists, i.e. if there exists a morphism $g : b \to a$ with $f \circ g = 1b$.

- section if a left inverse of f exists, i.e. if there exists a morphism $g : b \to a$ with $g \circ f = 1a$.

Every retraction is an epimorphism, and every section is a monomorphism. Furthermore, the following three statements are equivalent:

- f is a monomorphism and a retraction;

- f is an epimorphism and a section;

- f is an isomorphism.

35.4 Functors

Main article: Functor

Functors are structure-preserving maps between categories. They can be thought of as morphisms in the category of all (small) categories.

A (**covariant**) functor F from a category C to a category D, written $F : C \to D$, consists of:

- for each object x in C, an object $F(x)$ in D; and

- for each morphism $f : x \to y$ in C, a morphism $F(f) : F(x) \to F(y)$,

such that the following two properties hold:

- For every object x in C, $F(1_x) = 1_{F(x)}$;

- For all morphisms $f : x \to y$ and $g : y \to z$, $F(g \circ f) = F(g) \circ F(f)$.

A **contravariant** functor $F : C \to D$, is like a covariant functor, except that it "turns morphisms around" ("reverses all the arrows"). More specifically, every morphism $f : x \to y$ in C must be assigned to a morphism $F(f) : F(y) \to F(x)$ in D. In other words, a contravariant functor acts as a covariant functor from the opposite category C^{op} to D.

35.5 Natural transformations

Main article: Natural transformation

A *natural transformation* is a relation between two functors. Functors often describe "natural constructions" and natural transformations then describe "natural homomorphisms" between two such constructions. Sometimes two quite different constructions yield "the same" result; this is expressed by a natural isomorphism between the two functors.

If F and G are (covariant) functors between the categories C and D, then a natural transformation η from F to G associates to every object X in C a morphism $\eta X : F(X) \to G(X)$ in D such that for every morphism $f : X \to Y$ in C, we have $\eta Y \circ F(f) = G(f) \circ \eta X$; this means that the following diagram is commutative:

The two functors F and G are called *naturally isomorphic* if there exists a natural transformation from F to G such that ηX is an isomorphism for every object X in C.

35.6 Other concepts

35.6.1 Universal constructions, limits, and colimits

Main articles: Universal property and Limit (category theory)

$$F(X) \xrightarrow{\quad F(f) \quad} F(Y)$$

$$\eta_X \downarrow \qquad\qquad\qquad \downarrow \eta_Y$$

$$G(X) \xrightarrow{\quad G(f) \quad} G(Y)$$

Commutative diagram defining natural transformations

Using the language of category theory, many areas of mathematical study can be categorized. Categories include sets, groups and topologies.

Each category is distinguished by properties that all its objects have in common, such as the empty set or the product of two topologies, yet in the definition of a category, objects are considered to be atomic, i.e., we *do not know* whether an object A is a set, a topology, or any other abstract concept. Hence, the challenge is to define special objects without referring to the internal structure of those objects. To define the empty set without referring to elements, or the product topology without referring to open sets, one can characterize these objects in terms of their relations to other objects, as given by the morphisms of the respective categories. Thus, the task is to find *universal properties* that uniquely determine the objects of interest.

Indeed, it turns out that numerous important constructions can be described in a purely categorical way. The central concept which is needed for this purpose is called categorical *limit*, and can be dualized to yield the notion of a *colimit*.

35.6.2   Equivalent categories

Main articles: Equivalence of categories and Isomorphism of categories

It is a natural question to ask: under which conditions can two categories be considered to be "essentially the same", in the sense that theorems about one category can readily be transformed into theorems about the other category? The major tool one employs to describe such a situation is called *equivalence of categories*, which is given by appropriate functors between two categories. Categorical equivalence has found numerous applications in mathematics.

35.6.3 Further concepts and results

The definitions of categories and functors provide only the very basics of categorical algebra; additional important topics are listed below. Although there are strong interrelations between all of these topics, the given order can be considered as a guideline for further reading.

- The functor category D^C has as objects the functors from C to D and as morphisms the natural transformations of such functors. The Yoneda lemma is one of the most famous basic results of category theory; it describes representable functors in functor categories.

- Duality: Every statement, theorem, or definition in category theory has a *dual* which is essentially obtained by "reversing all the arrows". If one statement is true in a category C then its dual will be true in the dual category C^{op}. This duality, which is transparent at the level of category theory, is often obscured in applications and can lead to surprising relationships.

- Adjoint functors: A functor can be left (or right) adjoint to another functor that maps in the opposite direction. Such a pair of adjoint functors typically arises from a construction defined by a universal property; this can be seen as a more abstract and powerful view on universal properties.

35.6.4 Higher-dimensional categories

Many of the above concepts, especially equivalence of categories, adjoint functor pairs, and functor categories, can be situated into the context of *higher-dimensional categories*. Briefly, if we consider a morphism between two objects as a "process taking us from one object to another", then higher-dimensional categories allow us to profitably generalize this by considering "higher-dimensional processes".

For example, a (strict) 2-category is a category together with "morphisms between morphisms", i.e., processes which allow us to transform one morphism into another. We can then "compose" these "bimorphisms" both horizontally and vertically, and we require a 2-dimensional "exchange law" to hold, relating the two composition laws. In this context, the standard example is **Cat**, the 2-category of all (small) categories, and in this example, bimorphisms of morphisms are simply natural transformations of morphisms in the usual sense. Another basic example is to consider a 2-category with a single object; these are essentially monoidal categories. Bicategories are a weaker notion of 2-dimensional categories in which the composition of morphisms is not strictly associative, but only associative "up to" an isomorphism.

This process can be extended for all natural numbers n, and these are called *n*-categories. There is even a notion of *ω-category* corresponding to the ordinal number ω.

Higher-dimensional categories are part of the broader mathematical field of higher-dimensional algebra, a concept introduced by Ronald Brown. For a conversational introduction to these ideas, see John Baez, 'A Tale of *n*-categories' (1996).

35.7 Historical notes

In 1942–45, Samuel Eilenberg and Saunders Mac Lane introduced categories, functors, and natural transformations as part of their work in topology, especially algebraic topology. Their work was an important part of the transition from intuitive and geometric homology to axiomatic homology theory. Eilenberg and Mac Lane later wrote that their goal was to understand natural transformations; in order to do that, functors had to be defined, which required categories.

Stanisław Ulam, and some writing on his behalf, have claimed that related ideas were current in the late 1930s in Poland. Eilenberg was Polish, and studied mathematics in Poland in the 1930s. Category theory is also, in some sense, a continuation of the work of Emmy Noether (one of Mac Lane's teachers) in formalizing abstract processes; Noether realized that in order to understand a type of mathematical structure, one needs to understand the processes preserving that structure. In order to achieve this understanding, Eilenberg and Mac Lane proposed an axiomatic formalization of the relation between structures and the processes preserving them.

The subsequent development of category theory was powered first by the computational needs of homological algebra, and later by the axiomatic needs of algebraic geometry, the field most resistant to being grounded in either axiomatic set theory or the Russell-Whitehead view of united foundations. General category theory, an extension of universal algebra having many new features allowing for semantic flexibility and higher-order logic, came later; it is now applied throughout mathematics.

Certain categories called topoi (singular *topos*) can even serve as an alternative to axiomatic set theory as a foundation of mathematics. A topos can also be considered as a specific type of category with two additional topos axioms. These foundational applications of category theory have been worked out in fair detail as a basis for, and justification of, constructive mathematics. Topos theory is a form of abstract sheaf theory, with geometric origins, and leads to ideas such as pointless topology.

Categorical logic is now a well-defined field based on type theory for intuitionistic logics, with applications in functional programming and domain theory, where a cartesian closed category is taken as a non-syntactic description of a lambda calculus. At the very least, category theoretic language clarifies what exactly these related areas have in common (in some abstract sense).

Category theory has been applied in other fields as well. For example, John Baez has shown a link between Feynman diagrams in Physics and monoidal categories.[6] Another application of category theory, more specifically: topos theory, has been made in mathematical music theory, see for example the book *The Topos of Music, Geometric Logic of* Concepts, Theory, and Performance *by Guerino Mazzola*.

More recent efforts to introduce undergraduates to categories as a foundation for mathematics include those of William Lawvere and Rosebrugh (2003) and Lawvere and Stephen Schanuel (1997) and Mirroslav Yotov (2012).

35.8   See also

- Group theory

- Domain theory

- Enriched category theory

- Glossary of category theory

- Higher category theory

- Higher-dimensional algebra

- Important publications in category theory

- Outline of category theory

- Timeline of category theory and related mathematics

35.9   Notes

[1] Awodey 2006

[2] Geroch, Robert (1985). *Mathematical physics* ([Repr.] ed.). Chicago: University of Chicago Press. p. 7. ISBN 0-226-28862-5. Retrieved 20 August 2012. Note that theorem 3 is actually easier for categories in general than it is for the special case of sets. This phenomenon is by no means rare.

[3] Mac Lane 1998, p. 18: "As Eilenberg-Mac Lane first observed, 'category' has been defined in order to be able to define 'functor' and 'functor' has been defined in order to be able to define 'natural transformation' "

[4] Some authors compose in the opposite order, writing fg or $f \circ g$ for $g \circ f$. Computer scientists using category theory very commonly write $f ; g$ for $g \circ f$

[5] Note that a morphism that is both epic and monic is not necessarily an isomorphism! An elementary counterexample: in the category consisting of two objects A and B, the identity morphisms, and a single morphism f from A to B, f is both epic and monic but is not an isomorphism.

[6] Baez, J.C.; Stay, M. (2009). "Physics, topology, logic and computation: A Rosetta stone" (PDF). arXiv:0903.0340.

35.10 References

- Adámek, Jiří; Herrlich, Horst; Strecker, George E. (1990). *Abstract and concrete categories*. John Wiley & Sons. ISBN 0-471-60922-6.

- Awodey, Steve (2006). *Category Theory*. Oxford Logic Guides **49**. Oxford University Press. ISBN 978-0-19-151382-4.

- Barr, Michael; Wells, Charles (2012), *Category Theory for Computing Science*, Reprints in Theory and Applications of Categories **22** (3rd ed.).

- Barr, Michael; Wells, Charles (2005), *Toposes, Triples and Theories*, Reprints in Theory and Applications of Categories **12** (revised ed.), MR 2178101.

- Borceux, Francis (1994). *Handbook of categorical algebra*. Encyclopedia of Mathematics and its Applications 50-52. Cambridge University Press.

- Bucur, Ion; Deleanu, Aristide (1968). *Introduction to the theory of categories and functors*. Wiley.

- Freyd, Peter J. (1964). *Abelian Categories*. New York: Harper and Row.

- Freyd, Peter J.; Scedrov, Andre (1990). *Categories, allegories*. North Holland Mathematical Library **39**. North Holland. ISBN 978-0-08-088701-2.

- Goldblatt, Robert (2006) [1979]. *Topoi: The Categorial Analysis of Logic*. Studies in logic and the foundations of mathematics **94** (Reprint, revised ed.). Dover Publications. ISBN 978-0-486-45026-1.

- Hatcher, William S. (1982). "Ch. 8". *The logical foundations of mathematics*. Foundations & philosophy of science & technology (2nd ed.). Pergamon Press.

- Herrlich, Horst; Strecker, George E. (2007), *Category Theory* (3rd ed.), Heldermann Verlag Berlin, ISBN 978-3-88538-001-6.

- Kashiwara, Masaki; Schapira, Pierre (2006). *Categories and Sheaves*. Grundlehren der Mathematischen Wissenschaften **332**. Springer. ISBN 978-3-540-27949-5.

- Lawvere, F. William; Rosebrugh, Robert (2003). *Sets for Mathematics*. Cambridge University Press. ISBN 978-0-521-01060-3.

- Lawvere, F. W.; Schanuel, Stephen Hoel (2009) [1997]. *Conceptual Mathematics: A First Introduction to Categories* (2nd ed.). Cambridge University Press. ISBN 978-0-521-89485-2.

- Leinster, Tom (2004). *Higher operads, higher categories*. London Math. Society Lecture Note Series **298**. Cambridge University Press. ISBN 978-0-521-53215-0.

- Leinster, Tom (2014). *Basic Category Theory*. Cambridge University Press.

- Lurie, Jacob (2009). *Higher topos theory*. Annals of Mathematics Studies **170**. Princeton, NJ: Princeton University Press. arXiv:math.CT/0608040. ISBN 978-0-691-14049-0. MR 2522659.

- Mac Lane, Saunders (1998). *Categories for the Working Mathematician*. Graduate Texts in Mathematics **5** (2nd ed.). Springer-Verlag. ISBN 0-387-98403-8. MR 1712872.

- Mac Lane, Saunders; Birkhoff, Garrett (1999) [1967]. *Algebra* (2nd ed.). Chelsea. ISBN 0-8218-1646-2.

- Martini, A.; Ehrig, H.; Nunes, D. (1996). "Elements of basic category theory". *Technical Report* (Technical University Berlin) **96** (5).

- May, Peter (1999). *A Concise Course in Algebraic Topology*. University of Chicago Press. ISBN 0-226-51183-9.

- Guerino, Mazzola (2002). *The Topos of Music, Geometric Logic of Concepts, Theory, and Performance*. Birkhäuser. ISBN 3-7643-5731-2.

- Pedicchio, Maria Cristina; Tholen, Walter, eds. (2004). *Categorical foundations. Special topics in order, topology, algebra, and sheaf theory*. Encyclopedia of Mathematics and Its Applications **97**. Cambridge: Cambridge University Press. ISBN 0-521-83414-7. Zbl 1034.18001.

- Pierce, Benjamin C. (1991). *Basic Category Theory for Computer Scientists*. MIT Press. ISBN 978-0-262-66071-6.

- Schalk, A.; Simmons, H. (2005). *An introduction to Category Theory in four easy movements* (PDF). Notes for a course offered as part of the MSc. in Mathematical Logic, Manchester University.

- Simpson, Carlos. *Homotopy theory of higher categories*. arXiv:1001.4071., draft of a book.

- Taylor, Paul (1999). *Practical Foundations of Mathematics*. Cambridge Studies in Advanced Mathematics **59**. Cambridge University Press. ISBN 978-0-521-63107-5.

- Turi, Daniele (1996–2001). "Category Theory Lecture Notes" (PDF). Retrieved 11 December 2009. Based on Mac Lane 1998.

35.11   Further reading

- Jean-Pierre Marquis (2008). *From a Geometrical Point of View: A Study of the History and Philosophy of Category Theory*. Springer Science & Business Media. ISBN 978-1-4020-9384-5.

35.12   External links

- Theory and Application of Categories, an electronic journal of category theory, full text, free, since 1995.

- nLab, a wiki project on mathematics, physics and philosophy with emphasis on the n-categorical point of view.

- André Joyal, CatLab, a wiki project dedicated to the exposition of categorical mathematics.

- Category Theory, a web page of links to lecture notes and freely available books on category theory.

- Hillman, Chris, *A Categorical Primer*, CiteSeerX: 10.1.1.24.3264, a formal introduction to category theory.

- Adamek, J.; Herrlich, H.; Stecker, G. "Abstract and Concrete Categories-The Joy of Cats" (PDF).

- Category Theory entry by Jean-Pierre Marquis in the *Stanford Encyclopedia of Philosophy* with an extensive bibliography.

- List of academic conferences on category theory

- Baez, John (1996). "The Tale of n-categories". — An informal introduction to higher order categories.

- WildCats is a category theory package for Mathematica. Manipulation and visualization of objects, morphisms, categories, functors, natural transformations, universal properties.

- The catsters's channel on YouTube, a channel about category theory.

- Category Theory at PlanetMath.org.

- Video archive of recorded talks relevant to categories, logic and the foundations of physics.

- Interactive Web page which generates examples of categorical constructions in the category of finite sets.

- Category Theory for the Sciences, an instruction on category theory as a tool throughout the sciences.

Chapter 36

Set theory

This article is about the branch of mathematics. For musical set theory, see Set theory (music).

Set theory is the branch of mathematical logic that studies sets, which informally are collections of objects. Although

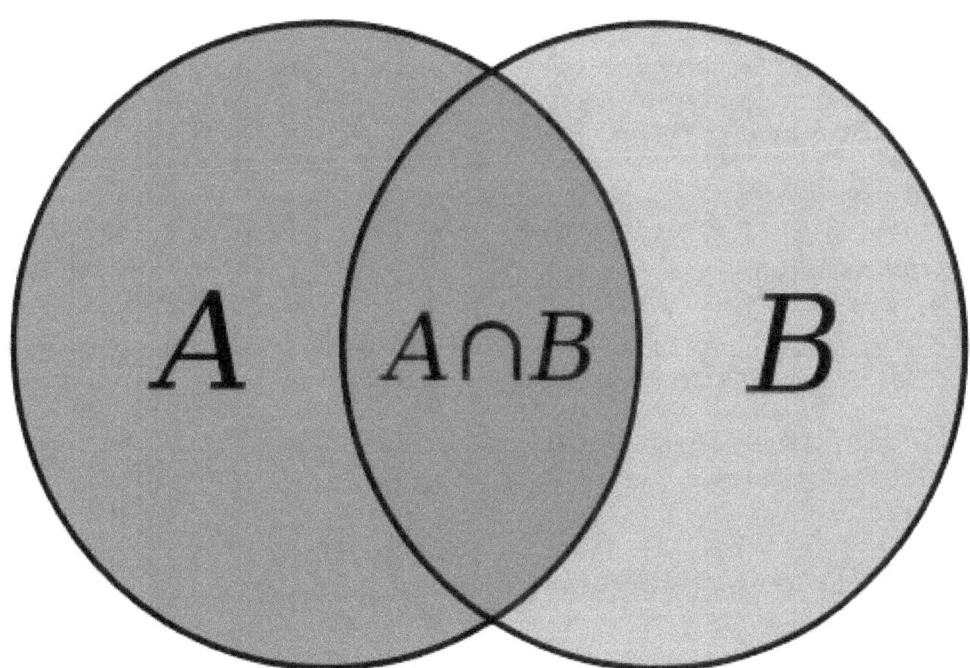

A Venn diagram illustrating the intersection of two sets.

any type of object can be collected into a set, set theory is applied most often to objects that are relevant to mathematics. The language of set theory can be used in the definitions of nearly all mathematical objects.

The modern study of set theory was initiated by Georg Cantor and Richard Dedekind in the 1870s. After the discovery of paradoxes in naive set theory, numerous axiom systems were proposed in the early twentieth century, of which the Zermelo–Fraenkel axioms, with the axiom of choice, are the best-known.

Set theory is commonly employed as a foundational system for mathematics, particularly in the form of Zermelo–Fraenkel set theory with the axiom of choice. Beyond its foundational role, set theory is a branch of mathematics in its own right, with an active research community. Contemporary research into set theory includes a diverse collection of topics, ranging from the structure of the real number line to the study of the consistency of large cardinals.

36.1 History

Mathematical topics typically emerge and evolve through interactions among many researchers. Set theory, however, was founded by a single paper in 1874 by Georg Cantor: "On a Characteristic Property of All Real Algebraic Numbers".[1][2]

Since the 5th century BC, beginning with Greek mathematician Zeno of Elea in the West and early Indian mathematicians in the East, mathematicians had struggled with the concept of infinity. Especially notable is the work of Bernard Bolzano in the first half of the 19th century.[3] Modern understanding of infinity began in 1867–71, with Cantor's work on number theory. An 1872 meeting between Cantor and Richard Dedekind influenced Cantor's thinking and culminated in Cantor's 1874 paper.

Cantor's work initially polarized the mathematicians of his day. While Karl Weierstrass and Dedekind supported Cantor, Leopold Kronecker, now seen as a founder of mathematical constructivism, did not. Cantorian set theory eventually became widespread, due to the utility of Cantorian concepts, such as one-to-one correspondence among sets, his proof that there are more real numbers than integers, and the "infinity of infinities" ("Cantor's paradise") resulting from the power set operation. This utility of set theory led to the article "Mengenlehre" contributed in 1898 by Arthur Schoenflies to Klein's encyclopedia.

The next wave of excitement in set theory came around 1900, when it was discovered that Cantorian set theory gave rise to several contradictions, called antinomies or paradoxes. Bertrand Russell and Ernst Zermelo independently found the simplest and best known paradox, now called Russell's paradox: consider "the set of all sets that are not members of themselves", which leads to a contradiction since it must be a member of itself, and not a member of itself. In 1899 Cantor had himself posed the question "What is the cardinal number of the set of all sets?", and obtained a related paradox. Russell used his paradox as a theme in his 1903 review of continental mathematics in his *The Principles of Mathematics*.

In 1906 English readers were treated to *Theory of Sets of Points*[4] by William Henry Young and his wife Grace Chisholm Young, published by Cambridge University Press.

The momentum of set theory was such that debate on the paradoxes did not lead to its abandonment. The work of Zermelo in 1908 and Abraham Fraenkel in 1922 resulted in the set of axioms ZFC, which became the most commonly used set of axioms for set theory. The work of analysts such as Henri Lebesgue demonstrated the great mathematical utility of set theory, which has since become woven into the fabric of modern mathematics. Set theory is commonly used as a foundational system, although in some areas category theory is thought to be a preferred foundation.

36.2 Basic concepts and notation

Main articles: Set (mathematics) and Algebra of sets

Set theory begins with a fundamental binary relation between an object o and a set A. If o is a **member** (or **element**) of A, write $o \in A$. Since sets are objects, the membership relation can relate sets as well.

A derived binary relation between two sets is the subset relation, also called **set inclusion**. If all the members of set A are also members of set B, then A is a **subset** of B, denoted $A \subseteq B$. For example, $\{1,2\}$ is a subset of $\{1,2,3\}$, and so is $\{2\}$ but $\{1,4\}$ is not. From this definition, it is clear that a set is a subset of itself; for cases where one wishes to rule this out, the term **proper subset** is defined. A is called a **proper subset** of B if and only if A is a subset of B, but B is **not** a subset of A. Note also that 1 and 2 and 3 are members (elements) of set $\{1,2,3\}$, but are *not* subsets, and the subsets in turn are *not* as such members of the set.

Just as arithmetic features binary operations on numbers, set theory features binary operations on sets. The:

- **Union** of the sets A and B, denoted $A \cup B$, is the set of all objects that are a member of A, or B, or both. The union of $\{1, 2, 3\}$ and $\{2, 3, 4\}$ is the set $\{1, 2, 3, 4\}$.

- **Intersection** of the sets A and B, denoted $A \cap B$, is the set of all objects that are members of both A and B. The intersection of $\{1, 2, 3\}$ and $\{2, 3, 4\}$ is the set $\{2, 3\}$.

- **Set difference** of U and A, denoted $U \setminus A$, is the set of all members of U that are not members of A. The set difference $\{1,2,3\} \setminus \{2,3,4\}$ is $\{1\}$, while, conversely, the set difference $\{2,3,4\} \setminus \{1,2,3\}$ is $\{4\}$. When A is a subset of U, the set difference $U \setminus A$ is also called the **complement** of A in U. In this case, if the choice of U is clear from the context, the notation A^c is sometimes used instead of $U \setminus A$, particularly if U is a universal set as in the study of Venn diagrams.

- **Symmetric difference** of sets A and B, denoted $A \triangle B$ or $A \ominus B$, is the set of all objects that are a member of exactly one of A and B (elements which are in one of the sets, but not in both). For instance, for the sets $\{1,2,3\}$ and $\{2,3,4\}$, the symmetric difference set is $\{1,4\}$. It is the set difference of the union and the intersection, $(A \cup B) \setminus (A \cap B)$ or $(A \setminus B) \cup (B \setminus A)$.

- **Cartesian product** of A and B, denoted $A \times B$, is the set whose members are all possible ordered pairs (a,b) where a is a member of A and b is a member of B. The cartesian product of $\{1, 2\}$ and $\{$red, white$\}$ is $\{(1, \text{red}), (1, \text{white}), (2, \text{red}), (2, \text{white})\}$.

- **Power set** of a set A is the set whose members are all possible subsets of A. For example, the power set of $\{1, 2\}$ is $\{\ \{\}, \{1\}, \{2\}, \{1,2\}\ \}$.

Some basic sets of central importance are the empty set (the unique set containing no elements), the set of natural numbers, and the set of real numbers.

36.3   Some ontology

Main article: von Neumann universe

A set is pure if all of its members are sets, all members of its members are sets, and so on. For example, the set $\{\{\}\}$ containing only the empty set is a nonempty pure set. In modern set theory, it is common to restrict attention to the **von Neumann universe** of pure sets, and many systems of axiomatic set theory are designed to axiomatize the pure sets only. There are many technical advantages to this restriction, and little generality is lost, because essentially all mathematical concepts can be modeled by pure sets. Sets in the von Neumann universe are organized into a cumulative hierarchy, based on how deeply their members, members of members, etc. are nested. Each set in this hierarchy is assigned (by transfinite recursion) an ordinal number α, known as its **rank**. The rank of a pure set X is defined to be the least upper bound of all successors of ranks of members of X. For example, the empty set is assigned rank 0, while the set $\{\{\}\}$ containing only the empty set is assigned rank 1. For each ordinal α, the set $V\alpha$ is defined to consist of all pure sets with rank less than α. The entire von Neumann universe is denoted V.

36.4   Axiomatic set theory

Elementary set theory can be studied informally and intuitively, and so can be taught in primary schools using Venn diagrams. The intuitive approach tacitly assumes that a set may be formed from the class of all objects satisfying any particular defining condition. This assumption gives rise to paradoxes, the simplest and best known of which are Russell's paradox and the Burali-Forti paradox. Axiomatic set theory was originally devised to rid set theory of such paradoxes.[5]

The most widely studied systems of axiomatic set theory imply that all sets form a cumulative hierarchy. Such systems come in two flavors, those whose ontology consists of:

- *Sets alone*. This includes the most common axiomatic set theory, **Zermelo–Fraenkel set theory (ZFC)**, which includes the axiom of choice. Fragments of ZFC include:

- Zermelo set theory, which replaces the axiom schema of replacement with that of separation;
- General set theory, a small fragment of Zermelo set theory sufficient for the Peano axioms and finite sets;
- Kripke–Platek set theory, which omits the axioms of infinity, powerset, and choice, and weakens the axiom schemata of separation and replacement.

- *Sets and proper classes.* These include Von Neumann–Bernays–Gödel set theory, which has the same strength as ZFC for theorems about sets alone, and Morse-Kelley set theory and Tarski–Grothendieck set theory, both of which are stronger than ZFC.

The above systems can be modified to allow **urelements**, objects that can be members of sets but that are not themselves sets and do not have any members.

The systems of **New Foundations NFU** (allowing urelements) and **NF** (lacking them) are not based on a cumulative hierarchy. NF and NFU include a "set of everything," relative to which every set has a complement. In these systems urelements matter, because NF, but not NFU, produces sets for which the axiom of choice does not hold.

Systems of constructive set theory, such as CST, CZF, and IZF, embed their set axioms in intuitionistic instead of classical logic. Yet other systems accept classical logic but feature a nonstandard membership relation. These include rough set theory and fuzzy set theory, in which the value of an atomic formula embodying the membership relation is not simply **True** or **False**. The Boolean-valued models of ZFC are a related subject.

An enrichment of ZFC called Internal Set Theory was proposed by Edward Nelson in 1977.

36.5 Applications

Many mathematical concepts can be defined precisely using only set theoretic concepts. For example, mathematical structures as diverse as graphs, manifolds, rings, and vector spaces can all be defined as sets satisfying various (axiomatic) properties. Equivalence and order relations are ubiquitous in mathematics, and the theory of mathematical relations can be described in set theory.

Set theory is also a promising foundational system for much of mathematics. Since the publication of the first volume of *Principia Mathematica*, it has been claimed that most or even all mathematical theorems can be derived using an aptly designed set of axioms for set theory, augmented with many definitions, using first or second order logic. For example, properties of the natural and real numbers can be derived within set theory, as each number system can be identified with a set of equivalence classes under a suitable equivalence relation whose field is some infinite set.

Set theory as a foundation for mathematical analysis, topology, abstract algebra, and discrete mathematics is likewise uncontroversial; mathematicians accept that (in principle) theorems in these areas can be derived from the relevant definitions and the axioms of set theory. Few full derivations of complex mathematical theorems from set theory have been formally verified, however, because such formal derivations are often much longer than the natural language proofs mathematicians commonly present. One verification project, Metamath, includes human-written, computer-verified derivations of more than 12,000 theorems starting from ZFC set theory, first order logic and propositional logic.

36.6 Areas of study

Set theory is a major area of research in mathematics, with many interrelated subfields.

36.6.1 Combinatorial set theory

Main article: Infinitary combinatorics

Combinatorial set theory concerns extensions of finite combinatorics to infinite sets. This includes the study of cardinal arithmetic and the study of extensions of Ramsey's theorem such as the Erdős–Rado theorem.

36.6.2   Descriptive set theory

Main article: Descriptive set theory

Descriptive set theory is the study of subsets of the real line and, more generally, subsets of Polish spaces. It begins with the study of pointclasses in the Borel hierarchy and extends to the study of more complex hierarchies such as the projective hierarchy and the Wadge hierarchy. Many properties of Borel sets can be established in ZFC, but proving these properties hold for more complicated sets requires additional axioms related to determinacy and large cardinals.

The field of effective descriptive set theory is between set theory and recursion theory. It includes the study of lightface pointclasses, and is closely related to hyperarithmetical theory. In many cases, results of classical descriptive set theory have effective versions; in some cases, new results are obtained by proving the effective version first and then extending ("relativizing") it to make it more broadly applicable.

A recent area of research concerns Borel equivalence relations and more complicated definable equivalence relations. This has important applications to the study of invariants in many fields of mathematics.

36.6.3   Fuzzy set theory

Main article: Fuzzy set theory

In set theory as Cantor defined and Zermelo and Fraenkel axiomatized, an object is either a member of a set or not. In fuzzy set theory this condition was relaxed by Lotfi A. Zadeh so an object has a *degree of membership* in a set, a number between 0 and 1. For example, the degree of membership of a person in the set of "tall people" is more flexible than a simple yes or no answer and can be a real number such as 0.75.

36.6.4   Inner model theory

Main article: Inner model theory

An **inner model** of Zermelo–Fraenkel set theory (ZF) is a transitive class that includes all the ordinals and satisfies all the axioms of ZF. The canonical example is the constructible universe L developed by Gödel. One reason that the study of inner models is of interest is that it can be used to prove consistency results. For example, it can be shown that regardless of whether a model V of ZF satisfies the continuum hypothesis or the axiom of choice, the inner model L constructed inside the original model will satisfy both the generalized continuum hypothesis and the axiom of choice. Thus the assumption that ZF is consistent (has at least one model) implies that ZF together with these two principles is consistent.

The study of inner models is common in the study of determinacy and large cardinals, especially when considering axioms such as the axiom of determinacy that contradict the axiom of choice. Even if a fixed model of set theory satisfies the axiom of choice, it is possible for an inner model to fail to satisfy the axiom of choice. For example, the existence of sufficiently large cardinals implies that there is an inner model satisfying the axiom of determinacy (and thus not satisfying the axiom of choice).[6]

36.6.5   Large cardinals

Main article: Large cardinal property

A **large cardinal** is a cardinal number with an extra property. Many such properties are studied, including inaccessible cardinals, measurable cardinals, and many more. These properties typically imply the cardinal number must be very large, with the existence of a cardinal with the specified property unprovable in Zermelo-Fraenkel set theory.

36.6.6 Determinacy

Main article: Determinacy

Determinacy refers to the fact that, under appropriate assumptions, certain two-player games of perfect information are determined from the start in the sense that one player must have a winning strategy. The existence of these strategies has important consequences in descriptive set theory, as the assumption that a broader class of games is determined often implies that a broader class of sets will have a topological property. The axiom of determinacy (AD) is an important object of study; although incompatible with the axiom of choice, AD implies that all subsets of the real line are well behaved (in particular, measurable and with the perfect set property). AD can be used to prove that the Wadge degrees have an elegant structure.

36.6.7 Forcing

Main article: Forcing (mathematics)

Paul Cohen invented the method of forcing while searching for a model of ZFC in which the continuum hypothesis fails, or a model of ZF in which the axiom of choice fails. Forcing adjoins to some given model of set theory additional sets in order to create a larger model with properties determined (i.e. "forced") by the construction and the original model. For example, Cohen's construction adjoins additional subsets of the natural numbers without changing any of the cardinal numbers of the original model. Forcing is also one of two methods for proving relative consistency by finitistic methods, the other method being Boolean-valued models.

36.6.8 Cardinal invariants

Main article: Cardinal invariant

A **cardinal invariant** is a property of the real line measured by a cardinal number. For example, a well-studied invariant is the smallest cardinality of a collection of meagre sets of reals whose union is the entire real line. These are invariants in the sense that any two isomorphic models of set theory must give the same cardinal for each invariant. Many cardinal invariants have been studied, and the relationships between them are often complex and related to axioms of set theory.

36.6.9 Set-theoretic topology

Main article: Set-theoretic topology

Set-theoretic topology studies questions of general topology that are set-theoretic in nature or that require advanced methods of set theory for their solution. Many of these theorems are independent of ZFC, requiring stronger axioms for their proof. A famous problem is the normal Moore space question, a question in general topology that was the subject of intense research. The answer to the normal Moore space question was eventually proved to be independent of ZFC.

36.7 Objections to set theory as a foundation for mathematics

From set theory's inception, some mathematicians have objected to it as a foundation for mathematics. The most common objection to set theory, one Kronecker voiced in set theory's earliest years, starts from the constructivist view that mathematics is loosely related to computation. If this view is granted, then the treatment of infinite sets, both in naive and in axiomatic set theory, introduces into mathematics methods and objects that are not computable even in principle.

Ludwig Wittgenstein condemned set theory. He wrote that "set theory is wrong", since it builds on the "nonsense" of fictitious symbolism, has "pernicious idioms", and that it is nonsensical to talk about "all numbers".[7] Wittgenstein's views about the foundations of mathematics were later criticised by Georg Kreisel and Paul Bernays, and investigated by Crispin Wright, among others.

Category theorists have proposed topos theory as an alternative to traditional axiomatic set theory. Topos theory can interpret various alternatives to that theory, such as constructivism, finite set theory, and computable set theory.[8] Topoi also give a natural setting for forcing and discussions of the independence of choice from ZF, as well as providing the framework for pointless topology and Stone spaces.[9]

An active area of research is the univalent foundations arising from homotopy type theory. Here, sets may be defined as certain kinds of types, with universal properties of sets arising from higher inductive types. Principles such as the axiom of choice and the law of the excluded middle appear in a spectrum of different forms, some of which can be proven, others which correspond to the classical notions; this allows for a detailed discussion of the effect of these axioms on mathematics.[10][11]

36.8   See also

- Glossary of set theory

- Category theory

- List of set theory topics

- Relational model – borrows from set theory

36.9   Notes

[1] Cantor, Georg (1874), "Ueber eine Eigenschaft des Inbegriffes aller reellen algebraischen Zahlen", *J. Reine Angew. Math.* **77**: 258–262, doi:10.1515/crll.1874.77.258

[2] Johnson, Philip (1972), *A History of Set Theory*, Prindle, Weber & Schmidt, ISBN 0-87150-154-6

[3] Bolzano, Bernard (1975), Berg, Jan, ed., *Einleitung zur Größenlehre und erste Begriffe der allgemeinen Größenlehre*, Bernard-Bolzano-Gesamtausgabe, edited by Eduard Winter et al., Vol. II, A, 7, Stuttgart, Bad Cannstatt: Friedrich Frommann Verlag, p. 152, ISBN 3-7728-0466-7

[4] William Henry Young & Grace Chisholm Young (1906) Theory of Sets of Points, link from Internet Archive

[5] In his 1925, John von Neumann observed that "set theory in its first, "naive" version, due to Cantor, led to contradictions. These are the well-known antinomies of the set of all sets that do not contain themselves (Russell), of the set of all transfinte ordinal numbers (Burali-Forti), and the set of all finitely definable real numbers (Richard)." He goes on to observe that two "tendencies" were attempting to "rehabilitate" set theory. Of the first effort, exemplified by Bertrand Russell, Julius König, Hermann Weyl and L. E. J. Brouwer, von Neumann called the "overall effect of their activity . . . devastating". With regards to the axiomatic method employed by second group composed of Zermelo, Abraham Fraenkel and Arthur Moritz Schoenflies, von Neumann worried that "We see only that the known modes of inference leading to the antinomies fail, but who knows where there are not others?" and he set to the task, "in the spirit of the second group", to "produce, by means of a finite number of purely formal operations . . . all the sets that we want to see formed" but not allow for the antinomies. (All quotes from von Neumann 1925 reprinted in van Heijenoort, Jean (1967, third printing 1976), "From Frege to Gödel: A Source Book in Mathematical Logic, 1979–1931", Harvard University Press, Cambridge MA, ISBN 0-674-32449-8 (pbk). A synopsis of the history, written by van Heijenoort, can be found in the comments that precede von Neumann's 1925.

[6] Jech, Thomas (2003), *Set Theory*, Springer Monographs in Mathematics (Third Millennium ed.), Berlin, New York: Springer-Verlag, p. 642, ISBN 978-3-540-44085-7, Zbl 1007.03002

[7] Wittgenstein, Ludwig (1975). *Philosophical Remarks, §129, §174*. Oxford: Basil Blackwell. ISBN 0631191305.

[8] Ferro, A.; Omodeo, E. G.; Schwartz, J. T. (1980), "Decision procedures for elementary sublanguages of set theory. I. Multi-level syllogistic and some extensions", *Comm. Pure Appl. Math.* **33** (5): 599–608, doi:10.1002/cpa.3160330503

[9] Saunders Mac Lane and Ieke Moerdijk (1992) *Sheaves in Geometry and Logic: a First Introduction to Topos Theory.* Springer Verlag.

[10] homotopy type theory in *nLab*

[11] *Homotopy Type Theory: Univalent Foundations of Mathematics.* The Univalent Foundations Program. Institute for Advanced Study.

36.10 Further reading

- Devlin, Keith, 1993. *The Joy of Sets* (2nd ed.). Springer Verlag. ISBN 0-387-94094-4

- Ferreirós, Jose, 2007 (1999). *Labyrinth of Thought: A history of set theory and its role in modern mathematics.* Basel, Birkhäuser. ISBN 978-3-7643-8349-7

- Johnson, Philip, 1972. *A History of Set Theory.* Prindle, Weber & Schmidt ISBN 0-87150-154-6

- Kunen, Kenneth, 1980. *Set Theory: An Introduction to Independence Proofs.* North-Holland, ISBN 0-444-85401-0.

- Potter, Michael, 2004. *Set Theory and Its Philosophy: A Critical Introduction.* Oxford University Press.

- Tiles, Mary, 2004 (1989). *The Philosophy of Set Theory: An Historical Introduction to Cantor's Paradise.* Dover Publications. ISBN 978-0-486-43520-6

36.11 External links

- Foreman, Matthew, Akihiro Kanamori, eds. *Handbook of Set Theory.* 3 vols., 2010. Each chapter surveys some aspect of contemporary research in set theory. Does not cover established elementary set theory, on which see Devlin (1993).

- Hazewinkel, Michiel, ed. (2001), "Axiomatic set theory", *Encyclopedia of Mathematics*, Springer, ISBN 978-1-55608-010-4

- Hazewinkel, Michiel, ed. (2001), "Set theory", *Encyclopedia of Mathematics*, Springer, ISBN 978-1-55608-010-4

- Jech, Thomas (2002). "Set Theory", *Stanford Encyclopedia of Philosophy.*

- Schoenflies, Arthur (1898). Mengenlehre in Klein's encyclopedia.

- Online books, and library resources in your library and in other libraries about set theory

Georg Cantor

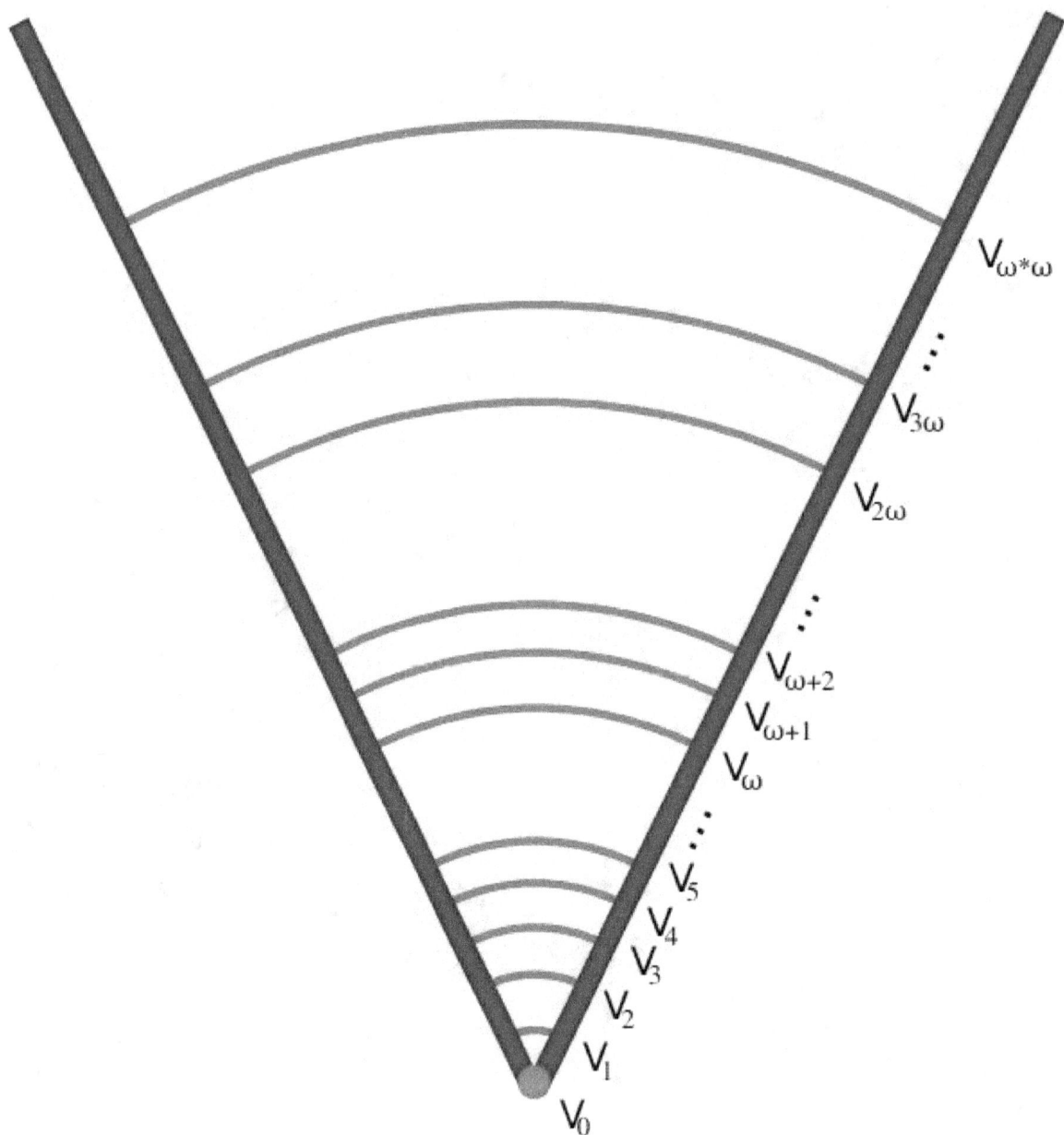

An initial segment of the von Neumann hierarchy.

Chapter 37

Type theory

In mathematics, logic, and computer science, a **type theory** is any of a class of formal systems, some of which can serve as alternatives to set theory as a foundation for all mathematics. In type theory, every "term" has a "type" and operations are restricted to terms of a certain type.

Type theory is closely related to (and in some cases overlaps with) type systems, which are a programming language feature used to reduce bugs. The types of type theory were created to avoid paradoxes in a variety of formal logics and rewrite systems and sometimes "type theory" is used to refer to this broader application.

Two well-known type theories that can serve as mathematical foundations are Alonzo Church's typed λ-calculi and Per Martin-Löf's intuitionistic type theory.

37.1 History

Main article: History of type theory

The types of type theory were invented by Bertrand Russell in response to his discovery that Gottlob Frege's version of naive set theory was afflicted with Russell's paradox. This theory of types features prominently in Whitehead and Russell's *Principia Mathematica*. It avoids Russell's paradox by first creating a hierarchy of types, then assigning each mathematical (and possibly other) entity to a type. Objects of a given type are built exclusively from objects of preceding types (those lower in the hierarchy), thus preventing loops.

The common usage of "type theory" is when those types are used with a term rewrite system. The most famous early example is Alonzo Church's lambda calculus. *Church's Theory of Types*[1] helped the formal system avoid the Kleene–Rosser paradox that afflicted the original untyped lambda calculus. Church demonstrated that it could serve as a foundation of mathematics and it was referred to as a higher-order logic.

Some other type theories include Per Martin-Löf's intuitionistic type theory, which has been the foundation used in some areas of constructive mathematics and for the proof assistant Agda. Thierry Coquand's calculus of constructions and its derivatives are the foundation used by Coq and others. The field is an area of active research, as demonstrated by homotopy type theory.

37.2 Basic concepts

In a system of type theory, each **term** has a **type** and operations are restricted to terms of a certain type. A typing judgment $M : A$ describes that the term M has type A. For example, nat may be a type representing the natural numbers and $0, 1, 2, \ldots$ may be inhabitants of that type. The judgement that 2 has type nat is written as $2 : \text{nat}$.

A function in type theory is denoted with an arrow $\to$. The function addOne (commonly called successor), has the judgement addOne : nat $\rightarrow$ nat . Calling or "applying" a function to an argument is usually written without parentheses, so addOne 2 instead of addOne(2) . (This allows for consistent currying.)

Type theories also contain rules for rewriting terms. These are called **conversion rules** or, if the rule only works in one direction, a **reduction rule**. For example, $2 + 1$ and 3 are syntactically different terms, but the first reduces to the latter. This reduction is denoted as $2 + 1 \twoheadrightarrow 3$.

37.3   Difference from set theory

There are many different set theories and many different systems of type theory, so what follows are generalizations.

- Set theory is built on top of logic. It requires a separate system like Frege's underneath it. In type theory, concepts like "and" and "or" can be encoded as types in the type theory itself.

- In set theory, an element can belong to multiple sets, either to a subset or superset. In type theory, terms (generally) belong to only one type. (Where a subset would be used, type theory creates a new type, called a dependent sum type, with new terms. Union is similarly achieved by a new sum type and new terms.)

- In set theory, sets can contain unrelated elements, e.g., apples and real numbers. In type theory, types that combine unrelated types do so by creating new terms.

- Set theory usually encodes numbers as sets. (0 is the empty set, 1 is a set containing the empty set, etc.) Type theory can encode numbers as functions using Church encoding or more naturally as inductive types, which are a type with well-behaved constant terms.

- Set theory allows set builder notation.

- Type theory has a simple connection to constructive mathematics through the BHK interpretation.

37.4   Optional features

37.4.1   Normalization

The term $2+1$ reduces to 3 . Since 3 cannot be reduced further, it is called a **normal form**. A system of type theory is said to be **strongly normalizing** if all terms have a normal form and any order of reductions reaches it. **Weakly normalizing** systems have a normal form but some orders of reductions may loop forever and never reach it.

For a normalizing system, some borrow the word **element** from set theory and use it to refer to all closed terms that can reduce to the same normal form. A **closed term** is one without parameters. (A term like $x + 1$ with its parameter x is called an **open term**.) Thus, $2 + 1$ and $3 + 0$ may be different terms but they're both from the element 3 .

A similar idea that works for open and closed terms is convertibility. Two terms are **convertible** if there exists a term that they both reduce to. For example, $2 + 1$ and $1 + 2$ are convertible. As are $x + (1 + 1)$ and $x + 2$. However, $x + 1$ and $1 + x$ (where x is a free variable) are not because both are in normal form and they are not the same. Confluent and weakly normalizing systems can test if two terms are convertible by checking if they both reduce to the same normal form.

37.4.2   Dependent types

Main article: dependent types

A **dependent type** is a type that depends on a term or on another type. Thus, the type returned by a function may depend upon the argument to the function.

For example, a list of nat s of length 4 may be a different type than a list of nat s of length 5. In a type theory with dependent types, it is possible to define a function that take a parameter "n" and returns a list containing "n" zeros. Calling the function with 4 would produce a term with a different type than if the function was called with 5.

Dependent types play a central role in intuitionistic type theory and in the design of functional programming languages like Idris, ATS, Agda and Epigram.

37.4.3   Equality types (or "identity types")

Many systems of type theory have a type that represents equality of types and terms. This type is different from convertibility, and is often denoted **propositional equality**.

In intuitionistic type theory, the equality type is known as I for identity. There is a type $I\ A\ a\ b$ when A is a type and a and b are both terms of type A. A term of type $I\ A\ a\ b$ is interpreted as meaning that a is equal to b.

In practice, it is possible to build a type I nat 3 4 but there will not exist a term of that type. In intuitionistic type theory, new terms of equality start with reflexivity. If 3 is a term of type nat, then there exists a term of type I nat 3 3. More complicated equalities can be created by creating a reflexive term and then doing a reduction on one side. So if $2+1$ is a term of type nat, then there is a term of type I nat $(2+1)\ (2+1)$ and, by reduction, generate a term of type I nat $(2+1)\ 3$. Thus, in this system, the equality type denotes that two values of the same type are convertible by reductions.

Having a type for equality is important because it can be manipulated inside the system. There is usually no judgement to say two terms are *not* equal; instead, as in the Brouwer–Heyting–Kolmogorov interpretation, we map $\neg(a = b)$ to $(a = b) \to \bot$, where $\bot$ is the bottom type having no values. There exists a term with type $(I\ \text{nat}\ 3\ 4) \to \bot$, but not one of type $(I\ \text{nat}\ 3\ 3) \to \bot$.

Homotopy type theory differs from intuitionistic type theory mostly by its handling of the equality type.

37.4.4   Inductive types

Main article: Inductive type

A system of type theory requires some basic terms and types to operate on. Some systems build them out of functions using Church encoding. Other systems have **inductive types**: a set of base types and a set of type constructors that generate types with well-behaved properties. For example, certain recursive functions called on inductive types are guaranteed to terminate.

Coinductive type are infinite data types created by giving a function that generates the next element(s). See Coinduction and Corecursion.

Induction induction is a feature for declaring an inductive type and a family of types that depends on the inductive type.

Induction recursion allows a wider range of well-behaved types but requires that the type and the recursive functions that operate on them be defined at the same time.

37.4.5   Universe types

Types were created to prevent paradoxes, such as Russell's paradox. However, the motives that lead to those paradoxes – being able to say things about all types – still exist. So many type theories have a "universe type", which contains all other types.

In systems where you might want to say something about universe types, there is a hierarchy of universe types, each containing the one below it in the hierarchy. The hierarchy is defined as being infinite, but statements must only refer to a finite number of universe levels.

Type universes are particularly tricky in type theory. The initial proposal of intuitionistic type theory suffered from Girard's paradox.

37.4.6 Computational component

Many systems of type theory, such as the simply-typed lambda calculus, intuitionistic type theory, and the calculus of constructions, are also programming languages. That is, they are said to have a "computational component". The computation is the reduction of terms of the language using rewriting rules.

A system of type theory that has a well-behaved computational component also has a simple connection to constructive mathematics through the BHK interpretation.

Non-constructive mathematics in these systems is possible by adding operators on continuations such as call with current continuation. However, these operators tend to break desirable properties such as canonicity and parametricity.

37.5 Systems of type theory

37.5.1 Major

- Simply typed lambda calculus which is a higher-order logic
- Intuitionistic type theory
- System F
- LF is often used to define other type theories
- Calculus of constructions and its derivatives

37.5.2 Minor

- Automath
- some forms of combinatory logic
- ST type theory
- others defined in the lambda cube
- others under the name typed lambda calculus
- others under the name pure type system

37.5.3 Active

- Homotopy type theory is being researched

37.6 Practical impact

37.6.1 Programming languages

Main article: Type system

There is extensive overlap and interaction between the fields of type theory and type systems. Type systems are a programming language feature designed to identify bugs. Any static program analysis, such as the type checking algorithms in the semantic analysis phase of compiler, has a connection to type theory.

A prime example is Agda, a programming language which uses intuitionistic type theory for its type system. The programming language ML was developed for manipulating type theories (see LCF) and its own type system was heavily influenced by them.

37.6.2 Mathematical foundations

The first computer proof assistant, called Automath, used type theory to encode mathematics on a computer. Martin-Löf specifically developed intuitionistic type theory to encode *all* mathematics - to serve as a new foundation for mathematics. There is current research into mathematical foundations using homotopy type theory.

Mathematicians working in category theory already had difficulty working with the widely accepted foundation of Zermelo–Fraenkel set theory. This led to proposals such as Lawvere's Elementary Theory of the Category of Sets (ETCS).[2] Homotopy type theory continues in this line using type theory. Researchers are exploring connections between dependent types (especially the identity type) and algebraic topology (specifically homotopy).

37.6.3 Proof assistants

Main article: Proof assistant

Much of the current research into type theory is driven by proof checkers, interactive proof assistants, and automated theorem provers. Most of these systems use a type theory as the mathematical foundation for encoding proofs. This is not surprising, given the close connection between type theory and programming languages.

- LF is used by Twelf, often to define other type theories

- Multiple type theories falling under higher-order logic are used by the HOL family of provers and PVS

- Intuitionistic type theory is used by Agda which is both a programming language and proof assistant.

- Computational Type Theory is used by NuPRL

- The calculus of constructions and its derivatives are used by Coq and Matita.

Multiple type theories are supported by LEGO and Isabelle. Isabelle also supports foundations besides type theories, such as ZFC. Mizar is an example of a proof system that only supports set theory.

37.6.4 Linguistics

Type theory is also widely in use in formal theories of semantics of natural languages, especially Montague grammar and its descendants. In particular, categorial grammars and pregroup grammars make extensive use of type constructors to define the types (*noun*, *verb*, etc.) of words.

The most common construction takes the basic types e and t for individuals and truth-values, respectively, and defines the set of types recursively as follows:

- if a and b are types, then so is $\langle a, b \rangle$.

- Nothing except the basic types, and what can be constructed from them by means of the previous clause are types.

A complex type $\langle a, b \rangle$ is the type of functions from entities of type a to entities of type b . Thus one has types like $\langle e, t \rangle$ which are interpreted as elements of the set of functions from entities to truth-values, i.e. indicator functions of sets of entities. An expression of type $\langle \langle e, t \rangle, t \rangle$ is a function from sets of entities to truth-values, i.e. a (indicator function of a) set of sets. This latter type is standardly taken to be the type of natural language quantifiers, like *everybody* or *nobody* (Montague 1973, Barwise and Cooper 1981).

37.6.5 Social sciences

Gregory Bateson introduced a theory of logical types into the social sciences; his notions of double bind and logical levels are based on Russell's theory of types.

37.7 Relation to category theory

Although the initial motivation for category theory was far removed from foundationalism, the two fields turned out to have deep connections. As John Lane Bell writes: "In fact categories can *themselves* be viewed as type theories of a certain kind; this fact alone indicates that type theory is much more closely related to category theory than it is to set theory." In brief, a category can be viewed as a type theory by regarding its objects as types (or sorts), i.e. "Roughly speaking, a category may be thought of as a type theory shorn of its syntax." A number of significant results follow in this way:[3]

- cartesian closed categories correspond to the typed λ-calculus (Lambek, 1970)

- C-monoids (categories with products and exponentials and a single, nonterminal object) correspond to the untyped λ-calculus (observed independently by Lambek and Dana Scott around 1980)

- locally cartesian closed categories correspond to Martin-Löf type theories (Seely, 1984)

The interplay, known as categorical logic, has been a subject of active research since then; see the monograph of Jacobs (1999) for instance.

37.8 See also

- Data type for concrete types of data in programming

- Domain theory

- Type (model theory)

- Type system for a more practical discussion of type systems for programming languages

37.9 References

- W. Farmer, *The seven virtues of simple type theory*, Journal of Applied Logic, Vol. 6, No. 3. (September 2008), pp. 267–286.

[1] Alonzo Church, *A formulation of the simple theory of types*, The Journal of Symbolic Logic 5(2):56–68 (1940)

[2] ETCS in *nLab*

[3] John L. Bell (2012). "Types, Sets and Categories". In Akihiro Kanamory. *Handbook of the History of Logic. Volume 6. Sets and Extensions in the Twentieth Century* (PDF). Elsevier. ISBN 978-0-08-093066-4.

37.10 Further reading

- Constable, Robert L., 2002, "Naïve Computational Type Theory," in H. Schwichtenberg and R. Steinbruggen (eds.), *Proof and System-Reliability*: 213–259. Intended as a type theory counterpart of Paul Halmos's (1960) *Naïve Set Theory*

- Andrews B., Peter (2002). *An Introduction to Mathematical Logic and Type Theory: To Truth Through Proof, 2nd ed.* Kluwer Academic Publishers. ISBN 978-1-4020-0763-7.

- Jacobs, Bart (1999). *Categorical Logic and Type Theory.* Studies in Logic and the Foundations of Mathematics 141. North Holland, Elsevier. ISBN 0-444-50170-3. Covers type theory in depth, including polymorphic and dependent type extensions. Gives categorical semantics.

- Collins, Jordan E. (2012). *A History of the Theory of Types: Developments After the Second Edition of 'Principia Mathematica'.* LAP Lambert Academic Publishing. ISBN 978-3-8473-2963-3. Provides a historical survey of the developments of the theory of types with a focus on the decline of the theory as a foundation of mathematics over the four decades following the publication of the second edition of 'Principia Mathematica'.

- Cardelli, Luca, 1997, "Type Systems," in Allen B. Tucker, ed., *The Computer Science and Engineering Handbook.* CRC Press: 2208–2236.

- Thompson, Simon, 1991. *Type Theory and Functional Programming.* Addison–Wesley. ISBN 0-201-41667-0.

- J. Roger Hindley, *Basic Simple Type Theory*, Cambridge University Press, 2008, ISBN 0-521-05422-2 (also 1995, 1997). A good introduction to simple type theory for computer scientists; the system described is not exactly Church's STT though. Book review

- Stanford Encyclopedia of Philosophy: Type Theory" – by Thierry Coquand.

- Fairouz D. Kamareddine, Twan Laan, Rob P. Nederpelt, *A modern perspective on type theory: from its origins until today*, Springer, 2004, ISBN 1-4020-2334-0

- José Ferreirós, José Ferreirós Domínguez, *Labyrinth of thought: a history of set theory and its role in modern mathematics*, Edition 2, Springer, 2007, ISBN 3-7643-8349-6, chapter X "Logic and Type Theory in the Interwar Period"

37.11   External links

- Computational type theory at Scholarpedia, curated by Robert L. Constable.

- The TYPES Forum — moderated e-mail forum focusing on type theory in computer science, operating since 1987.

- The Nuprl Book: "Introduction to Type Theory."

- Types Project lecture notes of summer schools 2005–2008
    - The 2005 summer school has introductory lectures

Chapter 38

Proof theory

Proof theory is a branch of mathematical logic that represents proofs as formal mathematical objects, facilitating their analysis by mathematical techniques. Proofs are typically presented as inductively-defined data structures such as plain lists, boxed lists, or trees, which are constructed according to the axioms and rules of inference of the logical system. As such, proof theory is syntactic in nature, in contrast to model theory, which is semantic in nature. Together with model theory, axiomatic set theory, and recursion theory, proof theory is one of the so-called *four pillars* of the foundations of mathematics.[1]

Some of the major areas of proof theory include structural proof theory, ordinal analysis, provability logic, reverse mathematics, proof mining, automated theorem proving, and proof complexity. Much research also focuses on applications in computer science, linguistics, and philosophy.

38.1   History

Although the formalisation of logic was much advanced by the work of such figures as Gottlob Frege, Giuseppe Peano, Bertrand Russell, and Richard Dedekind, the story of modern proof theory is often seen as being established by David Hilbert, who initiated what is called Hilbert's program in the foundations of mathematics. The central idea of this program was that if we could give finitary proofs of consistency for all the sophisticated formal theories needed by mathematicians, then we could ground these theories by means of a metamathematical argument, which shows that all of their purely universal assertions (more technically their provable Π_1^0 sentences) are finitarily true; once so grounded we do not care about the non-finitary meaning of their existential theorems, regarding these as pseudo-meaningful stipulations of the existence of ideal entities.

The failure of the program was induced by Kurt Gödel's incompleteness theorems, which showed that any ω-consistent theory that is sufficiently strong to express certain simple arithmetic truths, cannot prove its own consistency, which on Gödel's formulation is a Π_1^0 sentence. However, modified versions of Hilbert's program emerged and research has been carried out on related topics. This has led, in particular, to:

- Refinement of Gödel's result, particularly J. Barkley Rosser's refinement, weakening the above requirement of ω-consistency to simple consistency;

- Axiomatisation of the core of Gödel's result in terms of a modal language, provability logic;

- Transfinite iteration of theories, due to Alan Turing and Solomon Feferman;

- The recent discovery of self-verifying theories, systems strong enough to talk about themselves, but too weak to carry out the diagonal argument that is the key to Gödel's unprovability argument.

In parallel to the rise and fall of Hilbert's program, the foundations of structural proof theory were being founded. Jan Łukasiewicz suggested in 1926 that one could improve on Hilbert systems as a basis for the axiomatic presentation of logic if one allowed the drawing of conclusions from assumptions in the inference rules of the logic. In response to this Stanisław Jaśkowski (1929) and Gerhard Gentzen (1934) independently provided such systems, called calculi of natural deduction, with Gentzen's approach introducing the idea of symmetry between the grounds for asserting propositions, expressed in introduction rules, and the consequences of accepting propositions in the elimination rules, an idea that has proved very important in proof theory.[2] Gentzen (1934) further introduced the idea of the sequent calculus, a calculus advanced in a similar spirit that better expressed the duality of the logical connectives,[3] and went on to make fundamental advances in the formalisation of intuitionistic logic, and provide the first combinatorial proof of the consistency of Peano arithmetic. Together, the presentation of natural deduction and the sequent calculus introduced the fundamental idea of analytic proof to proof theory.

38.2   Structural proof theory

Main article: Structural proof theory

Structural proof theory is the subdiscipline of proof theory that studies the specifics of proof calculi. The three most well-known styles of proof calculi are:

- The Hilbert calculi

- The natural deduction calculi

- The sequent calculi

Each of these can give a complete and axiomatic formalization of propositional or predicate logic of either the classical or intuitionistic flavour, almost any modal logic, and many substructural logics, such as relevance logic or linear logic. Indeed it is unusual to find a logic that resists being represented in one of these calculi.

Proof theorists are typically interested in proof calculi that support a notion of analytic proof. The notion of analytic proof was introduced by Gentzen for the sequent calculus; there the analytic proofs are those that are cut-free. Much of the interest in cut-free proofs comes from the subformula property: every formula in the end sequent of a cut-free proof is a subformula of one of the premises. This allows one to show consistency of the sequent calculus easily; if the empty sequent were derivable it would have to be a subformula of some premise, which it is not. Gentzen's midsequent theorem, the Craig interpolation theorem, and Herbrand's theorem also follow as corollaries of the cut-elimination theorem.

Gentzen's natural deduction calculus also supports a notion of analytic proof, as shown by Dag Prawitz. The definition is slightly more complex: we say the analytic proofs are the normal forms, which are related to the notion of normal form in term rewriting. More exotic proof calculi such as Jean-Yves Girard's proof nets also support a notion of analytic proof.

Structural proof theory is connected to type theory by means of the Curry-Howard correspondence, which observes a structural analogy between the process of normalisation in the natural deduction calculus and beta reduction in the typed lambda calculus. This provides the foundation for the intuitionistic type theory developed by Per Martin-Löf, and is often extended to a three way correspondence, the third leg of which are the cartesian closed categories.

Other research topics in structural theory include analytic tableau, which apply the central idea of analytic proof from structural proof theory to provide decision procedures and semi-decision procedures for a wide range of logics, and the proof theory of substructural logics.

38.3   Ordinal analysis

Main article: Ordinal analysis

Ordinal analysis is a powerful technique for providing combinatorial consistency proofs for subsystems of arithmetic, analysis, and set theory. Gödel's second incompleteness theorem is often interpreted as demonstrating that finitistic consistency proofs are impossible for theories of sufficient strength. Ordinal analysis allows one to measure precisely the infinitary content of the consistency of theories. For a consistent recursively axiomatized theory T, one can prove in finitistic arithmetic that the well-foundedness of a certain transfinite ordinal implies the consistency of T. Gödel's second incompleteness theorem implies that the well-foundedness of such an ordinal cannot be proved in the theory T.

Consequences of ordinal analysis include (1) consistency of subsystems of classical second order arithmetic and set theory relative to constructive theories, (2) combinatorial independence results, and (3) classifications of provably total recursive functions and provably well-founded ordinals.

Ordinal analysis was originated by Genzten, who proved the consistency of Peano Arithmetic using transfinite induction up to ordinal ε_0. Ordinal analysis has been extended to many fragments of first and second order arithmetic and set theory. One major challenge has been the ordinal analysis of impredicative theories. The first breakthrough in this direction was Takeuti's proof of the consistency of $\Pi 1$
1-CA_0 using the method of ordinal diagrams

38.4 Provability logic

Main article: Provability logic

Provability logic is a modal logic, in which the box operator is interpreted as 'it is provable that'. The point is to capture the notion of a proof predicate of a reasonably rich formal theory. As basic axioms of the provability logic GL, which captures provable in Peano Arithmetic, one takes modal analogues of the Hilbert-Bernays derivability conditions and Löb's theorem (if it is provable that the provability of A implies A, then A is provable).

Some of the basic results concerning the incompleteness of Peano Arithmetic and related theories have analogues in provability logic. For example, it is a theorem in GL that if a contradiction is not provable then it is not provable that a contradiction is not provable (Gödel's second incompleteness theorem). There are also modal analogues of the fixed-point theorem. Robert Solovay proved that the modal logic GL is complete with respect to Peano Arithmetic. That is, the propositional theory of provability in Peano Arithmetic is completely represented by the modal logic GL. This straightforwardly implies that propositional reasoning about provability in Peano Arithmetic is complete and decidable.

Other research in provability logic has focused on first-order provability logic, polymodal provability logic (with one modality representing provability in the object theory and another representing provability in the meta-theory), and inter-pretability logics intended to capture the interaction between provability and interpretability. Some very recent research has involved applications of graded provability algebras to the ordinal analysis of arithmetical theories.

38.5 Reverse mathematics

Main article: Reverse mathematics

Reverse mathematics is a program in mathematical logic that seeks to determine which axioms are required to prove theorems of mathematics. It was founded by Harvey Friedman. Its defining method can be described as "going backwards from the theorems to the axioms", in contrast to the ordinary mathematical practice of deriving theorems from axioms. The reverse mathematics program was foreshadowed by results in set theory such as the classical theorem that the axiom of choice and Zorn's lemma are equivalent over ZF set theory. The goal of reverse mathematics, however, is to study possible axioms of ordinary theorems of mathematics rather than possible axioms for set theory.

In reverse mathematics, one starts with a framework language and a base theory—a core axiom system—that is too weak to prove most of the theorems one might be interested in, but still powerful enough to develop the definitions necessary to state these theorems. For example, to study the theorem "Every bounded sequence of real numbers has a supremum" it is necessary to use a base system which can speak of real numbers and sequences of real numbers.

For each theorem that can be stated in the base system but is not provable in the base system, the goal is to determine the particular axiom system (stronger than the base system) that is necessary to prove that theorem. To show that a system S

is required to prove a theorem T, two proofs are required. The first proof shows T is provable from S; this is an ordinary mathematical proof along with a justification that it can be carried out in the system S. The second proof, known as a **reversal**, shows that T itself implies S; this proof is carried out in the base system. The reversal establishes that no axiom system S' that extends the base system can be weaker than S while still proving T.

One striking phenomenon in reverse mathematics is the robustness of the *Big Five* axiom systems. In order of increasing strength, these systems are named by the initialisms RCA_0, WKL_0, ACA_0, ATR_0, and $\Pi^1_1\text{-}CA_0$. Nearly every theorem of ordinary mathematics that has been reverse mathematically analyzed has been proven equivalent to one of these five systems. Much recent research has focused on combinatorial principles that do not fit neatly into this framework, like RT^2_2 (Ramsey's theorem for pairs).

Research in reverse mathematics often incorporates methods and techniques from recursion theory as well as proof theory.

38.6   Functional interpretations

Functional interpretations are interpretations of non-constructive theories in functional ones. Functional interpretations usually proceed in two stages. First, one "reduces" a classical theory C to an intuitionistic one I. That is, one provides a constructive mapping that translates the theorems of C to the theorems of I. Second, one reduces the intuitionistic theory I to a quantifier free theory of functionals F. These interpretations contribute to a form of Hilbert's program, since they prove the consistency of classical theories relative to constructive ones. Successful functional interpretations have yielded reductions of infinitary theories to finitary theories and impredicative theories to predicative ones.

Functional interpretations also provide a way to extract constructive information from proofs in the reduced theory. As a direct consequence of the interpretation one usually obtains the result that any recursive function whose totality can be proven either in I or in C is represented by a term of F. If one can provide an additional interpretation of F in I, which is sometimes possible, this characterization is in fact usually shown to be exact. It often turns out that the terms of F coincide with a natural class of functions, such as the primitive recursive or polynomial-time computable functions. Functional interpretations have also been used to provide ordinal analyses of theories and classify their provably recursive functions.

The study of functional interpretations began with Kurt Gödel's interpretation of intuitionistic arithmetic in a quantifier-free theory of functionals finite type. This interpretation is commonly known as the Dialectica interpretation. Together with the double-negation interpretation of classical logic in intuitionistic logic, it provides a reduction of classical arithmetic to intuitionistic arithmetic.

38.7   Formal and informal proof

Main article: Formal proof

The *informal* proofs of everyday mathematical practice are unlike the *formal* proofs of proof theory. They are rather like high-level sketches that would allow an expert to reconstruct a formal proof at least in principle, given enough time and patience. For most mathematicians, writing a fully formal proof is too pedantic and long-winded to be in common use.

Formal proofs are constructed with the help of computers in interactive theorem proving. Significantly, these proofs can be checked automatically, also by computer. (Checking formal proofs is usually simple, whereas *finding* proofs (automated theorem proving) is generally hard.) An informal proof in the mathematics literature, by contrast, requires weeks of peer review to be checked, and may still contain errors.

38.8   Proof-theoretic semantics

Main articles: proof-theoretic semantics and logical harmony

In linguistics, type-logical grammar, categorial grammar and Montague grammar apply formalisms based on structural proof theory to give a formal natural language semantics.

38.9   See also

- Intermediate logic
- Model theory
- Proof (truth)
- Proof techniques

38.10   Notes

[1] E.g., Wang (1981), pp. 3–4, and Barwise (1978).

[2] Prawitz (2006, p. 98).

[3] Girard, Lafont, and Taylor (1988).

38.11   References

- J. Avigad, E.H. Reck (2001). "Clarifying the nature of the infinite": the development of metamathematics and proof theory. Carnegie-Mellon Technical Report CMU-PHIL-120.

- J. Barwise (ed., 1978). Handbook of Mathematical Logic. North-Holland.

- S. Buss. Handbook of Proof Theory. Elsevier.

- A. S. Troelstra, H. Schwichtenberg (1996). *Basic Proof Theory*. In series *Cambridge Tracts in Theoretical Computer Science*, Cambridge University Press, ISBN 0-521-77911-1.

- G. Gentzen (1935/1969). Investigations into logical deduction. In M. E. Szabo, editor, *Collected Papers of Gerhard Gentzen*. North-Holland. Translated by Szabo from "Untersuchungen über das logische Schliessen", Mathematisches Zeitschrift 39: 176-210, 405-431.

- Hazewinkel, Michiel, ed. (2001), "Proof theory", *Encyclopedia of Mathematics*, Springer, ISBN 978-1-55608-010-4

- Luis Moreno & Bharath Sriraman (2005).*Structural Stability and Dynamic Geometry: Some Ideas on Situated Proof*. *International Reviews on Mathematical Education. Vol. 37, no.3, pp. 130–139*

- Prawitz, Dag (2006) [1965]. *Natural deduction: A proof-theoretical study*. Mineola, New York: Dover Publications. ISBN 978-0-486-44655-4.

- J. von Plato (2008). The Development of Proof Theory. Stanford Encyclopedia of Philosophy.

- Wang, Hao (1981). *Popular Lectures on Mathematical Logic*. Van Nostrand Reinhold Company. ISBN 0-442-23109-1.

Chapter 39

Computability theory

For the concept of computability, see Computability.

Computability theory, also called **recursion theory**, is a branch of mathematical logic, of computer science, and of the theory of computation that originated in the 1930s with the study of computable functions and Turing degrees.

The basic questions addressed by recursion theory are "What does it mean for a function on the natural numbers to be computable?" and "How can noncomputable functions be classified into a hierarchy based on their level of noncomputability?". The answers to these questions have led to a rich theory that is still being actively researched. The field has since grown to include the study of generalized computability and definability. Invention of the central combinatorial object of recursion theory, namely the Universal Turing Machine, predates and predetermines the invention of modern computers. Historically, the study of algorithmically undecidable sets and functions was motivated by various problems in mathematics that turned to be undecidable; for example, word problem for groups and the like. There are several applications of the theory to other branches of mathematics that do not necessarily concentrate on undecidability. The early applications include the celebrated Higman's embedding theorem that provides a link between recursion theory and group theory, results of Michael O. Rabin and Anatoly Maltsev on algorithmic presentations of algebras, and the negative solution to Hilbert's Tenth Problem. The more recent applications include algorithmic randomness, results of Soare et al. who applied recursion-theoretic methods to solve a problem in algebraic geometry,[1] and the very recent work of Slaman et al. on normal numbers that solves a problem in analytic number theory.

Recursion theory overlaps with proof theory, effective descriptive set theory, model theory, and abstract algebra. Arguably, computational complexity theory is a child of recursion theory; both theories share the same technical tool, namely the Turing Machine. Recursion theorists in mathematical logic often study the theory of relative computability, reducibility notions and degree structures described in this article. This contrasts with the theory of subrecursive hierarchies, formal methods and formal languages that is common in the study of computability theory in computer science. There is a considerable overlap in knowledge and methods between these two research communities; however, no firm line can be drawn between them. For instance, parametrized complexity was invented by a complexity theorist Michael Fellows and a recursion theorist Rod Downey.

39.1 Computable and uncomputable sets

Recursion theory originated in the 1930s, with work of Kurt Gödel, Alonzo Church, Alan Turing, Stephen Kleene and Emil Post.[2]

The fundamental results the researchers obtained established Turing computability as the correct formalization of the informal idea of effective calculation. These results led Stephen Kleene (1952) to coin the two names "Church's thesis" (Kleene 1952:300) and "Turing's Thesis" (Kleene 1952:376). Nowadays these are often considered as a single hypothesis, the **Church–Turing thesis**, which states that any function that is computable by an algorithm is a computable function. Although initially skeptical, by 1946 Gödel argued in favor of this thesis:

"Tarski has stressed in his lecture (and I think justly) the great importance of the concept of general recursiveness (or Turing's computability). It seems to me that this importance is largely due to the fact that with this concept one has for the first time succeeded in giving an absolute notion to an interesting epistemological notion, i.e., one not depending on the formalism chosen.*"(Gödel 1946 in Davis 1965:84).[3]

With a definition of effective calculation came the first proofs that there are problems in mathematics that cannot be effectively decided. Church (1936a, 1936b) and Turing (1936), inspired by techniques used by Gödel (1931) to prove his incompleteness theorems, independently demonstrated that the Entscheidungsproblem is not effectively decidable. This result showed that there is no algorithmic procedure that can correctly decide whether arbitrary mathematical propositions are true or false.

Many problems in mathematics have been shown to be undecidable after these initial examples were established. In 1947, Markov and Post published independent papers showing that the word problem for semigroups cannot be effectively decided. Extending this result, Pyotr Novikov and William Boone showed independently in the 1950s that the word problem for groups is not effectively solvable: there is no effective procedure that, given a word in a finitely presented group, will decide whether the element represented by the word is the identity element of the group. In 1970, Yuri Matiyasevich proved (using results of Julia Robinson) Matiyasevich's theorem, which implies that Hilbert's tenth problem has no effective solution; this problem asked whether there is an effective procedure to decide whether a Diophantine equation over the integers has a solution in the integers. The list of undecidable problems gives additional examples of problems with no computable solution.

The study of which mathematical constructions can be effectively performed is sometimes called **recursive mathematics**: the *Handbook of Recursive Mathematics* (Ershov *et al.* 1998) covers many of the known results in this field.

39.2 Turing computability

The main form of computability studied in recursion theory was introduced by Turing (1936). A set of natural numbers is said to be a *computable set* (also called a *decidable*, *recursive*, or *Turing computable* set) if there is a Turing machine that, given a number n, halts with output 1 if n is in the set and halts with output 0 if n is not in the set. A function f from the natural numbers to themselves is a *recursive* or *(Turing) computable function* if there is a Turing machine that, on input n, halts and returns output $f(n)$. The use of Turing machines here is not necessary; there are many other models of computation that have the same computing power as Turing machines; for example the μ-recursive functions obtained from primitive recursion and the μ operator.

The terminology for recursive functions and sets is not completely standardized. The definition in terms of μ-recursive functions as well as a different definition of *rekursiv* functions by Gödel led to the traditional name *recursive* for sets and functions computable by a Turing machine. The word *decidable* stems from the German word *Entscheidungsproblem* which was used in the original papers of Turing and others. In contemporary use, the term "computable function" has various definitions: according to Cutland (1980), it is a partial recursive function (which can be undefined for some inputs), while according to Soare (1987) it is a total recursive (equivalently, general recursive) function. This article follows the second of these conventions. Soare (1996) gives additional comments about the terminology.

Not every set of natural numbers is computable. The halting problem, which is the set of (descriptions of) Turing machines that halt on input 0, is a well-known example of a noncomputable set. The existence of many noncomputable sets follows from the facts that there are only countably many Turing machines, and thus only countably many computable sets, but there are uncountably many sets of natural numbers.

Although the halting problem is not computable, it is possible to simulate program execution and produce an infinite list of the programs that do halt. Thus the halting problem is an example of a *recursively enumerable set*, which is a set that can be enumerated by a Turing machine (other terms for recursively enumerable include *computably enumerable* and *semidecidable*). Equivalently, a set is recursively enumerable if and only if it is the range of some computable function. The recursively enumerable sets, although not decidable in general, have been studied in detail in recursion theory.

39.3 Areas of research

Beginning with the theory of recursive sets and functions described above, the field of recursion theory has grown to include the study of many closely related topics. These are not independent areas of research: each of these areas draws ideas and results from the others, and most recursion theorists are familiar with the majority of them.

39.3.1 Relative computability and the Turing degrees

Main articles: Turing reduction and Turing degree

Recursion theory in mathematical logic has traditionally focused on *relative computability*, a generalization of Turing computability defined using oracle Turing machines, introduced by Turing (1939). An oracle Turing machine is a hypothetical device which, in addition to performing the actions of a regular Turing machine, is able to ask questions of an *oracle*, which is a particular set of natural numbers. The oracle machine may only ask questions of the form "Is n in the oracle set?". Each question will be immediately answered correctly, even if the oracle set is not computable. Thus an oracle machine with a noncomputable oracle will be able to compute sets that a Turing machine without an oracle cannot.

Informally, a set of natural numbers A is *Turing reducible* to a set B if there is an oracle machine that correctly tells whether numbers are in A when run with B as the oracle set (in this case, the set A is also said to be (*relatively*) *computable from B* and *recursive in B*). If a set A is Turing reducible to a set B and B is Turing reducible to A then the sets are said to have the same *Turing degree* (also called *degree of unsolvability*). The Turing degree of a set gives a precise measure of how uncomputable the set is.

The natural examples of sets that are not computable, including many different sets that encode variants of the halting problem, have two properties in common:

1. They are recursively enumerable, and

2. Each can be translated into any other via a many-one reduction. That is, given such sets A and B, there is a total computable function f such that $A = \{x : f(x) \in B\}$. These sets are said to be *many-one equivalent* (or *m-equivalent*).

Many-one reductions are "stronger" than Turing reductions: if a set A is many-one reducible to a set B, then A is Turing reducible to B, but the converse does not always hold. Although the natural examples of noncomputable sets are all many-one equivalent, it is possible to construct recursively enumerable sets A and B such that A is Turing reducible to B but not many-one reducible to B. It can be shown that every recursively enumerable set is many-one reducible to the halting problem, and thus the halting problem is the most complicated recursively enumerable set with respect to many-one reducibility and with respect to Turing reducibility. Post (1944) asked whether *every* recursively enumerable set is either computable or Turing equivalent to the halting problem, that is, whether there is no recursively enumerable set with a Turing degree intermediate between those two.

As intermediate results, Post defined natural types of recursively enumerable sets like the simple, hypersimple and hyper-hypersimple sets. Post showed that these sets are strictly between the computable sets and the halting problem with respect to many-one reducibility. Post also showed that some of them are strictly intermediate under other reducibility notions stronger than Turing reducibility. But Post left open the main problem of the existence of recursively enumerable sets of intermediate Turing degree; this problem became known as *Post's problem*. After ten years, Kleene and Post showed in 1954 that there are intermediate Turing degrees between those of the computable sets and the halting problem, but they failed to show that any of these degrees contains a recursively enumerable set. Very soon after this, Friedberg and Muchnik independently solved Post's problem by establishing the existence of recursively enumerable sets of intermediate degree. This groundbreaking result opened a wide study of the Turing degrees of the recursively enumerable sets which turned out to possess a very complicated and non-trivial structure.

There are uncountably many sets that are not recursively enumerable, and the investigation of the Turing degrees of all sets is as central in recursion theory as the investigation of the recursively enumerable Turing degrees. Many degrees with special properties were constructed: *hyperimmune-free degrees* where every function computable relative to that degree is majorized by a (unrelativized) computable function; *high degrees* relative to which one can compute a function f which dominates every computable function g in the sense that there is a constant c depending on g such that $g(x) < f(x)$ for all $x > c$; *random degrees* containing algorithmically random sets; *1-generic* degrees of 1-generic sets; and the degrees below the halting problem of limit-recursive sets.

The study of arbitrary (not necessarily recursively enumerable) Turing degrees involves the study of the Turing jump. Given a set A, the *Turing jump* of A is a set of natural numbers encoding a solution to the halting problem for oracle Turing machines running with oracle A. The Turing jump of any set is always of higher Turing degree than the original set, and a theorem of Friedburg shows that any set that computes the Halting problem can be obtained as the Turing jump of another set. Post's theorem establishes a close relationship between the Turing jump operation and the arithmetical hierarchy, which is a classification of certain subsets of the natural numbers based on their definability in arithmetic.

Much recent research on Turing degrees has focused on the overall structure of the set of Turing degrees and the set of Turing degrees containing recursively enumerable sets. A deep theorem of Shore and Slaman (1999) states that the function mapping a degree x to the degree of its Turing jump is definable in the partial order of the Turing degrees. A recent survey by Ambos-Spies and Fejer (2006) gives an overview of this research and its historical progression.

39.3.2 Other reducibilities

Main article: Reduction (recursion theory)

An ongoing area of research in recursion theory studies reducibility relations other than Turing reducibility. Post (1944) introduced several *strong reducibilities*, so named because they imply truth-table reducibility. A Turing machine implementing a strong reducibility will compute a total function regardless of which oracle it is presented with. *Weak reducibilities* are those where a reduction process may not terminate for all oracles; Turing reducibility is one example.

The strong reducibilities include:

One-one reducibility A is *one-one reducible* (or *1-reducible*) to B if there is a total computable injective function f such that each n is in A if and only if $f(n)$ is in B.

Many-one reducibility This is essentially one-one reducibility without the constraint that f be injective. A is *many-one reducible* (or *m-reducible*) to B if there is a total computable function f such that each n is in A if and only if $f(n)$ is in B.

Truth-table reducibility A is truth-table reducible to B if A is Turing reducible to B via an oracle Turing machine that computes a total function regardless of the oracle it is given. Because of compactness of Cantor space, this is equivalent to saying that the reduction presents a single list of questions (depending only on the input) to the oracle simultaneously, and then having seen their answers is able to produce an output without asking additional questions regardless of the oracle's answer to the initial queries. Many variants of truth-table reducibility have also been studied.

Further reducibilities (positive, disjunctive, conjunctive, linear and their weak and bounded versions) are discussed in the article Reduction (recursion theory).

The major research on strong reducibilities has been to compare their theories, both for the class of all recursively enumerable sets as well as for the class of all subsets of the natural numbers. Furthermore, the relations between the reducibilities has been studied. For example, it is known that every Turing degree is either a truth-table degree or is the union of infinitely many truth-table degrees.

Reducibilities weaker than Turing reducibility (that is, reducibilities that are implied by Turing reducibility) have also been studied. The most well known are arithmetical reducibility and hyperarithmetical reducibility. These reducibilities are closely connected to definability over the standard model of arithmetic.

39.3.3   Rice's theorem and the arithmetical hierarchy

Rice showed that for every nontrivial class C (which contains some but not all r.e. sets) the index set $E = \{e$: the eth r.e. set We is in $C\}$ has the property that either the halting problem or its complement is many-one reducible to E, that is, can be mapped using a many-one reduction to E (see Rice's theorem for more detail). But, many of these index sets are even more complicated than the halting problem. These type of sets can be classified using the arithmetical hierarchy. For example, the index set FIN of class of all finite sets is on the level Σ_2, the index set REC of the class of all recursive sets is on the level Σ_3, the index set COFIN of all cofinite sets is also on the level Σ_3 and the index set COMP of the class of all Turing-complete sets Σ_4. These hierarchy levels are defined inductively, Σ_{n+1} contains just all sets which are recursively enumerable relative to Σ_n; Σ_1 contains the recursively enumerable sets. The index sets given here are even complete for their levels, that is, all the sets in these levels can be many-one reduced to the given index sets.

39.3.4   Reverse mathematics

Main article: Reverse mathematics

The program of *reverse mathematics* asks which set-existence axioms are necessary to prove particular theorems of mathematics in subsystems of second-order arithmetic. This study was initiated by Harvey Friedman and was studied in detail by Stephen Simpson and others; Simpson (1999) gives a detailed discussion of the program. The set-existence axioms in question correspond informally to axioms saying that the powerset of the natural numbers is closed under various reducibility notions. The weakest such axiom studied in reverse mathematics is *recursive comprehension*, which states that the powerset of the naturals is closed under Turing reducibility.

39.3.5   Numberings

A numbering is an enumeration of functions; it has two parameters, e and x and outputs the value of the e-th function in the numbering on the input x. Numberings can be partial-recursive although some of its members are total recursive, that is, computable functions. Admissible numberings are those into which all others can be translated. A Friedberg numbering (named after its discoverer) is a one-one numbering of all partial-recursive functions; it is necessarily not an admissible numbering. Later research dealt also with numberings of other classes like classes of recursively enumerable sets. Goncharov discovered for example a class of recursively enumerable sets for which the numberings fall into exactly two classes with respect to recursive isomorphisms.

39.3.6   The priority method

> *For further explanation, see the section* Post's problem and the priority method *in the article* Turing degree.

Post's problem was solved with a method called the *priority method*; a proof using this method is called a *priority argument*. This method is primarily used to construct recursively enumerable sets with particular properties. To use this method, the desired properties of the set to be constructed are broken up into an infinite list of goals, known as *requirements*, so that satisfying all the requirements will cause the set constructed to have the desired properties. Each requirement is assigned to a natural number representing the priority of the requirement; so 0 is assigned to the most important priority, 1 to the second most important, and so on. The set is then constructed in stages, each stage attempting to satisfy one of more of the requirements by either adding numbers to the set or banning numbers from the set so that the final set will satisfy the requirement. It may happen that satisfying one requirement will cause another to become unsatisfied; the priority order is used to decide what to do in such an event.

Priority arguments have been employed to solve many problems in recursion theory, and have been classified into a hierarchy based on their complexity (Soare 1987). Because complex priority arguments can be technical and difficult to follow, it has traditionally been considered desirable to prove results without priority arguments, or to see if results proved with priority arguments can also be proved without them. For example, Kummer published a paper on a proof for the existence of Friedberg numberings without using the priority method.

39.3.7 The lattice of recursively enumerable sets

When Post defined the notion of a simple set as an r.e. set with an infinite complement not containing any infinite r.e. set, he started to study the structure of the recursively enumerable sets under inclusion. This lattice became a well-studied structure. Recursive sets can be defined in this structure by the basic result that a set is recursive if and only if the set and its complement are both recursively enumerable. Infinite r.e. sets have always infinite recursive subsets; but on the other hand, simple sets exist but do not have a coinfinite recursive superset. Post (1944) introduced already hypersimple and hyperhypersimple sets; later maximal sets were constructed which are r.e. sets such that every r.e. superset is either a finite variant of the given maximal set or is co-finite. Post's original motivation in the study of this lattice was to find a structural notion such that every set which satisfies this property is neither in the Turing degree of the recursive sets nor in the Turing degree of the halting problem. Post did not find such a property and the solution to his problem applied priority methods instead; Harrington and Soare (1991) found eventually such a property.

39.3.8 Automorphism problems

Another important question is the existence of automorphisms in recursion-theoretic structures. One of these structures is that one of recursively enumerable sets under inclusion modulo finite difference; in this structure, A is below B if and only if the set difference $B - A$ is finite. Maximal sets (as defined in the previous paragraph) have the property that they cannot be automorphic to non-maximal sets, that is, if there is an automorphism of the recursive enumerable sets under the structure just mentioned, then every maximal set is mapped to another maximal set. Soare (1974) showed that also the converse holds, that is, every two maximal sets are automorphic. So the maximal sets form an orbit, that is, every automorphism preserves maximality and any two maximal sets are transformed into each other by some automorphism. Harrington gave a further example of an automorphic property: that of the creative sets, the sets which are many-one equivalent to the halting problem.

Besides the lattice of recursively enumerable sets, automorphisms are also studied for the structure of the Turing degrees of all sets as well as for the structure of the Turing degrees of r.e. sets. In both cases, Cooper claims to have constructed nontrivial automorphisms which map some degrees to other degrees; this construction has, however, not been verified and some colleagues believe that the construction contains errors and that the question of whether there is a nontrivial automorphism of the Turing degrees is still one of the main unsolved questions in this area (Slaman and Woodin 1986, Ambos-Spies and Fejer 2006).

39.3.9 Kolmogorov complexity

Main article: Kolmogorov complexity

The field of Kolmogorov complexity and algorithmic randomness was developed during the 1960s and 1970s by Chaitin, Kolmogorov, Levin, Martin-Löf and Solomonoff (the names are given here in alphabetical order; much of the research was independent, and the unity of the concept of randomness was not understood at the time). The main idea is to consider a universal Turing machine U and to measure the complexity of a number (or string) x as the length of the shortest input p such that $U(p)$ outputs x. This approach revolutionized earlier ways to determine when an infinite sequence (equivalently, characteristic function of a subset of the natural numbers) is random or not by invoking a notion of randomness for finite objects. Kolmogorov complexity became not only a subject of independent study but is also applied to other subjects as a tool for obtaining proofs. There are still many open problems in this area. For that reason, a recent research conference in this area was held in January 2007[4] and a list of open problems[5] is maintained by Joseph Miller and Andre Nies.

39.3.10 Frequency computation

This branch of recursion theory analyzed the following question: For fixed m and n with $0 < m < n$, for which functions A is it possible to compute for any different n inputs $x_1, x_2, ..., x_n$ a tuple of n numbers $y_1, y_2, ..., y_n$ such that at least m of the equations $A(x_k) = y_k$ are true. Such sets are known as (m, n)-recursive sets. The first major result in this branch of Recursion Theory is Trakhtenbrot's result that a set is computable if it is (m, n)-recursive for some m, n with $2m > n$. On the other hand, Jockusch's semirecursive sets (which were already known informally before Jockusch introduced them 1968) are examples of a set which is (m, n)-recursive if and only if $2m < n + 1$. There are uncountably many of these sets and also some recursively enumerable but noncomputable sets of this type. Later, Degtev established a hierarchy of recursively enumerable sets that are $(1, n + 1)$-recursive but not $(1, n)$-recursive. After a long phase of research by Russian scientists, this subject became repopularized in the west by Beigel's thesis on bounded queries, which linked frequency computation to the above-mentioned bounded reducibilities and other related notions. One of the major results was Kummer's Cardinality Theory which states that a set A is computable if and only if there is an n such that some algorithm enumerates for each tuple of n different numbers up to n many possible choices of the cardinality of this set of n numbers intersected with A; these choices must contain the true cardinality but leave out at least one false one.

39.3.11   Inductive inference

This is the recursion-theoretic branch of learning theory. It is based on Gold's model of learning in the limit from 1967 and has developed since then more and more models of learning. The general scenario is the following: Given a class S of computable functions, is there a learner (that is, recursive functional) which outputs for any input of the form $(f(0),f(1),...,f(n))$ a hypothesis. A learner M learns a function f if almost all hypotheses are the same index e of f with respect to a previously agreed on acceptable numbering of all computable functions; M learns S if M learns every f in S. Basic results are that all recursively enumerable classes of functions are learnable while the class REC of all computable functions is not learnable. Many related models have been considered and also the learning of classes of recursively enumerable sets from positive data is a topic studied from Gold's pioneering paper in 1967 onwards.

39.3.12   Generalizations of Turing computability

Recursion theory includes the study of generalized notions of this field such as arithmetic reducibility, hyperarithmetical reducibility and α-recursion theory, as described by Sacks (1990). These generalized notions include reducibilities that cannot be executed by Turing machines but are nevertheless natural generalizations of Turing reducibility. These studies include approaches to investigate the analytical hierarchy which differs from the arithmetical hierarchy by permitting quantification over sets of natural numbers in addition to quantification over individual numbers. These areas are linked to the theories of well-orderings and trees; for example the set of all indices of recursive (nonbinary) trees without infinite branches is complete for level Π_1^1 of the analytical hierarchy. Both Turing reducibility and hyperarithmetical reducibility are important in the field of effective descriptive set theory. The even more general notion of degrees of constructibility is studied in set theory.

39.3.13   Continuous computability theory

Computability theory for digital computation is well developed. Computability theory is less well developed for analog computation that occurs in analog computers, analog signal processing, analog electronics, neural networks and continuous-time control theory, modelled by differential equations and continuous dynamical systems (Orponen 1997; Moore 1996).

39.4   Relationships between definability, proof and computability

There are close relationships between the Turing degree of a set of natural numbers and the difficulty (in terms of the arithmetical hierarchy) of defining that set using a first-order formula. One such relationship is made precise by Post's theorem. A weaker relationship was demonstrated by Kurt Gödel in the proofs of his completeness theorem and incompleteness theorems. Gödel's proofs show that the set of logical consequences of an effective first-order theory is a recursively enumerable set, and that if the theory is strong enough this set will be uncomputable. Similarly, Tarski's indefinability theorem can be interpreted both in terms of definability and in terms of computability.

Recursion theory is also linked to second order arithmetic, a formal theory of natural numbers and sets of natural numbers. The fact that certain sets are computable or relatively computable often implies that these sets can be defined in weak subsystems of second order arithmetic. The program of reverse mathematics uses these subsystems to measure the non-computability inherent in well known mathematical theorems. Simpson (1999) discusses many aspects of second-order arithmetic and reverse mathematics.

The field of proof theory includes the study of second-order arithmetic and Peano arithmetic, as well as formal theories of the natural numbers weaker than Peano arithmetic. One method of classifying the strength of these weak systems is by characterizing which computable functions the system can prove to be total (see Fairtlough and Wainer (1998)). For example, in primitive recursive arithmetic any computable function that is provably total is actually primitive recursive, while Peano arithmetic proves that functions like the Ackermann function, which are not primitive recursive, are total. Not every total computable function is provably total in Peano arithmetic, however; an example of such a function is provided by Goodstein's theorem.

39.5 Name of the subject

The field of mathematical logic dealing with computability and its generalizations has been called "recursion theory" since its early days. Robert I. Soare, a prominent researcher in the field, has proposed (Soare 1996) that the field should be called "computability theory" instead. He argues that Turing's terminology using the word "computable" is more natural and more widely understood than the terminology using the word "recursive" introduced by Kleene. Many contemporary researchers have begun to use this alternate terminology.[6] These researchers also use terminology such as *partial computable function* and *computably enumerable (c.e.) set* instead of *partial recursive function* and *recursively enumerable (r.e.) set*. Not all researchers have been convinced, however, as explained by Fortnow[7] and Simpson.[8] Some commentators argue that both the names *recursion theory* and *computability theory* fail to convey the fact that most of the objects studied in recursion theory are not computable.[9]

Rogers (1967) has suggested that a key property of recursion theory is that its results and structures should be invariant under computable bijections on the natural numbers (this suggestion draws on the ideas of the Erlangen program in geometry). The idea is that a computable bijection merely renames numbers in a set, rather than indicating any structure in the set, much as a rotation of the Euclidean plane does not change any geometric aspect of lines drawn on it. Since any two infinite computable sets are linked by a computable bijection, this proposal identifies all the infinite computable sets (the finite computable sets are viewed as trivial). According to Rogers, the sets of interest in recursion theory are the noncomputable sets, partitioned into equivalence classes by computable bijections of the natural numbers.

39.6 Professional organizations

The main professional organization for recursion theory is the *Association for Symbolic Logic*, which holds several research conferences each year. The interdisciplinary research Association *Computability in Europe (CiE)* also organizes a series of annual conferences.

39.7 See also

- Recursion (computer science)
- Computability logic
- Transcomputational problem

39.8 Notes

[1] Csima, Barbara F., et al. "Bounding prime models." The Journal of Symbolic Logic 69.04 (2004): 1117-1142.

[2] Many of these foundational papers are collected in *The Undecidable* (1965) edited by Martin Davis.

[3] The full paper can also be found at pages 150ff (with commentary by Charles Parsons at 144ff) in Feferman et al. editors 1990 *Kurt Gödel Volume II Publications 1938-1974*, Oxford University Press, New York, ISBN 978-0-19-514721-6. Both reprintings have the following footnote * added to the Davis volume by Gödel in 1965: "To be more precise: a function of integers is computable in any formal system containing arithmetic if and only if it is computable in arithmetic, where a function f is called computable in S if there is in S a computable term representing f (p. 150).

[4] Conference on Logic, Computability and Randomness, January 10–13, 2007.

[5] The homepage of Andre Nies has a list of open problems in Kolmogorov complexity

[6] Mathscinet searches for the titles like "computably enumerable" and "c.e." show that many papers have been published with this terminology as well as with the other one.

[7] Lance Fortnow, "Is it Recursive, Computable or Decidable?," 2004-2-15, accessed 2006-1-9.

[8] Stephen G. Simpson, "What is computability theory?," FOM email list, 1998-8-24, accessed 2006-1-9.

[9] Harvey Friedman, "Renaming recursion theory," FOM email list, 1998-8-28, accessed 2006-1-9.

39.9   References

Undergraduate level texts   • S. B. Cooper, 2004. *Computability Theory*, Chapman & Hall/CRC. ISBN 1-58488-237-9

  • N. Cutland, 1980. *Computability, An introduction to recursive function theory*, Cambridge University Press. ISBN 0-521-29465-7

  • Y. Matiyasevich, 1993. *Hilbert's Tenth Problem*, MIT Press. ISBN 0-262-13295-8

Advanced texts   • S. Jain, D. Osherson, J. Royer and A. Sharma, 1999. *Systems that learn, an introduction to learning theory, second edition*, Bradford Book. ISBN 0-262-10077-0

  • S. Kleene, 1952. *Introduction to Metamathematics*, North-Holland (11th printing; 6th printing added comments). ISBN 0-7204-2103-9

  • M. Lerman, 1983. *Degrees of unsolvability*, Perspectives in Mathematical Logic. Springer-Verlag. ISBN 3-540-12155-2.

  • Andre Nies, 2009. *Computability and Randomness*, Oxford University Press, 447 pages. ISBN 978-0-19-923076-1.

  • P. Odifreddi, 1989. *Classical Recursion Theory*, North-Holland. ISBN 0-444-87295-7

  • P. Odifreddi, 1999. *Classical Recursion Theory, Volume II*, Elsevier. ISBN 0-444-50205-X

  • H. Rogers, Jr., 1967. *The Theory of Recursive Functions and Effective Computability*, second edition 1987. MIT Press. ISBN 0-262-68052-1 (paperback), ISBN 0-07-053522-1

  • G Sacks, 1990. *Higher Recursion Theory*, Springer-Verlag. ISBN 3-540-19305-7

  • S. G. Simpson, 1999. *Subsystems of Second Order Arithmetic*, Springer-Verlag. ISBN 3-540-64882-8

  • R. I. Soare, 1987. *Recursively Enumerable Sets and Degrees*, Perspectives in Mathematical Logic. Springer-Verlag. ISBN 0-387-15299-7.

Survey papers and collections   • K. Ambos-Spies and P. Fejer, 2006. "Degrees of Unsolvability." Unpublished preprint.

  • H. Enderton, 1977. "Elements of Recursion Theory." *Handbook of Mathematical Logic*, edited by J. Barwise. North-Holland (1977), pp. 527–566. ISBN 0-7204-2285-X

  • Y. L. Ershov, S. S. Goncharov, A. Nerode, and J. B. Remmel, 1998. *Handbook of Recursive Mathematics*, North-Holland (1998). ISBN 0-7204-2285-X

- M. Fairtlough and S. Wainer, 1998. "Hierarchies of Provably Recursive Functions". In *Handbook of Proof Theory*, edited by S. Buss, Elsevier (1998).
- R. I. Soare, 1996. *Computability and recursion, Bulletin of Symbolic Logic* v. 2 pp. 284–321.

Research papers and collections
- Burgin, M. and Klinger, A. "Experience, Generations, and Limits in Machine Learning." *Theoretical Computer Science* v. 317, No. 1/3, 2004, pp. 71–91
- A. Church, 1936a. "An unsolvable problem of elementary number theory." *American Journal of Mathematics* v. 58, pp. 345–363. Reprinted in "The Undecidable", M. Davis ed., 1965.
- A. Church, 1936b. "A note on the Entscheidungsproblem." *Journal of Symbolic Logic* v. 1, n. 1, and v. 3, n. 3. Reprinted in "The Undecidable", M. Davis ed., 1965.
- M. Davis, ed., 1965. *The Undecidable—Basic Papers on Undecidable Propositions, Unsolvable Problems and Computable Functions*, Raven, New York. Reprint, Dover, 2004. ISBN 0-486-43228-9
- R. M. Friedberg, 1958. "Three theorems on recursive enumeration: I. Decomposition, II. Maximal Set, III. Enumeration without repetition." *The Journal of Symbolic Logic*, v. 23, pp. 309–316.
- Gold, E. Mark (1967). *Language Identification in the Limit* (PDF) **10**, Information and Control, pp. 447–474
- L. Harrington and R. I. Soare, 1991. "Post's Program and incomplete recursively enumerable sets", *Proceedings of the National Academy of Sciences of the USA*, volume 88, pages 10242—10246.
- C. Jockusch jr. "Semirecursive sets and positive reducibility", *Trans. Amer. Math. Soc.* **137** (1968) 420-436
- S. C. Kleene and E. L. Post, 1954. "The upper semi-lattice of degrees of recursive unsolvability." *Annals of Mathematics* v. 2 n. 59, 379–407.
- Moore, C. (1996). "Recursion theory on the reals and continuous-time computation". *Theoretical Computer Science*. CiteSeerX: 10.1.1.6.5519.
- J. Myhill, 1956. "The lattice of recursively enumerable sets." *The Journal of Symbolic Logic*, v. 21, pp. 215–220.
- Orponen, P. (1997). "A survey of continuous-time computation theory". *Advances in algorithms, languages, and complexity*. CiteSeerX: 10.1.1.53.1991.
- E. Post, 1944, "Recursively enumerable sets of positive integers and their decision problems", *Bulletin of the American Mathematical Society*, volume 50, pages 284–316.
- E. Post, 1947. "Recursive unsolvability of a problem of Thue." *Journal of Symbolic Logic* v. 12, pp. 1–11. Reprinted in "The Undecidable", M. Davis ed., 1965.
- Shore, Richard A.; Slaman, Theodore A. (1999), "Defining the Turing jump" (PDF), *Mathematical Research Letters* **6**: 711–722, doi:10.4310/mrl.1999.v6.n6.a10, ISSN 1073-2780, MR 1739227
- T. Slaman and W. H. Woodin, 1986. "Definability in the Turing degrees." *Illinois J. Math.* v. 30 n. 2, pp. 320–334.
- R. I. Soare, 1974. "Automorphisms of the lattice of recursively enumerable sets, Part I: Maximal sets." *Annals of Mathematics*, v. 100, pp. 80–120.
- A. Turing, 1937. "On computable numbers, with an application to the Entscheidungsproblem." *Proceedings of the London Mathematics Society*, ser. 2 v. 42, pp. 230–265. Corrections *ibid.* v. 43 (1937) pp. 544–546. Reprinted in "The Undecidable", M. Davis ed., 1965. PDF from comlab.ox.ac.uk
- A. Turing, 1939. "Systems of logic based on ordinals." *Proceedings of the London Mathematics Society*, ser. 2 v. 45, pp. 161–228. Reprinted in "The Undecidable", M. Davis ed., 1965.

39.10   External links

- Association for Symbolic Logic homepage
- Computability in Europe homepage
- Webpage on Recursion Theory Course at Graduate Level with approximately 100 pages of lecture notes
- German language lecture notes on inductive inference

Chapter 40

Combinatorics

Not to be confused with combinatoriality.

Combinatorics is a branch of mathematics concerning the study of finite or countable discrete structures. Aspects of combinatorics include counting the structures of a given kind and size (enumerative combinatorics), deciding when certain criteria can be met, and constructing and analyzing objects meeting the criteria (as in combinatorial designs and matroid theory), finding "largest", "smallest", or "optimal" objects (extremal combinatorics and combinatorial optimization), and studying combinatorial structures arising in an algebraic context, or applying algebraic techniques to combinatorial problems (algebraic combinatorics).

Combinatorial problems arise in many areas of pure mathematics, notably in algebra, probability theory, topology, and geometry,[1] and combinatorics also has many applications in mathematical optimization, computer science, ergodic theory and statistical physics. Many combinatorial questions have historically been considered in isolation, giving an *ad hoc* solution to a problem arising in some mathematical context. In the later twentieth century, however, powerful and general theoretical methods were developed, making combinatorics into an independent branch of mathematics in its own right. One of the oldest and most accessible parts of combinatorics is graph theory, which also has numerous natural connections to other areas. Combinatorics is used frequently in computer science to obtain formulas and estimates in the analysis of algorithms.

A mathematician who studies combinatorics is called a **combinatorialist** or a **combinatorist**.

40.1 History

Main article: History of combinatorics

Basic combinatorial concepts and enumerative results appeared throughout the ancient world. In 6th century BCE, ancient Indian physician Sushruta asserts in Sushruta Samhita that 63 combinations can be made out of 6 different tastes, taken one at a time, two at a time, etc., thus computing all $2^6 - 1$ possibilities. Greek historian Plutarch discusses an argument between Chrysippus (3rd century BCE) and Hipparchus (2nd century BCE) of a rather delicate enumerative problem, which was later shown to be related to Schröder numbers.[2][3] In the *Ostomachion*, Archimedes (3rd century BCE) considers a tiling puzzle.

In the Middle Ages, combinatorics continued to be studied, largely outside of the European civilization. The Indian mathematician Mahāvīra (c. 850) provided formulae for the number of permutations and combinations,[4][5] and these formulas may have been familiar to Indian mathematicians as early as the 6th century CE.[6] The philosopher and astronomer Rabbi Abraham ibn Ezra (c. 1140) established the symmetry of binomial coefficients, while a closed formula was obtained later by the talmudist and mathematician Levi ben Gerson (better known as Gersonides), in 1321.[7] The arithmetical triangle— a graphical diagram showing relationships among the binomial coefficients— was presented by mathematicians in treatises dating as far back as the 10th century, and would eventually become known as Pascal's triangle. Later, in Medieval England, campanology provided examples of what is now known as Hamiltonian cycles in certain Cayley graphs on permutations.[8]

During the Renaissance, together with the rest of mathematics and the sciences, combinatorics enjoyed a rebirth. Works of Pascal, Newton, Jacob Bernoulli and Euler became foundational in the emerging field. In modern times, the works of J. J. Sylvester (late 19th century) and Percy MacMahon (early 20th century) laid the foundation for enumerative and algebraic combinatorics. Graph theory also enjoyed an explosion of interest at the same time, especially in connection with the four color problem.

In the second half of 20th century, combinatorics enjoyed a rapid growth, which led to establishment of dozens of new journals and conferences in the subject.[9] In part, the growth was spurred by new connections and applications to other fields, ranging from algebra to probability, from functional analysis to number theory, etc. These connections shed the boundaries between combinatorics and parts of mathematics and theoretical computer science, but at the same time led to a partial fragmentation of the field.

40.2 Approaches and subfields of combinatorics

40.2.1 Enumerative combinatorics

Main article: Enumerative combinatorics

Enumerative combinatorics is the most classical area of combinatorics, and concentrates on counting the number of certain combinatorial objects. Although counting the number of elements in a set is a rather broad mathematical problem, many of the problems that arise in applications have a relatively simple combinatorial description. Fibonacci numbers is the basic example of a problem in enumerative combinatorics. The twelvefold way provides a unified framework for counting permutations, combinations and partitions.

40.2.2 Analytic combinatorics

Main article: Analytic combinatorics

Analytic combinatorics concerns the enumeration of combinatorial structures using tools from complex analysis and probability theory. In contrast with enumerative combinatorics, which uses explicit combinatorial formulae and generating functions to describe the results, analytic combinatorics aims at obtaining asymptotic formulae.

40.2.3 Partition theory

Main article: Partition theory

Partition theory studies various enumeration and asymptotic problems related to integer partitions, and is closely related to q-series, special functions and orthogonal polynomials. Originally a part of number theory and analysis, it is now considered a part of combinatorics or an independent field. It incorporates the bijective approach and various tools in analysis, analytic number theory, and has connections with statistical mechanics.

40.2.4 Graph theory

Main article: Graph theory

Graphs are basic objects in combinatorics. The questions range from counting (e.g., the number of graphs on n vertices with k edges) to structural (e.g., which graphs contain Hamiltonian cycles) to algebraic questions (e.g., given a graph G and two numbers x and y, does the Tutte polynomial $TG(x,y)$ have a combinatorial interpretation?). It should be noted that while there are very strong connections between graph theory and combinatorics, these two are sometimes thought of as separate subjects.[10]

40.2.5 Design theory

Main article: Combinatorial design

Design theory is a study of combinatorial designs, which are collections of subsets with certain intersection properties. Block designs are combinatorial designs of a special type. This area is one of the oldest parts of combinatorics, such as in Kirkman's schoolgirl problem proposed in 1850. The solution of the problem is a special case of a Steiner system, which systems play an important role in the classification of finite simple groups. The area has further connections to coding theory and geometric combinatorics.

40.2.6 Finite geometry

Main article: Finite geometry

Finite geometry is the study of geometric systems having only a finite number of points. Structures analogous to those found in continuous geometries (Euclidean plane, real projective space, etc.) but defined combinatorially are the main items studied. This area provides a rich source of examples for design theory. It should not be confused with discrete geometry (combinatorial geometry).

40.2.7 Order theory

Main article: Order theory

Order theory is the study of partially ordered sets, both finite and infinite. Various examples of partial orders appear in algebra, geometry, number theory and throughout combinatorics and graph theory. Notable classes and examples of partial orders include lattices and Boolean algebras.

40.2.8 Matroid theory

Main article: Matroid theory

Matroid theory abstracts part of geometry. It studies the properties of sets (usually, finite sets) of vectors in a vector space that do not depend on the particular coefficients in a linear dependence relation. Not only the structure but also enumerative properties belong to matroid theory. Matroid theory was introduced by Hassler Whitney and studied as a part of the order theory. It is now an independent field of study with a number of connections with other parts of combinatorics.

40.2.9 Extremal combinatorics

Main article: Extremal combinatorics

Extremal combinatorics studies extremal questions on set systems. The types of questions addressed in this case are about the largest possible graph which satisfies certain properties. For example, the largest triangle-free graph on $2n$ vertices is a complete bipartite graph $Kn.n$. Often it is too hard even to find the extremal answer $f(n)$ exactly and one can only give an asymptotic estimate.

Ramsey theory is another part of extremal combinatorics. It states that any sufficiently large configuration will contain some sort of order. It is an advanced generalization of the pigeonhole principle.

40.2.10 Probabilistic combinatorics

Main article: Probabilistic method

In probabilistic combinatorics, the questions are of the following type: what is the probability of a certain property for a random discrete object, such as a random graph? For instance, what is the average number of triangles in a random graph? Probabilistic methods are also used to determine the existence of combinatorial objects with certain prescribed properties (for which explicit examples might be difficult to find), simply by observing that the probability of randomly selecting an object with those properties is greater than 0. This approach (often referred to as *the* probabilistic method) proved highly effective in applications to extremal combinatorics and graph theory. A closely related area is the study of finite Markov chains, especially on combinatorial objects. Here again probabilistic tools are used to estimate the mixing time.

Often associated with Paul Erdős, who did the pioneer work on the subject, probabilistic combinatorics was traditionally viewed as a set of tools to study problems in other parts of combinatorics. However, with the growth of applications to analysis of algorithms in computer science, as well as classical probability, additive and probabilistic number theory, the area recently grew to become an independent field of combinatorics.

40.2.11 Algebraic combinatorics

Main article: Algebraic combinatorics

Algebraic combinatorics is an area of mathematics that employs methods of abstract algebra, notably group theory and representation theory, in various combinatorial contexts and, conversely, applies combinatorial techniques to problems in algebra. Algebraic combinatorics is continuously expanding its scope, in both topics and techniques, and can be seen as the area of mathematics where the interaction of combinatorial and algebraic methods is particularly strong and significant.

40.2.12 Combinatorics on words

Main article: Combinatorics on words

Combinatorics on words deals with formal languages. It arose independently within several branches of mathematics, including number theory, group theory and probability. It has applications to enumerative combinatorics, fractal analysis, theoretical computer science, automata theory and linguistics. While many applications are new, the classical Chomsky–Schützenberger hierarchy of classes of formal grammars is perhaps the best known result in the field.

40.2.13 Geometric combinatorics

Main article: Geometric combinatorics

Geometric combinatorics is related to convex and discrete geometry, in particular polyhedral combinatorics. It asks, for example, how many faces of each dimension can a convex polytope have. Metric properties of polytopes play an important role as well, e.g. the Cauchy theorem on rigidity of convex polytopes. Special polytopes are also considered, such as permutohedra, associahedra and Birkhoff polytopes. We should note that combinatorial geometry is an old fashioned name for discrete geometry.

40.2.14   Topological combinatorics

Main article: Topological combinatorics

Combinatorial analogs of concepts and methods in topology are used to study graph coloring, fair division, partitions, partially ordered sets, decision trees, necklace problems and discrete Morse theory. It should not be confused with combinatorial topology which is an older name for algebraic topology.

40.2.15   Arithmetic combinatorics

Main article: Arithmetic combinatorics

Arithmetic combinatorics arose out of the interplay between number theory, combinatorics, ergodic theory and harmonic analysis. It is about combinatorial estimates associated with arithmetic operations (addition, subtraction, multiplication, and division). *Additive combinatorics* refers to the special case when only the operations of addition and subtraction are involved. One important technique in arithmetic combinatorics is the ergodic theory of dynamical systems.

40.2.16   Infinitary combinatorics

Main article: Infinitary combinatorics

Infinitary combinatorics, or combinatorial set theory, is an extension of ideas in combinatorics to infinite sets. It is a part of set theory, an area of mathematical logic, but uses tools and ideas from both set theory and extremal combinatorics.

Gian-Carlo Rota used the name *continuous combinatorics*[11] to describe probability and measure theory, since there are many analogies between *counting* and *measure*.

40.3   Related fields

40.3.1   Combinatorial optimization

Combinatorial optimization is the study of optimization on discrete and combinatorial objects. It started as a part of combinatorics and graph theory, but is now viewed as a branch of applied mathematics and computer science, related to operations research, algorithm theory and computational complexity theory.

40.3.2   Coding theory

Coding theory started as a part of design theory with early combinatorial constructions of error-correcting codes. The main idea of the subject is to design efficient and reliable methods of data transmission. It is now a large field of study, part of information theory.

40.3.3   Discrete and computational geometry

Discrete geometry (also called combinatorial geometry) also began a part of combinatorics, with early results on convex polytopes and kissing numbers. With the emergence of applications of discrete geometry to computational geometry, these two fields partially merged and became a separate field of study. There remain many connections with geometric and topological combinatorics, which themselves can be viewed as outgrowths of the early discrete geometry.

40.3.4   Combinatorics and dynamical systems

Combinatorial aspects of dynamical systems is another emerging field. Here dynamical systems can be defined on combinatorial objects. See for example graph dynamical system.

40.3.5   Combinatorics and physics

There are increasing interactions between combinatorics and physics, particularly statistical physics. Examples include an exact solution of the Ising model, and a connection between the Potts model on one hand, and the chromatic and Tutte polynomials on the other hand.

40.4   See also

- Combinatorial biology
- Combinatorial chemistry
- Combinatorial data analysis
- Combinatorial game theory
- Combinatorial group theory
- List of combinatorics topics
- Phylogenetics

40.5   Notes

[1] Björner and Stanley, p. 2

[2] Stanley, Richard P.: "Hipparchus, Plutarch, Schröder, and Hough". *American Mathematical Monthly* **104** (1997), no. 4, 344–350.

[3] Habsieger, Laurent; Kazarian, Maxim; and Lando, Sergei: "On the Second Number of Plutarch", *American Mathematical Monthly* **105** (1998), no. 5, 446.

[4] O'Connor, John J.; Robertson, Edmund F., "Combinatorics", *MacTutor History of Mathematics archive*, University of St Andrews.

[5] Puttaswamy, Tumkur K. (2000). "The Mathematical Accomplishments of Ancient Indian Mathematicians", in Selin, Helaine, *Mathematics Across Cultures: The History of Non-Western Mathematics*, Netherlands: Kluwer Academic Publishers, p. 417. ISBN 978-1-4020-0260-1

[6] Biggs, Norman L. (1979). "The Roots of Combinatorics". *Historia Mathematica* **6**: 109–136.

[7] Maistrov, L. E. (1974), *Probability Theory: A Historical Sketch*, Academic Press, p. 35, ISBN 9781483218632. (Translation from 1967 Russian ed.)

[8] White, Arthur T.: "Ringing the Cosets", *American Mathematical Monthly*, **94** (1987), no. 8, 721–746; White, Arthur T.: "Fabian Stedman: The First Group Theorist?", *American Mathematical Monthly*, **103** (1996), no. 9, 771–778.

[9] See Journals in Combinatorics and Graph Theory

[10] Sanders, Daniel P.; *2-Digit MSC Comparison*

[11] *Continuous and profinite combinatorics*

40.6 References

- Björner, Anders; and Stanley, Richard P.; (2010); *A Combinatorial Miscellany*

- Bóna, Miklós; (2011); *A Walk Through Combinatorics (3rd Edition).* ISBN 978-981-4335-23-2, ISBN 978-981-4460-00-2(pbk)

- Graham, Ronald L.; Groetschel, Martin; and Lovász, László; eds. (1996); *Handbook of Combinatorics*, Volumes 1 and 2. Amsterdam, NL, and Cambridge, MA: Elsevier (North-Holland) and MIT Press. ISBN 0-262-07169-X

- Lindner, Charles C.; and Rodger, Christopher A.; eds. (1997); *Design Theory*, CRC-Press; 1st. edition (October 31, 1997). ISBN 0-8493-3986-3.

- Riordan, John (1958); *An Introduction to Combinatorial Analysis*, New York, NY: Wiley & Sons (republished)

- Stanley, Richard P. (1997, 1999); *Enumerative Combinatorics*, Volumes 1 and 2, Cambridge University Press. ISBN 0-521-55309-1, ISBN 0-521-56069-1

- van Lint, Jacobus H.; and Wilson, Richard M.; (2001); *A Course in Combinatorics*, 2nd Edition, Cambridge University Press. ISBN 0-521-80340-3

40.7 External links

- Hazewinkel, Michiel, ed. (2001), "Combinatorial analysis", *Encyclopedia of Mathematics*, Springer, ISBN 978-1-55608-010-4

- Combinatorial Analysis – an article in Encyclopædia Britannica Eleventh Edition

- Combinatorics, a MathWorld article with many references.

- Combinatorics, from a *MathPages.com* portal.

- The Hyperbook of Combinatorics, a collection of math articles links.

- The Two Cultures of Mathematics by W. T. Gowers, article on problem solving vs theory building.

Plain Bob Minor

An example of change ringing (with six bells), one of the earliest nontrivial results in Graph Theory.

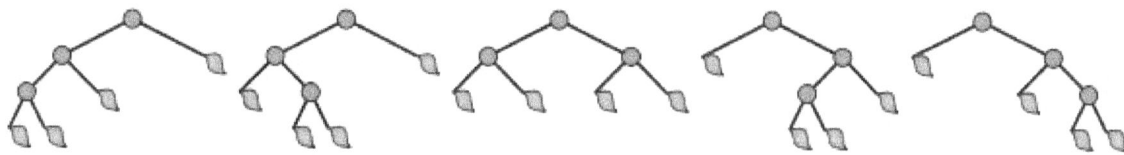

Five binary trees on three vertices, an example of Catalan numbers.

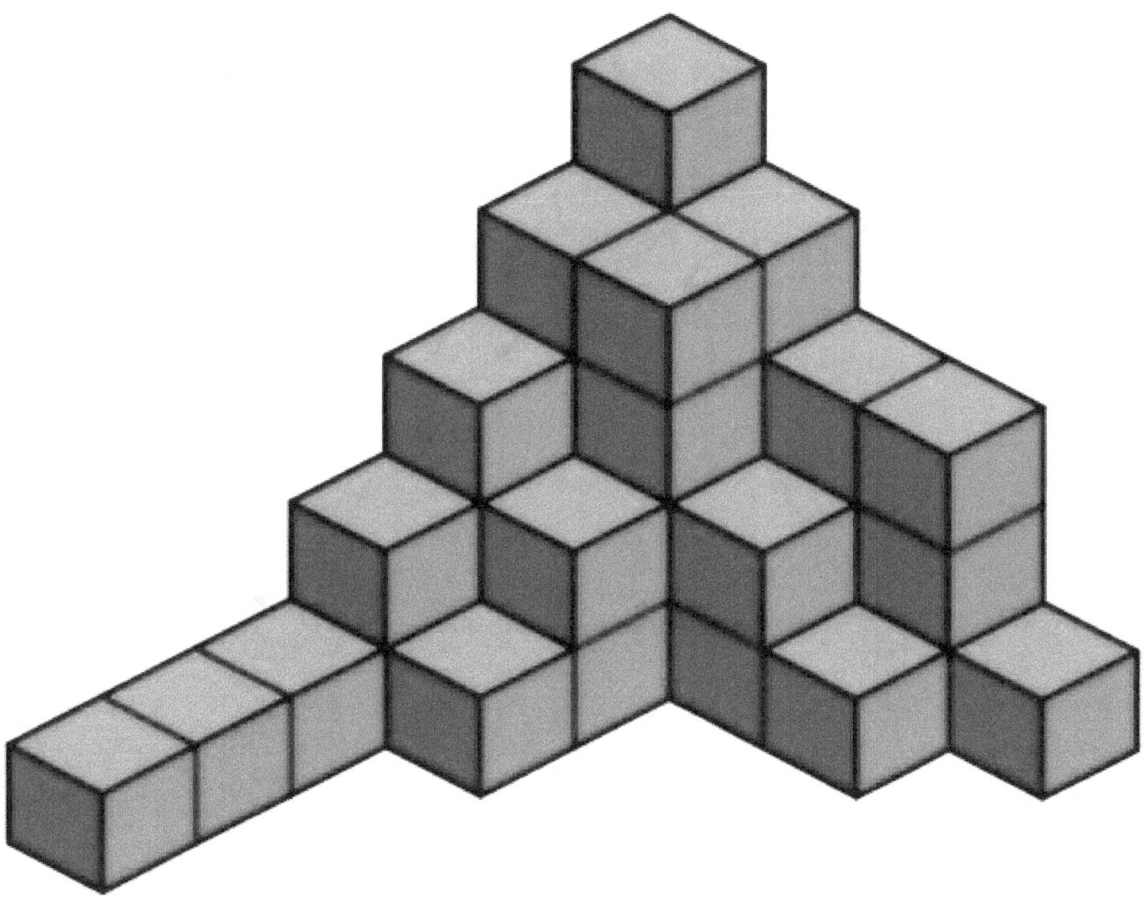

A plane partition.

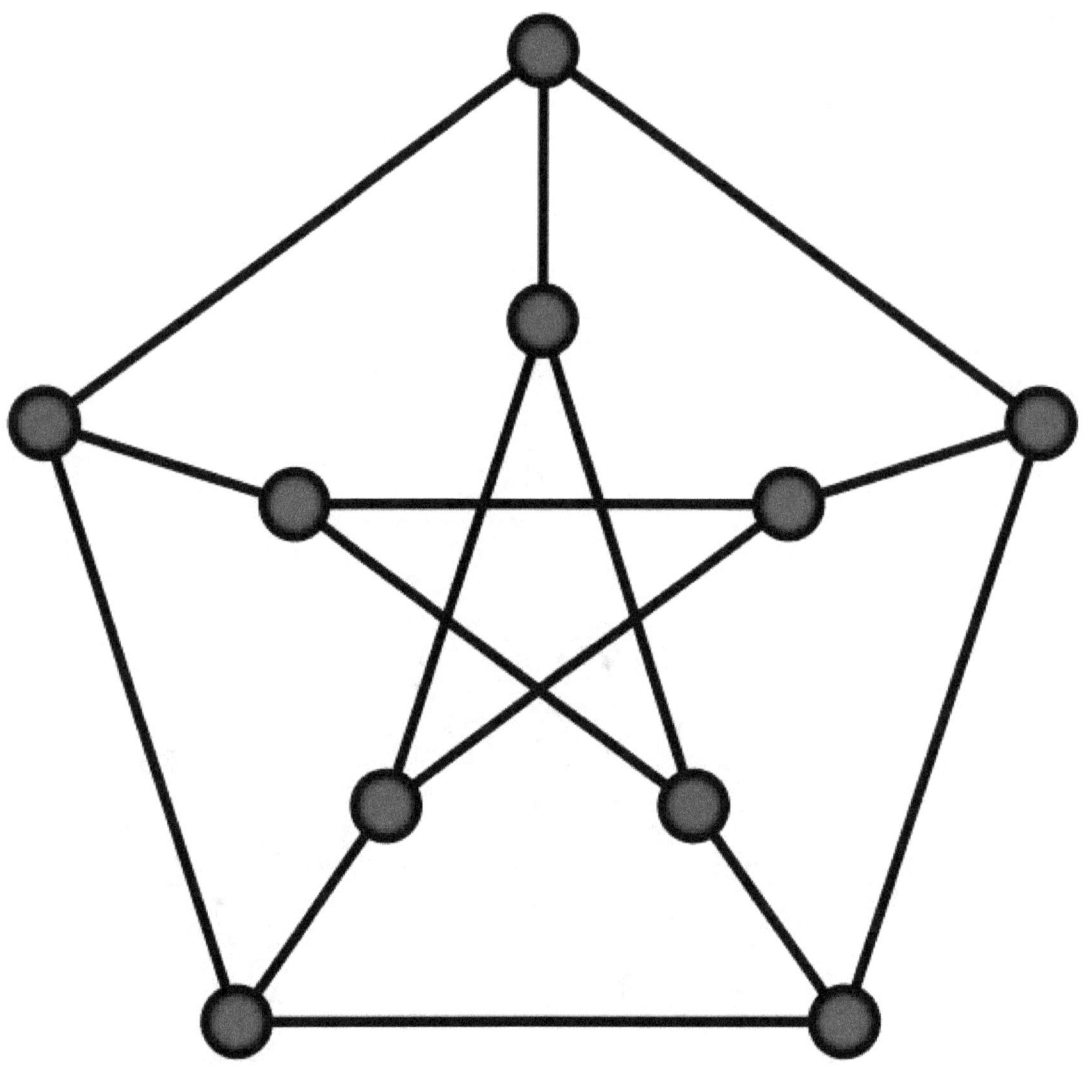

Petersen graph.

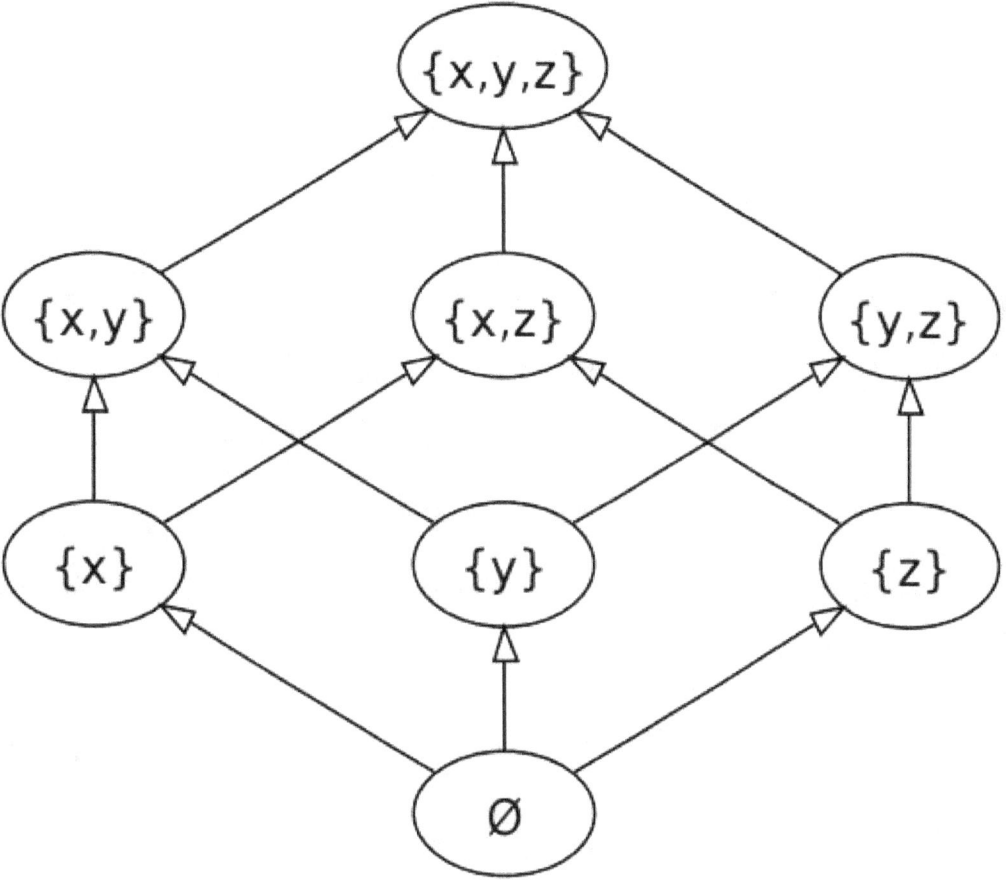

Hasse diagram of the powerset of {x,y,z} ordered by inclusion.

Self-avoiding walk in a square grid graph.

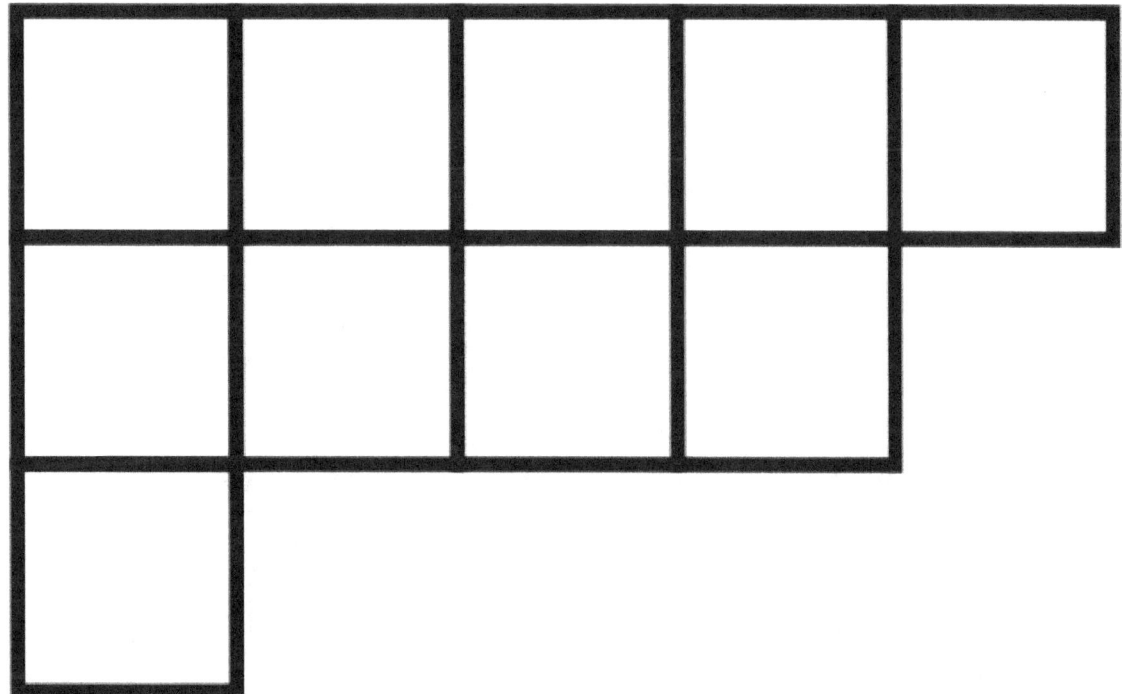

Young diagram of a partition (5,4,1).

0

Construction of a Thue–Morse infinite word.

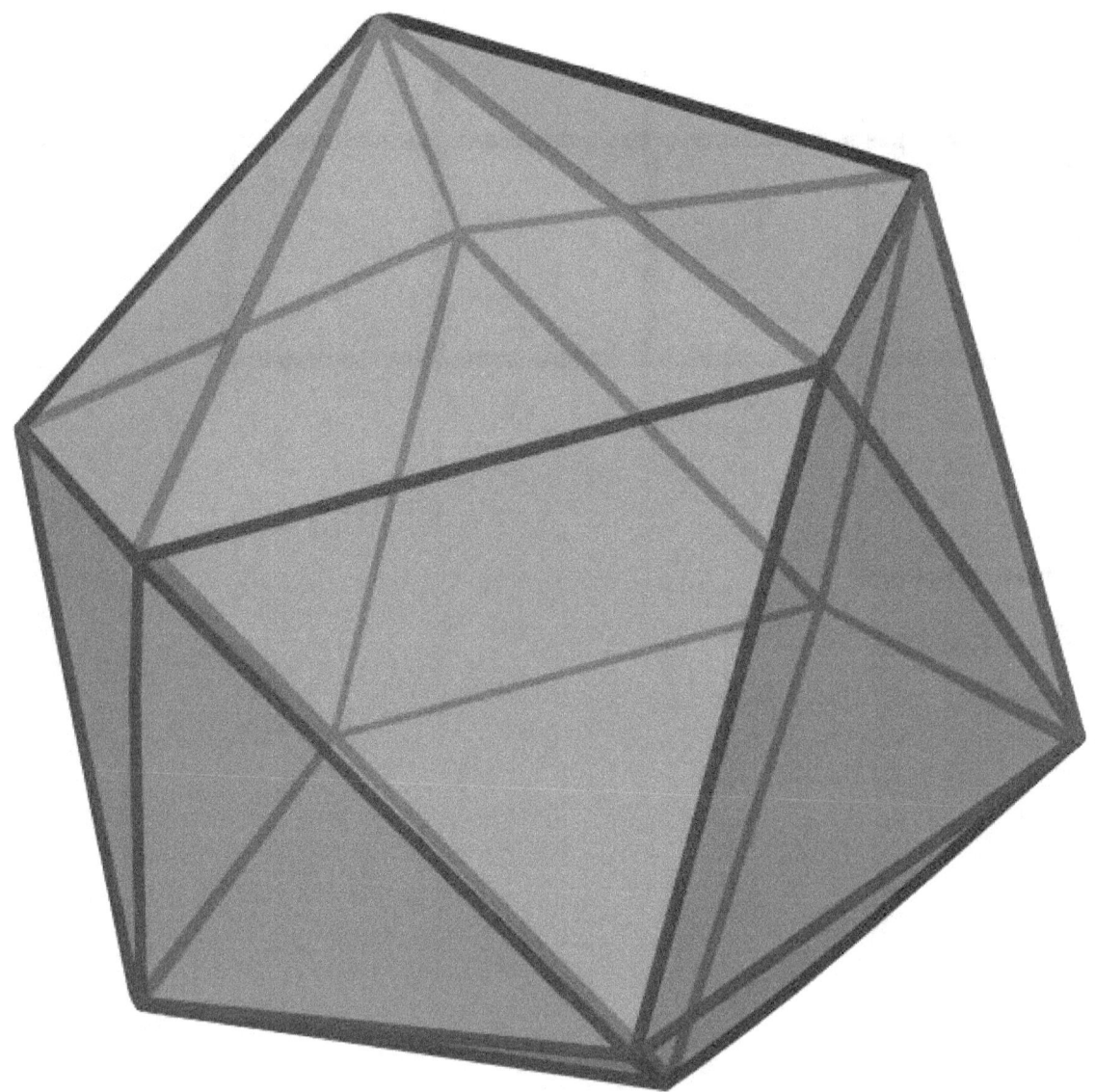

An icosahedron.

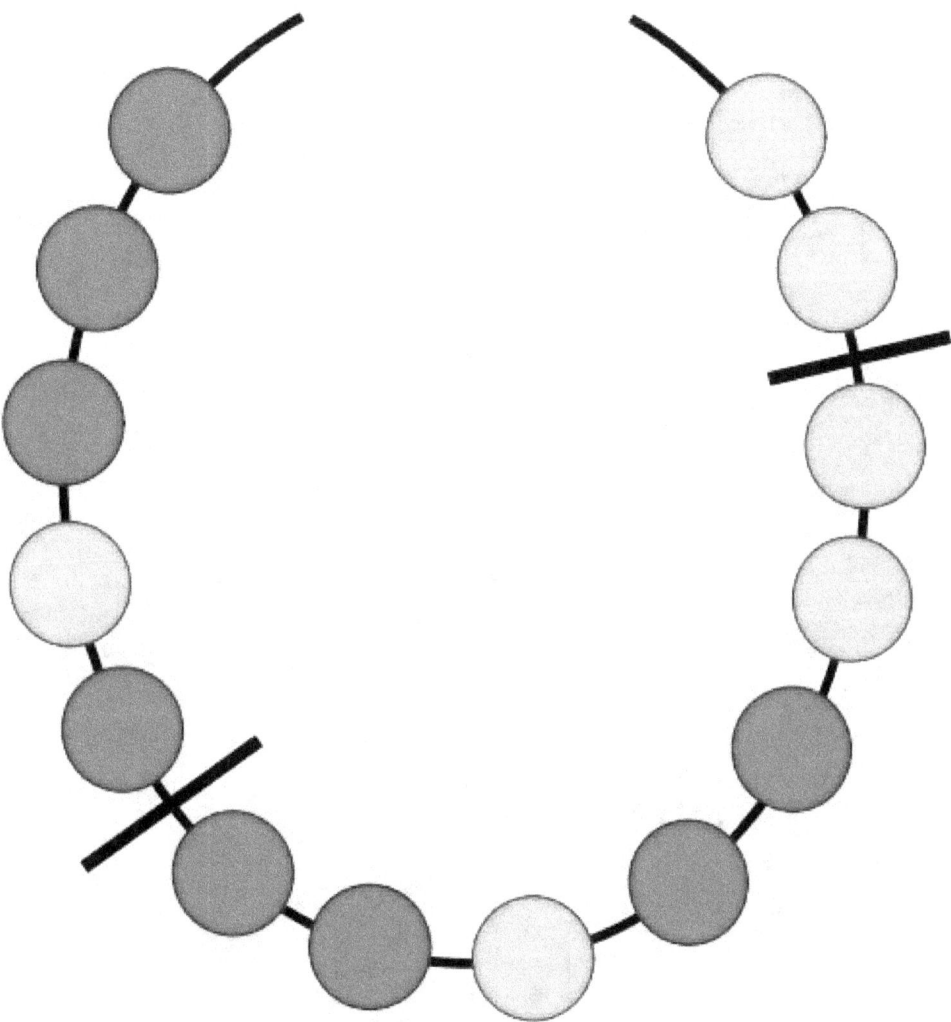

Splitting a necklace with two cuts.

Kissing spheres are connected to both coding theory and discrete geometry.

Chapter 41

Theory of computation

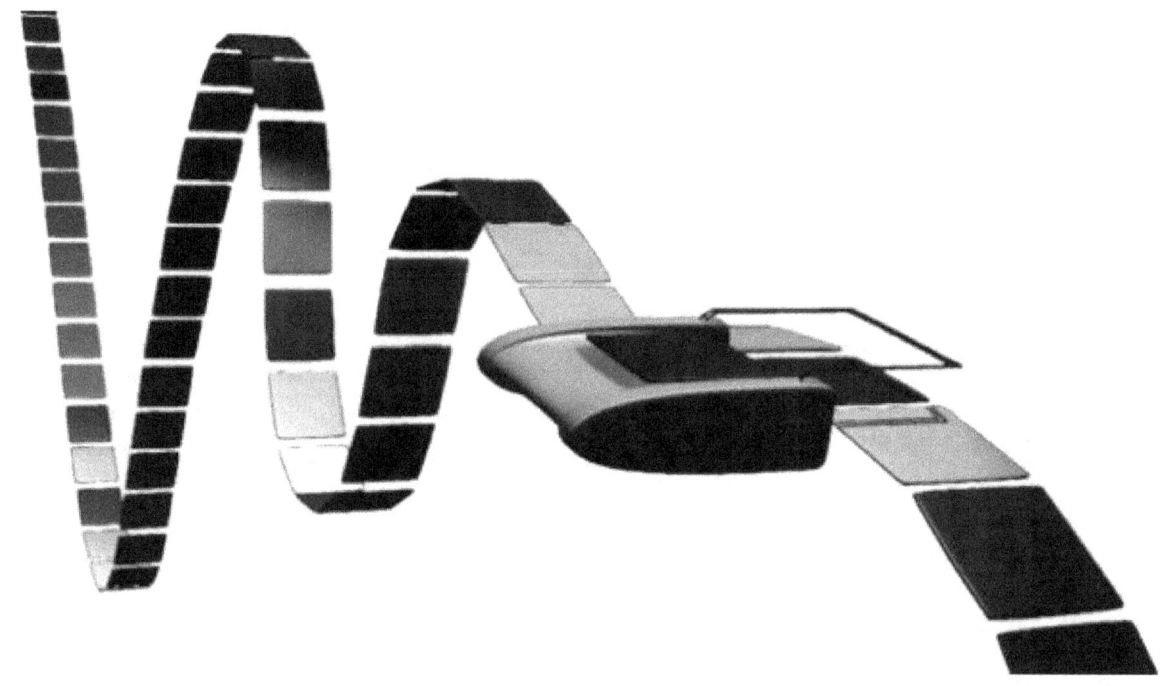

An artistic representation of a Turing machine. Turing machines are frequently used as theoretical models for computing.

In theoretical computer science and mathematics, the **theory of computation** is the branch that deals with how efficiently problems can be solved on a model of computation, using an algorithm. The field is divided into three major branches: automata theory and language, computability theory, and computational complexity theory, which are linked by the question: *"What are the fundamental capabilities and limitations of computers?."*[1]

In order to perform a rigorous study of computation, computer scientists work with a mathematical abstraction of computers called a model of computation. There are several models in use, but the most commonly examined is the Turing machine.[2] Computer scientists study the Turing machine because it is simple to formulate, can be analyzed and used to prove results, and because it represents what many consider the most powerful possible "reasonable" model of computation (see Church–Turing thesis).[3] It might seem that the potentially infinite memory capacity is an unrealizable attribute, but any decidable problem[4] solved by a Turing machine will always require only a finite amount of memory. So in principle, any problem that can be solved (decided) by a Turing machine can be solved by a computer that has a bounded amount of memory.

41.1 History

The theory of computation can be considered the creation of models of all kinds in the field of computer science. Therefore, mathematics and logic are used. In the last century it became an independent academic discipline and was separated from mathematics.

Some pioneers of the theory of computation were Alonzo Church, Kurt Gödel, Alan Turing, Stephen Kleene, John von Neumann and Claude Shannon.

41.2 Branches

41.2.1 Automata theory

Main article: Automata theory

Automata theory is the study of abstract machines (or more appropriately, abstract 'mathematical' machines or systems) and the computational problems that can be solved using these machines. These abstract machines are called automata. Automata comes from the Greek word (Αυτόματα) which means that something is doing something by itself. Automata theory is also closely related to formal language theory,[5] as the automata are often classified by the class of formal languages they are able to recognize. An automaton can be a finite representation of a formal language that may be an infinite set. Automata are used as theoretical models for computing machines, and are used for proofs about computability.

41.2.2 Formal Language Theory

Main article: Formal language
Language theory is a branch of mathematics concerned with describing languages as a set of operations over an alphabet. It is closely linked with automata theory, as automata are used to generate and recognize formal languages. There are several classes of formal languages, each allowing more complex language specification than the one before it, i.e. Chomsky hierarchy,[6] and each corresponding to a class of automata which recognizes it. Because automata are used as models for computation, formal languages are the preferred mode of specification for any problem that must be computed.

41.2.3 Computability theory

Main article: Computability theory

Computability theory deals primarily with the question of the extent to which a problem is solvable on a computer. The statement that the halting problem cannot be solved by a Turing machine[7] is one of the most important results in computability theory, as it is an example of a concrete problem that is both easy to formulate and impossible to solve using a Turing machine. Much of computability theory builds on the halting problem result.

Another important step in computability theory was Rice's theorem, which states that for all non-trivial properties of partial functions, it is undecidable whether a Turing machine computes a partial function with that property.[8]

Computability theory is closely related to the branch of mathematical logic called recursion theory, which removes the restriction of studying only models of computation which are reducible to the Turing model.[9] Many mathematicians and computational theorists who study recursion theory will refer to it as computability theory.

41.2.4 Computational complexity theory

Main article: Computational complexity theory
Complexity theory considers not only whether a problem can be solved at all on a computer, but also how efficiently the

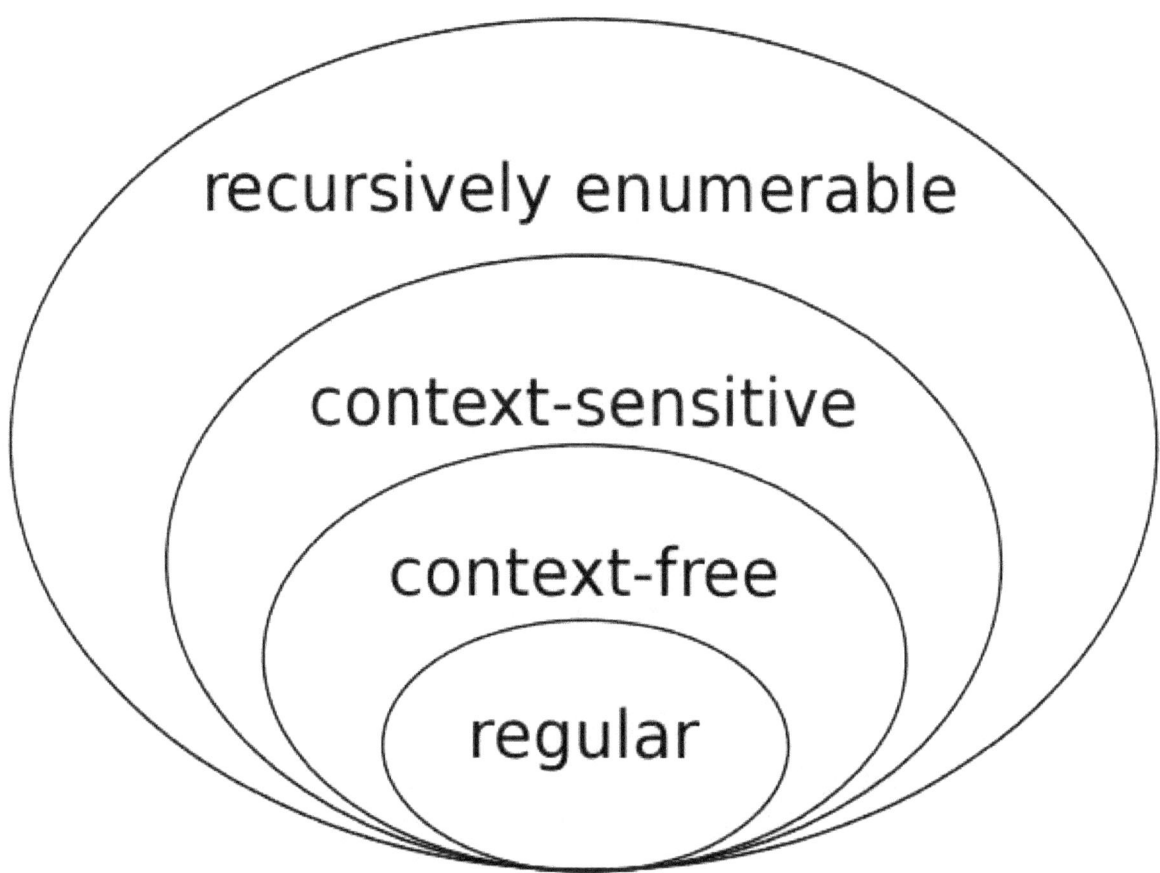

Set inclusions described by the Chomsky hierarchy

problem can be solved. Two major aspects are considered: time complexity and space complexity, which are respectively how many steps does it take to perform a computation, and how much memory is required to perform that computation.

In order to analyze how much time and space a given algorithm requires, computer scientists express the time or space required to solve the problem as a function of the size of the input problem. For example, finding a particular number in a long list of numbers becomes harder as the list of numbers grows larger. If we say there are n numbers in the list, then if the list is not sorted or indexed in any way we may have to look at every number in order to find the number we're seeking. We thus say that in order to solve this problem, the computer needs to perform a number of steps that grows linearly in the size of the problem.

To simplify this problem, computer scientists have adopted Big O notation, which allows functions to be compared in a way that ensures that particular aspects of a machine's construction do not need to be considered, but rather only the asymptotic behavior as problems become large. So in our previous example we might say that the problem requires $O(n)$ steps to solve.

Perhaps the most important open problem in all of computer science is the question of whether a certain broad class of problems denoted NP can be solved efficiently. This is discussed further at Complexity classes P and NP, and P versus NP problem is one of the seven Millennium Prize Problems stated by the Clay Mathematics Institute in 2000. The Official Problem Description was given by Turing Award winner Stephen Cook.

41.3 Models of computation

Main article: Model of computation

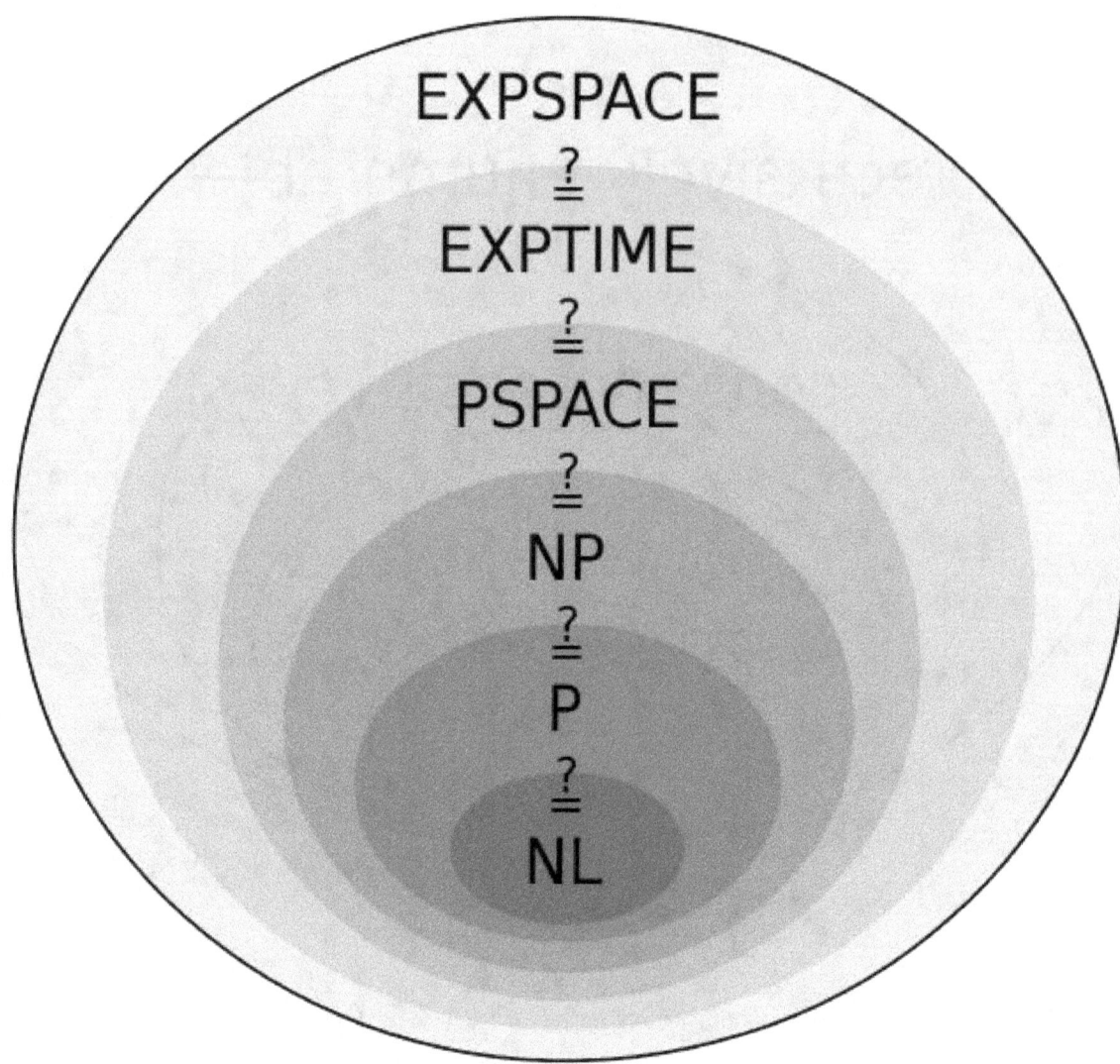

A representation of the relation among complexity classes

Aside from a Turing machine, other equivalent (See: Church–Turing thesis) models of computation are in use.

Lambda calculus A computation consists of an initial lambda expression (or two if you want to separate the function and its input) plus a finite sequence of lambda terms, each deduced from the preceding term by one application of Beta reduction.

Combinatory logic is a concept which has many similarities to λ -calculus, but also important differences exist (e.g. fixed point combinator **Y** has normal form in combinatory logic but not in λ -calculus). Combinatory logic was developed with great ambitions: understanding the nature of paradoxes, making foundations of mathematics more economic (conceptually), eliminating the notion of variables (thus clarifying their role in mathematics).

μ-recursive functions a computation consists of a mu-recursive function, *i.e.* its defining sequence, any input value(s) and a sequence of recursive functions appearing in the defining sequence with inputs and outputs. Thus, if in the defining sequence of a recursive function $f(x)$ the functions $g(x)$ and $h(x, y)$ appear, then terms of the form 'g(5)=7' or 'h(3,2)=10' might appear. Each entry in this sequence needs to be an application of a basic function or follow from the entries above by using composition, primitive recursion or μ recursion. For instance if $f(x) =$

$h(x, g(x))$, then for 'f(5)=3' to appear, terms like 'g(5)=6' and 'h(5,6)=3' must occur above. The computation terminates only if the final term gives the value of the recursive function applied to the inputs.

Markov algorithm a string rewriting system that uses grammar-like rules to operate on strings of symbols.

Register machine is a theoretically interesting idealization of a computer. There are several variants. In most of them, each register can hold a natural number (of unlimited size), and the instructions are simple (and few in number), e.g. only decrementation (combined with conditional jump) and incrementation exist (and halting). The lack of the infinite (or dynamically growing) external store (seen at Turing machines) can be understood by replacing its role with Gödel numbering techniques: the fact that each register holds a natural number allows the possibility of representing a complicated thing (e.g. a sequence, or a matrix etc.) by an appropriate huge natural number — unambiguity of both representation and interpretation can be established by number theoretical foundations of these techniques.

In addition to the general computational models, some simpler computational models are useful for special, restricted applications. Regular expressions, for example, specify string patterns in many contexts, from office productivity software to programming languages. Another formalism mathematically equivalent to regular expressions, Finite automata are used in circuit design and in some kinds of problem-solving. Context-free grammars specify programming language syntax. Non-deterministic pushdown automata are another formalism equivalent to context-free grammars. Primitive recursive functions are a defined subclass of the recursive functions.

Different models of computation have the ability to do different tasks. One way to measure the power of a computational model is to study the class of formal languages that the model can generate; in such a way to the Chomsky hierarchy of languages is obtained.

41.4 References

[1] Michael Sipser (2013). *Introduction to the Theory of Computation 3rd*. Cengage Learning. ISBN 978-1-133-18779-0. central areas of the theory of computation: automata, computability, and complexity. (Page 1)

[2] Andrew Hodges (2012). *Alan Turing: The Enigma (THE CENTENARY EDITION)*. Princeton University Press. ISBN 978-0-691-15564-7.

[3] Rabin, Michael O. (June 2012). *Turing, Church, Gödel, Computability, Complexity and Randomization: A Personal View*.

[4] Donald Monk (1976). *Mathematical Logic*. Springer-Verlag. ISBN 9780387901701.

[5] Hopcroft, John E. and Jeffrey D. Ullman (2006). *Introduction to Automata Theory, Languages, and Computation*. 3rd ed. Reading, MA: Addison-Wesley. ISBN 978-0-321-45536-9.

[6] Chomsky hierarchy (1956). "Three models for the description of language". *Information Theory, IRE Transactions on* (IEEE) **2** (3): 113–124. doi:10.1109/TIT.1956.1056813. Retrieved 6 January 2015.

[7] Alan Turing (1937). "On computable numbers, with an application to the Entscheidungsproblem". *Proceedings of the London Mathematical Society* (IEEE) **2** (42): 230–265. doi:10.1112/plms/s2-42.1.230. Retrieved 6 January 2015.

[8] Henry Gordon Rice (1953). "Classes of Recursively Enumerable Sets and Their Decision Problems". *Transactions of the American Mathematical Society* (American Mathematical Society) **74** (2): 358–366. doi:10.2307/1990888. JSTOR 1990888.

[9] Martin Davis (2004). *The undecidable: Basic papers on undecidable propositions, unsolvable problems and computable functions (Dover Ed)*. Dover Publications. ISBN 978-0486432281.

41.5 Further reading

Textbooks aimed at computer scientists

(There are many textbooks in this area; this list is by necessity incomplete.)

- Hopcroft, John E., and Jeffrey D. Ullman (2006). *Introduction to Automata Theory, Languages, and Computation.* *3rd ed* Reading, MA: Addison-Wesley. ISBN 978-0-321-45536-9 One of the standard references in the field.

- Michael Sipser (2013). *Introduction to the Theory of Computation* (3rd ed.). Cengage Learning. ISBN 978-1-133-18779-0.

- Eitan Gurari (1989). *An Introduction to the Theory of Computation.* Computer Science Press. ISBN 0-7167-8182-4.

- Hein, James L. (1996) *Theory of Computation.* Sudbury, MA: Jones & Bartlett. ISBN 978-0-86720-497-1 A gentle introduction to the field, appropriate for second-year undergraduate computer science students.

- Taylor, R. Gregory (1998). *Models of Computation and Formal Languages.* New York: Oxford University Press. ISBN 978-0-19-510983-2 An unusually readable textbook, appropriate for upper-level undergraduates or beginning graduate students.

- Lewis, F. D. (2007). *Essentials of theoretical computer science* A textbook covering the topics of formal languages, automata and grammars. The emphasis appears to be on presenting an overview of the results and their applications rather than providing proofs of the results.

- Martin Davis, Ron Sigal, Elaine J. Weyuker, *Computability, complexity, and languages: fundamentals of theoretical computer science*, 2nd ed., Academic Press, 1994, ISBN 0-12-206382-1. Covers a wider range of topics than most other introductory books, including program semantics and quantification theory. Aimed at graduate students.

Books on computability theory from the (wider) mathematical perspective

- Hartley Rogers, Jr (1987). *Theory of Recursive Functions and Effective Computability*, MIT Press. ISBN 0-262-68052-1

- S. Barry Cooper (2004). *Computability Theory.* Chapman and Hall/CRC. ISBN 1-58488-237-9..

- Carl H. Smith, *A recursive introduction to the theory of computation*, Springer, 1994, ISBN 0-387-94332-3. A shorter textbook suitable for graduate students in Computer Science.

Historical perspective

- Richard L. Epstein and Walter A. Carnielli (2000). *Computability: Computable Functions, Logic, and the Foundations of Mathematics, with Computability: A Timeline (2nd ed.).* Wadsworth/Thomson Learning. ISBN 0-534-54644-7..

41.6 External links

- Theory of Computation at MIT

- Theory of Computation at Harvard

- Computability Logic - A theory of interactive computation. The main web source on this subject.

Chapter 42

Cryptography

"Secret code" redirects here. For the Aya Kamiki album, see Secret Code.
"Cryptology" redirects here. For the David S. Ware album, see Cryptology (album).

Cryptography or **cryptology**: from Greek κρυπτός *kryptós*, "hidden, secret"; and γράφειν *graphein*, "writing", or

German Lorenz cipher machine, used in World War II to encrypt very-high-level general staff messages

-λογία *-logia*, "study", respectively[1] is the practice and study of techniques for secure communication in the presence of third parties (called adversaries).[2] More generally, it is about constructing and analyzing protocols that block adversaries;[3] various aspects in information security such as data confidentiality, data integrity, authentication, and non-repudiation[4] are central to modern cryptography. Modern cryptography exists at the intersection of the disciplines of mathematics, computer science, and electrical engineering. Applications of cryptography include ATM cards, computer passwords, and electronic commerce.

Cryptography prior to the modern age was effectively synonymous with *encryption*, the conversion of information from a readable state to apparent nonsense. The originator of an encrypted message shared the decoding technique needed to recover the original information only with intended recipients, thereby precluding unwanted persons from doing the same. Since World War I and the advent of the computer, the methods used to carry out cryptology have become increasingly complex and its application more widespread.

Modern cryptography is heavily based on mathematical theory and computer science practice; cryptographic algorithms are designed around computational hardness assumptions, making such algorithms hard to break in practice by any adversary. It is theoretically possible to break such a system, but it is infeasible to do so by any known practical means. These schemes are therefore termed computationally secure; theoretical advances, e.g., improvements in integer factorization algorithms, and faster computing technology require these solutions to be continually adapted. There exist information-theoretically secure schemes that provably cannot be broken even with unlimited computing power—an example is the one-time pad—but these schemes are more difficult to implement than the best theoretically breakable but computationally secure mechanisms.

The growth of cryptographic technology has raised a number of legal issues in the information age. Cryptography's potential for use as a tool for espionage and sedition has led many governments to classify it as a weapon and to limit or even prohibit its use and export.[5] In some jurisdictions where the use of cryptography is legal, laws permit investigators to compel the disclosure of encryption keys for documents relevant to an investigation.[6] Cryptography also plays a major role in digital rights management and piracy of digital media.[7]

42.1 Terminology

Until modern times, cryptography referred almost exclusively to *encryption*, which is the process of converting ordinary information (called plaintext) into unintelligible text (called ciphertext).[8] Decryption is the reverse, in other words, moving from the unintelligible ciphertext back to plaintext. A *cipher* (or *cypher*) is a pair of algorithms that create the encryption and the reversing decryption. The detailed operation of a cipher is controlled both by the algorithm and in each instance by a "key". This is a secret (ideally known only to the communicants), usually a short string of characters, which is needed to decrypt the ciphertext. Formally, a "cryptosystem" is the ordered list of elements of finite possible plaintexts, finite possible cyphertexts, finite possible keys, and the encryption and decryption algorithms which correspond to each key. Keys are important both formally and in actual practice, as ciphers without variable keys can be trivially broken with only the knowledge of the cipher used and are therefore useless (or even counter-productive) for most purposes. Historically, ciphers were often used directly for encryption or decryption without additional procedures such as authentication or integrity checks.

In colloquial use, the term "code" is often used to mean any method of encryption or concealment of meaning. However, in cryptography, *code* has a more specific meaning. It means the replacement of a unit of plaintext (i.e., a meaningful word or phrase) with a code word (for example, "wallaby" replaces "attack at dawn"). Codes are no longer used in serious cryptography—except incidentally for such things as unit designations (e.g., Bronco Flight or Operation Overlord)—since properly chosen ciphers are both more practical and more secure than even the best codes and also are better adapted to computers.

Cryptanalysis is the term used for the study of methods for obtaining the meaning of encrypted information without access to the key normally required to do so; i.e., it is the study of how to crack encryption algorithms or their implementations.

Some use the terms *cryptography* and *cryptology* interchangeably in English, while others (including US military practice generally) use *cryptography* to refer specifically to the use and practice of cryptographic techniques and *cryptology* to refer to the combined study of cryptography and cryptanalysis.[9][10] English is more flexible than several other languages in which *cryptology* (done by cryptologists) is always used in the second sense above. RFC 2828 advises that steganography is sometimes included in cryptology.[11]

The study of characteristics of languages that have some application in cryptography or cryptology (e.g. frequency data, letter combinations, universal patterns, etc.) is called cryptolinguistics.

42.2   History of cryptography and cryptanalysis

Main article: History of cryptography

Before the modern era, cryptography was concerned solely with message confidentiality (i.e., encryption)—conversion of messages from a comprehensible form into an incomprehensible one and back again at the other end, rendering it unreadable by interceptors or eavesdroppers without secret knowledge (namely the key needed for decryption of that message). Encryption attempted to ensure secrecy in communications, such as those of spies, military leaders, and diplomats. In recent decades, the field has expanded beyond confidentiality concerns to include techniques for message integrity checking, sender/receiver identity authentication, digital signatures, interactive proofs and secure computation, among others.

42.2.1   Classic cryptography

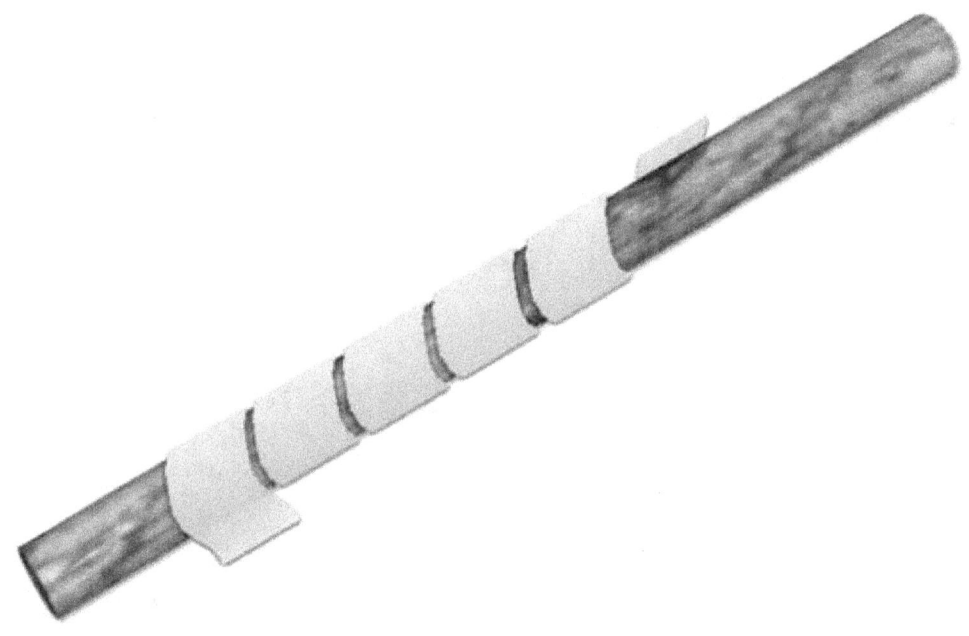

Reconstructed ancient Greek scytale, an early cipher device

The earliest forms of secret writing required little more than writing implements since most people could not read. More literacy, or literate opponents, required actual cryptography. The main classical cipher types are transposition ciphers, which rearrange the order of letters in a message (e.g., 'hello world' becomes 'ehlol owrdl' in a trivially simple rearrangement scheme), and substitution ciphers, which systematically replace letters or groups of letters with other letters or groups of letters (e.g., 'fly at once' becomes 'gmz bu podf' by replacing each letter with the one following it in the Latin alphabet). Simple versions of either have never offered much confidentiality from enterprising opponents. An early substitution cipher was the Caesar cipher, in which each letter in the plaintext was replaced by a letter some fixed number of positions further down the alphabet. Suetonius reports that Julius Caesar used it with a shift of three to communicate with his generals. Atbash is an example of an early Hebrew cipher. The earliest known use of cryptography is some carved ciphertext on stone in Egypt (ca 1900 BCE), but this may have been done for the amusement of literate observers rather than as a way of concealing information.

The Greeks of Classical times are said to have known of ciphers (e.g., the scytale transposition cipher claimed to have been used by the Spartan military).[12] Steganography (i.e., hiding even the existence of a message so as to keep it confidential) was also first developed in ancient times. An early example, from Herodotus, was a message tattooed on a slave's shaved head and concealed under the regrown hair.[8] More modern examples of steganography include the use of invisible ink, microdots, and digital watermarks to conceal information.

In India, the 2000-year old Kamasutra of Vātsyāyana speaks of two different kinds of ciphers called Kautiliyam and Mulavediya. In the Kautiliyam, the cipher letter substitutions are based on phonetic relations, such as vowels becoming consonants. In the Mulavediya, the cipher alphabet consists of pairing letters and using the reciprocal ones.[8]

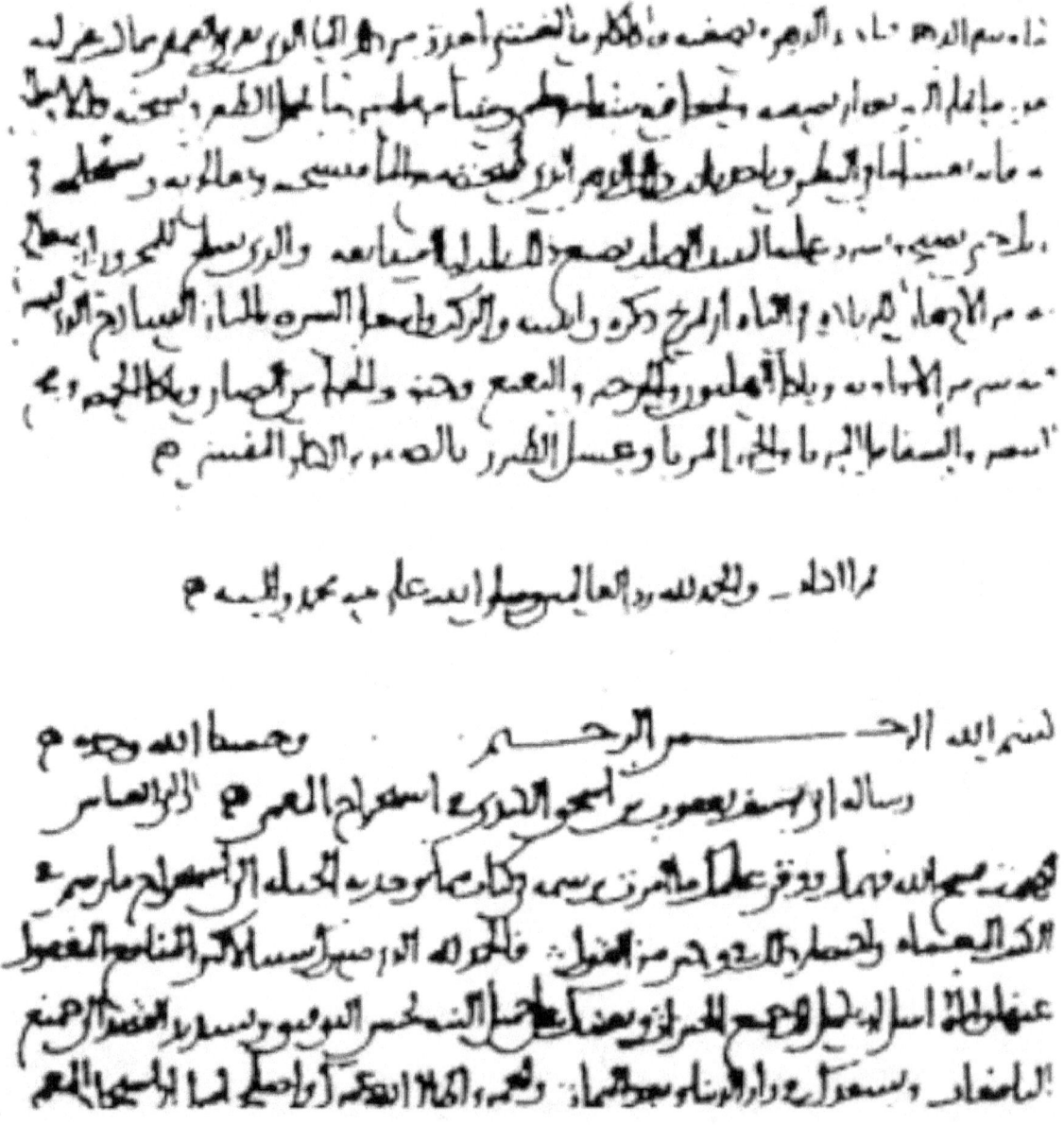

First page of a book by Al-Kindi which discusses encryption of messages

Ciphertexts produced by a classical cipher (and some modern ciphers) always reveal statistical information about the plaintext, which can often be used to break them. After the discovery of frequency analysis, perhaps by the Arab mathematician and polymath Al-Kindi (also known as *Alkindus*) in the 9th century,[13] nearly all such ciphers became more or less readily breakable by any informed attacker. Such classical ciphers still enjoy popularity today, though mostly as puzzles (see cryptogram). Al-Kindi wrote a book on cryptography entitled *Risalah fi Istikhraj al-Mu'amma* (*Manuscript for the Deciphering Cryptographic Messages*), which described the first known use frequency analysis cryptanalysis techniques.[13][14]

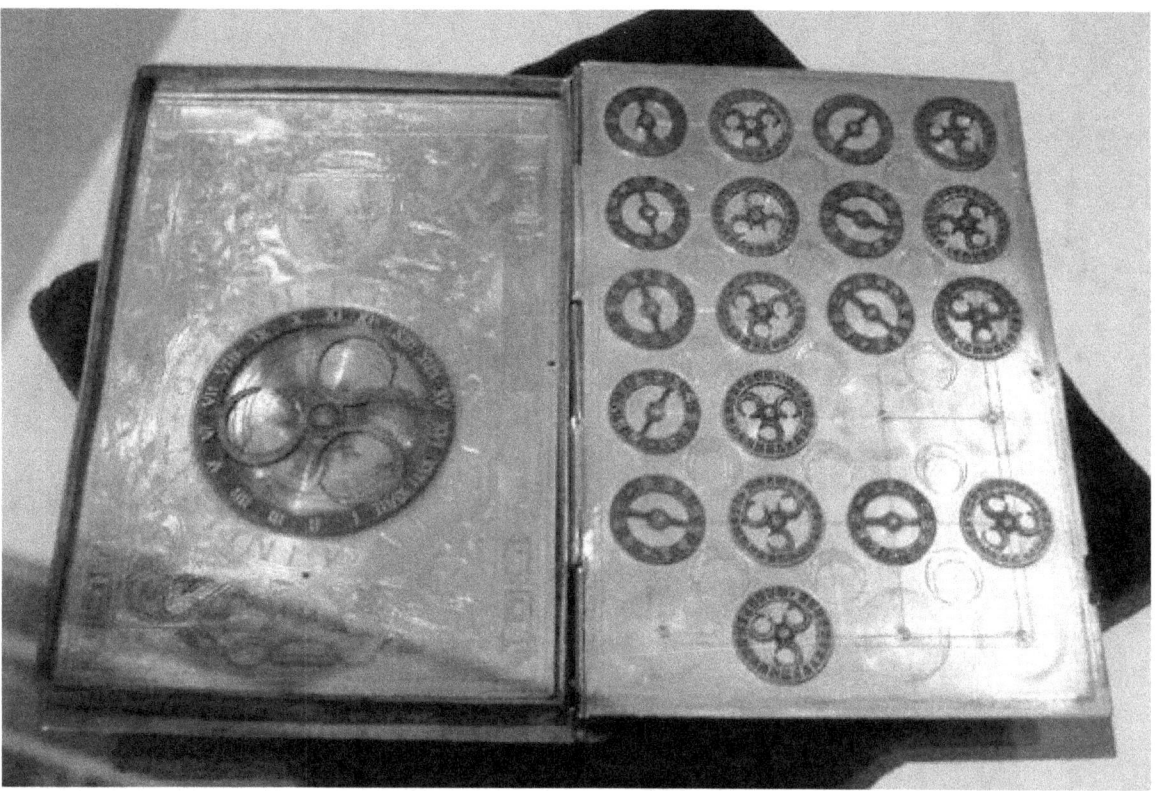

16th-century book-shaped French cipher machine, with arms of Henri II of France

Essentially all ciphers remained vulnerable to cryptanalysis using the frequency analysis technique until the development of the polyalphabetic cipher, most clearly by Leon Battista Alberti around the year 1467, though there is some indication that it was already known to Al-Kindi.[14] Alberti's innovation was to use different ciphers (i.e., substitution alphabets) for various parts of a message (perhaps for each successive plaintext letter at the limit). He also invented what was probably the first automatic cipher device, a wheel which implemented a partial realization of his invention. In the polyalphabetic Vigenère cipher, encryption uses a *key word*, which controls letter substitution depending on which letter of the key word is used. In the mid-19th century Charles Babbage showed that the Vigenère cipher was vulnerable to Kasiski examination, but this was first published about ten years later by Friedrich Kasiski.[15]

Although frequency analysis can be a powerful and general technique against many ciphers, encryption has still often been effective in practice, as many a would-be cryptanalyst was unaware of the technique. Breaking a message without using frequency analysis essentially required knowledge of the cipher used and perhaps of the key involved, thus making espionage, bribery, burglary, defection, etc., more attractive approaches to the cryptanalytically uninformed. It was finally explicitly recognized in the 19th century that secrecy of a cipher's algorithm is not a sensible nor practical safeguard of message security; in fact, it was further realized that any adequate cryptographic scheme (including ciphers) should remain secure even if the adversary fully understands the cipher algorithm itself. Security of the key used should alone be sufficient for a good cipher to maintain confidentiality under an attack. This fundamental principle was first explicitly stated in 1883 by Auguste Kerckhoffs and is generally called Kerckhoffs's Principle; alternatively and more bluntly, it was restated by Claude Shannon, the inventor of information theory and the fundamentals of theoretical cryptography, as *Shannon's Maxim*—'the enemy knows the system'.

Different physical devices and aids have been used to assist with ciphers. One of the earliest may have been the scytale

Enciphered letter from Gabriel de Luetz d'Aramon, French Ambassador to the Ottoman Empire, after 1546, with partial decipherment

of ancient Greece, a rod supposedly used by the Spartans as an aid for a transposition cipher (see image above). In medieval times, other aids were invented such as the cipher grille, which was also used for a kind of steganography. With the invention of polyalphabetic ciphers came more sophisticated aids such as Alberti's own cipher disk, Johannes Trithemius' tabula recta scheme, and Thomas Jefferson's multi cylinder (not publicly known, and reinvented independently by Bazeries around 1900). Many mechanical encryption/decryption devices were invented early in the 20th century, and several patented, among them rotor machines—famously including the Enigma machine used by the German government and military from the late 1920s and during World War II.[16] The ciphers implemented by better quality examples of these machine designs brought about a substantial increase in cryptanalytic difficulty after WWI.[17]

42.2.2 Computer era

Cryptanalysis of the new mechanical devices proved to be both difficult and laborious. In the United Kingdom, cryptanalytic efforts at Bletchley Park during WWII spurred the development of more efficient means for carrying out repetitious tasks. This culminated in the development of the Colossus, the world's first fully electronic, digital, programmable computer, which assisted in the decryption of ciphers generated by the German Army's Lorenz SZ40/42 machine.

Just as the development of digital computers and electronics helped in cryptanalysis, it made possible much more complex ciphers. Furthermore, computers allowed for the encryption of any kind of data representable in any binary format, unlike classical ciphers which only encrypted written language texts; this was new and significant. Computer use has thus supplanted linguistic cryptography, both for cipher design and cryptanalysis. Many computer ciphers can be characterized by their operation on binary bit sequences (sometimes in groups or blocks), unlike classical and mechanical schemes, which generally manipulate traditional characters (i.e., letters and digits) directly. However, computers have also assisted cryptanalysis, which has compensated to some extent for increased cipher complexity. Nonetheless, good modern ciphers have stayed ahead of cryptanalysis; it is typically the case that use of a quality cipher is very efficient (i.e., fast and requiring few resources, such as memory or CPU capability), while breaking it requires an effort many orders of magnitude larger, and vastly larger than that required for any classical cipher, making cryptanalysis so inefficient and impractical as to be effectively impossible.

Extensive open academic research into cryptography is relatively recent; it began only in the mid-1970s. In recent times, IBM personnel designed the algorithm that became the Federal (i.e., US) Data Encryption Standard; Whitfield Diffie and Martin Hellman published their key agreement algorithm;[18] and the RSA algorithm was published in Martin Gardner's *Scientific American* column. Since then, cryptography has become a widely used tool in communications, computer networks, and computer security generally. Some modern cryptographic techniques can only keep their keys secret if certain mathematical problems are intractable, such as the integer factorization or the discrete logarithm problems, so there are deep connections with abstract mathematics. There are very few cryptosystems that are proven to be unconditionally secure. The one-time pad is one. There are a few important ones that are proven secure under certain unproven assumptions. For example, the infeasibility of factoring extremely large integers is the basis for believing that RSA is secure, and some other systems, but even there, the proof is usually lost due to practical considerations. There are systems similar to RSA, such as one by Michael O. Rabin that is provably secure provided factoring $n = pq$ is impossible, but the more practical system RSA has never been proved secure in this sense. The discrete logarithm problem is the basis for believing some other cryptosystems are secure, and again, there are related, less practical systems that are provably secure relative to the discrete log problem.[19]

As well as being aware of cryptographic history, cryptographic algorithm and system designers must also sensibly consider probable future developments while working on their designs. For instance, continuous improvements in computer processing power have increased the scope of brute-force attacks, so when specifying key lengths, the required key lengths are similarly advancing.[20] The potential effects of quantum computing are already being considered by some cryptographic system designers; the announced imminence of small implementations of these machines may be making the need for this preemptive caution rather more than merely speculative.[4]

Essentially, prior to the early 20th century, cryptography was chiefly concerned with linguistic and lexicographic patterns. Since then the emphasis has shifted, and cryptography now makes extensive use of mathematics, including aspects of information theory, computational complexity, statistics, combinatorics, abstract algebra, number theory, and finite mathematics generally. Cryptography is also a branch of engineering, but an unusual one since it deals with active, intelligent, and malevolent opposition (see cryptographic engineering and security engineering); other kinds of engineering (e.g., civil or chemical engineering) need deal only with neutral natural forces. There is also active research examining the relationship between cryptographic problems and quantum physics (see quantum cryptography and quantum computer).

42.3   Modern cryptography

The modern field of cryptography can be divided into several areas of study. The chief ones are discussed here; see Topics in Cryptography for more.

42.3.1   Symmetric-key cryptography

Main article: Symmetric-key algorithm
 Symmetric-key cryptography refers to encryption methods in which both the sender and receiver share the same key (or, less commonly, in which their keys are different, but related in an easily computable way). This was the only kind of encryption publicly known until June 1976.[18]

Symmetric key ciphers are implemented as either block ciphers or stream ciphers. A block cipher enciphers input in blocks of plaintext as opposed to individual characters, the input form used by a stream cipher.

The Data Encryption Standard (DES) and the Advanced Encryption Standard (AES) are block cipher designs which have been designated cryptography standards by the US government (though DES's designation was finally withdrawn after the AES was adopted).[21] Despite its deprecation as an official standard, DES (especially its still-approved and much more secure triple-DES variant) remains quite popular; it is used across a wide range of applications, from ATM encryption[22] to e-mail privacy[23] and secure remote access.[24] Many other block ciphers have been designed and released, with considerable variation in quality. Many have been thoroughly broken, such as FEAL.[4][25]

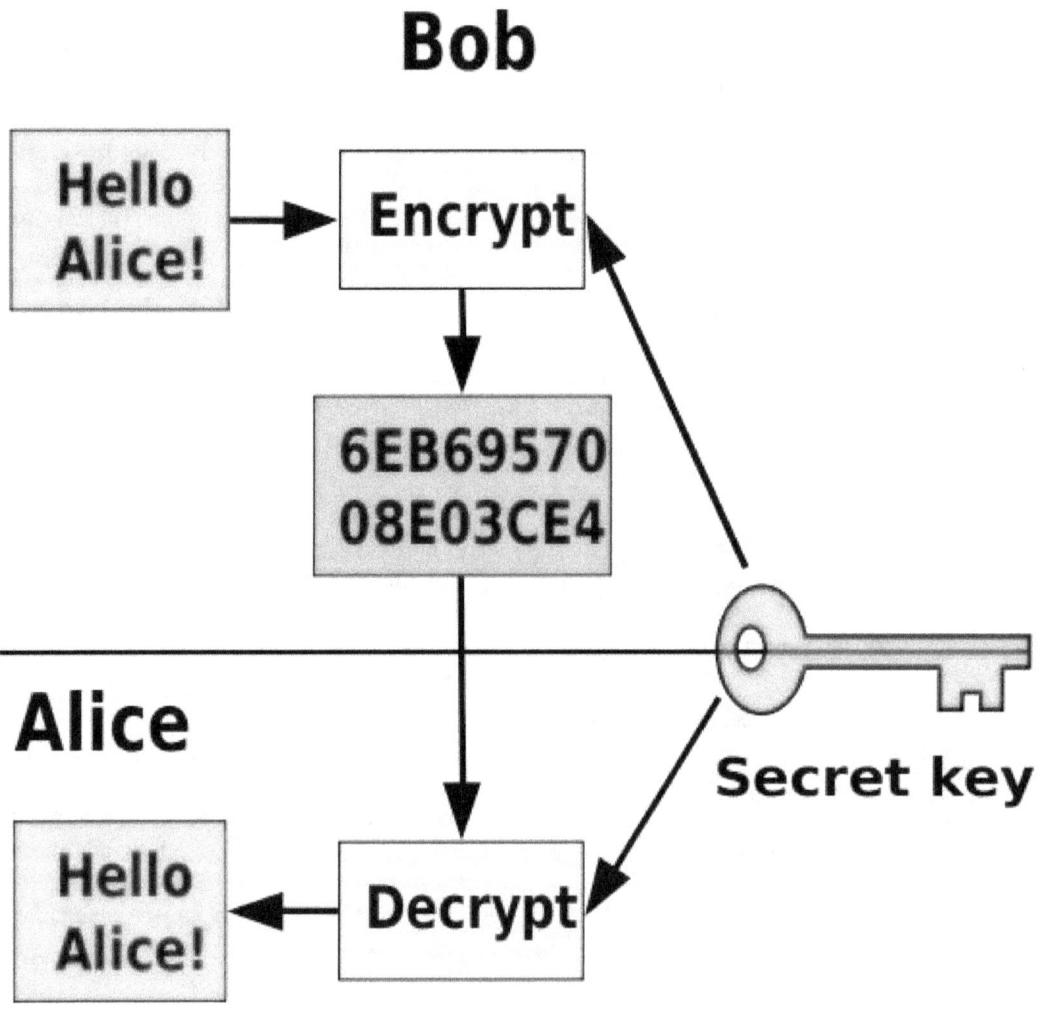

Symmetric-key cryptography, where a single key is used for encryption and decryption

Stream ciphers, in contrast to the 'block' type, create an arbitrarily long stream of key material, which is combined with the plaintext bit-by-bit or character-by-character, somewhat like the one-time pad. In a stream cipher, the output stream is created based on a hidden internal state which changes as the cipher operates. That internal state is initially set up using the secret key material. RC4 is a widely used stream cipher; see Category:Stream ciphers.[4] Block ciphers can be used as stream ciphers; see Block cipher modes of operation.

Cryptographic hash functions are a third type of cryptographic algorithm. They take a message of any length as input, and output a short, fixed length hash which can be used in (for example) a digital signature. For good hash functions, an attacker cannot find two messages that produce the same hash. MD4 is a long-used hash function which is now broken; MD5, a strengthened variant of MD4, is also widely used but broken in practice. The US National Security Agency developed the Secure Hash Algorithm series of MD5-like hash functions: SHA-0 was a flawed algorithm that the agency withdrew; SHA-1 is widely deployed and more secure than MD5, but cryptanalysts have identified attacks against it; the SHA-2 family improves on SHA-1, but it isn't yet widely deployed; and the US standards authority thought it "prudent" from a security perspective to develop a new standard to "significantly improve the robustness of NIST's overall hash algorithm toolkit."[26] Thus, a hash function design competition was meant to select a new U.S. national standard, to be

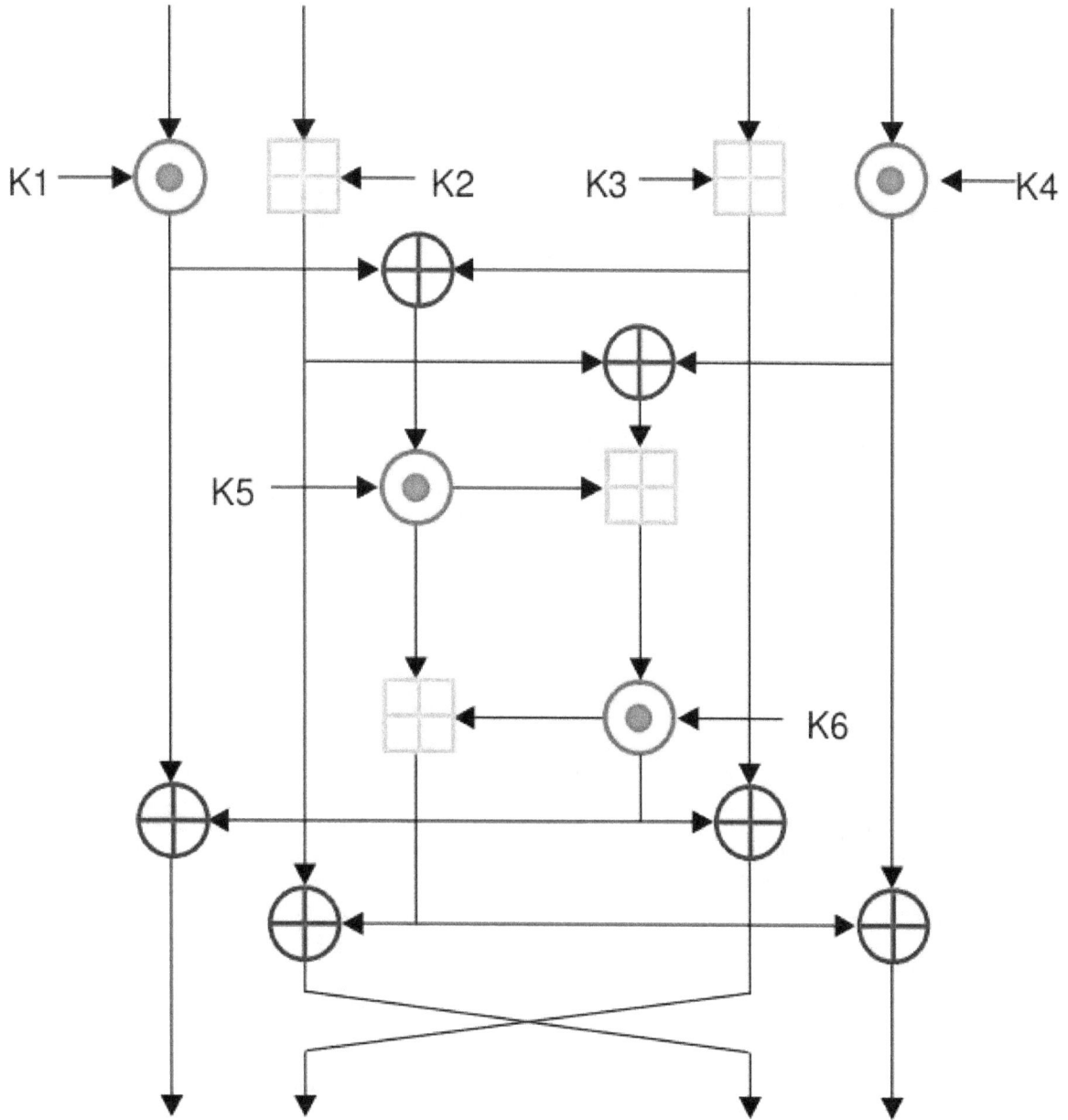

One round (out of 8.5) of the IDEA cipher, used in some versions of PGP for high-speed encryption of, for instance, e-mail

called SHA-3, by 2012. The competition ended on October 2, 2012 when the NIST announced that Keccak would be the new SHA-3 hash algorithm.[27]

Message authentication codes (MACs) are much like cryptographic hash functions, except that a secret key can be used to authenticate the hash value upon receipt;[4] this additional complication blocks an attack scheme against bare digest algorithms, and so has been thought worth the effort.

42.3.2   Public-key cryptography

Main article: Public-key cryptography

 Symmetric-key cryptosystems use the same key for encryption and decryption of a message, though a message or group of messages may have a different key than others. A significant disadvantage of symmetric ciphers is the key management

Bob

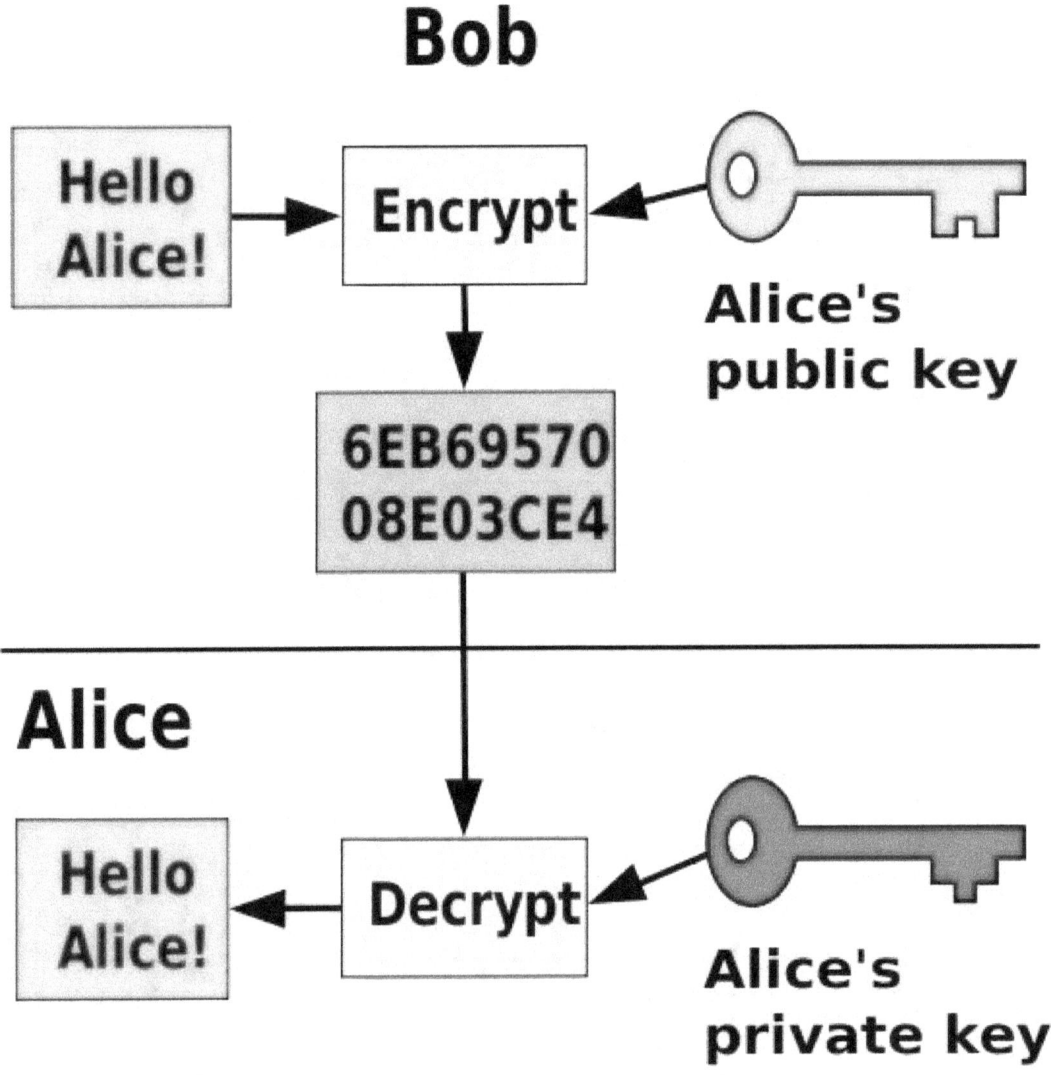

Public-key cryptography, where different keys are used for encryption and decryption

necessary to use them securely. Each distinct pair of communicating parties must, ideally, share a different key, and perhaps each ciphertext exchanged as well. The number of keys required increases as the square of the number of network members, which very quickly requires complex key management schemes to keep them all consistent and secret. The difficulty of securely establishing a secret key between two communicating parties, when a secure channel does not already exist between them, also presents a chicken-and-egg problem which is a considerable practical obstacle for cryptography users in the real world.

In a groundbreaking 1976 paper, Whitfield Diffie and Martin Hellman proposed the notion of *public-key* (also, more generally, called *asymmetric key*) cryptography in which two different but mathematically related keys are used—a *public* key and a *private* key.[28] A public key system is so constructed that calculation of one key (the 'private key') is computationally infeasible from the other (the 'public key'), even though they are necessarily related. Instead, both keys are generated secretly, as an interrelated pair.[29] The historian David Kahn described public-key cryptography as "the most revolutionary new concept in the field since polyalphabetic substitution emerged in the Renaissance".[30]

In public-key cryptosystems, the public key may be freely distributed, while its paired private key must remain secret. In

Whitfield Diffie and Martin Hellman, authors of the first published paper on public-key cryptography

a public-key encryption system, the *public key* is used for encryption, while the *private* or *secret key* is used for decryption. While Diffie and Hellman could not find such a system, they showed that public-key cryptography was indeed possible by presenting the Diffie–Hellman key exchange protocol, a solution that is now widely used in secure communications to allow two parties to secretly agree on a shared encryption key.[18]

Diffie and Hellman's publication sparked widespread academic efforts in finding a practical public-key encryption system. This race was finally won in 1978 by Ronald Rivest, Adi Shamir, and Len Adleman, whose solution has since become known as the RSA algorithm.[31]

The Diffie–Hellman and RSA algorithms, in addition to being the first publicly known examples of high quality public-key algorithms, have been among the most widely used. Others include the Cramer–Shoup cryptosystem, ElGamal encryption, and various elliptic curve techniques. See Category:Asymmetric-key cryptosystems.

To much surprise, a document published in 1997 by the Government Communications Headquarters (GCHQ), a British intelligence organization, revealed that cryptographers at GCHQ had anticipated several academic developments.[32] Reportedly, around 1970, James H. Ellis had conceived the principles of asymmetric key cryptography. In 1973, Clifford Cocks invented a solution that essentially resembles the RSA algorithm.[32][33] And in 1974, Malcolm J. Williamson is claimed to have developed the Diffie–Hellman key exchange.[34]

Padlock icon from the Firefox Web browser, which indicates that TLS, a public-key cryptography system, is in use.

Public-key cryptography can also be used for implementing digital signature schemes. A digital signature is reminiscent of an ordinary signature; they both have the characteristic of being easy for a user to produce, but difficult for anyone else to forge. Digital signatures can also be permanently tied to the content of the message being signed; they cannot then be 'moved' from one document to another, for any attempt will be detectable. In digital signature schemes, there are two algorithms: one for *signing*, in which a secret key is used to process the message (or a hash of the message, or both), and one for *verification*, in which the matching public key is used with the message to check the validity of the signature. RSA and DSA are two of the most popular digital signature schemes. Digital signatures are central to the operation of public key infrastructures and many network security schemes (e.g., SSL/TLS, many VPNs, etc.).[25]

Public-key algorithms are most often based on the computational complexity of "hard" problems, often from number theory. For example, the hardness of RSA is related to the integer factorization problem, while Diffie–Hellman and DSA are related to the discrete logarithm problem. More recently, *elliptic curve cryptography* has developed, a system in which security is based on number theoretic problems involving elliptic curves. Because of the difficulty of the underlying problems, most public-key algorithms involve operations such as modular multiplication and exponentiation, which are much more computationally expensive than the techniques used in most block ciphers, especially with typical key sizes. As a result, public-key cryptosystems are commonly hybrid cryptosystems, in which a fast high-quality symmetric-key encryption algorithm is used for the message itself, while the relevant symmetric key is sent with the message, but encrypted using a public-key algorithm. Similarly, hybrid signature schemes are often used, in which a cryptographic hash function is computed, and only the resulting hash is digitally signed.[4]

42.3.3 Cryptanalysis

Main article: Cryptanalysis
The goal of cryptanalysis is to find some weakness or insecurity in a cryptographic scheme, thus permitting its subversion or evasion.

It is a common misconception that every encryption method can be broken. In connection with his WWII work at Bell Labs, Claude Shannon proved that the one-time pad cipher is unbreakable, provided the key material is truly random, never reused, kept secret from all possible attackers, and of equal or greater length than the message.[35] Most ciphers, apart from the one-time pad, can be broken with enough computational effort by brute force attack, but the amount of effort needed may be exponentially dependent on the key size, as compared to the effort needed to make use of the cipher. In such cases, effective security could be achieved if it is proven that the effort required (i.e., "work factor", in Shannon's terms) is beyond the ability of any adversary. This means it must be shown that no efficient method (as opposed to the time-consuming brute force method) can be found to break the cipher. Since no such proof has been found to date, the one-time-pad remains the only theoretically unbreakable cipher.

There are a wide variety of cryptanalytic attacks, and they can be classified in any of several ways. A common distinction turns on what an attacker knows and what capabilities are available. In a ciphertext-only attack, the cryptanalyst has access only to the ciphertext (good modern cryptosystems are usually effectively immune to ciphertext-only attacks). In a known-plaintext attack, the cryptanalyst has access to a ciphertext and its corresponding plaintext (or to many such pairs). In a chosen-plaintext attack, the cryptanalyst may choose a plaintext and learn its corresponding ciphertext (perhaps many times); an example is gardening, used by the British during WWII. Finally, in a chosen-ciphertext attack, the cryptanalyst may be able to *choose* ciphertexts and learn their corresponding plaintexts.[4] Also important, often overwhelmingly so, are mistakes (generally in the design or use of one of the protocols involved; see Cryptanalysis of the Enigma for some historical examples of this).

Cryptanalysis of symmetric-key ciphers typically involves looking for attacks against the block ciphers or stream ciphers that are more efficient than any attack that could be against a perfect cipher. For example, a simple brute force attack against DES requires one known plaintext and 2^{55} decryptions, trying approximately half of the possible keys, to reach a point at which chances are better than even that the key sought will have been found. But this may not be enough assurance; a linear cryptanalysis attack against DES requires 2^{43} known plaintexts and approximately 2^{43} DES operations.[36] This is a considerable improvement on brute force attacks.

Public-key algorithms are based on the computational difficulty of various problems. The most famous of these is integer factorization (e.g., the RSA algorithm is based on a problem related to integer factoring), but the discrete logarithm problem is also important. Much public-key cryptanalysis concerns numerical algorithms for solving these computational problems, or some of them, efficiently (i.e., in a practical time). For instance, the best known algorithms for solving the elliptic curve-based version of discrete logarithm are much more time-consuming than the best known algorithms for factoring, at least for problems of more or less equivalent size. Thus, other things being equal, to achieve an equivalent strength of attack resistance, factoring-based encryption techniques must use larger keys than elliptic curve techniques.

For this reason, public-key cryptosystems based on elliptic curves have become popular since their invention in the mid-1990s.

While pure cryptanalysis uses weaknesses in the algorithms themselves, other attacks on cryptosystems are based on actual use of the algorithms in real devices, and are called *side-channel attacks*. If a cryptanalyst has access to, for example, the amount of time the device took to encrypt a number of plaintexts or report an error in a password or PIN character, he may be able to use a timing attack to break a cipher that is otherwise resistant to analysis. An attacker might also study the pattern and length of messages to derive valuable information; this is known as traffic analysis[37] and can be quite useful to an alert adversary. Poor administration of a cryptosystem, such as permitting too short keys, will make any system vulnerable, regardless of other virtues. And, of course, social engineering, and other attacks against the personnel who work with cryptosystems or the messages they handle (e.g., bribery, extortion, blackmail, espionage, torture, ...) may be the most productive attacks of all.

42.3.4   Cryptographic primitives

Much of the theoretical work in cryptography concerns cryptographic *primitives*—algorithms with basic cryptographic properties—and their relationship to other cryptographic problems. More complicated cryptographic tools are then built from these basic primitives. These primitives provide fundamental properties, which are used to develop more complex tools called *cryptosystems* or *cryptographic protocols*, which guarantee one or more high-level security properties. Note however, that the distinction between cryptographic *primitives* and cryptosystems, is quite arbitrary; for example, the RSA algorithm is sometimes considered a cryptosystem, and sometimes a primitive. Typical examples of cryptographic primitives include pseudorandom functions, one-way functions, etc.

42.3.5   Cryptosystems

One or more cryptographic primitives are often used to develop a more complex algorithm, called a cryptographic system, or *cryptosystem*. Cryptosystems (e.g., El-Gamal encryption) are designed to provide particular functionality (e.g., public key encryption) while guaranteeing certain security properties (e.g., chosen-plaintext attack (CPA) security in the random oracle model). Cryptosystems use the properties of the underlying cryptographic primitives to support the system's security properties. Of course, as the distinction between primitives and cryptosystems is somewhat arbitrary, a sophisticated cryptosystem can be derived from a combination of several more primitive cryptosystems. In many cases, the cryptosystem's structure involves back and forth communication among two or more parties in space (e.g., between the sender of a secure message and its receiver) or across time (e.g., cryptographically protected backup data). Such cryptosystems are sometimes called *cryptographic protocols*.

Some widely known cryptosystems include RSA encryption, Schnorr signature, El-Gamal encryption, PGP, etc. More complex cryptosystems include electronic cash[38] systems, signcryption systems, etc. Some more 'theoretical' cryptosystems include interactive proof systems,[39] (like zero-knowledge proofs),[40] systems for secret sharing,[41][42] etc.

Until recently, most security properties of most cryptosystems were demonstrated using empirical techniques or using ad hoc reasoning. Recently, there has been considerable effort to develop formal techniques for establishing the security of cryptosystems; this has been generally called *provable security*. The general idea of provable security is to give arguments about the computational difficulty needed to compromise some security aspect of the cryptosystem (i.e., to any adversary).

The study of how best to implement and integrate cryptography in software applications is itself a distinct field (see Cryptographic engineering and Security engineering).

42.4   Legal issues

See also: Cryptography laws in different nations

42.4.1 Prohibitions

Cryptography has long been of interest to intelligence gathering and law enforcement agencies. Secret communications may be criminal or even treasonous. Because of its facilitation of privacy, and the diminution of privacy attendant on its prohibition, cryptography is also of considerable interest to civil rights supporters. Accordingly, there has been a history of controversial legal issues surrounding cryptography, especially since the advent of inexpensive computers has made widespread access to high quality cryptography possible.

In some countries, even the domestic use of cryptography is, or has been, restricted. Until 1999, France significantly restricted the use of cryptography domestically, though it has since relaxed many of these rules. In China and Iran, a license is still required to use cryptography.[5] Many countries have tight restrictions on the use of cryptography. Among the more restrictive are laws in Belarus, Kazakhstan, Mongolia, Pakistan, Singapore, Tunisia, and Vietnam.[43]

In the United States, cryptography is legal for domestic use, but there has been much conflict over legal issues related to cryptography. One particularly important issue has been the export of cryptography and cryptographic software and hardware. Probably because of the importance of cryptanalysis in World War II and an expectation that cryptography would continue to be important for national security, many Western governments have, at some point, strictly regulated export of cryptography. After World War II, it was illegal in the US to sell or distribute encryption technology overseas; in fact, encryption was designated as auxiliary military equipment and put on the United States Munitions List.[44] Until the development of the personal computer, asymmetric key algorithms (i.e., public key techniques), and the Internet, this was not especially problematic. However, as the Internet grew and computers became more widely available, high-quality encryption techniques became well known around the globe.

42.4.2 Export controls

Main article: Export of cryptography

In the 1990s, there were several challenges to US export regulation of cryptography. After the source code for Philip Zimmermann's Pretty Good Privacy (PGP) encryption program found its way onto the Internet in June 1991, a complaint by RSA Security (then called RSA Data Security, Inc.) resulted in a lengthy criminal investigation of Zimmermann by the US Customs Service and the FBI, though no charges were ever filed.[45][46] Daniel J. Bernstein, then a graduate student at UC Berkeley, brought a lawsuit against the US government challenging some aspects of the restrictions based on free speech grounds. The 1995 case Bernstein v. United States ultimately resulted in a 1999 decision that printed source code for cryptographic algorithms and systems was protected as free speech by the United States Constitution.[47]

In 1996, thirty-nine countries signed the Wassenaar Arrangement, an arms control treaty that deals with the export of arms and "dual-use" technologies such as cryptography. The treaty stipulated that the use of cryptography with short key-lengths (56-bit for symmetric encryption, 512-bit for RSA) would no longer be export-controlled.[48] Cryptography exports from the US became less strictly regulated as a consequence of a major relaxation in 2000;[49] there are no longer very many restrictions on key sizes in US-exported mass-market software. Since this relaxation in US export restrictions, and because most personal computers connected to the Internet include US-sourced web browsers such as Firefox or Internet Explorer, almost every Internet user worldwide has potential access to quality cryptography via their browsers (e.g., via Transport Layer Security). The Mozilla Thunderbird and Microsoft Outlook E-mail client programs similarly can transmit and receive emails via TLS, and can send and receive email encrypted with S/MIME. Many Internet users don't realize that their basic application software contains such extensive cryptosystems. These browsers and email programs are so ubiquitous that even governments whose intent is to regulate civilian use of cryptography generally don't find it practical to do much to control distribution or use of cryptography of this quality, so even when such laws are in force, actual enforcement is often effectively impossible.

42.4.3 NSA involvement

See also: Clipper chip

Another contentious issue connected to cryptography in the United States is the influence of the National Security Agency on cipher development and policy. The NSA was involved with the design of DES during its development at IBM and its consideration by the National Bureau of Standards as a possible Federal Standard for cryptography.[50] DES was designed to be resistant to differential cryptanalysis,[51] a powerful and general cryptanalytic technique known to the NSA and IBM, that became publicly known only when it was rediscovered in the late 1980s.[52] According to Steven Levy, IBM discovered differential cryptanalysis,[46] but kept the technique secret at the NSA's request. The technique became publicly known only when Biham and Shamir re-discovered and announced it some years later. The entire affair illustrates the difficulty of determining what resources and knowledge an attacker might actually have.

Another instance of the NSA's involvement was the 1993 Clipper chip affair, an encryption microchip intended to be part of the Capstone cryptography-control initiative. Clipper was widely criticized by cryptographers for two reasons. The cipher algorithm (called Skipjack) was then classified (declassified in 1998, long after the Clipper initiative lapsed). The classified cipher caused concerns that the NSA had deliberately made the cipher weak in order to assist its intelligence efforts. The whole initiative was also criticized based on its violation of Kerckhoffs's Principle, as the scheme included a special escrow key held by the government for use by law enforcement, for example in wiretaps.[46]

42.4.4   Digital rights management

Main article: Digital rights management

Cryptography is central to digital rights management (DRM), a group of techniques for technologically controlling use of copyrighted material, being widely implemented and deployed at the behest of some copyright holders. In 1998, U.S. President Bill Clinton signed the Digital Millennium Copyright Act (DMCA), which criminalized all production, dissemination, and use of certain cryptanalytic techniques and technology (now known or later discovered); specifically, those that could be used to circumvent DRM technological schemes.[53] This had a noticeable impact on the cryptography research community since an argument can be made that *any* cryptanalytic research violated, or might violate, the DMCA. Similar statutes have since been enacted in several countries and regions, including the implementation in the EU Copyright Directive. Similar restrictions are called for by treaties signed by World Intellectual Property Organization member-states.

The United States Department of Justice and FBI have not enforced the DMCA as rigorously as had been feared by some, but the law, nonetheless, remains a controversial one. Niels Ferguson, a well-respected cryptography researcher, has publicly stated that he will not release some of his research into an Intel security design for fear of prosecution under the DMCA.[54] Both Alan Cox (longtime number 2 in Linux kernel development) and Edward Felten (and some of his students at Princeton) have encountered problems related to the Act. Dmitry Sklyarov was arrested during a visit to the US from Russia, and jailed for five months pending trial for alleged violations of the DMCA arising from work he had done in Russia, where the work was legal. In 2007, the cryptographic keys responsible for Blu-ray and HD DVD content scrambling were discovered and released onto the Internet. In both cases, the MPAA sent out numerous DMCA takedown notices, and there was a massive Internet backlash[7] triggered by the perceived impact of such notices on fair use and free speech.

42.4.5   Forced disclosure of encryption keys

Main article: Key disclosure law

In the United Kingdom, the Regulation of Investigatory Powers Act gives UK police the powers to force suspects to decrypt files or hand over passwords that protect encryption keys. Failure to comply is an offense in its own right, punishable on conviction by a two-year jail sentence or up to five years in cases involving national security.[6] Successful prosecutions have occurred under the Act; the first, in 2009,[55] resulted in a term of 13 months' imprisonment.[56] Similar forced disclosure laws in Australia, Finland, France, and India compel individual suspects under investigation to hand over encryption keys or passwords during a criminal investigation.

In the United States, the federal criminal case of United States v. Fricosu addressed whether a search warrant can compel a person to reveal an encryption passphrase or password.[57] The Electronic Frontier Foundation (EFF) argued that this is a violation of the protection from self-incrimination given by the Fifth Amendment.[58] In 2012, the court ruled that under the All Writs Act, the defendant was required to produce an unencrypted hard drive for the court.[59]

In many jurisdictions, the legal status of forced disclosure remains unclear.

42.5 See also

- List of cryptographers
- Encyclopedia of Cryptography and Security
- List of important publications in cryptography
- List of multiple discoveries (see "RSA")
- List of unsolved problems in computer science
- Outline of cryptography
- Global surveillance
- Strong cryptography
- A Syllabical and Steganographical table - first cryptography chart

42.6 References

[1] Liddell, Henry George; Scott, Robert; Jones, Henry Stuart; McKenzie, Roderick (1984). *A Greek-English Lexicon*. Oxford University Press.

[2] Rivest, Ronald L. (1990). "Cryptology". In J. Van Leeuwen. *Handbook of Theoretical Computer Science* 1. Elsevier.

[3] Bellare, Mihir; Rogaway, Phillip (21 September 2005). "Introduction". *Introduction to Modern Cryptography*. p. 10.

[4] Menezes, A. J.; van Oorschot, P. C.; Vanstone, S. A. *Handbook of Applied Cryptography*. ISBN 0-8493-8523-7.

[5] "Overview per country". *Crypto Law Survey*. February 2013. Retrieved 26 March 2015.

[6] "UK Data Encryption Disclosure Law Takes Effect". *PC World*. 1 October 2007. Retrieved 26 March 2015.

[7] Doctorow, Cory (2 May 2007). "Digg users revolt over AACS key". *Boing Boing*. Retrieved 26 March 2015.

[8] Kahn, David (1967). *The Codebreakers*. ISBN 0-684-83130-9.

[9] Oded Goldreich, *Foundations of Cryptography, Volume 1: Basic Tools*, Cambridge University Press, 2001, ISBN 0-521-79172-3

[10] "Cryptology (definition)". *Merriam-Webster's Collegiate Dictionary* (11th ed.). Merriam-Webster. Retrieved 26 March 2015.

[11] "RFC 2828 - Internet Security Glossary". *Internet Engineering Task Force*. May 2000. Retrieved 26 March 2015.

[12] Íāshchenko, V. V. (2002). *Cryptography: an introduction*. AMS Bookstore. p. 6. ISBN 0-8218-2986-6.

[13] Singh, Simon (2000). *The Code Book*. New York: Anchor Books. pp. 14–20. ISBN 9780385495325.

[14] Al-Kadi, Ibrahim A. (April 1992). "The origins of cryptology: The Arab contributions". *Cryptologia* 16 (2): 97–126.

[15] Schrödel, Tobias (October 2008). "Breaking Short Vigenère Ciphers".*Cryptologia*32(4): 334–337.doi:10.1080/0161119087.

[16] Hakim, Joy (1995). *A History of US: War, Peace and all that Jazz*. New York: Oxford University Press. ISBN 0-19-509514-6.

[17] Gannon, James (2001). *Stealing Secrets, Telling Lies: How Spies and Codebreakers Helped Shape the Twentieth Century*. Washington, D.C.: Brassey's. ISBN 1-57488-367-4.

[18] Diffie, Whitfield; Hellman, Martin (November 1976). "New Directions in Cryptography" (pdf). *IEEE Transactions on Information Theory*. IT-22: 644–654.

[19] *Cryptography: Theory and Practice*, Third Edition (Discrete Mathematics and Its Applications), 2005, by Douglas R. Stinson, Chapman and Hall/CRC

[20] Blaze, Matt; Diffie, Whitefield; Rivest, Ronald L.; Schneier, Bruce; Shimomura, Tsutomu; Thompson, Eric; Wiener, Michael (January 1996). "Minimal key lengths for symmetric ciphers to provide adequate commercial security". Fortify. Retrieved 26 March 2015.

[21] "FIPS PUB 197: The official Advanced Encryption Standard" (PDF). *Computer Security Resource Center*. National Institute of Standards and Technology. Retrieved 26 March 2015.

[22] "NCUA letter to credit unions" (PDF). *National Credit Union Administration*. July 2004. Retrieved 26 March 2015.

[23] "RFC 2440 - Open PGP Message Format". *Internet Engineering Task Force*. November 1998. Retrieved 26 March 2015.

[24] Golen, Pawel (19 July 2002). "SSH". *WindowSecurity*. Retrieved 26 March 2015.

[25] Schneier, Bruce (1996). *Applied Cryptography* (2nd ed.). Wiley. ISBN 0-471-11709-9.

[26] "Notices". *Federal Register* **72** (212). 2 November 2007.
Archived 28 February 2008 at the Wayback Machine

[27] "NIST Selects Winner of Secure Hash Algorithm (SHA-3) Competition". *Tech Beat*. National Institute of Standards and Technology. October 2, 2012. Retrieved 26 March 2015.

[28] Diffie, Whitfield; Hellman, Martin (8 June 1976). "Multi-user cryptographic techniques". *AFIPS Proceedings* **45**: 109–112.

[29] Ralph Merkle was working on similar ideas at the time and encountered publication delays, and Hellman has suggested that the term used should be Diffie–Hellman–Merkle aysmmetric key cryptography.

[30] Kahn, David (Fall 1979). "Cryptology Goes Public". *Foreign Affairs* **58** (1): 153.

[31] Rivest, Ronald L.; Shamir, A.; Adleman, L. (1978). "A Method for Obtaining Digital Signatures and Public-Key Cryptosystems". *Communications of the ACM* (Association for Computing Machinery) **21** (2): 120–126.
Archived November 16, 2001 at the Wayback Machine
Previously released as an MIT "Technical Memo" in April 1977, and published in Martin Gardner's*Scientific American* lrecreationscolumn

[32] Wayner, Peter (24 December 1997). "British Document Outlines Early Encryption Discovery". *New York Times*. Retrieved 26 March 2015.

[33] Cocks, Clifford (20 November 1973). "A Note on 'Non-Secret Encryption'" (PDF). *CESG Research Report*.

[34] Singh, Simon (1999). *The Code Book*. Doubleday. pp. 279–292.

[35] Shannon, Claude; Weaver, Warren (1963). *The Mathematical Theory of Communication*. University of Illinois Press. ISBN 0-252-72548-4.

[36] Junod, Pascal (2001). "On the Complexity of Matsui's Attack" (PDF). *Selected Areas in Cryptography*.

[37] Song, Dawn; Wagner, David A.; Tian, Xuqing (2001). "Timing Analysis of Keystrokes and Timing Attacks on SSH" (PDF). *Tenth USENIX Security Symposium*.

[38] Brands, S. (1994). "Untraceable Off-line Cash in Wallets with Observers". *Advances in Cryptology—Proceedings of CRYPTO* (Springer-Verlag).

[39] Babai, László (1985). "Trading group theory for randomness". *Proceedings of the Seventeenth Annual Symposium on the Theory of Computing* (Association for Computing Machinery).

[40] Goldwasser, S.; Micali, S.; Rackoff, C. (1989). "The Knowledge Complexity of Interactive Proof Systems". *SIAM Journal on Computing* **18** (1): 186–208.

[41] Blakley, G. (June 1979). "Safeguarding cryptographic keys". *Proceedings of AFIPS 1979* **48**: 313–317.

[42] Shamir, A. (1979). "How to share a secret". *Communications of the ACM* (Association for Computing Machinery) **22**: 612–613.

[43] "6.5.1 WHAT ARE THE CRYPTOGRAPHIC POLICIES OF SOME COUNTRIES?". RSA Laboratories. Retrieved 26 March 2015.

[44] Rosenoer, Jonathan (1995). "CRYPTOGRAPHY & SPEECH". *CyberLaw*.
Archived December 1, 2005 at the Wayback Machine

[45] "Case Closed on Zimmermann PGP Investigation". *IEEE Computer Society's Technical Committee on Security and Privacy*. 14 February 1996. Retrieved 26 March 2015.

[46] Levy, Steven (2001). *Crypto: How the Code Rebels Beat the Government—Saving Privacy in the Digital Age*. Penguin Books. p. 56. ISBN 0-14-024432-8. OCLC 244148644 48066852 48846639.

[47] "Bernstein v USDOJ". *Electronic Privacy Information Center*. United States Court of Appeals for the Ninth Circuit. 6 May 1999. Retrieved 26 March 2015.

[48] "DUAL-USE LIST - CATEGORY 5 – PART 2 – "INFORMATION SECURITY"" (DOC). *Wassenaar Arrangement*. Retrieved 26 March 2015.

[49] "6.4 UNITED STATES CRYPTOGRAPHY EXPORT/IMPORT LAWS". *RSA Laboratories*. Retrieved 26 March 2015.

[50] Schneier, Bruce (15 June 2000). "The Data Encryption Standard (DES)". *Crypto-Gram*. Retrieved 26 March 2015.

[51] Coppersmith, D. (May 1994). "The Data Encryption Standard (DES) and its strength against attacks" (PDF). *IBM Journal of Research and Development* **38** (3): 243. doi:10.1147/rd.383.0243. Retrieved 26 March 2015.

[52] Biham, E.; Shamir, A. (1991). "Differential cryptanalysis of DES-like cryptosystems" (PDF). *Journal of Cryptology* (Springer-Verlag) **4** (1): 3–72. Retrieved 26 March 2015.

[53] "The Digital Millennium Copyright Act of 1998" (PDF). *United States Copyright Office*. Retrieved 26 March 2015.

[54] Ferguson, Niels (15 August 2001). "Censorship in action: why I don't publish my HDCP results".
Archived December 1, 2001 at the Wayback Machine

[55] Williams, Christopher (11 August 2009). "Two convicted for refusal to decrypt data". *The Register*. Retrieved 26 March 2015.

[56] Williams, Christopher (24 November 2009). "UK jails schizophrenic for refusal to decrypt files". *The Register*. Retrieved 26 March 2015.

[57] Ingold, John (January 4, 2012). "Password case reframes Fifth Amendment rights in context of digital world". *The Denver Post*. Retrieved 26 March 2015.

[58] Leyden, John (13 July 2011). "US court test for rights not to hand over crypto keys". *The Register*. Retrieved 26 March 2015.

[59] "ORDER GRANTING APPLICATION UNDER THE ALL WRITS ACT REQUIRING DEFENDANT FRICOSU TO ASSIST IN THE EXECUTION OF PREVIOUSLY ISSUED SEARCH WARRANTS" (PDF). United States District Court for the District of Colorado. Retrieved 26 March 2015.

42.7 Further reading

Further information: Books on cryptography

- Becket, B (1988). *Introduction to Cryptology*. Blackwell Scientific Publications. ISBN 0-632-01836-4. OCLC 16832704. Excellent coverage of many classical ciphers and cryptography concepts and of the "modern" DES and RSA systems.

- *Cryptography and Mathematics* by Bernhard Esslinger, 200 pages, part of the free open-source package CrypTool, PDF download at the Wayback Machine (archived July 22, 2011). CrypTool is the most widespread e-learning program about cryptography and cryptanalysis, open source.

- *In Code: A Mathematical Journey* by Sarah Flannery (with David Flannery). Popular account of Sarah's award-winning project on public-key cryptography, co-written with her father.

- James Gannon, *Stealing Secrets, Telling Lies: How Spies and Codebreakers Helped Shape the Twentieth Century*, Washington, D.C., Brassey's, 2001, ISBN 1-57488-367-4.

- Oded Goldreich, Foundations of Cryptography, in two volumes, Cambridge University Press, 2001 and 2004.

- *Introduction to Modern Cryptography* by Jonathan Katz and Yehuda Lindell.

- *Alvin's Secret Code* by Clifford B. Hicks (children's novel that introduces some basic cryptography and cryptanalysis).

- Ibrahim A. Al-Kadi, "The Origins of Cryptology: the Arab Contributions," Cryptologia, vol. 16, no. 2 (April 1992), pp. 97–126.

- Christof Paar, Jan Pelzl, Understanding Cryptography, A Textbook for Students and Practitioners. Springer, 2009. (Slides, online cryptography lectures and other information are available on the companion web site.) Very accessible introduction to practical cryptography for non-mathematicians.

- *Introduction to Modern Cryptography* by Phillip Rogaway and Mihir Bellare, a mathematical introduction to theoretical cryptography including reduction-based security proofs. PDF download.

- Johann-Christoph Woltag, 'Coded Communications (Encryption)' in Rüdiger Wolfrum (ed) Max Planck Encyclopedia of Public International Law (Oxford University Press 2009). *"Max Planck Encyclopedia of Public International Law"., giving an overview of international law issues regarding cryptography.

- Jonathan Arbib & John Dwyer, Discrete Mathematics for Cryptography, 1st Edition ISBN 978-1-907934-01-8.

- Stallings, William (March 2013). *Cryptography and Network Security: Principles and Practice* (6th ed.). Prentice Hall. ISBN 978-0133354690.

42.8   External links

- The dictionary definition of cryptography at Wiktionary

- Media related to Cryptography at Wikimedia Commons

-

- Cryptography on *In Our Time* at the BBC. (listen now)

- Crypto Glossary and Dictionary of Technical Cryptography

- NSA's CryptoKids.

- Overview and Applications of Cryptology by the CrypTool Team; PDF; 3.8 MB—July 2008

- A Course in Cryptography by Raphael Pass & Abhi Shelat - offered at Cornell in the form of lecture notes.

- Cryptocorner.com by Chuck Easttom - A generalized resource on all aspects of cryptology.

- For more on the use of cryptographic elements in fiction, see: Dooley, John F., William and Marilyn Ingersoll Professor of Computer Science, Knox College (23 August 2012). "Cryptology in Fiction".

- The George Fabyan Collection at the Library of Congress has early editions of works of seventeenth-century English literature, publications relating to cryptography.

Variants of the Enigma machine, used by Germany's military and civil authorities from the late 1920s through World War II, implemented a complex electro-mechanical polyalphabetic cipher. Breaking and reading of the Enigma cipher at Poland's Cipher Bureau, for 7 years before the war, and subsequent decryption at Bletchley Park, was important to Allied victory.[8]

Poznań monument (center) to Polish cryptologists whose breaking of Germany's Enigma machine ciphers, beginning in 1932, altered the course of World War II

Chapter 43

Graph theory

This article is about sets of vertices connected by edges. For graphs of mathematical functions, see Graph of a function. For other uses, see Graph (disambiguation).

  In mathematics and computer science, **graph theory** is the study of *graphs*, which are mathematical structures used

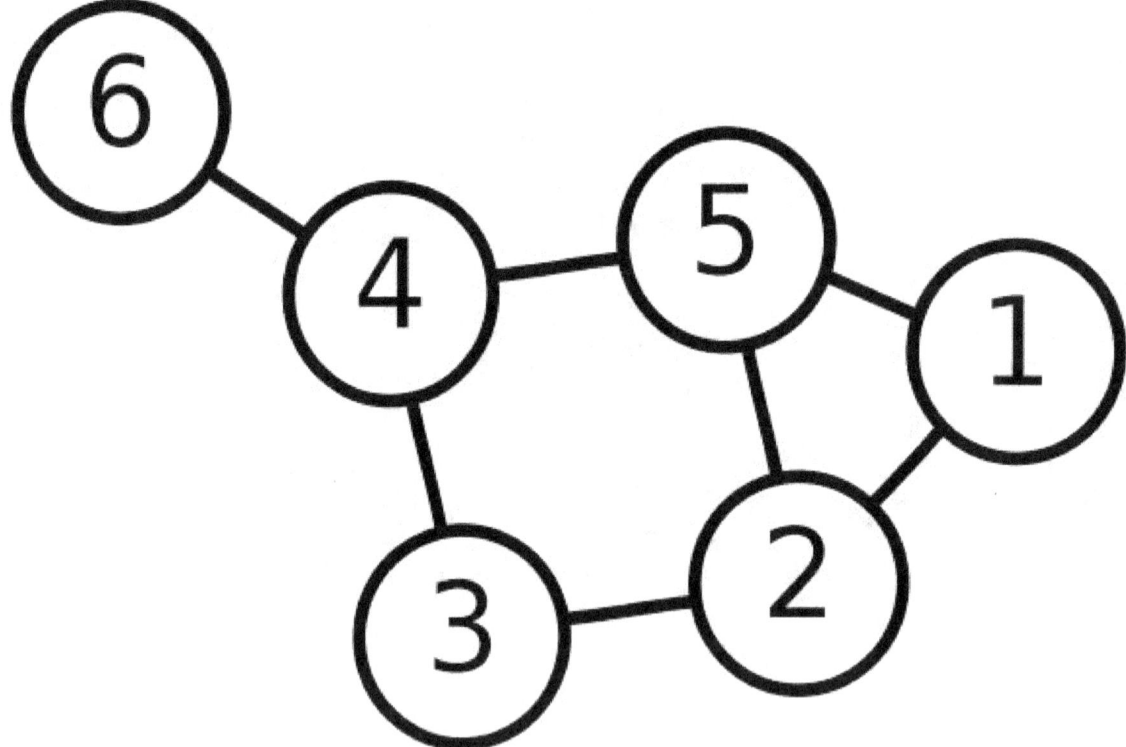

A drawing of a graph

to model pairwise relations between objects. A "graph" in this context is made up of "vertices" or "nodes" and lines called *edges* that connect them. A graph may be *undirected*, meaning that there is no distinction between the two vertices associated with each edge, or its edges may be *directed* from one vertex to another; see graph (mathematics) for more detailed definitions and for other variations in the types of graph that are commonly considered. Graphs are one of the prime objects of study in discrete mathematics.

Refer to the glossary of graph theory for basic definitions in graph theory.

43.1   Definitions

Definitions in graph theory vary. The following are some of the more basic ways of defining graphs and related mathematical structures.

43.1.1   Graph

In the most common sense of the term,[1] a **graph** is an ordered pair $G = (V, E)$ comprising a set V of **vertices** or **nodes** together with a set E of **edges** or **lines**, which are 2-element subsets of V (i.e., an edge is related with two vertices, and the relation is represented as an unordered pair of the vertices with respect to the particular edge). To avoid ambiguity, this type of graph may be described precisely as undirected and simple.

Other senses of *graph* stem from different conceptions of the edge set. In one more generalized notion,[2] V is a set together with a relation of **incidence** that associates with each edge two vertices. In another generalized notion, E is a multiset of unordered pairs of (not necessarily distinct) vertices. Many authors call this type of object a multigraph or pseudograph.

All of these variants and others are described more fully below.

The vertices belonging to an edge are called the **ends**, **endpoints**, or **end vertices** of the edge. A vertex may exist in a graph and not belong to an edge.

V and E are usually taken to be finite, and many of the well-known results are not true (or are rather different) for **infinite graphs** because many of the arguments fail in the infinite case. The **order** of a graph is $|V|$ (the number of vertices). A graph's **size** is $|E|$, the number of edges. The **degree** or **valency** of a vertex is the number of edges that connect to it, where an edge that connects to the vertex at both ends (a loop) is counted twice.

For an edge $\{u, v\}$, graph theorists usually use the somewhat shorter notation uv.

43.2   Applications

Graphs can be used to model many types of relations and processes in physical, biological,[4] social and information systems. Many practical problems can be represented by graphs.

In computer science, graphs are used to represent networks of communication, data organization, computational devices, the flow of computation, etc. For instance, the link structure of a website can be represented by a directed graph, in which the vertices represent web pages and directed edges represent links from one page to another. A similar approach can be taken to problems in travel, biology, computer chip design, and many other fields. The development of algorithms to handle graphs is therefore of major interest in computer science. The transformation of graphs is often formalized and represented by graph rewrite systems. Complementary to graph transformation systems focusing on rule-based in-memory manipulation of graphs are graph databases geared towards transaction-safe, persistent storing and querying of graph-structured data.

Graph-theoretic methods, in various forms, have proven particularly useful in linguistics, since natural language often lends itself well to discrete structure. Traditionally, syntax and compositional semantics follow tree-based structures, whose expressive power lies in the principle of compositionality, modeled in a hierarchical graph. More contemporary approaches such as head-driven phrase structure grammar model the syntax of natural language using typed feature structures, which are directed acyclic graphs. Within lexical semantics, especially as applied to computers, modeling word meaning is easier when a given word is understood in terms of related words; semantic networks are therefore important in computational linguistics. Still other methods in phonology (e.g. optimality theory, which uses lattice graphs) and morphology (e.g. finite-state morphology, using finite-state transducers) are common in the analysis of language as a graph. Indeed, the usefulness of this area of mathematics to linguistics has borne organizations such as TextGraphs, as well as various 'Net' projects, such as WordNet, VerbNet, and others.

Graph theory is also used to study molecules in chemistry and physics. In condensed matter physics, the three-dimensional structure of complicated simulated atomic structures can be studied quantitatively by gathering statistics on graph-theoretic

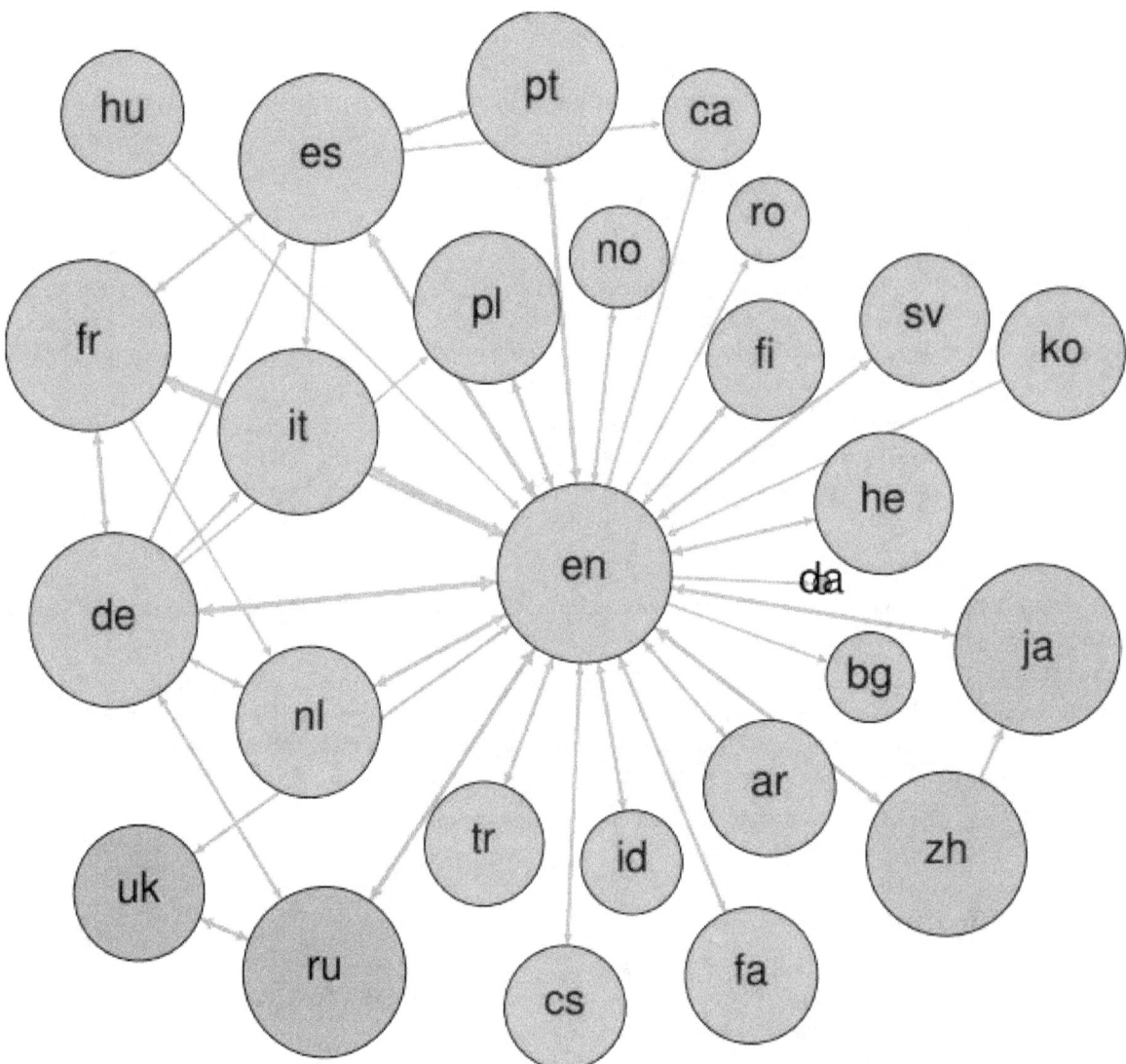

The network graph formed by Wikipedia editors (edges) contributing to different Wikipedia language versions (nodes) during one month in summer 2013.[3]

properties related to the topology of the atoms. In chemistry a graph makes a natural model for a molecule, where vertices represent atoms and edges bonds. This approach is especially used in computer processing of molecular structures, ranging from chemical editors to database searching. In statistical physics, graphs can represent local connections between interacting parts of a system, as well as the dynamics of a physical process on such systems. Graphs are also used to represent the micro-scale channels of porous media, in which the vertices represent the pores and the edges represent the smaller channels connecting the pores.

Graph theory is also widely used in sociology as a way, for example, to measure actors' prestige or to explore rumor spreading, notably through the use of social network analysis software. Under the umbrella of social networks are many different types of graphs.[5] Acquaintanceship and friendship graphs describe whether people know each other. Influence graphs model whether certain people can influence the behavior of others. Finally, collaboration graphs model whether two people work together in a particular way, such as acting in a movie together.

Likewise, graph theory is useful in biology and conservation efforts where a vertex can represent regions where certain species exist (or inhabit) and the edges represent migration paths, or movement between the regions. This information is important when looking at breeding patterns or tracking the spread of disease, parasites or how changes to the movement can affect other species.

In mathematics, graphs are useful in geometry and certain parts of topology such as knot theory. Algebraic graph theory has close links with group theory.

A graph structure can be extended by assigning a weight to each edge of the graph. Graphs with weights, or weighted graphs, are used to represent structures in which pairwise connections have some numerical values. For example, if a graph represents a road network, the weights could represent the length of each road.

43.3   History

The Königsberg Bridge problem

The paper written by Leonhard Euler on the *Seven Bridges of Königsberg* and published in 1736 is regarded as the first paper in the history of graph theory.[6] This paper, as well as the one written by Vandermonde on the *knight problem*, carried on with the *analysis situs* initiated by Leibniz. Euler's formula relating the number of edges, vertices, and faces of a convex polyhedron was studied and generalized by Cauchy[7] and L'Huillier,[8] and is at the origin of topology.

More than one century after Euler's paper on the bridges of Königsberg and while Listing introduced topology, Cayley was led by the study of particular analytical forms arising from differential calculus to study a particular class of graphs, the *trees*.[9] This study had many implications in theoretical chemistry. The involved techniques mainly concerned the enumeration of graphs having particular properties. Enumerative graph theory then rose from the results of Cayley and the fundamental results published by Pólya between 1935 and 1937 and the generalization of these by De Bruijn in 1959. Cayley linked his results on trees with the contemporary studies of chemical composition.[10] The fusion of the ideas coming from mathematics with those coming from chemistry is at the origin of a part of the standard terminology of graph theory.

In particular, the term "graph" was introduced by Sylvester in a paper published in 1878 in *Nature*, where he draws an analogy between "quantic invariants" and "co-variants" of algebra and molecular diagrams:[11]

> "[...] Every invariant and co-variant thus becomes expressible by a *graph* precisely identical with a Kekuléan diagram or chemicograph. [...] I give a rule for the geometrical multiplication of graphs, *i.e.* for constructing a *graph* to the product of in- or co-variants whose separate graphs are given. [...]" (italics as in the original).

The first textbook on graph theory was written by Dénes Kőnig, and published in 1936.[12] Another book by Frank Harary, published in 1969, was "considered the world over to be the definitive textbook on the subject",[13] and enabled mathematicians, chemists, electrical engineers and social scientists to talk to each other. Harary donated all of the royalties to fund the Pólya Prize.[14]

One of the most famous and stimulating problems in graph theory is the four color problem: "Is it true that any map drawn in the plane may have its regions colored with four colors, in such a way that any two regions having a common border have different colors?" This problem was first posed by Francis Guthrie in 1852 and its first written record is in a letter of De Morgan addressed to Hamilton the same year. Many incorrect proofs have been proposed, including those by Cayley, Kempe, and others. The study and the generalization of this problem by Tait, Heawood, Ramsey and Hadwiger led to the study of the colorings of the graphs embedded on surfaces with arbitrary genus. Tait's reformulation generated a new class of problems, the *factorization problems*, particularly studied by Petersen and Kőnig. The works of Ramsey on colorations and more specially the results obtained by Turán in 1941 was at the origin of another branch of graph theory, *extremal graph theory*.

The four color problem remained unsolved for more than a century. In 1969 Heinrich Heesch published a method for solving the problem using computers.[15] A computer-aided proof produced in 1976 by Kenneth Appel and Wolfgang Haken makes fundamental use of the notion of "discharging" developed by Heesch.[16][17] The proof involved checking the properties of 1,936 configurations by computer, and was not fully accepted at the time due to its complexity. A simpler proof considering only 633 configurations was given twenty years later by Robertson, Seymour, Sanders and Thomas.[18]

The autonomous development of topology from 1860 and 1930 fertilized graph theory back through the works of Jordan, Kuratowski and Whitney. Another important factor of common development of graph theory and topology came from the use of the techniques of modern algebra. The first example of such a use comes from the work of the physicist Gustav Kirchhoff, who published in 1845 his Kirchhoff's circuit laws for calculating the voltage and current in electric circuits.

The introduction of probabilistic methods in graph theory, especially in the study of Erdős and Rényi of the asymptotic probability of graph connectivity, gave rise to yet another branch, known as *random graph theory*, which has been a fruitful source of graph-theoretic results.

43.4   Graph drawing

Main article: Graph drawing

Graphs are represented visually by drawing a dot or circle for every vertex, and drawing an arc between two vertices if they are connected by an edge. If the graph is directed, the direction is indicated by drawing an arrow.

A graph drawing should not be confused with the graph itself (the abstract, non-visual structure) as there are several ways to structure the graph drawing. All that matters is which vertices are connected to which others by how many edges and not the exact layout. In practice it is often difficult to decide if two drawings represent the same graph. Depending on the problem domain some layouts may be better suited and easier to understand than others.

The pioneering work of W. T. Tutte was very influential in the subject of graph drawing. Among other achievements, he introduced the use of linear algebraic methods to obtain graph drawings.

Graph drawing also can be said to encompass problems that deal with the crossing number and its various generalizations. The crossing number of a graph is the minimum number of intersections between edges that a drawing of the graph in the plane must contain. For a planar graph, the crossing number is zero by definition.

Drawings on surfaces other than the plane are also studied.

43.5   Graph-theoretic data structures

Main article: Graph (abstract data type)

There are different ways to store graphs in a computer system. The data structure used depends on both the graph structure and the algorithm used for manipulating the graph. Theoretically one can distinguish between list and matrix structures but in concrete applications the best structure is often a combination of both. List structures are often preferred for sparse graphs as they have smaller memory requirements. Matrix structures on the other hand provide faster access for some applications but can consume huge amounts of memory.

List structures include the incidence list, an array of pairs of vertices, and the adjacency list, which separately lists the neighbors of each vertex: Much like the incidence list, each vertex has a list of which vertices it is adjacent to.

Matrix structures include the incidence matrix, a matrix of 0's and 1's whose rows represent vertices and whose columns represent edges, and the adjacency matrix, in which both the rows and columns are indexed by vertices. In both cases a 1 indicates two adjacent objects and a 0 indicates two non-adjacent objects. The Laplacian matrix is a modified form of the adjacency matrix that incorporates information about the degrees of the vertices, and is useful in some calculations such as Kirchhoff's theorem on the number of spanning trees of a graph. The distance matrix, like the adjacency matrix, has both its rows and columns indexed by vertices, but rather than containing a 0 or a 1 in each cell it contains the length of a shortest path between two vertices.

43.6   Problems in graph theory

43.6.1   Enumeration

There is a large literature on graphical enumeration: the problem of counting graphs meeting specified conditions. Some of this work is found in Harary and Palmer (1973).

43.6.2   Subgraphs, induced subgraphs, and minors

A common problem, called the subgraph isomorphism problem, is finding a fixed graph as a subgraph in a given graph. One reason to be interested in such a question is that many graph properties are *hereditary* for subgraphs, which means that a graph has the property if and only if all subgraphs have it too. Unfortunately, finding maximal subgraphs of a certain kind is often an NP-complete problem.

- Finding the largest complete subgraph is called the clique problem (NP-complete).

A similar problem is finding induced subgraphs in a given graph. Again, some important graph properties are hereditary with respect to induced subgraphs, which means that a graph has a property if and only if all induced subgraphs also have it. Finding maximal induced subgraphs of a certain kind is also often NP-complete. For example,

- Finding the largest edgeless induced subgraph, or independent set, called the independent set problem (NP-complete).

Still another such problem, the *minor containment problem*, is to find a fixed graph as a minor of a given graph. A minor or **subcontraction** of a graph is any graph obtained by taking a subgraph and contracting some (or no) edges. Many graph properties are hereditary for minors, which means that a graph has a property if and only if all minors have it too. A famous example:

- A graph is planar if it contains as a minor neither the complete bipartite graph $K_{3,3}$ (See the Three-cottage problem) nor the complete graph K_5 .

Another class of problems has to do with the extent to which various species and generalizations of graphs are determined by their *point-deleted subgraphs*, for example:

- The reconstruction conjecture.

43.6.3 Graph coloring

Many problems have to do with various ways of coloring graphs, for example:

- The four-color theorem

- The strong perfect graph theorem

- The Erdős–Faber–Lovász conjecture(unsolved)

- The total coloring conjecture, also called Behzad's conjecture) (unsolved)

- The list coloring conjecture (unsolved)

- The Hadwiger conjecture (graph theory) (unsolved)

43.6.4 Subsumption and unification

Constraint modeling theories concern families of directed graphs related by a partial order. In these applications, graphs are ordered by specificity, meaning that more constrained graphs—which are more specific and thus contain a greater amount of information—are subsumed by those that are more general. Operations between graphs include evaluating the direction of a subsumption relationship between two graphs, if any, and computing graph unification. The unification of two argument graphs is defined as the most general graph (or the computation thereof) that is consistent with (i.e. contains all of the information in) the inputs, if such a graph exists; efficient unification algorithms are known.

For constraint frameworks which are strictly compositional, graph unification is the sufficient satisfiability and combination function. Well-known applications include automatic theorem proving and modeling the elaboration of linguistic structure.

43.6.5 Route problems

- Hamiltonian path and cycle problems

- Minimum spanning tree

- Route inspection problem (also called the "Chinese Postman Problem")

- Seven Bridges of Königsberg

- Shortest path problem

- Steiner tree

- Three-cottage problem

- Traveling salesman problem (NP-hard)

43.6.6   Network flow

There are numerous problems arising especially from applications that have to do with various notions of flows in networks, for example:

- Max flow min cut theorem

43.6.7   Visibility problems

- Museum guard problem

43.6.8   Covering problems

Covering problems in graphs are specific instances of subgraph-finding problems, and they tend to be closely related to the clique problem or the independent set problem.

- Set cover problem

- Vertex cover problem

43.6.9   Decomposition problems

Decomposition, defined as partitioning the edge set of a graph (with as many vertices as necessary accompanying the edges of each part of the partition), has a wide variety of question. Often, it is required to decompose a graph into subgraphs isomorphic to a fixed graph; for instance, decomposing a complete graph into Hamiltonian cycles. Other problems specify a family of graphs into which a given graph should be decomposed, for instance, a family of cycles, or decomposing a complete graph Kn into $n - 1$ specified trees having, respectively, 1, 2, 3, ..., $n - 1$ edges.

Some specific decomposition problems that have been studied include:

- Arboricity, a decomposition into as few forests as possible

- Cycle double cover, a decomposition into a collection of cycles covering each edge exactly twice

- Edge coloring, a decomposition into as few matchings as possible

- Graph factorization, a decomposition of a regular graph into regular subgraphs of given degrees

43.6.10   Graph classes

Many problems involve characterizing the members of various classes of graphs. Some examples of such questions are below:

- Enumerating the members of a class

- Characterizing a class in terms of forbidden substructures

- Ascertaining relationships among classes (e.g., does one property of graphs imply another)

- Finding efficient algorithms to decide membership in a class

- Finding representations for members of a class.

43.7 See also

- Gallery of named graphs
- Glossary of graph theory
- List of graph theory topics
- Publications in graph theory

43.7.1 Related topics

- Algebraic graph theory
- Citation graph
- Conceptual graph
- Data structure
- Disjoint-set data structure
- Dual-phase evolution
- Entitative graph
- Existential graph
- Graph algebras
- Graph automorphism
- Graph coloring
- Graph database
- Graph data structure
- Graph drawing
- Graph equation
- Graph rewriting
- Graph sandwich problem
- Graph property
- Intersection graph
- Logical graph
- Loop
- Network theory
- Null graph
- Pebble motion problems
- Percolation
- Perfect graph

- Quantum graph
- Random regular graphs
- Semantic networks
- Spectral graph theory
- Strongly regular graphs
- Symmetric graphs
- Transitive reduction
- Tree data structure

43.7.2 Algorithms

- Bellman–Ford algorithm
- Dijkstra's algorithm
- Ford–Fulkerson algorithm
- Kruskal's algorithm
- Nearest neighbour algorithm
- Prim's algorithm
- Depth-first search
- Breadth-first search

43.7.3 Subareas

- Algebraic graph theory
- Geometric graph theory
- Extremal graph theory
- Probabilistic graph theory
- Topological graph theory

43.7.4 Related areas of mathematics

- Combinatorics
- Group theory
- Knot theory
- Ramsey theory

43.7.5 Generalizations

- Hypergraph
- Abstract simplicial complex

43.7.6 Prominent graph theorists

- Alon, Noga
- Berge, Claude
- Bollobás, Béla
- Bondy, Adrian John
- Brightwell, Graham
- Chudnovsky, Maria
- Chung, Fan
- Dirac, Gabriel Andrew
- Erdős, Paul
- Euler, Leonhard
- Faudree, Ralph
- Golumbic, Martin
- Graham, Ronald
- Harary, Frank
- Heawood, Percy John
- Kotzig, Anton
- Kőnig, Dénes
- Lovász, László
- Murty, U. S. R.
- Nešetřil, Jaroslav
- Rényi, Alfréd
- Ringel, Gerhard
- Robertson, Neil
- Seymour, Paul
- Szemerédi, Endre
- Thomas, Robin
- Thomassen, Carsten
- Turán, Pál
- Tutte, W. T.
- Whitney, Hassler

43.8   Notes

[1] See, for instance, Iyanaga and Kawada, **69 J**, p. 234 or Biggs, p. 4.

[2] See, for instance, Graham et al., p. 5.

[3] Hale, Scott A. (2013). "Multilinguals and Wikipedia Editing". arXiv:1312.0976 [cs.CY].

[4] Mashaghi, A. et al. (2004). "Investigation of a protein complex network". *European Physical Journal B* **41** (1): 113–121. doi:10.1140/epjb/e2004-00301-0.

[5] Rosen, Kenneth H. *Discrete mathematics and its applications* (7th ed.). New York: McGraw-Hill. ISBN 978-0-07-338309-5.

[6] Biggs, N.; Lloyd, E. and Wilson, R. (1986), *Graph Theory, 1736-1936*, Oxford University Press

[7] Cauchy, A.L. (1813), "Recherche sur les polyèdres - premier mémoire", *Journal de l'École Polytechnique*, 9 (Cahier 16): 66–86.

[8] L'Huillier, S.-A.-J. (1861), "Mémoire sur la polyèdrométrie", *Annales de Mathématiques* **3**: 169–189.

[9] Cayley, A. (1857), "On the theory of the analytical forms called trees", *Philosophical Magazine*, Series IV **13** (85): 172–176. doi:10.1017/CBO9780511703690.046.

[10] Cayley, A. (1875), "Ueber die Analytischen Figuren, welche in der Mathematik Bäume genannt werden und ihre Anwendung auf die Theorie chemischer Verbindungen",*Berichte der deutschen Chemischen Gesellschaft***8**(2): 1056–1059,doi:10.1002/cber..

[11] Joseph Sylvester, John (1878). "Chemistry and Algebra". *Nature* **17**: 284. doi:10.1038/017284a0.

[12] Tutte, W.T. (2001), *Graph Theory*, Cambridge University Press, p. 30, ISBN 978-0-521-79489-3.

[13] Gardner, Martin (1992), *Fractal Music, Hypercards, and more...Mathematical Recreations from Scientific American*, W. H. Freeman and Company, p. 203.

[14] Society for Industrial and Applied Mathematics (2002), "The George Polya Prize", *Looking Back, Looking Ahead: A SIAM History* (PDF), p. 26.

[15] Heinrich Heesch: Untersuchungen zum Vierfarbenproblem. Mannheim: Bibliographisches Institut 1969.

[16] Appel, K. and Haken, W. (1977), "Every planar map is four colorable. Part I. Discharging", *Illinois J. Math.* **21**: 429–490.

[17] Appel, K. and Haken, W. (1977), "Every planar map is four colorable. Part II. Reducibility", *Illinois J. Math.* **21**: 491–567.

[18] Robertson, N.; Sanders, D.; Seymour, P. and Thomas, R. (1997), "The four color theorem", *Journal of Combinatorial Theory Series B* **70**: 2–44, doi:10.1006/jctb.1997.1750.

43.9   References

• Berge, Claude (1958), *Théorie des graphes et ses applications*, Collection Universitaire de Mathématiques **II**, Paris: Dunod. English edition, Wiley 1961; Methuen & Co, New York 1962; Russian, Moscow 1961; Spanish, Mexico 1962; Roumanian, Bucharest 1969; Chinese, Shanghai 1963; Second printing of the 1962 first English edition, Dover, New York 2001.

• Biggs, N.; Lloyd, E.; Wilson, R. (1986), *Graph Theory, 1736–1936*, Oxford University Press.

• Bondy, J.A.; Murty, U.S.R. (2008), *Graph Theory*, Springer, ISBN 978-1-84628-969-9.

• Bondy, Riordan, O.M (2003), *Mathematical results on scale-free random graphs in "Handbook of Graphs and Networks" (S. Bornholdt and H.G. Schuster (eds))*, Wiley VCH, Weinheim, 1st ed..

• Chartrand, Gary (1985), *Introductory Graph Theory*, Dover, ISBN 0-486-24775-9.

• Gibbons, Alan (1985), *Algorithmic Graph Theory*, Cambridge University Press.

- Reuven Cohen, Shlomo Havlin (2010), *Complex Networks: Structure, Robustness and Function*, Cambridge University Press

- Golumbic, Martin (1980), *Algorithmic Graph Theory and Perfect Graphs*, Academic Press.

- Harary, Frank (1969), *Graph Theory*, Reading, MA: Addison-Wesley.

- Harary, Frank; Palmer, Edgar M. (1973), *Graphical Enumeration*, New York, NY: Academic Press.

- Mahadev, N.V.R.; Peled, Uri N. (1995), *Threshold Graphs and Related Topics*, North-Holland.

- Mark Newman (2010), *Networks: An Introduction*, Oxford University Press.

43.10 External links

- Graph theory with examples

- Hazewinkel, Michiel, ed. (2001), "Graph theory", *Encyclopedia of Mathematics*, Springer, ISBN 978-1-55608-010-4

- Graph theory tutorial

- A searchable database of small connected graphs

- Image gallery: graphs at the Wayback Machine (archived February 6, 2006)

- Concise, annotated list of graph theory resources for researchers

- rocs — a graph theory IDE

- The Social Life of Routers — non-technical paper discussing graphs of people and computers

- Graph Theory Software — tools to teach and learn graph theory

- Online books, and library resources in your library and in other libraries about graph theory

43.10.1 Online textbooks

- Phase Transitions in Combinatorial Optimization Problems, Section 3: Introduction to Graphs (2006) by Hartmann and Weigt

- Digraphs: Theory Algorithms and Applications 2007 by Jorgen Bang-Jensen and Gregory Gutin

- Graph Theory, by Reinhard Diestel

www.ingramcontent.com/pod-product-compliance
Lightning Source LLC
Chambersburg PA
CBHW080755180526

45168CB00006B/2222